Jeep Cherokee & Comanche Automotive Repair Manual

by Bob Henderson and John H Haynes

Member of the Guild of Motoring Writers

Models covered:

All Jeep Cherokee, Wagoneer and Comanche models
1984 through 2000
Does not include diesel engine information

(3G28 - 50010)

(1553)

ABCDE
FGHIJ
KL

3

Haynes Publishing Group
Sparkford Nr Yeovil
Somerset BA22 7JJ England

Haynes North America, Inc
861 Lawrence Drive
Newbury Park
California 91320 USA

Acknowledgements

We are grateful for the help and cooperation of the Chrysler Corporation for assistance with technical information and certain illustrations. Certain wiring diagrams originated exclusively for Haynes North America, Inc. by Valley Forge Technical Services. Technical writers who contributed to this project include Larry Warren, Mike Stubblefield and Ken Freund.

A book in the Haynes Automotive Repair Manual Series

Printed in the U.S.A.

ISBN 1 56392 400 5

Library of Congress Catalog Card Number 00-109302

Contents

Haynes photographer, mechanic and author and with Jeep Cherokee

About this manual

Its purpose

The purpose of this manual is to help you get the best value from your vehicle. It can do so in several ways. It can help you decide what work must be done, even if you choose to have it done by a dealer service department or a repair shop; it provides information and procedures for routine maintenance and servicing; and it offers diagnostic and repair procedures to follow when trouble occurs.

We hope you use the manual to tackle the work yourself. For many simpler jobs, doing it yourself may be quicker than arranging an appointment to get the vehicle into a shop and making the trips to leave it and pick it up. More importantly, a lot of money can be saved by avoiding the expense the shop must pass on to you to cover its labor and overhead costs. An added benefit is the sense of satisfaction and accomplishment that you feel after doing the job yourself.

Using the manual

The manual is divided into Chapters. Each Chapter is divided into numbered Sections, which are headed in bold type between horizontal lines. Each Section consists of consecutively numbered paragraphs.

At the beginning of each numbered Section you will be referred to any illustrations which apply to the procedures in that Section. The reference numbers used in illustration captions pinpoint the pertinent Section and the Step within that Section. That is, illustration 3.2 means the illustration refers to Section 3 and Step (or paragraph) 2 within that Section.

Procedures, once described in the text, are not normally repeated. When it's necessary to refer to another Chapter, the reference will be given as Chapter and Section number. Cross references given without use of the word "Chapter" apply to Sections and/or paragraphs in the same Chapter. For example, "see Section 8" means in the same Chapter.

References to the left or right side of the vehicle assume you are sitting in the driver's seat, facing forward.

Even though we have prepared this manual with extreme care, neither the publisher nor the author can accept responsibility for any errors in, or omissions from, the information given.

NOTE

A **Note** provides information necessary to properly complete a procedure or information which will make the procedure easier to understand.

CAUTION

A **Caution** provides a special procedure or special steps which must be taken while completing the procedure where the Caution is found. Not heeding a Caution can result in damage to the assembly being worked on.

WARNING

A **Warning** provides a special procedure or special steps which must be taken while completing the procedure where the Warning is found. Not heeding a Warning can result in personal injury.

Introduction to the Jeep Cherokee, Wagoneer and Comanche

Jeep Cherokee models are available in two and four door station wagon body styles. Comanche models are available in short and long bed pick-up styles.

The inline four, six-cylinder and V6 engines used in these vehicles are equipped with a carburetor or fuel injection, depending on model. The engine drives the rear wheels through either a four or five-speed manual or three or four-speed automatic transmission via a driveshaft and solid rear axle. A transfer case and driveshaft are used to drive the front axle on 4WD models.

The suspension features solid axles at the front and rear. The front axle is suspended by coil springs and shock absorbers and is located by suspension arms and a track bar. The rear axle is suspended by leaf springs and shock absorbers.

The steering box is mounted to the left of the engine and is connected to the steering arms through a series of rods which incorporates a damper. Power assist is optional on most models.

The brakes are disc at the front and drums at the rear, with power assist standard. Some later models are equipped with an optional Antilock Braking System (ABS).

Vehicle identification numbers

Modifications are a continuing and unpublicized process in vehicle manufacturing. Since spare parts manuals and lists are compiled on a numerical basis, the individual vehicle numbers are essential to correctly identify the component required.

Vehicle Identification Number (VIN)

This very important identification number is stamped on a plate attached to the left side of the dashboard just inside the windshield on the driver's side of the vehicle (see illustration). The VIN also appears on the Vehicle Certificate of Title and Registration. It contains information such as where and when the vehicle was manufactured, the model year and the body style.

Safety Certification label

The Safety Certification label is affixed to the left front door pillar. The plate contains the name of the manufacturer, the month and year of production, the Gross Vehicle Weight Rating (GVWR) and the certification statement.

Vehicle Identification Plate

This plate is located on the radiator support on the driver's side. It contains information on the vehicle model, emission certification, engine and transmission type as well as the paint code.

Engine identification number

The engine ID number on four-cylinder engines is located on a machined surface on the right side of the block between the number three and four cylinders (see illustration). On V6 engines, the ID number is located on a pad at the front of the block (see illustration). On inline six-cylinder engines, the ID number is located on a machined surface on the right side of the block between the number two and three cylinders (see illustration).

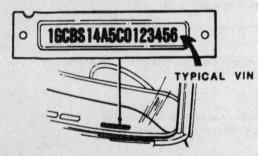

The Vehicle Identification Number (VIN) is visible from outside the vehicle through the driver's side of the windshield

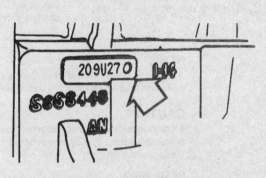

Four-cylinder engine ID number location

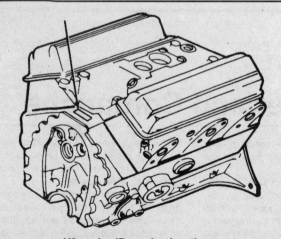

V6 engine ID number location

Inline six-cylinder engine ID number location

Transmission identification number

The ID number on T 4/5 manual transmissions is located on a tag attached to the rear of the case. On AX 4/5 manual transmissions, there are two identification codes: a model/code shipping date stamped on the shift tower and an eight digit code stamped on the bottom surface of the case (see illustration). On the BA 10/5 manual transmis-

sion, the ID plate is attached to the left side of the front case (see illustration). On three-speed automatic transmissions the ID numbers are stamped on the left edge of the case (see illustration). The ID plate on four-speed automatic transmissions is located on the right rear of the case (see illustration).

Transfer case identification number

On most models the transfer case iden-

tification plate is located on the left rear side of the case (see illustration).

Axle identification numbers

On most front axles the identification number is located on a tag attached to the differential housing cover (see illustration). On rear axles, the identification tag is on the left side of the housing and the build date and manufacturer number are stamped on the axle tube (see illustration).

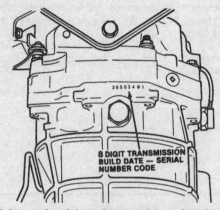

The build date and serial number are stamped on the bottom of the AX 4/5 manual transmission case

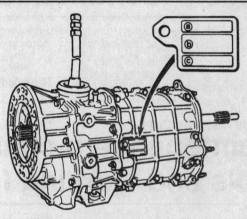

BA 10/5 manual transmission ID number location

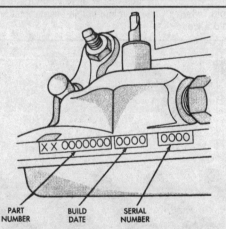

The three-speed automatic transmission numbers are stamped on the edge of the left side of the case

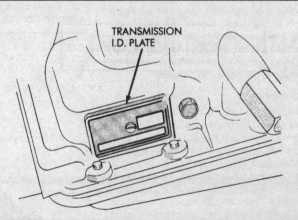

The four-speed automatic transmission ID plate is on the right rear side of the case

Typical transfer case ID tag

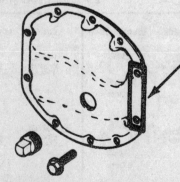

Front axle ID number location

Rear axle number locations

Buying parts

Replacement parts are available from many sources, which generally fall into one of two categories - authorized dealer parts departments and independent retail auto parts stores. Our advice concerning these parts is as follows:

Retail auto parts stores: Good auto parts stores will stock frequently needed components which wear out relatively fast, such as clutch components, exhaust systems, brake parts, tune-up parts, etc. These stores often supply new or reconditioned parts on an exchange basis, which can save a considerable amount of money. Discount auto parts stores are often very good places to buy materials and parts needed for general vehicle maintenance such as oil, grease, filters, spark plugs, belts, touch-up paint, bulbs, etc. They also usually sell tools and general accessories, have convenient hours, charge lower prices and can often be found not far from home.

Authorized dealer parts department: This is the best source for parts which are unique to the vehicle and not generally available elsewhere (such as major engine parts, transmission parts, trim pieces, etc.).

Warranty information: If the vehicle is still covered under warranty, be sure that any replacement parts purchased - regardless of the source - do not invalidate the warranty!

To be sure of obtaining the correct parts, have engine and chassis numbers available and, if possible, take the old parts along for positive identification.

Maintenance techniques, tools and working facilities

Maintenance techniques

There are a number of techniques involved in maintenance and repair that will be referred to throughout this manual. Application of these techniques will enable the home mechanic to be more efficient, better organized and capable of performing the various tasks properly, which will ensure that the repair job is thorough and complete.

Fasteners

Fasteners are nuts, bolts, studs and screws used to hold two or more parts together. There are a few things to keep in mind when working with fasteners. Almost all of them use a locking device of some type, either a lockwasher, locknut, locking tab or thread adhesive. All threaded fasteners should be clean and straight, with undamaged threads and undamaged corners on the hex head where the wrench fits. Develop the habit of replacing all damaged nuts and bolts with new ones. Special locknuts with nylon or fiber inserts can only be used once. If they are removed, they lose their locking ability and must be replaced with new ones.

Rusted nuts and bolts should be treated with a penetrating fluid to ease removal and prevent breakage. Some mechanics use turpentine in a spout-type oil can, which works quite well. After applying the rust penetrant, let it work for a few minutes before trying to loosen the nut or bolt. Badly rusted fasteners may have to be chiseled or sawed off or removed with a special nut breaker, available at tool stores.

If a bolt or stud breaks off in an assembly, it can be drilled and removed with a special tool commonly available for this purpose. Most automotive machine shops can perform this task, as well as other repair procedures, such as the repair of threaded holes that have been stripped out.

Flat washers and lockwashers, when removed from an assembly, should always be replaced exactly as removed. Replace any damaged washers with new ones. Never use a lockwasher on any soft metal surface (such as aluminum), thin sheet metal or plastic.

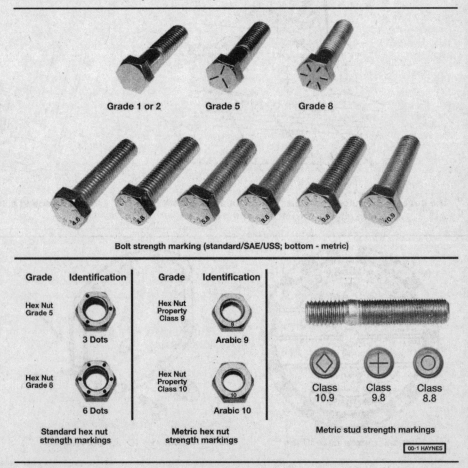

Grade 1 or 2   Grade 5   Grade 8

Bolt strength marking (standard/SAE/USS; bottom - metric)

Grade	Identification	Grade	Identification
Hex Nut Grade 5	3 Dots	Hex Nut Property Class 9	Arabic 9
Hex Nut Grade 8	6 Dots	Hex Nut Property Class 10	Arabic 10

Standard hex nut strength markings

Metric hex nut strength markings

Class 10.9   Class 9.8   Class 8.8

Metric stud strength markings

00-1 HAYNES

Fastener sizes

For a number of reasons, automobile manufacturers are making wider and wider use of metric fasteners. Therefore, it is important to be able to tell the difference between standard (sometimes called U.S. or SAE) and metric hardware, since they cannot be interchanged.

All bolts, whether standard or metric, are sized according to diameter, thread pitch and length. For example, a standard 1/2 - 13 x 1 bolt is 1/2 inch in diameter, has 13 threads per inch and is 1 inch long. An M12 - 1.75 x 25 metric bolt is 12 mm in diameter, has a thread pitch of 1.75 mm (the distance between threads) and is 25 mm long. The two bolts are nearly identical, and easily confused, but they are not interchangeable.

In addition to the differences in diameter, thread pitch and length, metric and standard bolts can also be distinguished by examining the bolt heads. To begin with, the distance across the flats on a standard bolt head is measured in inches, while the same dimension on a metric bolt is sized in millimeters (the same is true for nuts). As a result, a standard wrench should not be used on a metric bolt and a metric wrench should not be used on a standard bolt. Also, most standard bolts have slashes radiating out from the center of the head to denote the grade or strength of the bolt, which is an indication of the amount of torque that can be applied to it. The greater the number of slashes, the greater the strength of the bolt. Grades 0 through 5 are commonly used on automobiles. Metric bolts have a property class (grade) number, rather than a slash, molded into their heads to indicate bolt strength. In this case, the higher the number, the stronger the bolt. Property class numbers 8.8, 9.8 and 10.9 are commonly used on automobiles.

Strength markings can also be used to distinguish standard hex nuts from metric hex nuts. Many standard nuts have dots stamped into one side, while metric nuts are marked with a number. The greater the number of dots, or the higher the number, the greater the strength of the nut.

Metric studs are also marked on their ends according to property class (grade). Larger studs are numbered (the same as metric bolts), while smaller studs carry a geometric code to denote grade.

It should be noted that many fasteners, especially Grades 0 through 2, have no distinguishing marks on them. When such is the case, the only way to determine whether it is standard or metric is to measure the thread pitch or compare it to a known fastener of the same size.

Standard fasteners are often referred to as SAE, as opposed to metric. However, it should be noted that SAE technically refers to a non-metric fine thread fastener only. Coarse thread non-metric fasteners are referred to as USS sizes.

Since fasteners of the same size (both standard and metric) may have different

Metric thread sizes	Ft-lbs	Nm
M-6	6 to 9	9 to 12
M-8	14 to 21	19 to 28
M-10	28 to 40	38 to 54
M-12	50 to 71	68 to 96
M-14	80 to 140	109 to 154

Pipe thread sizes		
1/8	5 to 8	7 to 10
1/4	12 to 18	17 to 24
3/8	22 to 33	30 to 44
1/2	25 to 35	34 to 47

U.S. thread sizes		
1/4 - 20	6 to 9	9 to 12
5/16 - 18	12 to 18	17 to 24
5/16 - 24	14 to 20	19 to 27
3/8 - 16	22 to 32	30 to 43
3/8 - 24	27 to 38	37 to 51
7/16 - 14	40 to 55	55 to 74
7/16 - 20	40 to 60	55 to 81
1/2 - 13	55 to 80	75 to 108

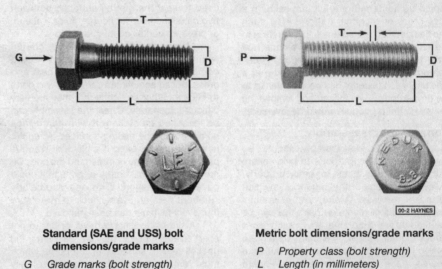

Standard (SAE and USS) bolt dimensions/grade marks

G   Grade marks (bolt strength)
L   Length (in inches)
T   Thread pitch (number of threads per inch)
D   Nominal diameter (in inches)

Metric bolt dimensions/grade marks

P   Property class (bolt strength)
L   Length (in millimeters)
T   Thread pitch (distance between threads in millimeters)
D   Diameter

strength ratings, be sure to reinstall any bolts, studs or nuts removed from your vehicle in their original locations. Also, when replacing a fastener with a new one, make sure that the new one has a strength rating equal to or greater than the original.

Tightening sequences and procedures

Most threaded fasteners should be tightened to a specific torque value (torque is the twisting force applied to a threaded component such as a nut or bolt). Overtightening the fastener can weaken it and cause it to break, while undertightening can cause it to eventually come loose. Bolts, screws and studs, depending on the material they are made of and their thread diameters, have

specific torque values, many of which are noted in the Specifications at the beginning of each Chapter. Be sure to follow the torque recommendations closely. For fasteners not assigned a specific torque, a general torque value chart is presented here as a guide. These torque values are for dry (unlubricated) fasteners threaded into steel or cast iron (not aluminum). As was previously mentioned, the size and grade of a fastener determine the amount of torque that can safely be applied to it. The figures listed here are approximate for Grade 2 and Grade 3 fasteners. Higher grades can tolerate higher torque values.

Fasteners laid out in a pattern, such as cylinder head bolts, oil pan bolts, differential cover bolts, etc., must be loosened or tightened in sequence to avoid warping the com-

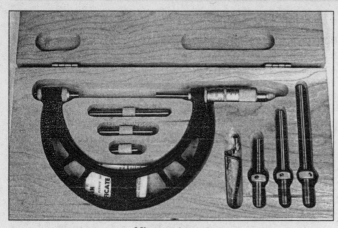

Micrometer set

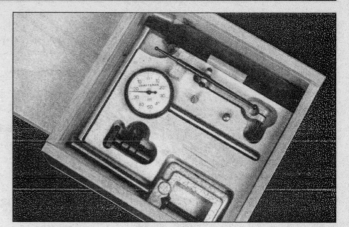

Dial indicator set

ponent. This sequence will normally be shown in the appropriate Chapter. If a specific pattern is not given, the following procedures can be used to prevent warping.

Initially, the bolts or nuts should be assembled finger-tight only. Next, they should be tightened one full turn each, in a criss-cross or diagonal pattern. After each one has been tightened one full turn, return to the first one and tighten them all one-half turn, following the same pattern. Finally, tighten each of them one-quarter turn at a time until each fastener has been tightened to the proper torque. To loosen and remove the fasteners, the procedure would be reversed.

Component disassembly

Component disassembly should be done with care and purpose to help ensure that the parts go back together properly. Always keep track of the sequence in which parts are removed. Make note of special characteristics or marks on parts that can be installed more than one way, such as a grooved thrust washer on a shaft. It is a good idea to lay the disassembled parts out on a clean surface in the order that they were removed. It may also be helpful to make sketches or take instant photos of components before removal.

When removing fasteners from a component, keep track of their locations. Sometimes threading a bolt back in a part, or putting the washers and nut back on a stud, can prevent mix-ups later. If nuts and bolts cannot be returned to their original locations, they should be kept in a compartmented box or a series of small boxes. A cupcake or muffin tin is ideal for this purpose, since each cavity can hold the bolts and nuts from a particular area (i.e. oil pan bolts, valve cover bolts, engine mount bolts, etc.). A pan of this type is especially helpful when working on assemblies with very small parts, such as the carburetor, alternator, valve train or interior dash and trim pieces. The cavities can be marked with paint or tape to identify the contents.

Whenever wiring looms, harnesses or connectors are separated, it is a good idea to identify the two halves with numbered pieces of masking tape so they can be easily reconnected.

Gasket sealing surfaces

Throughout any vehicle, gaskets are used to seal the mating surfaces between two parts and keep lubricants, fluids, vacuum or pressure contained in an assembly.

Many times these gaskets are coated with a liquid or paste-type gasket sealing compound before assembly. Age, heat and pressure can sometimes cause the two parts to stick together so tightly that they are very difficult to separate. Often, the assembly can be loosened by striking it with a soft-face hammer near the mating surfaces. A regular hammer can be used if a block of wood is placed between the hammer and the part. Do not hammer on cast parts or parts that could be easily damaged. With any particularly stubborn part, always recheck to make sure that every fastener has been removed.

Avoid using a screwdriver or bar to pry apart an assembly, as they can easily mar the gasket sealing surfaces of the parts, which must remain smooth. If prying is absolutely necessary, use an old broom handle, but keep in mind that extra clean up will be necessary if the wood splinters.

After the parts are separated, the old gasket must be carefully scraped off and the gasket surfaces cleaned. Stubborn gasket material can be soaked with rust penetrant or treated with a special chemical to soften it so it can be easily scraped off. A scraper can be fashioned from a piece of copper tubing by flattening and sharpening one end. Copper is recommended because it is usually softer than the surfaces to be scraped, which reduces the chance of gouging the part. Some gaskets can be removed with a wire brush, but regardless of the method used, the mating surfaces must be left clean and smooth. If for some reason the gasket surface is gouged, then a gasket sealer thick enough to fill scratches will have to be used during reassembly of the components. For most applications, a non-drying (or semi-drying) gasket sealer should be used.

Hose removal tips

Warning: *If the vehicle is equipped with air conditioning, do not disconnect any of the A/C hoses without first having the system depressurized by a dealer service department or a service station.*

Hose removal precautions closely parallel gasket removal precautions. Avoid scratching or gouging the surface that the hose mates against or the connection may leak. This is especially true for radiator hoses. Because of various chemical reactions, the rubber in hoses can bond itself to the metal spigot that the hose fits over. To remove a hose, first loosen the hose clamps that secure it to the spigot. Then, with slip-joint pliers, grab the hose at the clamp and rotate it around the spigot. Work it back and forth until it is completely free, then pull it off. Silicone or other lubricants will ease removal if they can be applied between the hose and the outside of the spigot. Apply the same lubricant to the inside of the hose and the outside of the spigot to simplify installation.

As a last resort (and if the hose is to be replaced with a new one anyway), the rubber can be slit with a knife and the hose peeled from the spigot. If this must be done, be careful that the metal connection is not damaged.

If a hose clamp is broken or damaged, do not reuse it. Wire-type clamps usually weaken with age, so it is a good idea to replace them with screw-type clamps whenever a hose is removed.

Tools

A selection of good tools is a basic requirement for anyone who plans to maintain and repair his or her own vehicle. For the owner who has few tools, the initial investment might seem high, but when compared to the spiraling costs of professional auto maintenance and repair, it is a wise one.

To help the owner decide which tools are needed to perform the tasks detailed in this manual, the following tool lists are offered: *Maintenance and minor repair, Repair/overhaul* and *Special.*

The newcomer to practical mechanics

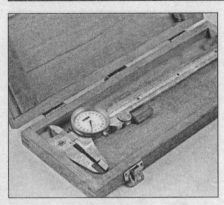

Dial caliper

Hand-operated vacuum pump

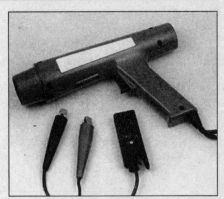

Timing light

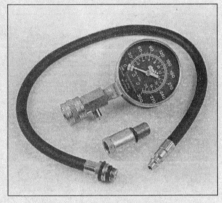

Compression gauge with spark plug
hole adapter

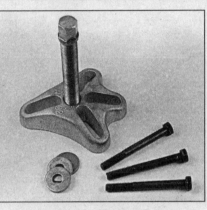

Damper/steering wheel puller

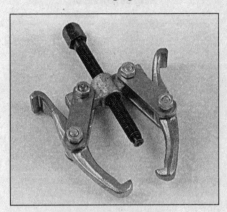

General purpose puller

Hydraulic lifter removal tool

Valve spring compressor

Valve spring compressor

Ridge reamer

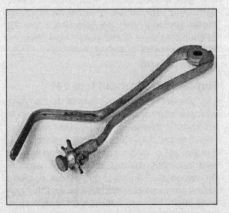

Piston ring groove cleaning tool

Ring removal/installation tool

Ring compressor

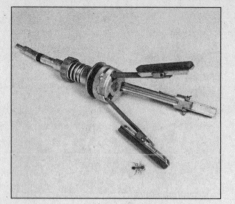

Cylinder hone

Brake hold-down spring tool

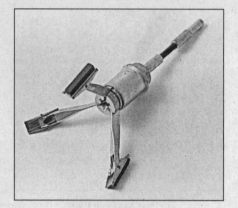

Brake cylinder hone

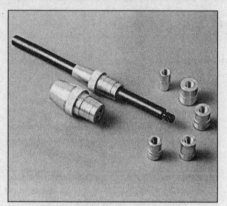

Clutch plate alignment tool

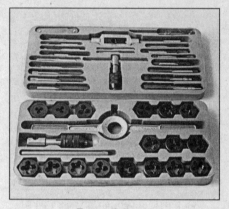

Tap and die set

should start off with the *maintenance and minor repair* tool kit, which is adequate for the simpler jobs performed on a vehicle. Then, as confidence and experience grow, the owner can tackle more difficult tasks, buying additional tools as they are needed. Eventually the basic kit will be expanded into the *repair and overhaul* tool set. Over a period of time, the experienced do-it-yourselfer will assemble a tool set complete enough for most repair and overhaul procedures and will add tools from the special category when it is felt that the expense is justified by the frequency of use.

Maintenance and minor repair tool kit

The tools in this list should be considered the minimum required for performance of routine maintenance, servicing and minor repair work. We recommend the purchase of combination wrenches (box-end and open-end combined in one wrench). While more expensive than open end wrenches, they offer the advantages of both types of wrench.

*Combination wrench set (1/4-inch to
    1 inch or 6 mm to 19 mm)
Adjustable wrench, 8 inch
Spark plug wrench with rubber insert
Spark plug gap adjusting tool
Feeler gauge set
Brake bleeder wrench
Standard screwdriver (5/16-inch x
    6 inch)*

*Phillips screwdriver (No. 2 x 6 inch)
Combination pliers - 6 inch
Hacksaw and assortment of blades
Tire pressure gauge
Grease gun
Oil can
Fine emery cloth
Wire brush
Battery post and cable cleaning tool
Oil filter wrench
Funnel (medium size)
Safety goggles
Jackstands (2)
Drain pan*

Note: *If basic tune-ups are going to be part of routine maintenance, it will be necessary to purchase a good quality stroboscopic timing light and combination tachometer/dwell meter. Although they are included in the list of special tools, it is mentioned here because they are absolutely necessary for tuning most vehicles properly.*

Repair and overhaul tool set

These tools are essential for anyone who plans to perform major repairs and are in addition to those in the maintenance and minor repair tool kit. Included is a comprehensive set of sockets which, though expensive, are invaluable because of their versatility, especially when various extensions and drives are available. We recommend the 1/2-inch drive over the 3/8-inch drive. Although the larger drive is bulky and more expensive,

it has the capacity of accepting a very wide range of large sockets. Ideally, however, the mechanic should have a 3/8-inch drive set and a 1/2-inch drive set.

*Socket set(s)
Reversible ratchet
Extension - 10 inch
Universal joint
Torque wrench (same size drive as
    sockets)
Ball peen hammer - 8 ounce
Soft-face hammer (plastic/rubber)
Standard screwdriver (1/4-inch x 6 inch)
Standard screwdriver (stubby -
    5/16-inch)
Phillips screwdriver (No. 3 x 8 inch)
Phillips screwdriver (stubby - No. 2)
Pliers - vise grip
Pliers - lineman's
Pliers - needle nose
Pliers - snap-ring (internal and external)
Cold chisel - 1/2-inch
Scribe
Scraper (made from flattened copper
    tubing)
Centerpunch
Pin punches (1/16, 1/8, 3/16-inch)
Steel rule/straightedge - 12 inch
Allen wrench set (1/8 to 3/8-inch or
    4 mm to 10 mm)
A selection of files
Wire brush (large)
Jackstands (second set)
Jack (scissor or hydraulic type)*

Note: *Another tool which is often useful is an electric drill with a chuck capacity of 3/8-inch and a set of good quality drill bits.*

Special tools

The tools in this list include those which are not used regularly, are expensive to buy, or which need to be used in accordance with their manufacturer's instructions. Unless these tools will be used frequently, it is not very economical to purchase many of them. A consideration would be to split the cost and use between yourself and a friend or friends. In addition, most of these tools can be obtained from a tool rental shop on a temporary basis.

This list primarily contains only those tools and instruments widely available to the public, and not those special tools produced by the vehicle manufacturer for distribution to dealer service departments. Occasionally, references to the manufacturer's special tools are included in the text of this manual. Generally, an alternative method of doing the job without the special tool is offered. However, sometimes there is no alternative to their use. Where this is the case, and the tool cannot be purchased or borrowed, the work should be turned over to the dealer service department or an automotive repair shop.

Valve spring compressor
Piston ring groove cleaning tool
Piston ring compressor
Piston ring installation tool
Cylinder compression gauge
Cylinder ridge reamer
Cylinder surfacing hone
Cylinder bore gauge
Micrometers and/or dial calipers
Hydraulic lifter removal tool
Balljoint separator
Universal-type puller
Impact screwdriver
Dial indicator set
Stroboscopic timing light (inductive pick-up)
Hand operated vacuum/pressure pump
Tachometer/dwell meter
Universal electrical multimeter
Cable hoist
Brake spring removal and installation tools
Floor jack

Buying tools

For the do-it-yourselfer who is just starting to get involved in vehicle maintenance and repair, there are a number of options available when purchasing tools. If maintenance and minor repair is the extent of the work to be done, the purchase of individual tools is satisfactory. If, on the other hand, extensive work is planned, it would be a good idea to purchase a modest tool set from one of the large retail chain stores. A set can usually be bought at a substantial savings over the individual tool prices, and they often come with a tool box. As additional tools are needed, add-on sets, individual tools and a larger tool box can be purchased to expand the tool selection. Building a tool set gradually allows the cost of the tools to be spread over a longer period of time and gives the mechanic the freedom to choose only those tools that will actually be used.

Tool stores will often be the only source of some of the special tools that are needed, but regardless of where tools are bought, try to avoid cheap ones, especially when buying screwdrivers and sockets, because they won't last very long. The expense involved in replacing cheap tools will eventually be greater than the initial cost of quality tools.

Care and maintenance of tools

Good tools are expensive, so it makes sense to treat them with respect. Keep them clean and in usable condition and store them properly when not in use. Always wipe off any dirt, grease or metal chips before putting them away. Never leave tools lying around in the work area. Upon completion of a job, always check closely under the hood for tools that may have been left there so they won't get lost during a test drive.

Some tools, such as screwdrivers, pliers, wrenches and sockets, can be hung on a panel mounted on the garage or workshop wall, while others should be kept in a tool box or tray. Measuring instruments, gauges, meters, etc. must be carefully stored where they cannot be damaged by weather or impact from other tools.

When tools are used with care and stored properly, they will last a very long time. Even with the best of care, though, tools will wear out if used frequently. When a tool is damaged or worn out, replace it. Subsequent jobs will be safer and more enjoyable if you do.

How to repair damaged threads

Sometimes, the internal threads of a nut or bolt hole can become stripped, usually from overtightening. Stripping threads is an all-too-common occurrence, especially when working with aluminum parts, because aluminum is so soft that it easily strips out.

Usually, external or internal threads are only partially stripped. After they've been cleaned up with a tap or die, they'll still work. Sometimes, however, threads are badly damaged. When this happens, you've got three choices:

1) *Drill and tap the hole to the next suitable oversize and install a larger diameter bolt, screw or stud.*
2) *Drill and tap the hole to accept a threaded plug, then drill and tap the plug to the original screw size. You can also buy a plug already threaded to the original size. Then you simply drill a hole to the specified size, then run the threaded plug into the hole with a bolt and jam nut. Once the plug is fully seated, remove the jam nut and bolt.*
3) *The third method uses a patented thread repair kit like Heli-Coil or Slimsert. These easy-to-use kits are designed to repair damaged threads in straight-through holes and blind holes. Both are available as kits which can handle a variety of sizes and thread patterns. Drill the hole, then tap it with the special included tap. Install the Heli-Coil and the hole is back to its original diameter and thread pitch.*

Regardless of which method you use, be sure to proceed calmly and carefully. A little impatience or carelessness during one of these relatively simple procedures can ruin your whole day's work and cost you a bundle if you wreck an expensive part.

Working facilities

Not to be overlooked when discussing tools is the workshop. If anything more than routine maintenance is to be carried out, some sort of suitable work area is essential.

It is understood, and appreciated, that many home mechanics do not have a good workshop or garage available, and end up removing an engine or doing major repairs outside. It is recommended, however, that the overhaul or repair be completed under the cover of a roof.

A clean, flat workbench or table of comfortable working height is an absolute necessity. The workbench should be equipped with a vise that has a jaw opening of at least four inches.

As mentioned previously, some clean, dry storage space is also required for tools, as well as the lubricants, fluids, cleaning solvents, etc. which soon become necessary.

Sometimes waste oil and fluids, drained from the engine or cooling system during normal maintenance or repairs, present a disposal problem. To avoid pouring them on the ground or into a sewage system, pour the used fluids into large containers, seal them with caps and take them to an authorized disposal site or recycling center. Plastic jugs, such as old antifreeze containers, are ideal for this purpose.

Always keep a supply of old newspapers and clean rags available. Old towels are excellent for mopping up spills. Many mechanics use rolls of paper towels for most work because they are readily available and disposable. To help keep the area under the vehicle clean, a large cardboard box can be cut open and flattened to protect the garage or shop floor.

Whenever working over a painted surface, such as when leaning over a fender to service something under the hood, always cover it with an old blanket or bedspread to protect the finish. Vinyl covered pads, made especially for this purpose, are available at auto parts stores.

Booster battery (jump) starting

Observe the following precautions when using a booster battery to start a vehicle:

a) Before connecting the booster battery, make sure the ignition switch is in the Off position.
b) Ensure that all electrical equipment (lights, heater, wipers etc.) are switched off.
c) Make sure that the booster battery is the same voltage as the discharged battery in the vehicle.
d) If the battery is being jump started from the battery in another vehicle, the two vehicles MUST NOT TOUCH each other.
e) Make sure the transaxle is in Neutral (manual transaxle) or Park (automatic transaxle).

f) Wear eye protection when jump starting a vehicle.

Connect one jumper lead between the positive (+) terminals of the two batteries. Connect the other jumper lead first to the negative (-) terminal of the booster battery, then to a good engine ground on the vehicle to be started (see illustration). Attach the lead at least 18 inches from the battery, if possible. Make sure that the jumper leads will not contact the fan, drivebelt of other moving parts of the engine.

Start the engine using the booster battery and allow the engine idle speed to stabilize. Disconnect the jumper leads in the reverse order of connection.

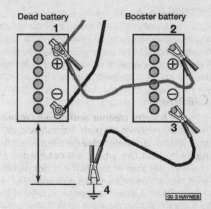

Make the booster battery cable connections in the numerical order shown (note that the negative cable of the booster battery is NOT attached to the negative terminal of the dead battery)

Jacking and towing

Jacking

The jack supplied with the vehicle should only be used for raising the vehicle when changing a tire or placing jackstands under the frame. **Warning:** *Never work under the vehicle or start the engine while this jack is being used as the only means of support.*

The vehicle should be on level ground with the hazard flashers on, the wheels blocked, the parking brake applied and the transmission in Park (automatic) or Reverse (manual). If a tire is being changed, loosen the lug nuts one-half turn and leave them in place until the wheel is raised off the ground.

Place the jack under the vehicle suspension in the indicated position (see illustration). Operate the jack with a slow, smooth motion until the wheel is raised off the ground. Remove the lug nuts, pull off the wheel, install the spare and thread the lug nuts back on with the bevelled sides facing in. Tighten them snugly, but wait until the vehicle is lowered to tighten them completely. Note that some spare tires are designed for temporary use only, don't exceed the recommended speed, mileage or other restrictions accompanying the spare.

Lower the vehicle, remove the jack and tighten the nuts (if loosened or removed) in a criss-cross pattern.

Towing

As a general rule, vehicles can be towed with all four wheels on the ground, provided that the driveshaft(s) are removed (see Chapter 8).

Equipment specifically designed for towing should be used and should be attached to the main structural members of the vehicle, not the bumper or brackets. Tow hooks are attached to the frame at both ends of the vehicle. However, they are for emergency use only and should not be used for highway towing. Stand clear of vehicles when using the tow hooks, tow straps and chains may break, causing serious injury.

Safety is a major consideration when towing and all applicable state and local laws must be obeyed. A safety chain must be used for all towing (in addition to the tow bar).

While towing, the parking brake should be released and the transmission and (if equipped) transfer case must be in Neutral. The steering must be unlocked (ignition switch in the Off position). Remember that power steering and power brakes will not work with the engine off.

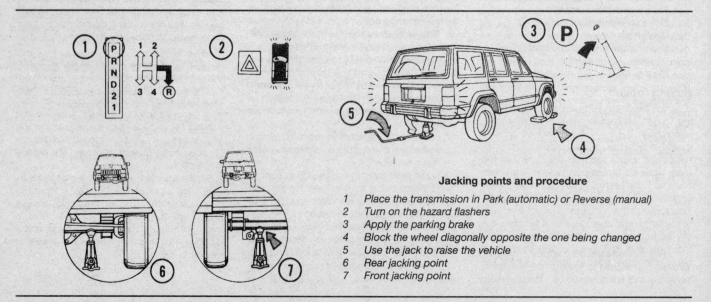

Jacking points and procedure

1   Place the transmission in Park (automatic) or Reverse (manual)
2   Turn on the hazard flashers
3   Apply the parking brake
4   Block the wheel diagonally opposite the one being changed
5   Use the jack to raise the vehicle
6   Rear jacking point
7   Front jacking point

Automotive chemicals and lubricants

A number of automotive chemicals and lubricants are available for use during vehicle maintenance and repair. They include a wide variety of products ranging from cleaning solvents and degreasers to lubricants and protective sprays for rubber, plastic and vinyl.

Cleaners

Carburetor cleaner and choke cleaner is a strong solvent for gum, varnish and carbon. Most carburetor cleaners leave a dry-type lubricant film which will not harden or gum up. Because of this film it is not recommended for use on electrical components.

Brake system cleaner is used to remove grease and brake fluid from the brake system, where clean surfaces are absolutely necessary. It leaves no residue and often eliminates brake squeal caused by contaminants.

Electrical cleaner removes oxidation, corrosion and carbon deposits from electrical contacts, restoring full current flow. It can also be used to clean spark plugs, carburetor jets, voltage regulators and other parts where an oil-free surface is desired.

Demoisturants remove water and moisture from electrical components such as alternators, voltage regulators, electrical connectors and fuse blocks. They are non-conductive, non-corrosive and non-flammable.

Degreasers are heavy-duty solvents used to remove grease from the outside of the engine and from chassis components. They can be sprayed or brushed on and, depending on the type, are rinsed off either with water or solvent.

Lubricants

Motor oil is the lubricant formulated for use in engines. It normally contains a wide variety of additives to prevent corrosion and reduce foaming and wear. Motor oil comes in various weights (viscosity ratings) from 0 to 50. The recommended weight of the oil depends on the season, temperature and the demands on the engine. Light oil is used in cold climates and under light load conditions. Heavy oil is used in hot climates and where high loads are encountered. Multi-viscosity oils are designed to have characteristics of both light and heavy oils and are available in a number of weights from 5W-20 to 20W-50.

Gear oil is designed to be used in differentials, manual transmissions and other areas where high-temperature lubrication is required.

Chassis and wheel bearing grease is a heavy grease used where increased loads and friction are encountered, such as for wheel bearings, balljoints, tie-rod ends and universal joints.

High-temperature wheel bearing grease is designed to withstand the extreme temperatures encountered by wheel bearings in disc brake equipped vehicles. It usually contains molybdenum disulfide (moly), which is a dry-type lubricant.

White grease is a heavy grease for metal-to-metal applications where water is a problem. White grease stays soft under both low and high temperatures (usually from -100 to +190-degrees F), and will not wash off or dilute in the presence of water.

Assembly lube is a special extreme pressure lubricant, usually containing moly, used to lubricate high-load parts (such as main and rod bearings and cam lobes) for initial start-up of a new engine. The assembly lube lubricates the parts without being squeezed out or washed away until the engine oiling system begins to function.

Silicone lubricants are used to protect rubber, plastic, vinyl and nylon parts.

Graphite lubricants are used where oils cannot be used due to contamination problems, such as in locks. The dry graphite will lubricate metal parts while remaining uncontaminated by dirt, water, oil or acids. It is electrically conductive and will not foul electrical contacts in locks such as the ignition switch.

Moly penetrants loosen and lubricate frozen, rusted and corroded fasteners and prevent future rusting or freezing.

Heat-sink grease is a special electrically non-conductive grease that is used for mounting electronic ignition modules where it is essential that heat is transferred away from the module.

Sealants

RTV sealant is one of the most widely used gasket compounds. Made from silicone, RTV is air curing, it seals, bonds, waterproofs, fills surface irregularities, remains flexible, doesn't shrink, is relatively easy to remove, and is used as a supplementary sealer with almost all low and medium temperature gaskets.

Anaerobic sealant is much like RTV in that it can be used either to seal gaskets or to form gaskets by itself. It remains flexible, is solvent resistant and fills surface imperfections. The difference between an anaerobic sealant and an RTV-type sealant is in the curing. RTV cures when exposed to air, while an anaerobic sealant cures only in the absence of air. This means that an anaerobic sealant cures only after the assembly of parts, sealing them together.

Thread and pipe sealant is used for sealing hydraulic and pneumatic fittings and vacuum lines. It is usually made from a Teflon compound, and comes in a spray, a paint-on liquid and as a wrap-around tape.

Chemicals

Anti-seize compound prevents seizing, galling, cold welding, rust and corrosion in fasteners. High-temperature anti-seize, usually made with copper and graphite lubricants, is used for exhaust system and exhaust manifold bolts.

Anaerobic locking compounds are used to keep fasteners from vibrating or working loose and cure only after installation, in the absence of air. Medium strength locking compound is used for small nuts, bolts and screws that may be removed later. High-strength locking compound is for large nuts, bolts and studs which aren't removed on a regular basis.

Oil additives range from viscosity index improvers to chemical treatments that claim to reduce internal engine friction. It should be noted that most oil manufacturers caution against using additives with their oils.

Gas additives perform several functions, depending on their chemical makeup. They usually contain solvents that help dissolve gum and varnish that build up on carburetor, fuel injection and intake parts. They also serve to break down carbon deposits that form on the inside surfaces of the combustion chambers. Some additives contain upper cylinder lubricants for valves and piston rings, and others contain chemicals to remove condensation from the gas tank.

Miscellaneous

Brake fluid is specially formulated hydraulic fluid that can withstand the heat and pressure encountered in brake systems. Care must be taken so this fluid does not come in contact with painted surfaces or plastics. An opened container should always be resealed to prevent contamination by water or dirt.

Weatherstrip adhesive is used to bond weatherstripping around doors, windows and trunk lids. It is sometimes used to attach trim pieces.

Undercoating is a petroleum-based, tar-like substance that is designed to protect metal surfaces on the underside of the vehicle from corrosion. It also acts as a sound-deadening agent by insulating the bottom of the vehicle.

Waxes and polishes are used to help protect painted and plated surfaces from the weather. Different types of paint may require the use of different types of wax and polish. Some polishes utilize a chemical or abrasive cleaner to help remove the top layer of oxidized (dull) paint on older vehicles. In recent years many non-wax polishes that contain a wide variety of chemicals such as polymers and silicones have been introduced. These non-wax polishes are usually easier to apply and last longer than conventional waxes and polishes.

Conversion factors

Length (distance)
Inches (in)	X 25.4	= Millimetres (mm)	X 0.0394	= Inches (in)
Feet (ft)	X 0.305	= Metres (m)	X 3.281	= Feet (ft)
Miles	X 1.609	= Kilometres (km)	X 0.621	= Miles

Volume (capacity)
Cubic inches (cu in; in³)	X 16.387	= Cubic centimetres (cc; cm³)	X 0.061	= Cubic inches (cu in; in³)
Imperial pints (Imp pt)	X 0.568	= Litres (l)	X 1.76	= Imperial pints (Imp pt)
Imperial quarts (Imp qt)	X 1.137	= Litres (l)	X 0.88	= Imperial quarts (Imp qt)
Imperial quarts (Imp qt)	X 1.201	= US quarts (US qt)	X 0.833	= Imperial quarts (Imp qt)
US quarts (US qt)	X 0.946	= Litres (l)	X 1.057	= US quarts (US qt)
Imperial gallons (Imp gal)	X 4.546	= Litres (l)	X 0.22	= Imperial gallons (Imp gal)
Imperial gallons (Imp gal)	X 1.201	= US gallons (US gal)	X 0.833	= Imperial gallons (Imp gal)
US gallons (US gal)	X 3.785	= Litres (l)	X 0.264	= US gallons (US gal)

Mass (weight)
Ounces (oz)	X 28.35	= Grams (g)	X 0.035	= Ounces (oz)
Pounds (lb)	X 0.454	= Kilograms (kg)	X 2.205	= Pounds (lb)

Force
Ounces-force (ozf; oz)	X 0.278	= Newtons (N)	X 3.6	= Ounces-force (ozf; oz)
Pounds-force (lbf; lb)	X 4.448	= Newtons (N)	X 0.225	= Pounds-force (lbf; lb)
Newtons (N)	X 0.1	= Kilograms-force (kgf; kg)	X 9.81	= Newtons (N)

Pressure
Pounds-force per square inch (psi; lbf/in²; lb/in²)	X 0.070	= Kilograms-force per square centimetre (kgf/cm²; kg/cm²)	X 14.223	= Pounds-force per square inch (psi; lbf/in²; lb/in²)
Pounds-force per square inch (psi; lbf/in²; lb/in²)	X 0.068	= Atmospheres (atm)	X 14.696	= Pounds-force per square inch (psi; lbf/in²; lb/in²)
Pounds-force per square inch (psi; lbf/in²; lb/in²)	X 0.069	= Bars	X 14.5	= Pounds-force per square inch (psi; lbf/in²; lb/in²)
Pounds-force per square inch (psi; lbf/in²; lb/in²)	X 6.895	= Kilopascals (kPa)	X 0.145	= Pounds-force per square inch (psi; lbf/in²; lb/in²)
Kilopascals (kPa)	X 0.01	= Kilograms-force per square centimetre (kgf/cm²; kg/cm²)	X 98.1	= Kilopascals (kPa)

Torque (moment of force)
Pounds-force inches (lbf in; lb in)	X 1.152	= Kilograms-force centimetre (kgf cm; kg cm)	X 0.868	= Pounds-force inches (lbf in; lb in)
Pounds-force inches (lbf in; lb in)	X 0.113	= Newton metres (Nm)	X 8.85	= Pounds-force inches (lbf in; lb in)
Pounds-force inches (lbf in; lb in)	X 0.083	= Pounds-force feet (lbf ft; lb ft)	X 12	= Pounds-force inches (lbf in; lb in)
Pounds-force feet (lbf ft; lb ft)	X 0.138	= Kilograms-force metres (kgf m; kg m)	X 7.233	= Pounds-force feet (lbf ft; lb ft)
Pounds-force feet (lbf ft; lb ft)	X 1.356	= Newton metres (Nm)	X 0.738	= Pounds-force feet (lbf ft; lb ft)
Newton metres (Nm)	X 0.102	= Kilograms-force metres (kgf m; kg m)	X 9.804	= Newton metres (Nm)

Vacuum
Inches mercury (in. Hg)	X 3.377	= Kilopascals (kPa)	X 0.2961	= Inches mercury
Inches mercury (in. Hg)	X 25.4	= Millimeters mercury (mm Hg)	X 0.0394	= Inches mercury

Power
Horsepower (hp)	X 745.7	= Watts (W)	X 0.0013	= Horsepower (hp)

Velocity (speed)
Miles per hour (miles/hr; mph)	X 1.609	= Kilometres per hour (km/hr; kph)	X 0.621	= Miles per hour (miles/hr; mph)

Fuel consumption*
Miles per gallon, Imperial (mpg)	X 0.354	= Kilometres per litre (km/l)	X 2.825	= Miles per gallon, Imperial (mpg)
Miles per gallon, US (mpg)	X 0.425	= Kilometres per litre (km/l)	X 2.352	= Miles per gallon, US (mpg)

Temperature
Degrees Fahrenheit = (°C x 1.8) + 32   Degrees Celsius (Degrees Centigrade; °C) = (°F - 32) x 0.56

*It is common practice to convert from miles per gallon (mpg) to litres/100 kilometres (l/100km),
where mpg (Imperial) x l/100 km = 282 and mpg (US) x l/100 km = 235

Safety first!

Regardless of how enthusiastic you may be about getting on with the job at hand, take the time to ensure that your safety is not jeopardized. A moment's lack of attention can result in an accident, as can failure to observe certain simple safety precautions. The possibility of an accident will always exist, and the following points should not be considered a comprehensive list of all dangers. Rather, they are intended to make you aware of the risks and to encourage a safety conscious approach to all work you carry out on your vehicle.

Essential DOs and DON'Ts

DON'T rely on a jack when working under the vehicle. Always use approved jackstands to support the weight of the vehicle and place them under the recommended lift or support points.

DON'T attempt to loosen extremely tight fasteners (i.e. wheel lug nuts) while the vehicle is on a jack - it may fall.

DON'T start the engine without first making sure that the transmission is in Neutral (or Park where applicable) and the parking brake is set.

DON'T remove the radiator cap from a hot cooling system - let it cool or cover it with a cloth and release the pressure gradually.

DON'T attempt to drain the engine oil until you are sure it has cooled to the point that it will not burn you.

DON'T touch any part of the engine or exhaust system until it has cooled sufficiently to avoid burns.

DON'T siphon toxic liquids such as gasoline, antifreeze and brake fluid by mouth, or allow them to remain on your skin.

DON'T inhale brake lining dust - it is potentially hazardous (see *Asbestos* below).

DON'T allow spilled oil or grease to remain on the floor - wipe it up before someone slips on it.

DON'T use loose fitting wrenches or other tools which may slip and cause injury.

DON'T push on wrenches when loosening or tightening nuts or bolts. Always try to pull the wrench toward you. If the situation calls for pushing the wrench away, push with an open hand to avoid scraped knuckles if the wrench should slip.

DON'T attempt to lift a heavy component alone - get someone to help you.

DON'T rush or take unsafe shortcuts to finish a job.

DON'T allow children or animals in or around the vehicle while you are working on it.

DO wear eye protection when using power tools such as a drill, sander, bench grinder, etc. and when working under a vehicle.

DO keep loose clothing and long hair well out of the way of moving parts.

DO make sure that any hoist used has a safe working load rating adequate for the job.

DO get someone to check on you periodically when working alone on a vehicle.

DO carry out work in a logical sequence and make sure that everything is correctly assembled and tightened.

DO keep chemicals and fluids tightly capped and out of the reach of children and pets.

DO remember that your vehicle's safety affects that of yourself and others. If in doubt on any point, get professional advice.

Asbestos

Certain friction, insulating, sealing, and other products - such as brake linings, brake bands, clutch linings, torque converters, gaskets, etc. - may contain asbestos. Extreme care must be taken to avoid inhalation of dust from such products, since it is hazardous to health. If in doubt, assume that they do contain asbestos.

Fire

Remember at all times that gasoline is highly flammable. Never smoke or have any kind of open flame around when working on a vehicle. But the risk does not end there. A spark caused by an electrical short circuit, by two metal surfaces contacting each other, or even by static electricity built up in your body under certain conditions, can ignite gasoline vapors, which in a confined space are highly explosive. Do not, under any circumstances, use gasoline for cleaning parts. Use an approved safety solvent.

Always disconnect the battery ground (-) cable at the battery before working on any part of the fuel system or electrical system. Never risk spilling fuel on a hot engine or exhaust component. It is strongly recommended that a fire extinguisher suitable for use on fuel and electrical fires be kept handy in the garage or workshop at all times. Never try to extinguish a fuel or electrical fire with water.

Fumes

Certain fumes are highly toxic and can quickly cause unconsciousness and even death if inhaled to any extent. Gasoline vapor falls into this category, as do the vapors from some cleaning solvents. Any draining or pouring of such volatile fluids should be done in a well ventilated area.

When using cleaning fluids and solvents, read the instructions on the container carefully. Never use materials from unmarked containers.

Never run the engine in an enclosed space, such as a garage. Exhaust fumes contain carbon monoxide, which is extremely poisonous. If you need to run the engine, always do so in the open air, or at least have the rear of the vehicle outside the work area.

If you are fortunate enough to have the use of an inspection pit, never drain or pour gasoline and never run the engine while the vehicle is over the pit. The fumes, being heavier than air, will concentrate in the pit with possibly lethal results.

The battery

Never create a spark or allow a bare light bulb near a battery. They normally give off a certain amount of hydrogen gas, which is highly explosive.

Always disconnect the battery ground (-) cable at the battery before working on the fuel or electrical systems.

If possible, loosen the filler caps or cover when charging the battery from an external source (this does not apply to sealed or maintenance-free batteries). Do not charge at an excessive rate or the battery may burst.

Take care when adding water to a non maintenance-free battery and when carrying a battery. The electrolyte, even when diluted, is very corrosive and should not be allowed to contact clothing or skin.

Always wear eye protection when cleaning the battery to prevent the caustic deposits from entering your eyes.

Household current

When using an electric power tool, inspection light, etc., which operates on household current, always make sure that the tool is correctly connected to its plug and that, where necessary, it is properly grounded. Do not use such items in damp conditions and, again, do not create a spark or apply excessive heat in the vicinity of fuel or fuel vapor.

Secondary ignition system voltage

A severe electric shock can result from touching certain parts of the ignition system (such as the spark plug wires) when the engine is running or being cranked, particularly if components are damp or the insulation is defective. In the case of an electronic ignition system, the secondary system voltage is much higher and could prove fatal.

Troubleshooting

Contents

This Section provides an easy reference guide to the more common problems that may occur during the operation of your vehicle. Various symptoms and their probable causes are grouped under headings denoting components or systems, such as Engine, Cooling system, etc. They also refer to the Chapter and/or Section that deals with the problem.

Remember that successful troubleshooting isn't a mysterious 'black art' practiced only by professional mechanics, it's simply the result of knowledge combined with an intelligent, systematic approach to a problem. Always use a process of elimination starting with the simplest solution and working through to the most complex, and never overlook the obvious. Anyone can run the gas tank dry or leave the lights on overnight, so don't assume that you're exempt from such oversights.

Finally, always establish a clear idea why a problem has occurred and take steps to ensure that it doesn't happen again. If the electrical system fails because of a poor connection, check all other connections in the system to make sure they don't fail as well. If a particular fuse continues to blow, find out why, don't just go on replacing fuses. Remember, failure of a small component can often be indicative of potential failure or incorrect functioning of a more important component or system.

Engine and performance

1 Engine will not rotate when attempting to start

1   Battery terminal connections loose or corroded. Check the cable terminals at the battery; tighten cable clamp and/or clean off corrosion as necessary (see Chapter 1).
2   Battery discharged or faulty. If the cable ends are clean and tight on the battery posts, turn the key to the On position and switch on the headlights or windshield wipers. If they won't run, the battery is discharged.
3   Automatic transmission not engaged in park (P) or Neutral (N).
4   Broken, loose or disconnected wires in the starting circuit. Inspect all wires and connectors at the battery, starter solenoid and ignition switch (on steering column).
5   Starter motor pinion jammed in flywheel ring gear. If manual transmission, place transmission in gear and rock the vehicle to manually turn the engine. Remove starter (Chapter 5) and inspect pinion and flywheel (Chapter 2) at earliest convenience.
6   Starter solenoid faulty (Chapter 5).
7   Starter motor faulty (Chapter 5).
8   Ignition switch faulty (Chapter 13).
9   Engine seized. Try to turn the crankshaft with a large socket and breaker bar on the pulley bolt.

2 Engine rotates but will not start

1   Fuel tank empty.
2   Battery discharged (engine rotates slowly). Check the operation of electrical components as described in previous Section.
3   Battery terminal connections loose or corroded. See previous Section.
4   Fuel not reaching carburetor or fuel injector. Check for clogged fuel filter or lines and defective fuel pump. Also make sure the tank vent lines aren't clogged (Chapter 4).
5   Choke not operating properly (Chapter 1).
6   Faulty distributor components. Check the cap and rotor (Chapter 1).
7   Low cylinder compression. Check as described in Chapter 2.
8   Valve clearances not properly adjusted (V6 engine) (Chapter 2B).
9   Water in fuel. Drain tank and fill with new fuel.
10 Defective ignition coil (Chapter 5).
11 Dirty or clogged carburetor jets or fuel injector. Carburetor out of adjustment. Check the float level (Chapter 4).
12 Wet or damaged ignition components (Chapters 1 and 5).
13 Worn, faulty or incorrectly gapped spark plugs (Chapter 1).
14 Broken, loose or disconnected wires in the starting circuit (see previous Section).
15 Loose distributor (changing ignition timing). Turn the distributor body as necessary to start the engine, then adjust the ignition timing as soon as possible (Chapter 1).
16 Broken, loose or disconnected wires at the ignition coil or faulty coil (Chapter 5).
17 Timing chain failure or wear affecting valve timing (Chapter 2).

3 Starter motor operates without turning engine

1   Starter pinion sticking. Remove the starter (Chapter 5) and inspect.
2   Starter pinion or flywheel/driveplate teeth worn or broken. Remove the inspection cover and inspect.

4 Engine hard to start when cold

1   Battery discharged or low. Check as described in Chapter 1.
2   Fuel not reaching the carburetor or fuel injectors. Check the fuel filter, lines and fuel pump (Chapters 1 and 4).
3   Choke inoperative (Chapters 1 and 4).
4   Defective spark plugs (Chapter 1).

5 Engine hard to start when hot

1   Air filter dirty (Chapter 1).
2   Fuel not reaching carburetor or fuel injectors (see Section 4). Check for a vapor lock situation, brought about by clogged fuel tank vent lines.
3   Bad engine ground connection.
4   Choke sticking (Chapter 1).
5   Defective pick-up coil in distributor (Chapter 5).
6   Float level too high (Chapter 4).

6 Starter motor noisy or engages roughly

1   Pinion or flywheel/driveplate teeth worn or broken. Remove the inspection cover on the left side of the engine and inspect.
2   Starter motor mounting bolts loose or missing.

7 Engine starts but stops immediately

1   Loose or damaged wire harness connections at distributor, coil or alternator.
2   Intake manifold vacuum leaks. Make sure all mounting bolts/nuts are tight and all vacuum hoses connected to the manifold are attached properly and in good condition.
3   Insufficient fuel flow (see Chapter 4).

8 Engine 'lopes' while idling or idles erratically

1   Vacuum leaks. Check mounting bolts at the intake manifold for tightness. Make sure that all vacuum hoses are connected and in good condition. Use a stethoscope or a length of fuel hose held against your ear to listen for vacuum leaks while the engine is running. A hissing sound will be heard. A soapy water solution will also detect leaks. Check the intake manifold gasket surfaces.
2   Leaking EGR valve or plugged PCV valve (see Chapters 1 and 6).
3   Air filter clogged (Chapter 1).
4   Fuel pump not delivering sufficient fuel (Chapter 4).
5   Leaking head gasket. Perform a cylinder compression check (Chapter 2).
6   Timing chain worn (Chapter 2).
7   Camshaft lobes worn (Chapter 2).
8   Valve clearance out of adjustment (V6 engine) (Chapter 2B). Valves burned or otherwise leaking (Chapter 2).
9   Ignition timing out of adjustment (Chapter 1).
10 Ignition system not operating properly (Chapters 1 and 5).
11 Thermostatic air cleaner not operating properly (Chapter 1).
12 Choke not operating properly (Chapters 1 and 4).
13 Dirty or clogged injectors. Carburetor dirty, clogged or out of adjustment. Check the float level (Chapter 4).
14 Idle speed out of adjustment (Chapter 1).

9   Engine misses at idle speed

1   Spark plugs faulty or not gapped properly (Chapter 1).
2   Faulty spark plug wires (Chapter 1).
3   Wet or damaged distributor components (Chapter 1).
4   Short circuits in ignition, coil or spark plug wires.
5   Sticking or faulty emissions systems (see Chapter 6).
6   Clogged fuel filter and/or foreign matter in fuel. Remove the fuel filter (Chapter 1) and inspect.
7   Vacuum leaks at intake manifold or hose connections. Check as described in Section 8.
8   Incorrect idle speed (Chapter 1) or idle mixture (Chapter 4).
9   Incorrect ignition timing (Chapter 1).
10   Low or uneven cylinder compression. Check as described in Chapter 2.
11   Choke not operating properly (Chapter 1).
12   Clogged or dirty fuel injectors (Chapter 4).

10   Excessively high idle speed

1   Sticking throttle linkage (Chapter 4).
2   Choke opened excessively at idle (Chapter 4).
3   Idle speed incorrectly adjusted (Chapter 1).
4   Valve clearances incorrectly adjusted (V6 engine) (Chapter 2B).

11   Battery will not hold a charge

1   Alternator drivebelt defective or not adjusted properly (Chapter 1).
2   Battery cables loose or corroded (Chapter 1).
3   Alternator not charging properly (Chapter 5).
4   Loose, broken or faulty wires in the charging circuit (Chapter 5).
5   Short circuit causing a continuous drain on the battery.
6   Battery defective internally.

12   Alternator light stays on

1   Fault in alternator or charging circuit (Chapter 5).
2   Alternator drivebelt defective or not properly adjusted (Chapter 1).

13   Alternator light fails to come on when key is turned on

1   Faulty bulb (Chapter 12).

2   Defective alternator (Chapter 5).
3   Fault in the printed circuit, dash wiring or bulb holder (Chapter 12).

14   Engine misses throughout driving speed range

1   Fuel filter clogged and/or impurities in the fuel system. Check fuel filter (Chapter 1) or clean system (Chapter 4).
2   Faulty or incorrectly gapped spark plugs (Chapter 1).
3   Incorrect ignition timing (Chapter 1).
4   Cracked distributor cap, disconnected distributor wires or damaged distributor components (Chapter 1).
5   Defective spark plug wires (Chapter 1).
6   Emissions system components faulty (Chapter 6).
7   Low or uneven cylinder compression pressures. Check as described in Chapter 2.
8   Weak or faulty ignition coil (Chapter 5).
9   Weak or faulty ignition system (Chapter 5).
10   Vacuum leaks at intake manifold or vacuum hoses (see Section 8).
11   Dirty or clogged carburetor or fuel injector (Chapter 4).
12   Leaky EGR valve (Chapter 6).
13   Carburetor out of adjustment (Chapter 4).
14   Idle speed out of adjustment (Chapter 1).

15   Hesitation or stumble during acceleration

1   Ignition timing incorrect (Chapter 1).
2   Ignition system not operating properly (Chapter 5).
3   Dirty or clogged carburetor or fuel injector (Chapter 4).
4   Low fuel pressure. Check for proper operation of the fuel pump and for restrictions in the fuel filter and lines (Chapter 4).
5   Carburetor out of adjustment (Chapter 4).

16   Engine stalls

1   Idle speed incorrect (Chapter 1).
2   Fuel filter clogged and/or water and impurities in the fuel system (Chapter 1).
3   Choke not operating properly (Chapter 1).
4   Damaged or wet distributor cap and wires.
5   Emissions system components faulty (Chapter 6).
6   Faulty or incorrectly gapped spark plugs (Chapter 1). Also check the spark plug wires (Chapter 1).
7   Vacuum leak at the carburetor, intake manifold or vacuum hoses. Check as described in Section 8.

17   Engine lacks power

1   Incorrect ignition timing (Chapter 1).
2   Excessive play in distributor shaft. At the same time check for faulty distributor cap, wires, etc. (Chapter 1).
3   Faulty or incorrectly gapped spark plugs (Chapter 1).
4   Air filter dirty (Chapter 1).
5   Faulty ignition coil (Chapter 5).
6   Brakes binding (Chapters 1 and 10).
7   Automatic transmission fluid level incorrect, causing slippage (Chapter 1).
8   Clutch slipping (Chapter 8).
9   Fuel filter clogged and/or impurities in the fuel system (Chapters 1 and 4).
10   EGR system not functioning properly (Chapter 6).
11   Use of sub-standard fuel. Fill tank with proper octane fuel.
12   Low or uneven cylinder compression pressures. Check as described in Chapter 2.
13   Air leak at carburetor or intake manifold (check as described in Section 8).
14   Dirty or clogged carburetor jets or malfunctioning choke (Chapters 1 and 4).
15   Clogged catalytic converter.

18   Engine backfires

1   EGR system not functioning properly (Chapter 6).
2   Ignition timing incorrect (Chapter 1).
3   Thermostatic air cleaner system not operating properly (Chapter 6).
4   Vacuum leak (refer to Section 8).
5   Valve clearances incorrect (V6 engine) (Chapter 2B).
6   Damaged valve springs or sticking valves (Chapter 2).
7   Intake air leak (see Section 8).
8   Carburetor float level out of adjustment (Chapter 4).

19   Engine surges while holding accelerator steady

1   Intake air leak (see Section 8).
2   Fuel pump not working properly (Chapter 4).

20   Pinging or knocking engine sounds when engine is under load

1   Incorrect grade of fuel. Fill tank with fuel of the proper octane rating.
2   Ignition timing incorrect (Chapter 1).
3   Carbon build-up in combustion chambers. Remove cylinder head(s) and clean combustion chambers (Chapter 2).
4   Incorrect spark plugs (Chapter 1).

21 Engine diesels (continues to run) after being turned off

1 Idle speed too high (Chapter 1).
2 Ignition timing incorrect (Chapter 1).
3 Incorrect spark plug heat range (Chapter 1).
4 Intake air leak (see Section 8).
5 Carbon build-up in combustion chambers. Remove the cylinder head(s) and clean the combustion chambers (Chapter 2).
6 Valves sticking (Chapter 2).
7 Valve clearance incorrect (V6 engine) (Chapter 2B).
8 EGR system not operating properly (Chapter 6).
9 Fuel shut-off system not operating properly (Chapter 6).
10 Check for causes of overheating (Section 27).

22 Low oil pressure

1 Improper grade of oil.
2 Oil pump worn or damaged (Chapter 2).
3 Engine overheating (refer to Section 27).
4 Clogged oil filter (Chapter 1).
5 Clogged oil strainer (Chapter 2).
6 Oil pressure gauge not working properly (Chapter 2).

23 Excessive oil consumption

1 Loose oil drain plug.
2 Loose bolts or damaged oil pan gasket (Chapter 2).
3 Loose bolts or damaged front cover gasket (Chapter 2).
4 Front or rear crankshaft oil seal leaking (Chapter 2).
5 Loose bolts or damaged rocker arm cover gasket (Chapter 2).
6 Loose oil filter (Chapter 1).
7 Loose or damaged oil pressure switch (Chapter 2).
8 Pistons and cylinders excessively worn (Chapter 2).
9 Piston rings not installed correctly on pistons (Chapter 2).
10 Worn or damaged piston rings (Chapter 2).
11 Intake and/or exhaust valve oil seals worn or damaged (Chapter 2).
12 Worn valve stems.
13 Worn or damaged valves/guides (Chapter 2).

24 Excessive fuel consumption

1 Dirty or clogged air filter element (Chapter 1).
2 Incorrect ignition timing (Chapter 1).
3 Incorrect idle speed (Chapter 1).
4 Low tire pressure or incorrect tire size (Chapter 11).

5 Fuel leakage. Check all connections, lines and components in the fuel system (Chapter 4).
6 Choke not operating properly (Chapter 1).
7 Dirty or clogged carburetor jets or fuel injectors (Chapter 4).

25 Fuel odor

1 Fuel leakage. Check all connections, lines and components in the fuel system (Chapter 4).
2 Fuel tank overfilled. Fill only to automatic shut-off.
3 Charcoal canister filter in Evaporative Emissions Control system clogged (Chapter 1).
4 Vapor leaks from Evaporative Emissions Control system lines (Chapter 6).

26 Miscellaneous engine noises

1 A strong dull noise that becomes more rapid as the engine accelerates indicates worn or damaged crankshaft bearings or an unevenly worn crankshaft. To pinpoint the trouble spot, remove the spark plug wire from one plug at a time and crank the engine over. If the noise stops, the cylinder with the removed plug wire indicates the problem area. Replace the bearing and/or service or replace the crankshaft (Chapter 2).
2 A similar (yet slightly higher pitched) noise to the crankshaft knocking described in the previous paragraph, that becomes more rapid as the engine accelerates, indicates worn or damaged connecting rod bearings (Chapter 2). The procedure for locating the problem cylinder is the same as described in Paragraph 1.
3 An overlapping metallic noise that increases in intensity as the engine speed increases, yet diminishes as the engine warms up indicates abnormal piston and cylinder wear (Chapter 2).To locate the problem cylinder, use the procedure described in Paragraph 1.
4 A rapid clicking noise that becomes faster as the engine accelerates indicates a worn piston pin or piston pin hole. This sound will happen each time the piston hits the highest and lowest points in the stroke (Chapter 2). The procedure for locating the problem piston is described in Paragraph 1.
5 A metallic clicking noise coming from the water pump indicates worn or damaged water pump bearings or pump. Replace the water pump with a new one (Chapter 3).
6 A rapid tapping sound or clicking sound that becomes faster as the engine speed increases indicates "valve tapping" or improperly adjusted valve clearances. This can be identified by holding one end of a section of hose to your ear and placing the other end at different spots along the rocker arm cover. The point where the sound is loudest

indicates the problem valve. Adjust the valve clearance (V6 engine) (Chapter 2B). If the problem persists, you likely have a collapsed valve lifter or other damaged valve train component. Changing the engine oil and adding a high viscosity oil treatment will sometimes cure a stuck lifter problem. If the problem still persists, the lifters, pushrods and rocker arms must be removed for inspection (see Chapter 2).
7 A steady metallic rattling or rapping sound coming from the area of the timing chain cover indicates a worn, damaged or out-of-adjustment timing chain. Service or replace the chain and related components (Chapter 2).

Cooling system

27 Overheating

1 Insufficient coolant in system (Chapter 1).
2 Drivebelt defective or not adjusted properly (Chapter 1).
3 Radiator core blocked or radiator grille dirty and restricted (Chapter 3).
4 Thermostat faulty (Chapter 3).
5 Fan not functioning properly (Chapter 3).
6 Radiator cap not maintaining proper pressure. Have cap pressure tested by gas station or repair shop.
7 Ignition timing incorrect (Chapter 1).
8 Defective water pump (Chapter 3).
9 Improper grade of engine oil.
10 Inaccurate temperature gauge (Chapter 12).

28 Overcooling

1 Thermostat faulty (Chapter 3).
2 Inaccurate temperature gauge (Chapter 12).

29 External coolant leakage

1 Deteriorated or damaged hoses. Loose clamps at hose connections (Chapter 1).
2 Water pump seals defective. If this is the case, water will drip from the weep hole in the water pump body (Chapter 3).
3 Leakage from radiator core or header tank. This will require the radiator to be professionally repaired (see Chapter 3 for removal procedures).
4 Engine drain plugs or water jacket freeze plugs leaking (see Chapters 1 and 2).
5 Leak from coolant temperature switch (Chapter 3).
6 Leak from damaged gaskets or small cracks (Chapter 2).
7 Damaged head gasket. This can be veri-

fied by checking the condition of the engine oil as noted in Section 30.

30 Internal coolant leakage

Note: *Internal coolant leaks can usually be detected by examining the oil. Check the dipstick and inside the rocker arm cover for water deposits and an oil consistency like that of a milkshake.*

1   Leaking cylinder head gasket. Have the system pressure tested or remove the cylinder head (Chapter 2) and inspect.
2   Cracked cylinder bore or cylinder head. Dismantle engine and inspect (Chapter 2).
3   Loose cylinder head bolts (tighten as described in Chapter 2).

31 Abnormal coolant loss

1   Overfilling system (Chapter 1).
2   Coolant boiling away due to overheating (see causes in Section 27).
3   Internal or external leakage (see Sections 29 and 30).
4   Faulty radiator cap. Have the cap pressure tested.
5   Cooling system being pressurized by engine compression. This could be due to a cracked head or block or leaking head gasket(s).

32 Poor coolant circulation

1   Inoperative water pump. A quick test is to pinch the top radiator hose closed with your hand while the engine is idling, then release it. You should feel a surge of coolant if the pump is working properly (Chapter 3).
2   Restriction in cooling system. Drain, flush and refill the system (Chapter 1). If necessary, remove the radiator (Chapter 3) and have it reverse flushed or professionally cleaned.
3   Loose water pump drivebelt (Chapter 1).
4   Thermostat sticking (Chapter 3).
5   Insufficient coolant (Chapter 1).

33 Corrosion

1   Excessive impurities in the water. Soft, clean water is recommended. Distilled or rainwater is satisfactory.
2   Insufficient antifreeze solution (refer to Chapter 1 for the proper ratio of water to antifreeze).
3   Infrequent flushing and draining of system. Regular flushing of the cooling system should be carried out at the specified intervals as described in (Chapter 1).

Clutch

Note: *All clutch related service information is located in Chapter 8, unless otherwise noted.*

34 Fails to release (pedal pressed to the floor, shift lever does not move freely in and out of Reverse)

1   Clutch contaminated with oil. Remove clutch plate and inspect.
2   Clutch plate warped, distorted or otherwise damaged.
3   Diaphragm spring fatigued. Remove clutch cover/pressure plate assembly and inspect.
4   Leakage of fluid from clutch hydraulic system. Inspect master cylinder, operating cylinder and connecting lines.
5   Air in clutch hydraulic system. Bleed the system.
6   Insufficient pedal stroke. Check and adjust as necessary.
7   Piston seal in operating cylinder deformed or damaged.
8   Lack of grease on pilot bushing.

35 Clutch slips (engine speed increases with no increase in vehicle speed)

1   Worn or oil soaked clutch plate.
2   Clutch plate not broken in. It may take 30 or 40 normal starts for a new clutch to seat.
3   Diaphragm spring weak or damaged. Remove clutch cover/pressure plate assembly and inspect.
4   Flywheel warped (Chapter 2).
5   Debris in master cylinder preventing the piston from returning to its normal position.
6   Clutch hydraulic line damaged.

36 Grabbing (chattering) as clutch is engaged

1   Oil on clutch plate. Remove and inspect. Repair any leaks.
2   Worn or loose engine or transmission mounts. They may move slightly when clutch is released. Inspect mounts and bolts.
3   Worn splines on transmission input shaft. Remove clutch components and inspect.
4   Warped pressure plate or flywheel. Remove clutch components and inspect.
5   Diaphragm spring fatigued. Remove clutch cover/pressure plate assembly and inspect.
6   Clutch linings hardened or warped.
7   Clutch lining rivets loose.

37 Squeal or rumble with clutch engaged (pedal released)

1   Improper pedal adjustment. Adjust pedal free play.
2   Release bearing binding on transmission shaft. Remove clutch components and check bearing. Remove any burrs or nicks, clean and relubricate before reinstallation.
3   Pilot bushing worn or damaged.
4   Clutch rivets loose.
5   Clutch plate cracked.
6   Fatigued clutch plate torsion springs. Replace clutch plate.

38 Squeal or rumble with clutch disengaged (pedal depressed)

1   Worn or damaged release bearing.
2   Worn or broken pressure plate diaphragm fingers.

39 Clutch pedal stays on floor when disengaged

Binding linkage or release bearing. Inspect linkage or remove clutch components as necessary.

Manual transmission

Note: *All manual transmission service information is located in Chapter 7, unless otherwise noted.*

40 Noisy in Neutral with engine running

1   Input shaft bearing worn.
2   Damaged main drive gear bearing.
3   Insufficient transmission oil (Chapter 1).
4   Transmission oil in poor condition. Drain and fill with proper grade oil. Check old oil for water and debris (Chapter 1).
5   Noise can be caused by variations in engine torque. Change the idle speed and see if noise disappears.

41 Noisy in all gears

1   Any of the above causes, and/or:
2   Worn or damaged output gear bearings or shaft.

42 Noisy in one particular gear

1   Worn, damaged or chipped gear teeth.
2   Worn or damaged synchronizer.

43 Slips out of gear

1 Transmission loose on clutch housing.
2 Stiff shift lever seal.
3 Shift linkage binding.
4 Broken or loose input gear bearing retainer.
5 Dirt between clutch lever and engine housing.
6 Worn linkage.
7 Damaged or worn check balls, fork rod ball grooves or check springs.
8 Worn mainshaft or countershaft bearings.
9 Loose engine mounts (Chapter 2).
10 Excessive gear end play.
11 Worn synchronizers.

44 Oil leaks

1 Excessive amount of lubricant in transmission (see Chapter 1 for correct checking procedures). Drain lubricant as required.
2 Rear oil seal or speedometer oil seal damaged.
3 To pinpoint a leak, first remove all built-up dirt and grime from the transmission. Degreasing agents and/or steam cleaning will achieve this. With the underside clean, drive the vehicle at low speeds so the air flow will not blow the leak far from its source. Raise the vehicle and determine where the leak is located.

45 Difficulty engaging gears

1 Clutch not releasing completely.
2 Loose or damaged shift linkage. Make a thorough inspection, replacing parts as necessary.
3 Insufficient transmission oil (Chapter 1).
4 Transmission oil in poor condition. Drain and fill with proper grade oil. Check oil for water and debris (Chapter 1).
5 Worn or damaged striking rod.
6 Sticking or jamming gears.

46 Noise occurs while shifting gears

1 Check for proper operation of the clutch (Chapter 8).
2 Faulty synchronizer assemblies. Measure baulk ring-to-gear clearance. Also, check for wear or damage to baulk rings or any parts of the synchromesh assemblies.

Automatic transmission

Note: *Due to the complexity of the automatic transmission, it's difficult for the home mechanic to properly diagnose and service. For problems other than the following, the vehicle should be taken to a reputable mechanic.*

47 Fluid leakage

1 Automatic transmission fluid is a deep red color, and fluid leaks should not be confused with engine oil which can easily be blown by air flow to the transmission.
2 To pinpoint a leak, first remove all built-up dirt and grime from the transmission. Degreasing agents and/or steam cleaning will achieve this. With the underside clean, drive the vehicle at low speeds so the air flow will not blow the leak far from its source. Raise the vehicle and determine where the leak is located. Common areas of leakage are:

a) *Fluid pan: tighten mounting bolts and/or replace pan gasket as necessary (Chapter 1).*
b) *Rear extension: tighten bolts and/or replace oil seal as necessary.*
c) *Filler pipe: replace the rubber oil seal where pipe enters transmission case.*
d) *Transmission oil lines: tighten fittings where lines enter transmission case and/or replace lines.*
e) *Vent pipe: transmission overfilled and/or water in fluid (see checking procedures, Chapter 1).*
f) *Speedometer connector: replace the O-ring where speedometer cable enters transmission case.*

48 General shift mechanism problems

Chapter 7 deals with checking and adjusting the shift linkage on automatic transmissions. Common problems which may be caused by out of adjustment linkage are:

a) *Engine starting in gears other than P (park) or N (Neutral).*
b) *Indicator pointing to a gear other than the one actually engaged.*
c) *Vehicle moves with transmission in P (Park) position.*

49 Transmission will not downshift with the accelerator pedal pressed to the floor

Chapter 7 deals with adjusting the TV linkage to enable the transmission to downshift properly.

50 Engine will start in gears other than Park or Neutral

Chapter 7 deals with adjusting the Neutral start switch installed on automatic transmissions.

51 Transmission slips, shifts rough, is noisy or has no drive in forward or Reverse gears

1 There are many probable causes for the above problems, but the home mechanic should concern himself only with one possibility; fluid level.
2 Before taking the vehicle to a shop, check the fluid level and condition as described in Chapter 1. Add fluid, if necessary, or change the fluid and filter if needed. If problems persist, have a professional diagnose the transmission.

Driveshaft

Note: *On 2WD models refer to Chapter 8, unless otherwise specified, for service information. On 4WD models, refer to Chapter 9.*

52 Leaks at front of driveshaft

Defective transmission rear seal. See Chapter 7 for replacement procedure. As this is done, check the splined yoke for burrs or roughness that could damage the new seal. Remove burrs with a fine file or whetstone.

53 Knock or clunk when transmission is under initial load (just after transmission is put into gear)

1 Loose or disconnected rear suspension components. Check all mounting bolts and bushings (Chapters 7 and 10).
2 Loose driveshaft bolts. Inspect all bolts and nuts and tighten them securely.
3 Worn or damaged universal joint bearings. Replace driveshaft (Chapter 8).
4 Worn sleeve yoke and mainshaft spline.

54 Metallic grating sound consistent with vehicle speed

Pronounced wear in the universal joint bearings. Replace U-joints or driveshafts, as necessary.

55 Vibration

Note: *Before blaming the driveshaft, make sure the tires are perfectly balanced and perform the following test.*

1 Install a tachometer inside the vehicle to monitor engine speed as the vehicle is driven. Drive the vehicle and note the engine speed at which the vibration (roughness) is most

pronounced. Now shift the transmission to a different gear and bring the engine speed to the same point.

2   If the vibration occurs at the same engine speed (rpm) regardless of which gear the transmission is in, the driveshaft is NOT at fault since the driveshaft speed varies.

3   If the vibration decreases or is eliminated when the transmission is in a different gear at the same engine speed, refer to the following probable causes.

4   Bent or dented driveshaft. Inspect and replace as necessary.

5   Undercoating or built-up dirt, etc. on the driveshaft. Clean the shaft thoroughly.

6   Worn universal joint bearings. Replace the U-joints or driveshaft as necessary.

7   Driveshaft and/or companion flange out of balance. Check for missing weights on the shaft. Remove driveshaft and reinstall 180~ from original position, then recheck. Have the driveshaft balanced if problem persists.

8   Loose driveshaft mounting bolts/nuts.

9   Defective center bearing, if so equipped.

10   Worn transmission rear bushing (Chapter 7).

56   Scraping noise

Make sure the dust cover on the sleeve yoke isn't rubbing on the transmission extension housing.

57   Whining or whistling noise

Defective center bearing, if so equipped.

Rear axle and differential

Note: *For differential servicing information on 2WD models, refer to Chapter 8, unless otherwise specified. On 4WD models, refer to Chapter 9.*

58   Noise, same when in drive as when vehicle is coasting

1   Road noise. No corrective action available.

2   Tire noise. Inspect tires and check tire pressures (Chapter 1).

3   Front wheel bearings loose, worn or damaged (Chapter 1).

4   Insufficient differential oil (Chapter 1).

5   Defective differential.

59   Knocking sound when starting or shifting gears

Defective or incorrectly adjusted differential.

60   Noise when turning

Defective differential.

61   Vibration

See probable causes under Driveshaft. Proceed under the guidelines listed for the driveshaft. If the problem persists, check the rear wheel bearings by raising the rear of the vehicle and spinning the wheels by hand. Listen for evidence of rough (noisy) bearings. Remove and inspect (Chapter 8).

62   Oil leaks

1   Pinion oil seal damaged (Chapter 8).

2   Axleshaft oil seals damaged (Chapter 8).

3   Differential cover leaking. Tighten mounting bolts or replace the gasket as required.

4   Loose filler or drain plug on differential (Chapter 1).

5   Clogged or damaged breather on differential.

Transfer case (4WD models)

Note: *Refer to Chapter 7C for 4WD system service and repair information.*

63   Gear jumping out of mesh

1   Incorrect control lever free play (Chapter 7C).

2   Interference between the control lever and the console.

3   Play or fatigue in the transfer case mounts.

4   Internal wear or incorrect adjustments.

64   Difficult shifting

1   Lack of oil.

2   Internal wear, damage or incorrect adjustment.

65   Noise

1   Lack of oil in transfer case.

2   Noise in 4H and 4L, but not in 2H indicates cause is in the front differential or front axle.

3   Noise in 2H, 4H and 4L indicates cause is in rear differential or rear axle.

4   Noise in 2H and 4H but not in 4L, or in 4L only, indicates internal wear or damage in transfer case.

Brakes

Note: *Before assuming a brake problem exists, make sure the tires are in good condition and inflated properly, the front end alignment is correct and the vehicle is not loaded with weight in an unequal manner. All service procedures for the brakes are included in Chapter 10, unless otherwise noted.*

66   Vehicle pulls to one side during braking

1   Defective, damaged or oil contaminated brake pad on one side. Inspect as described in Chapter 1. Refer to Chapter 10 if replacement is required.

2   Excessive wear of brake pad material or disc on one side. Inspect and repair as necessary.

3   Loose or disconnected front suspension components. Inspect and tighten all bolts securely (Chapters 1 and 11).

4   Defective caliper assembly. Remove caliper and inspect for stuck piston or damage.

5   Brake pad to rotor adjustment needed. Inspect automatic adjusting mechanism for proper operation.

6   Scored or out of round rotor.

7   Loose caliper mounting bolts.

8   Incorrect wheel bearing adjustment.

67   Noise (high-pitched squeal)

1   Front brake pads worn out. This noise comes from the wear sensor rubbing against the disc. Replace pads with new ones immediately!

2   Glazed or contaminated pads.

3   Dirty or scored rotor.

4   Bent support plate.

68   Excessive brake pedal travel

1   Partial brake system failure. Inspect entire system (Chapter 1) and correct as required.

2   Insufficient fluid in master cylinder. Check (Chapter 1) and add fluid, bleed system if necessary.

3   Air in system. Bleed system.

4   Excessive lateral rotor play.

5   Brakes out of adjustment. Check the operation of the automatic adjusters.

6   Defective proportioning valve. Replace valve and bleed system.

69   Brake pedal feels spongy when depressed

1   Air in brake lines. Bleed the brake system.

2   Deteriorated rubber brake hoses.

Inspect all system hoses and lines. Replace parts as necessary.

3   Master cylinder mounting nuts loose. Inspect master cylinder bolts (nuts) and tighten them securely.

4   Master cylinder faulty.

5   Incorrect shoe or pad clearance.

6   Defective check valve. Replace valve and bleed system.

7   Clogged reservoir cap vent hole.

8   Deformed rubber brake lines.

9   Soft or swollen caliper seals.

10   Poor quality brake fluid. Bleed entire system and fill with new approved fluid.

70   Excessive effort required to stop vehicle

1   Power brake booster not operating properly.

2   Excessively worn linings or pads. Check and replace if necessary.

3   One or more caliper pistons seized or sticking. Inspect and rebuild as required.

4   Brake pads or linings contaminated with oil or grease. Inspect and replace as required.

5   New pads or linings installed and not yet seated. It'll take a while for the new material to seat against the rotor or drum.

6   Worn or damaged master cylinder or caliper assemblies. Check particularly for frozen pistons.

7   Also see causes listed under Section 69.

71   Pedal travels to the floor with little resistance

Little or no fluid in the master cylinder reservoir caused by leaking caliper piston(s) or loose, damaged or disconnected brake lines. Inspect entire system and repair as necessary.

72   Brake pedal pulsates during brake application

1   Wheel bearings damaged, worn or out of adjustment (Chapter 1).

2   Caliper not sliding properly due to improper installation or obstructions. Remove and inspect.

3   Rotor not within specifications. Remove the rotor and check for excessive lateral runout and parallelism. Have the rotors resurfaced or replace them with new ones. Also make sure that all rotors are the same thickness.

4   Out of round rear brake drums. Remove the drums and have them turned or replace them with new ones.

73   Brakes drag (indicated by sluggish engine performance or wheels being very hot after driving)

1   Output rod adjustment incorrect at the brake pedal.

2   Obstructed master cylinder compensator. Disassemble master cylinder and clean.

3   Master cylinder piston seized in bore. Overhaul master cylinder.

4   Caliper assembly in need of overhaul.

5   Brake pads or shoes worn out.

6   Piston cups in master cylinder or caliper assembly deformed. Overhaul master cylinder.

7   Rotor not within specifications (Section 72).

8   Parking brake assembly will not release.

9   Clogged brake lines.

10   Wheel bearings out of adjustment (Chapter 1).

11   Brake pedal height improperly adjusted.

12   Wheel cylinder needs overhaul.

13   Improper shoe to drum clearance. Adjust as necessary.

74   Rear brakes lock up under light brake application

1   Tire pressures too high.

2   Tires excessively worn (Chapter 1).

75   Rear brakes lock up under heavy brake application

1   Tire pressures too high.

2   Tires excessively worn (Chapter 1).

3   Front brake pads contaminated with oil, mud or water. Clean or replace the pads.

4   Front brake pads excessively worn.

5   Defective master cylinder or caliper assembly.

Suspension and steering

Note: *All service procedures for the suspension and steering systems are included in Chapter 11, unless otherwise noted.*

76   Vehicle pulls to one side

1   Tire pressures uneven (Chapter 1).

2   Defective tire (Chapter 1).

3   Excessive wear in suspension or steering components (Chapter 1).

4   Front end alignment incorrect.

5   Front brakes dragging. Inspect as described in Section 73.

6   Wheel bearings improperly adjusted (Chapter 1).

7   Wheel lug nuts loose.

77   Shimmy, shake or vibration

1   Tire or wheel out of balance or out of round. Have them balanced on the vehicle.

2   Loose, worn or out of adjustment wheel bearings (Chapter 1).

3   Shock absorbers and/or suspension components worn or damaged. Check for worn bushings in the upper and lower links.

4   Wheel lug nuts loose.

5   Incorrect tire pressures.

6   Excessively worn or damaged tire.

7   Loosely mounted steering gear housing.

8   Steering gear improperly adjusted.

9   Loose, worn or damaged steering components.

10   Damaged idler arm.

11   Worn balljoint.

78   Excessive pitching and/or rolling around corners or during braking

1   Defective shock absorbers. Replace as a set.

2   Broken or weak leaf springs and/or suspension components.

3   Worn or damaged stabilizer bar or bushings.

79   Wandering or general instability

1   Improper tire pressures.

2   Worn or damaged upper and lower link or tension rod bushings.

3   Incorrect front end alignment.

4   Worn or damaged steering linkage or suspension components.

5   Improperly adjusted steering gear.

6   Out of balance wheels.

7   Loose wheel lug nuts.

8   Worn rear shock absorbers.

9   Fatigued or damaged rear leaf springs.

80   Excessively stiff steering

1   Lack of lubricant in power steering fluid reservoir, where appropriate (Chapter 1).

2   Incorrect tire pressures (Chapter 1).

3   Lack of lubrication at balljoints (Chapter 1).

4   Front end out of alignment.

5   Steering gear out of adjustment or lacking lubrication.

6   Improperly adjusted wheel bearings.

7   Worn or damaged steering gear.

8   Interference of steering column with turn signal switch.

9   Low tire pressures.

10   Worn or damaged balljoints.

11   Worn or damaged steering linkage.

12   See also Section 79.

81   Excessive play in steering

1   Loose wheel bearings (Chapter 1).
2   Excessive wear in suspension bushings (Chapter 1).
3   Steering gear improperly adjusted.
4   Incorrect front end alignment.
5   Steering gear mounting bolts loose.
6   Worn steering linkage.

82   Lack of power assistance

1   Steering pump drivebelt faulty or not adjusted properly (Chapter 1).
2   Fluid level low (Chapter 1).
3   Hoses or pipes restricting the flow. Inspect and replace parts as necessary.
4   Air in power steering system. Bleed system.
5   Defective power steering pump.

83   Steering wheel fails to return to straight-ahead position

1   Incorrect front end alignment.
2   Tire pressures low.
3   Steering gears improperly engaged.
4   Steering column out of alignment.
5   Worn or damaged balljoint.
6   Worn or damaged steering linkage.
7   Improperly lubricated idler arm.
8   Insufficient oil in steering gear.
9   Lack of fluid in power steering pump.

84   Steering effort not the same in both directions (power system)

1   Leaks in steering gear.
2   Clogged fluid passage in steering gear.

85   Noisy power steering pump

1   Insufficient oil in pump.
2   Clogged hoses or oil filter in pump.
3   Loose pulley.
4   Improperly adjusted drivebelt (Chapter 1).
5   Defective pump.

86   Miscellaneous noises

1   Improper tire pressures.
2   Insufficiently lubricated balljoint or steering linkage.
3   Loose or worn steering gear, steering linkage or suspension components.
4   Defective shock absorber.
5   Defective wheel bearing.
6   Worn or damaged suspension bushings.
7   Damaged leaf spring.
8   Loose wheel lug nuts.
9   Worn or damaged rear axleshaft spline.
10   Worn or damaged rear shock absorber mounting bushing.
11   Incorrect rear axle end play.
12   See also causes of noises at the rear axle and driveshaft.

87   Excessive tire wear (not specific to one area)

1   Incorrect tire pressures.
2   Tires out of balance. Have them balanced on the vehicle.
3   Wheels damaged. Inspect and replace as necessary.
4   Suspension or steering components worn (Chapter 1).

88   Excessive tire wear on outside edge

1   Incorrect tire pressure.
2   Excessive speed in turns.
3   Front end alignment incorrect (excessive toe-in).

89   Excessive tire wear on inside edge

1   Incorrect tire pressure.
2   Front end alignment incorrect (toe-out).
3   Loose or damaged steering components (Chapter 1).

90   Tire tread worn in one place

1   Tires out of balance. Have them balanced on the vehicle.
2   Damaged or buckled wheel. Inspect and replace if necessary.
3   Defective tire.

Chapter 1
Tune-up and routine maintenance

Contents

Specification

Recommended lubricants and fluids

Note: *Listed here are manufacturer recommendations at the time this manual was written. Manufacturers occasionally upgrade their fluid and lubricant specifications, so check with your local auto parts store for current recommendations.*

Engine oil type	API grade SJ or better multigrade and fuel efficient oil
Engine oil viscosity	See accompanying chart

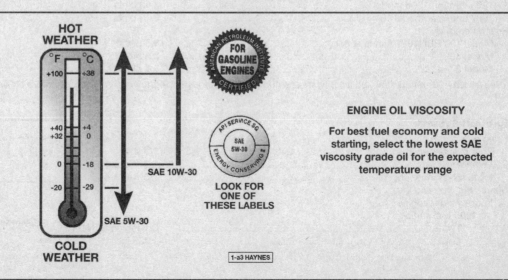

ENGINE OIL VISCOSITY

For best fuel economy and cold starting, select the lowest SAE viscosity grade oil for the expected temperature range

1-a3 HAYNES

Recommended lubricants and fluids

Automatic transmission fluid*	
AW-4 transmission	Mercon ATF
727 and 999 transmissions	Dexron II ATF
Manual transmission lubricant	SAE 75W-90 GL-5 gear lubricant

* The fluid type should be indicated on the dipstick

Differential lubricant	
Front differential	
Normal use	SAE 75W or SAE 80W-90 GL5 gear lubricant
Trailer towing	SAE 80W GL-5 gear lubricant
Limited slip differential	Add 2 oz. of Friction Modifier Additive
Rear differential	
Normal use	SAE 75W or SAE 80W-90 GL5 gear lubricant
Trailer towing with Class III hitch (5000 lb.)	SAE 75W-140 synthetic gear lubricant
Vehicles originally equipped with a trailer towing package	API GL5 80W-140 gear lubricant
With limited slip differential (all models)	Add 2 oz. of Friction Modifier Additive
Track Lock differential models 35,194 RBI and 8-1/4	Add 5 oz. of Friction Modifier Additive
Transfer case lubricant	Dexron II, or III, or Mercon automatic transmission fluid
Chassis grease	NLGI No. 2 chassis grease
Engine coolant	Mixture of water and ethylene glycol-base antifreeze
Brake fluid	DOT-3 brake fluid
Clutch fluid	DOT-3 brake fluid
Power steering fluid	Jeep power steering fluid or equivalent
Manual steering box lubricant	SAE 75W-90 GL-5 gear lubricant
Wheel bearing grease (2WD)	NLGI No. 2 moly-base wheel bearing grease

Capacities

Engine oil (with filter change, approximate)	
Four-cylinder engine	4 qts
V6 engine	4 qts
Inline six-cylinder engine	6 qts
Cooling system (approximate)	
Four-cylinder engine	10 qts
V6 engine and inline six-cylinder engines	12.5 qts
Fuel tank	
Standard	13.5 gal
Optional	20 gal
Automatic transmission (approximate)	4 qts (when draining pan and replacing filter only)
Manual transmission (approximate)	
4-speed	7.5 pints
5-speed	7.2 pints
Transfer case (approximate)	
Selec-trac	3.0 pints
Command-trac	2.2 pints
Front differential (4x4 models only)	
1984 through 1994 all models	2.5 pints
1995 through 1996 model 30	3.13 pints
1997 and later model 181 FBI	3.13 pints
Rear differential	
Model 194 RBI (1997 and later only)	3.5 pints*
Model 35	3.5 pints*
Model 8-1/4	4.8 pints*

* See above "rear differential" lubricant types and additives for various driving/towing applications before refilling.

Ignition system

Firing order	
Four-cylinder engine	1-3-4-2
V6 engine	1-2-3-4-5-6
Inline six-cylinder engine	1-5-3-6-2-4
Spark plug type and gap	
Four-cylinder engine	
1985 and earlier	
Type	Champion RFN14LY
Gap	0.035 in

FOUR-CYLINDER ENGINE

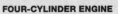

Cylinder location and distributor rotation

Spark plug type and gap (continued)
   Four-cylinder engine
      1986 and later
         Type... Champion RC12LYC
         Gap.. 0.035 in
   V6 engine
      Type... Champion RV12YC
      Gap.. 0.045 in
   Inline six-cylinder engine
      1990 and earlier
         Type... Champion RC9YC
         Gap.. 0.035 in
      1991 and later
         Type... Champion RC12LYC
         Gap.. 0.035 in
Ignition timing
   Four-cylinder engine (1984 through 1986 models only)
      Below 4000 feet... 12-degrees BTDC
      Above 4000 feet... 19-degrees BTDC
   V6 engine
      Automatic transmission .... 12-degrees BTDC
      Manual transmission
         California models.. 10-degrees BTDC
         All others.. 8-degrees BTDC
   Inline six-cylinder engine .... Timing not adjustable

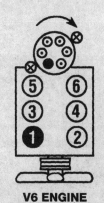

V6 ENGINE

The blackened terminal shown on the distributor cap indicates the Number One spark plug wire position

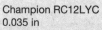

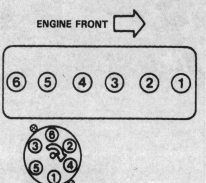

INLINE SIX-CYLINDER ENGINE

Cylinder location and distributor rotation

General

Engine idle speed (carbureted-models only)
   Four-cylinder engine
      Automatic transmission (in Drive) .... 700 rpm
      Manual transmission... 750 rpm
   V6 engine
      Automatic transmission (in Drive) .... 700 ± 50 rpm
      Manual transmission... 700 ± 50 rpm

Drivebelt tension (with special gauge)

Conventional V-belts
   New .... 120 to 150 lbs
   Used .... 90 to 115 lbs
Serpentine belt
   New .... 180 to 200 lbs
   Used .... 140 to 160 lbs

Brakes

Brake pad wear limit.. 1/8 in
Brake shoe wear limit .... 1/16 in

Torque specifications

	Ft-lbs
Differential (axle) fill plug	10 to 20
Engine oil drain plug	20
Wheel lug nuts	75
Manual transmission check/fill plug	15 to 25
Manual transmission drain plug	15 to 25
Transfer case drain/fill plug	20
Automatic transmission oil pan bolts	10
Carburetor mounting nuts	13 to 19
Throttle body mounting nuts	16
Carburetor-mounted fuel filter nut	18
Spark plugs	
Four-cylinder engine	27
V6 engine	22
Inline six-cylinder engine	27

Engine compartment component checking points (inline six-cylinder engine shown)

1	Crankcase Ventilation (CCV) system hose and fitting	
2	Clutch fluid reservoir	
3	Brake fluid reservoir	
4	Windshield washer reservoir	
5	Air cleaner housing	
6	Power steering fluid reservoir	
7	Drivebelt	
8	CCV fresh air hose	
9	Engine oil filler cap	
10	Engine oil dipstick	
11	Radiator hose	
12	Battery	
13	Air conditioner hose	
14	Coolant pressure bottle	

Typical engine compartment under side components (4WD vehicle shown)

1	Front driveshaft slip joint grease fitting	4	Exhaust pipe	8	Front driveaxle
2	Transmission	5	Brake hose	9	Front disc brake caliper
3	Engine oil drain plug	6	Steering linkage	10	Front driveshaft universal joint
		7	Steering damper		

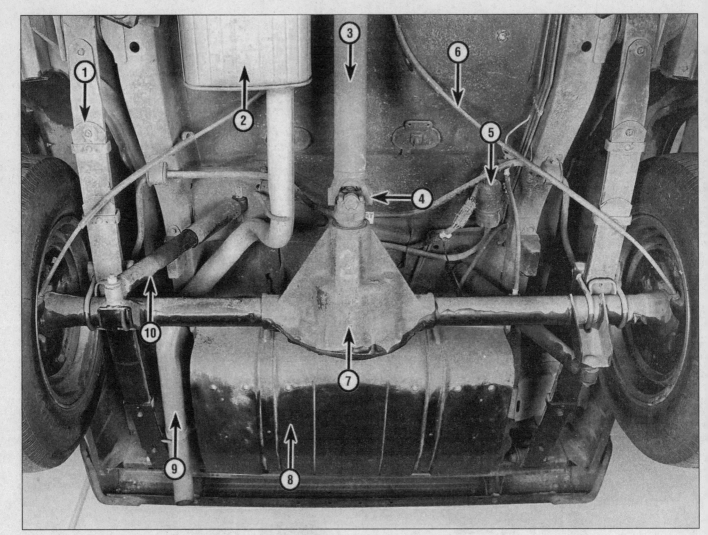

Typical rear under side vehicle components

1	Rear leaf spring	4	Universal joint	6	Parking brake cable	8	Fuel tank
2	Muffler	5	Fuel filter	7	Rear axle	9	Exhaust pipe
3	Driveshaft					10	Shock absorber

1 Jeep Cherokee/Wagoneer/Comanche Maintenance schedule

The following maintenance intervals are based on the assumption that the vehicle owner will be doing the maintenance or service work, as opposed to having a dealer service department do the work. Although the time/mileage intervals are loosely based on factory recommendations, most have been shortened to ensure, for example, that such items as lubricants and fluids are checked/changed at intervals that promote maximum engine/driveline service life. Also, subject to the preference of the individual owner interested in keeping his or her vehicle in peak condition at all times, and with the vehicle's ultimate resale in mind, many of the maintenance procedures may be performed more often than recommended in the following schedule. We encourage such owner initiative.

When the vehicle is new it should be serviced initially by a factory authorized dealer service department to protect the factory warranty. In many cases the initial maintenance check is done at no cost to the owner (check with your dealer service department for more information).

Every 250 miles or weekly, whichever comes first

Check the engine oil level (Section 4)
Check the engine coolant level (Section 4)
Check the windshield washer fluid level (Section 4)
Check the brake and clutch fluid levels (Section 4)
Check the tires and tire pressures (Section 5)

Every 3000 miles or 3 months, whichever comes first

All items listed above plus:

Check the automatic transmission fluid level (Section 6)
Check the power steering fluid level (Section 7)
Check and service the battery (Section 8)
Check the cooling system (Section 9)
Inspect and replace, if necessary, all underhood hoses (Section 10)
Inspect and replace, if necessary, the windshield wiper blades (Section 11)

Every 7500 miles or 12 months, whichever comes first

All items listed above plus:

Change the engine oil and filter (Section 12)*
Lubricate the chassis components (Section 13)
Inspect the suspension and steering components (Section 14)*
Inspect the exhaust system (Section 15)*
Check the manual transmission lubricant level (Section 16)*
Check the differential lubricant level (Section 17)*
Check the transfer case lubricant level (4WD models) (Section 18)
Rotate the tires (Section 19)
Check the brakes (Section 20)*
Inspect the fuel system (Section 21)

Check the carburetor choke operation (Section 22)
Check the carburetor/throttle body mounting nut torque (Section 23)
Check the throttle linkage (Section 24)
Check the thermostatically-controlled air cleaner (Section 25)
Check the engine drivebelts (Section 26)
Check the seat belts (Section 27)
Check the starter safety switch (Section 28)
Check the spare tire and jack (Section 29)

Every 30,000 miles or 24 months, whichever comes first

All items listed above plus:

Check and adjust, if necessary, the engine idle speed (Section 30)
Replace the fuel filter (Section 31) (1984 through 1996 models)
Replace the air and PCV filters (Section 32)
Check and adjust, if necessary, the ignition timing (Section 33)
Change the automatic transmission fluid (Section 34)**
Change the manual transmission lubricant (Section 35)
Change the differential lubricant (Section 36)
Change the transfer case lubricant (4WD models) (Section 37)
Check and repack the front wheel bearings (2WD models) (Section 38)
Service the cooling system (drain, flush and refill) (Section 39)
Inspect and replace, if necessary, the PCV valve (Section 40)
Inspect the evaporative emissions control system (Section 41)
Check the EGR system (Section 42)
Replace the spark plugs (Section 43)
Inspect the spark plug wires, distributor cap and rotor (Sections 44 and 45)*
* Replace the wires, cap and rotor at 60,000 miles

Every 82,500 miles or 82 months, whichever comes first

Replace the oxygen sensor and emissions timer (49-state models) (Section 46)

This item is affected by "severe" operating conditions as described below. If your vehicle is operated under severe conditions, perform all maintenance as indicated with an asterisk (*) at 3000 mile/3 month intervals. Severe conditions are indicated if you mainly operate your vehicle under one or more of the following:

Operating in dusty areas
Towing a trailer
Idling for extended periods and/or low speed operation
Operating when outside temperatures remain below freezing and when most trips are less than four miles

**If operated under one or more of the following conditions, change the automatic transmission fluid every 12,000 miles:*

In heavy city traffic where the outside temperature regularly reaches 90-degrees F (32-degrees C) or higher
In hilly or mountainous terrain
Frequent trailer pulling

2   Introduction

This Chapter is designed to help the home mechanic maintain the Jeep Cherokee, Wagoneer or Comanche with the goals of maximum performance, economy, safety and reliability in mind.

Included is a master maintenance schedule (page 1-7), followed by procedures dealing specifically with each item on the schedule. Visual checks, adjustments, component replacement and other helpful items are included. Refer to the accompanying illustrations of the engine compartment and the underside of the vehicle for the locations of various components.

Servicing your vehicle in accordance with the mileage/time maintenance schedule and the step-by-step procedures will result in a planned maintenance program that should produce a long and reliable service life. Keep in mind that it is a comprehensive plan, so maintaining some items but not others at the specified intervals will not produce the same results.

As you service your vehicle, you will discover that many of the procedures can, and should, be grouped together because of the nature of the particular procedure you're performing or because of the close proximity of two otherwise unrelated components to one another.

For example, if the vehicle is raised for chassis lubrication, you should inspect the exhaust, suspension, steering and fuel systems while you're under the vehicle. When you're rotating the tires, it makes good sense to check the brakes since the wheels are already removed. Finally, let's suppose you have to borrow or rent a torque wrench. Even if you only need it to tighten the spark plugs, you might as well check the torque of as many critical fasteners as time allows.

The first step in this maintenance program is to prepare yourself before the actual work begins. Read through all the procedures you're planning to do, then gather up all the parts and tools needed. If it looks like you might run into problems during a particular job, seek advice from a mechanic or an experienced do-it-yourselfer.

3   Tune-up general information

The term tune-up is used in this manual to represent a combination of individual operations rather than one specific procedure.

If, from the time the vehicle is new, the routine maintenance schedule is followed closely and frequent checks are made of fluid levels and high wear items, as suggested throughout this manual, the engine will be kept in relatively good running condition and the need for additional work will be minimized.

More likely than not, however, there will be times when the engine is running poorly

due to lack of regular maintenance. This is even more likely if a used vehicle, which has not received regular and frequent maintenance checks, is purchased. In such cases, an engine tune-up will be needed outside of the regular routine maintenance intervals.

The first step in any tune-up or diagnostic procedure to help correct a poor running engine is a cylinder compression check. A compression check (see Chapter 2 Part D) will help determine the condition of internal engine components and should be used as a guide for tune-up and repair procedures. If, for instance, a compression check indicates serious internal engine wear, a conventional tune-up will not improve the performance of the engine and would be a waste of time and money. Because of its importance, the compression check should be done by someone with the right equipment and the knowledge to use it properly.

The following procedures are those most often needed to bring a generally poor running engine back into a proper state of tune.

Minor tune-up

Check all engine related fluids (Section 4)
Clean, inspect and test the battery
   (Section 8)
Check and adjust the drivebelts
   (Section 26)
Replace the spark plugs (Section 43)
Inspect the distributor cap and rotor
   (Section 45)
Inspect the spark plug and coil wires
   (Section 44)
Check and adjust the ignition timing
   (Section 33)
Check the PCV valve or CCV
   hose(Section 40)
Check the air and PCV filters (Section 32)
Check the cooling system (Section 9)
Check all underhood hoses (Section 10)

Major tune-up

All items listed under Minor tune-up plus . . .
Check the EGR system (Section 42)
Check the ignition system (Chapter 5)

Check the charging system (Chapter 5)
Check the fuel system (Section 21)
Replace the air and PCV filters (Section 32)
Replace the distributor cap and rotor
   (Section 45)
Replace the spark plug wires (Section 44)

4   Fluid level checks

Note: *The following are fluid level checks to be done on a 250 mile or weekly basis. Additional fluid level checks can be found in specific maintenance procedures which follow. Regardless of intervals, be alert to fluid leaks under the vehicle which would indicate a fault to be corrected immediately.*

1   Fluids are an essential part of the lubrication, cooling, brake, clutch and windshield washer systems. Because the fluids gradually become depleted and/or contaminated during normal operation of the vehicle, they must be periodically replenished. See Recommended lubricants and fluids at the beginning of this Chapter before adding fluid to any of the following components. **Note:** *The vehicle must be on level ground when fluid levels are checked.*

Engine oil

Refer to illustrations 4.4 and 4.6

2   The engine oil level is checked with a dipstick that extends through a tube and into the oil pan at the bottom of the engine.

3   The oil level should be checked before the vehicle has been driven, or about 15 minutes after the engine has been shut off. If the oil is checked immediately after driving the vehicle, some of the oil will remain in the upper engine components, resulting in an inaccurate reading on the dipstick.

4   Pull the dipstick from the tube and wipe all the oil from the end with a clean rag or paper towel. Insert the clean dipstick all the way back into the tube, then pull it out again. Note the oil at the end of the dipstick. Add oil as necessary to keep the level between the ADD mark and the FULL mark on the dipstick **(see illustration).**

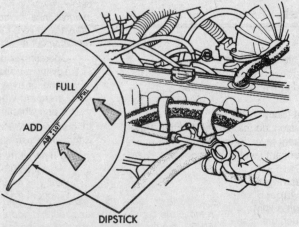

4.4 Checking the oil level on an inline six-cylinder engine (on V6 engines, the dipstick is on the driver's side of the engine; on four-cylinder engines, it's on the passenger's side) - the oil level should be at or near the FULL mark - if it isn't, add enough oil to bring the level to or near the FULL mark (it takes one quart to raise the level from ADD to FULL)

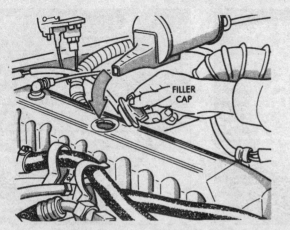

4.6 The twist-off oil filler cap is located on the rocker cover (except on V6 models, which have it mounted on top of a filler tube on the driver's side of the engine) - always make sure the area around this opening is clean before unscrewing the cap to prevent dirt from contaminating the engine

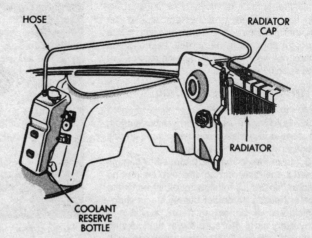

4.8a On four-cylinder and V6 engines, the coolant reserve bottle allows for contraction and expansion of the coolant - it can be checked visually with the engine hot or cold

5   Do not overfill the engine by adding too much oil since this may result in oil fouled spark plugs, oil leaks or oil seal failures.

6   Oil is added to the engine after removing a cap **(see illustration)**. An oil can spout or funnel may help to reduce spills.

7   Checking the oil level is an important preventive maintenance step. A consistently low oil level indicates oil leakage through damaged seals, defective gaskets or past worn rings or valve guides. If the oil looks milky in color or has water droplets in it, the cylinder head gasket(s) may be blown or the head(s) or block may be cracked. The engine should be checked immediately. The condition of the oil should also be checked. Whenever you check the oil level, slide your thumb and index finger up the dipstick before wiping off the oil. If you see small dirt or metal particles clinging to the dipstick, the oil should be changed (Section 12).

Engine coolant

Refer to illustrations 4.8a and 4.8b

Warning: *Do not allow antifreeze to come in contact with your skin or painted surfaces of the vehicle. Flush contaminated areas immediately with plenty of water. Don't store new coolant or leave old coolant lying around where it's accessible to children or pets, they're attracted by its sweet taste. Ingestion of even a small amount of coolant can be fatal! Wipe up garage floor and drip pan coolant spills immediately. Keep antifreeze containers covered and repair leaks in your cooling system immediately.*

8   All vehicles covered by this manual are equipped with a pressurized coolant recovery system. A white plastic coolant reserve or pressure bottle located in the engine compartment is connected by a hose to the radiator or radiator filler neck **(see illustrations)**. If the engine overheats on four-cylinder or V6 engines, coolant escapes through a valve in the radiator cap and travels through the hose

4.8b On inline six-cylinder engines, check the coolant level at the coolant pressure bottle: the engine must be off and cold, since the bottle is pressurized along with the rest of the cooling system - remove the cap and look inside the reservoir - the coolant level is correct if it's between the top of the post (FULL) and the notch in the post (ADD)

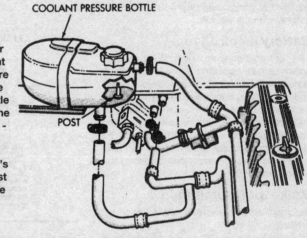

into the reservoir. As the engine cools, the coolant is automatically drawn back into the cooling system to maintain the correct level. On inline six-cylinder engines there is no radiator cap. Instead, the cap on the pressure bottle is pressurized. The bottle is part of the cooling system, and coolant flows through it whenever the engine is running. The cap should only be removed when the engine is off and cold.

9   The coolant level in the reservoir should be checked regularly. **Warning:** *Do not remove the radiator cap or pressure bottle cap to check the coolant level when the engine is warm. On four-cylinder or V6 engines, the coolant level in the reservoir should be kept between the FULL and ADD marks on the side of the reservoir. On inline six-cylinder engines, remove the pressure bottle cap with the engine off and cold. Look down into the bottle and make sure the coolant is at the top of the post (full) or no lower than the notch (add). If it is necessary to add coolant, allow the engine to cool, then remove the cap from the reservoir and add a 50/50 mixture of ethylene glycol-based*

antifreeze and water.

10   Drive the vehicle and recheck the coolant level. If only a small amount of coolant is required to bring the system up to the proper level, water can be used. However, repeated additions of water will dilute the antifreeze and water solution. In order to maintain the proper ratio of antifreeze and water, always top up the coolant level with the correct mixture. An empty plastic milk jug o bleach bottle makes an excellent container for mixing coolant. Do not use rust inhibitors or additives.

11   If the coolant level drops consistently, there may be a leak in the system. Inspect the radiator, hoses, filler cap, drain plugs and water pump (see Section 9). If no leaks are noted, have the radiator cap pressure tested by a service station.

12   If you have to remove the radiator cap, wait until the engine has cooled, then wrap a thick cloth around the cap and turn it to the first stop. If coolant or steam escapes, let the engine cool down longer, then remove the cap.

13   Check the condition of the coolant as

well. It should be relatively clear. If it's brown or rust colored, the system should be drained, flushed and refilled. Even if the coolant appears to be normal, the corrosion inhibitors wear out, so it must be replaced at the specified intervals.

Windshield washer fluid

Refer to illustration 4.14

14   Fluid for the windshield washer system is located in a plastic reservoir in the engine compartmeht **(see illustration)**.

15   In milder climates, plain water can be used in the reservoir, but it should be kept no more than 2/3 full to allow for expansion if the water freezes. In colder climates, use windshield washer system antifreeze, available at any auto parts store, to lower the freezing point of the fluid. Mix the antifreeze with water in accordance with the manufacturer's directions on the container. **Caution:** *Don't use cooling system antifreeze, it will damage the vehicle's paint.*

16   To help prevent icing in cold weather, warm the windshield with the defroster before using the washer.

Battery electrolyte

17   All vehicles with which this manual is concerned are equipped with a battery which is permanently sealed (except for vent holes) and has no filler caps. Water doesn't have to be added to these batteries at any time. If a maintenance-type battery is installed, the caps on the top of the battery should be removed periodically to check for a low water level. This check is most critical during the warm summer months.

Brake and clutch fluid

Refer to illustrations 4.18, 4.19a, 4.19b and 4.19c

18   The brake master cylinder is mounted on the front of the power booster unit in the engine compartment. The clutch master cylinder used on manual transmissions is mounted adjacent to it on the firewall **(see illustration)**.

4.14 The windshield washer fluid reservoir is located on the driver's side of the engine compartment - keep the level at or near the FULL line; fluid can be added after flipping up the cap

4.18 The clutch master cylinder is located next to the brake booster - maintain the fluid level between the MAX and MIN marks

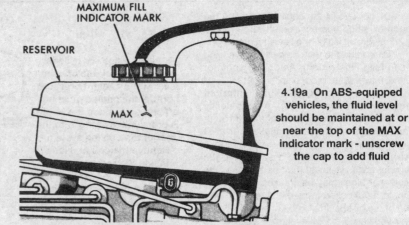

4.19a On ABS-equipped vehicles, the fluid level should be maintained at or near the top of the MAX indicator mark - unscrew the cap to add fluid

19   If the vehicle is equipped with ABS brakes, the fluid inside the reservoir is readily visible. The level should be at the MAX mark **(see illustration)**. On non-ABS vehicles, use a screwdriver to pry the clip free; then remove the cover **(see illustrations)**. Be sure to wipe the top of the reservoir cap or cover with a clean rag to prevent contamination of the brake and/or clutch system before removing the cover.

20   When adding fluid, pour it carefully into the reservoir to avoid spilling it onto sur-

4.19b On vehicles without ABS brakes, pry the clip off the brake master cylinder cover with a screwdriver

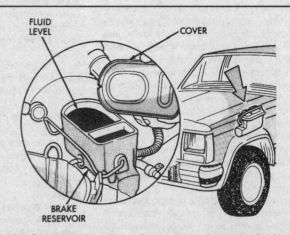

4.19c On reservoirs with a clip-on cover, the brake fluid level should be kept 1/4-inch below the top edge of the reservoir

rounding painted surfaces. Be sure the specified fluid is used, since mixing different types of brake fluid can cause damage to the system. See Recommended lubricants and fluids at the front of this Chapter or your owner's manual. **Warning:** *Brake fluid can harm your eyes and damage painted surfaces, so use extreme caution when handling or pouring it. Do not use brake fluid that has been standing open or is more than one year old. Brake fluid absorbs moisture from the air. Excess moisture can cause a dangerous loss of braking effectiveness.*

21   At this time the fluid and master cylinder can be inspected for contamination. The system should be drained and refilled if deposits, dirt particles or water droplets are seen in the fluid.

22   After filling the reservoir to the proper level, make sure the cover or cap is on tight to prevent fluid leakage.

23   The brake fluid level in the brake master cylinder will drop slightly as the pads and the brake shoes at each wheel wear down during normal operation. If the master cylinder requires repeated additions to keep it at the proper level, it's an indication of leakage in the brake system, which should be corrected immediately. Check all brake lines and connections (see Section 20 for more information).

24   If, upon checking the master cylinder fluid level, you discover one or both reservoirs empty or nearly empty, the brake system should be bled (Chapter 9).

5   Tire and tire pressure checks

Refer to illustrations 5.2, 5.3, 5.4a, 5.4b and 5.8

1   Periodic inspection of the tires may spare you the inconvenience of being stranded with a flat tire. It can also provide you with vital information regarding possible problems in the steering and suspension systems before major damage occurs.

2   The original tires on this vehicle are equipped with 1/2-inch side bands that will appear when tread depth reaches 1/16-inch, but they don't appear until the tires are worn out. Tread wear can be monitored with a simple, inexpensive device known as a tread depth indicator **(see illustration)**.

3   Note any abnormal tread wear **(see illustration)**. Tread pattern irregularities such as cupping, flat spots and more wear on one side than the other are indications of front

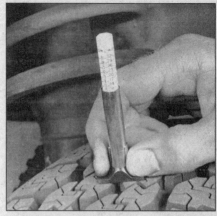

5.2 A tire tread depth indicator should be used to monitor tire wear - they are available at auto parts stores and service stations and cost very little

1

end alignment and/or balance problems. If any of these conditions are noted, take the vehicle to a tire shop or service station to correct the problem.

4   Look closely for cuts, punctures and embedded nails or tacks. Sometimes a tire will hold air pressure for a short time or leak

INCORRECT TOE-IN OR EXTREME CAMBER

UNDERINFLATION

CUPPING

Cupping may be caused by:
- **Underinflation and/or mechanical irregularities such as out-of-balance condition of wheel and/or tire, and bent or damaged wheel.**
- **Loose or worn steering tie-rod or steering idler arm.**
- **Loose, damaged or worn front suspension parts.**

OVERINFLATION

FEATHERING DUE TO MISALIGNMENT

5.3 This chart will help you determine the condition of your tires, the probable cause(s) of abnormal wear and the corrective action necessary

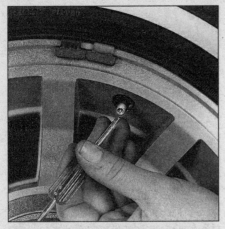

5.4a If a tire loses air on a steady basis, check the valve core first to make sure it's snug (special inexpensive wrenches are commonly available at auto parts stores)

5.4b If the valve core is tight, raise the corner of the vehicle with the low tire and spray a soapy water solution onto the tread as the tire is turned - slow leaks will cause small bubbles to appear

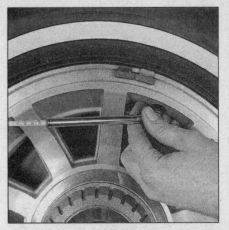

5.8 To extend the life of the tires, check the air pressure at least once a week with an accurate gauge (don't forget the spare!)

down very slowly after a nail has embedded itself in the tread. If a slow leak persists, check the valve stem core to make sure it's tight **(see illustration)**. Examine the tread for an object that may have embedded itself in the tire or for a -plug" that may have begun to leak (radial tire punctures are repaired with a plug that's installed in a puncture). If a puncture is suspected, it can be easily verified by spraying a solution of soapy water onto the puncture area **(see illustration)**. The soapy solution will bubble if there's a leak. Unless the puncture is unusually large, a tire shop or service station can usually repair the tire.

5   Carefully inspect the inner sidewall of each tire for evidence of brake fluid leakage. If you see any, inspect the brakes immediately.

6   Correct air pressure adds miles to the lifespan of the tires, improves mileage and enhances overall ride quality. Tire pressure cannot be accurately estimated by looking at a tire, especially if it's a radial. The correct tire pressures are located on a label on the inside of the glove box door. A tire pressure gauge is essential. Keep an accurate gauge in the vehicle. The pressure gauges attached to the nozzles of air hoses at gas stations are often inaccurate.

7   Always check tire pressure when the tires are cold. Cold, in this case, means the vehicle has not been driven over a mile in the three hours preceding a tire pressure check. A pressure rise of four to eight pounds is not uncommon once the tires are warm.

8   Unscrew the valve cap protruding from the wheel or hubcap and push the gauge firmly onto the valve stem **(see illustration)**. Note the reading on the gauge and compare the figure to the recommended tire pressure shown on the placard on the driver's side door pillar. Be sure to reinstall the valve cap to keep dirt and moisture out of the valve stem mechanism. Check all four tires and, if necessary, add enough air to bring them up to the recommended pressure.

9   Don't forget to keep the spare tire

6.6 The automatic transmission dipstick is located in a long tube located at the rear of the engine compartment

inflated to the specified pressure (refer to your owner's manual or the tire sidewall). Note that the pressure recommended for the compact spare is higher than for the tires on the vehicle.

6   Automatic transmission fluid level check

Refer to illustration 6.6

1   The automatic transmission fluid level should be carefully maintained. Low fluid level can lead to slipping or loss of drive, while overfilling can cause foaming and loss of fluid.

2   Warm the transmission by driving the vehicle at least 15 miles. With the parking brake set, start the engine, then move the shift lever through all the gear ranges, ending in Neutral. The fluid level must be checked with the vehicle level and the engine running at idle.

3   With the transmission at normal operating temperature, remove the dipstick from the filler tube. The dipstick is located at the rear of the engine compartment.

4   Wipe the fluid from the dipstick with a clean rag and push it back into the filler tube until the cap seats.

5   Pull the dipstick out again and note the fluid level.

6   The level should be between the ADD and FULL marks **(see illustration)**. If additional fluid is required, add it directly into the tube using a funnel. It takes about one pint to raise the level from the ADD mark to the FULL mark with a warm transmission, so add the fluid a little at a time and keep checking the level until it's correct.

7   The condition of the fluid should also be checked along with the level. If the fluid at the end of the dipstick is a dark reddish-brown color, or if it smells burned, it should be changed. If you are in doubt about the condition of the fluid, purchase some new fluid and compare the two for color and smell.

7   Power steering fluid level check

Refer to illustration 7.6

1   Unlike manual steering, the power steering system relies on fluid which may, over a

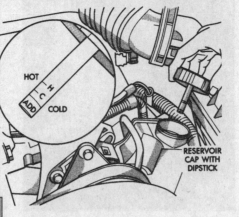

7.6 The power steering fluid filler cap/dipstick is located at the front of the engine - the dipstick has marks for checking the fluid at different temperatures

HOT

COLD

RESERVOIR CAP WITH DIPSTICK

8.1 Tools and materials required for battery maintenance

1   *Face shield/safety goggles - When removing corrosion with a brush, the acidic particles can easily fly up into your eyes*

2   *Baking soda - A solution of baking soda and water can be used to neutralize corrosion*

3   *Petroleum jelly - A layer of this on the battery posts will help prevent corrosion*

4   *Battery post/cable cleaner - This wire brush cleaning tool will remove all traces of corrosion from the battery posts and cable clamps*

5   *Treated felt washers - Placing one of these on each post, directly under the cable clamps, will help prevent corrosion*

6   *Puller - Sometimes the cable clamps are very difficult to pull off the posts, even after the nut/bolt has been completely loosened. This tool pulls the clamp straight up and off the post without damage*

7   *Battery post/cable cleaner - Here is another cleaning tool which is a slightly different version of number 4 above, but it does the same thing*

8   *Rubber gloves - Another safety item to consider when servicing the battery; remember that's acid inside the battery!*

period of time, require replenishing.

2   The fluid reservoir for the power steering pump is located on the pump body at the front of the engine.

3   For the check, the front wheels should be pointed straight ahead and the engine should be off.

4   Use a clean rag to wipe off the reservoir cap and the area around the cap. This will help prevent any foreign matter from entering the reservoir during the check.

5   Twist off the cap and check the temperature of the fluid at the end of the dipstick with your finger.

6   Wipe off the fluid with a clean rag, reinsert the dipstick, then withdraw it and read the fluid level. The level should be at the H (Hot) mark if the fluid was hot to the touch **(see illustration)**. It should be at the C (Cold) mark if the fluid was cool to the touch. Never allow the fluid level to drop below the ADD mark.

7   If additional fluid is required, pour the specified type directly into the reservoir, using a funnel to prevent spills.

8   If the reservoir requires frequent fluid additions, all power steering hoses, hose connections and the power steering pump should be carefully checked for leaks.

8   Battery check and maintenance

Refer to illustrations 8.1, 8.8a, 8.8b, 8.8c and 8.8d

Warning: *Certain precautions must be followed when checking and servicing the battery. Hydrogen gas, which is highly flammable, is always present in the battery cells, so keep lighted tobacco and all other open flames and sparks away from the battery. The electrolyte in the battery cells is actually dilute sulfuric acid, which will cause injury if splashed on your skin or in your eyes. It will also ruin clothes and painted surfaces. When removing the battery cables, always detach the negative cable first and hook it up last!*

1   Battery maintenance is an important procedure which will help ensure that you're

not stranded because of a dead battery. Several tools are required for this procedure **(see illustration)**.

2   Before servicing the battery, always turn the engine and all accessories off and disconnect the cable from the negative terminal.

3   A sealed (sometimes called maintenance-free) battery is standard equipment on these vehicles. The cell caps cannot be removed, no electrolyte checks are required and water cannot be added to the cells. However, if an aftermarket battery that requires regular maintenance has been installed, the following procedure can be used.

4   Check the electrolyte level in each of the battery cells. It must be above the plates. There's usually a split-ring indicator in each cell to indicate the correct level. If the level is low, add distilled water only, then install the cell caps. **Caution:** *Overfilling the cells may cause electrolyte to spill over during periods of heavy charging, causing corrosion and damage to nearby components.*

5   If the positive terminal and cable clamp on your vehicle's battery is equipped with a rubber protector, make sure that it's not torn or damaged. It should completely cover the terminal.

6   The external condition of the battery should be checked periodically. Look for damage such as a cracked case.

7   Check the tightness of the battery cable clamps to ensure good electrical connections and inspect the entire length of each cable, looking for cracked or abraded insulation and frayed conductors.

8   If corrosion (visible as white, fluffy deposits) is evident, remove the cables from the terminals, clean them with a battery brush and reinstall them **(see illustrations)**. Corrosion can be kept to a minimum by installing specially treated washers available at auto parts stores or by applying a layer of petroleum jelly or grease to the terminals and cable clamps after they are assembled.

9   Make sure that the battery carrier is in good condition and that the hold-down clamp bolt is tight. If the battery is removed

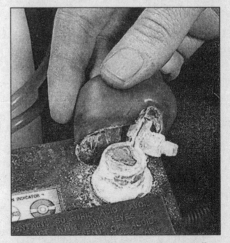

8.8a Battery terminal corrosion usually appears as light, fluffy powder

1

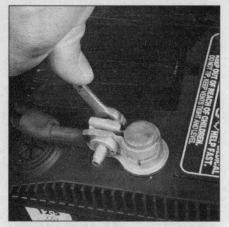

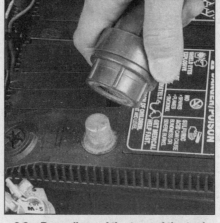

8.8b Removing a cable from the battery post with a wrench - sometimes, if corrosion has caused deterioration of the nut hex, a pair of special battery pliers is required for this procedure (always remove the ground cable first and hook it up last!)

8.8c Regardless of the type of the tool used to clean the battery post, a clean, shiny surface should be the result

8.8d When cleaning the cable clamps, all corrosion must be removed (the inside of the clamp is tapered to match the taper on the post, so don't remove too much material)

Check for a chafed area that could fail prematurely.

Check for a soft area indicating the hose has deteriorated inside.

Overtightening the clamp on a hardened hose will damage the hose and cause a leak.

Check each hose for swelling and oil-soaked ends. Cracks and breaks can be located by squeezing the hose.

9.4 Hoses, like drivebelts, have a habit of failing at the worst possible time - to prevent the inconvenience of a blown radiator or heater hose, inspect them carefully as shown here

(see Chapter 5 for the removal and installation procedure), make sure that no parts remain in the bottom of the carrier when it's reinstalled. When reinstalling the hold-down clamp, don't overtighten the bolt.

10   Corrosion on the carrier, battery case and surrounding areas can be removed with a solution of water and baking soda. Apply the mixture with a small brush, let it work, then rinse it off with plenty of clean water.

11   Any metal parts of the vehicle damaged by corrosion should be coated with a zinc-based primer, then painted.

12   Additional information on the battery, charging and jump starting can be found in Chapter 5 and at the front of this manual.

9   Cooling system check

Refer to illustration 9.4

1   Many major engine failures can be attributed to a faulty cooling system. If the vehicle is equipped with an automatic transmission, the cooling system also cools the transmission fluid and thus plays an important role in prolonging transmission life.

2   The cooling system should be checked with the engine cold. Do this before the vehicle is driven for the day or after it has been shut off for at least three hours.

3   On four-cylinder and V6 engines, remove the radiator cap by turning it to the left until it reaches a stop. If you hear a hissing sound (indicating there is still pressure in the system), wait until this stops. Now press down on the cap with the palm of your hand and continue turning to the left until the cap can be pulled off. Thoroughly clean the cap, inside and out, with clean water. Also clean the filler neck on the radiator. All traces of corrosion should be removed. On inline six-cylinder engines, unscrew the cap on the coolant pressure bottle and clean the cap. The coolant inside the radiator or pressure bottle should be relatively transparent. If it is rust colored, the system should be drained

and refilled (Section 39). If the coolant level is not up to the top, add additional anti-freeze/coolant mixture (see Section 4).

4   Carefully check the large upper and lower radiator hoses along with the smaller diameter heater hoses which run from the engine to the firewall. On some models the heater return hose runs directly to the radiator. On inline six-cylinder vehicles, also inspect the hoses attached to the coolant pressure bottle. Inspect each hose along its entire length, replacing any hose which is cracked, swollen or shows signs of deterioration. Cracks may become more apparent if the hose is squeezed **(see illustration)**. Regardless of condition, it's a good idea to replace hoses with new ones every two years.

5   Make sure that all hose connections are tight. A leak in the cooling system will usually show up as white or rust colored deposits on the areas adjoining the leak. If wire-type clamps are used at the ends of the hoses, it may be a good idea to replace them with more secure screw-type clamps.

6   Use compressed air or a soft brush to remove bugs, leaves, etc. from the front of the radiator or air conditioning condenser. Be careful not to damage the delicate cooling fins or cut yourself on them.

7   Every other inspection, or at the first indication of cooling system problems, have the cap and system pressure tested. If you don't have a pressure tester, most gas stations and repair shops will do this for a minimal charge.

10   Underhood hose check and replacement

General

1   **Caution:** *Replacement of air conditioning hoses must be left to a dealer service department or air conditioning shop that has*

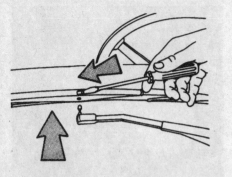

11.6a To remove the 1995 and earlier front wiper blade, insert a small, narrow screwdriver under the wiper release lever. Lift the lever and slide the wiper blade off the wiper arm pivot

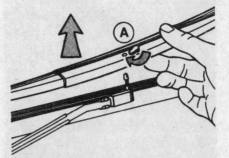

11.6b To remove the 1995 and earlier rear wiper blade, gently turn the pivot in the direction shown by the arrow. Slide the wiper blade off the wiper arm

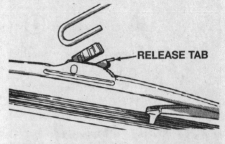

RELEASE TAB

11.6c On 1996 and later wiper blades, use a narrow screwdriver to compress the plastic release spring. Keeping the tab compressed in the direction of the arrow, slide the wiper blade out of the wiper arm hook, then slide the new wiper blade into the hook

the equipment to depressurize the system safely. Never remove air conditioning components or hoses until the system has been depressurized.

2   High temperatures in the engine compartment can cause the deterioration of the rubber and plastic hoses used for engine, accessory and emission systems operation. Periodic inspection should be made for cracks, loose clamps, material hardening and leaks. Information specific to the cooling system hoses can be found in Section 9.

3   Some, but not all, hoses are secured to the fittings with clamps. Where clamps are used, check to be sure they haven't lost their tension, allowing the hose to leak. If clamps aren't used, make sure the hose has not expanded and/or hardened where it slips over the fitting, allowing it to leak.

Vacuum hoses

4   It's quite common for vacuum hoses, especially those in the emissions system, to be color coded or identified by colored stripes molded into them. Various systems require hoses with different wall thicknesses, collapse resistance and temperature resistance. When replacing hoses, be sure the new ones are made of the same material.

5   Often the only effective way to check a hose is to remove it completely from the vehicle. If more than one hose is removed, be sure to label the hoses and fittings to ensure correct installation.

6   When checking vacuum hoses, be sure to include any plastic T-fittings in the check. Inspect the fittings for cracks and the hose where it fits over the fitting for distortion, which could cause leakage.

7   A small piece of vacuum hose (1/4-inch inside diameter) can be used as a stethoscope to detect vacuum leaks. Hold one end of the hose to your ear and probe around vacuum hoses and fittings, listening for the -hissing" sound characteristic of a vacuum leak.
Warning: When probing with the vacuum hose stethoscope, be very careful not to come into contact with moving engine components such as the drivebelt, cooling fan, etc.

Fuel hoses

Warning: There are certain precautions which must be taken when inspecting or servicing fuel system components. Work in a well ventilated area and do not allow open flames (cigarettes, appliance pilot lights, etc.) or bare light bulbs near the work area. Mop up any spills immediately and do not store fuel soaked rags where they could ignite. On vehicles equipped with fuel injection, the fuel system is under pressure, so if any fuel lines are to be disconnected, the pressure in the system must be relieved first (see Chapter 4 for more information).

8   Check all rubber fuel lines for deterioration and chafing. Check especially for cracks in areas where the hose bends and just before fittings, such as where a hose attaches to the fuel filter.

9   High quality fuel line, usually identified by the word Fluroelastomer printed on the hose, should be used for fuel line replacement. Never, under any circumstances, use unreinforced vacuum line, clear plastic tubing or water hose for fuel lines.

10   Spring-type clamps are commonly used on fuel lines. These clamps often lose their tension over a period of time, and can be -sprung" during removal. Replace all spring-type clamps with screw clamps whenever a hose is replaced.

Metal lines

11   Sections of metal line are often used for fuel line between the fuel pump and carburetor or fuel injection unit. Check carefully to be sure the line has not been bent or crimped and that cracks have not started in the line.

12   If a section of metal fuel line must be replaced, only seamless steel tubing should be used, since copper and aluminum tubing don't have the strength necessary to withstand normal engine vibration.

13   Check the metal brake lines where they enter the master cylinder and brake proportioning unit (if used) for cracks in the lines or loose fittings. Any sign of brake fluid leakage calls for an immediate, thorough inspection of the brake system.

11   Wiper blade inspection and replacement

Refer to illustrations 11.6a, 11.6b and 11.6c

1   The windshield wiper and blade assembly should be inspected periodically for damage, loose components and cracked or worn blade elements.

2   Road film can build up on the wiper blades and affect their efficiency, so they should be washed regularly with a mild detergent solution.

3   The action of the wiping mechanism can loosen the bolts, nuts and fasteners, so they should be checked and tightened, as necessary, at the same time the wiper blades are checked.

4   If the wiper blade elements (sometimes called inserts) are cracked, worn or warped, the blade/arm assemblies should be replaced with new ones.

5   Pull the wiper blade/arm assembly away from the glass.

6   Two types of wiper blade retaining systems are used. 1995 and earlier models employ a pin-pivot system. 1996 and later models use a plastic spring system **(see illustrations)**.

7   Compare the new blade/arm assembly to the old one for length, design, etc.

8   Reinstall the blade assembly on the arm, wet the windshield and check for proper operation.

12   Engine oil and filter change

Refer to illustrations 12.3, 12.9, 12.14, 12.16 and 12.18

1   Frequent oil changes are the most important preventive maintenance procedures that can be done by the home mechanic. As engine oil ages, it becomes diluted and contaminated, which leads to premature engine wear.

2   Although sorne sources recommend oil filter changes every other oil change, we feel

1

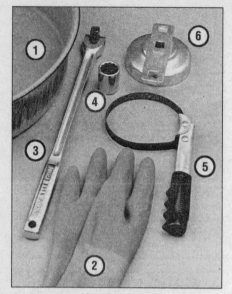

12.9 To avoid rounding off the corners of the drain plug, use the correct size box-end wrench or six-point socket to remove it

12.14 The oil filter is usually on very tight and will require a special wrench for removal - DO NOT use the wrench to tighten the new filter

12.3 These tools are required when changing the engine oil and filter

1 **Drain pan** - It should be fairly shallow in depth, but wide in order to prevent spills
2 **Rubber gloves** - When removing the drain plug and filter it is inevitable that you will get oil on your hands (the gloves will prevent burns)
3 **Breaker bar** - Sometimes the oil drain plug is pretty tight and a long breaker bar is needed to loosen it
4 **Socket** - To be used with the breaker bar or a ratchet (must be the correct size to fit the drain plug)
5 **Filter wrench** - This is a metal band-type wrench, which requires clearance around the filter to be effective
6 **Filter wrench** - This type fits on the bottom of the filter and can be turned with a ratchet or breaker bar (different size wrenches are available for different types of filters)

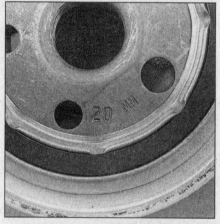

12.16 Some later inline six-cylinder engines use a metric 20 mm filter - if the old filter is so marked, make sure the new one is, too

12.18 Lubricate the gasket with clean engine oil before installing the filter on the engine

that the minimal cost of an oil filter and the relative ease with which it is installed dictate that a new filter be installed every time the oil is changed.

3   Gather together all necessary tools and materials before beginning this procedure **(see illustration)**.

4   You should have plenty of clean rags and newspapers handy to mop up any spills. Access to the underside of the vehicle is greatly improved if the vehicle can be lifted on a hoist, driven onto ramps or supported by jackstands. **Warning:** *Do not work under a vehicle which is supported only by a bumper, hydraulic or scissors-type jack.*

5   If this is your first oil change, get under the vehicle and familiarize yourself with the locations of the oil drain plug and the oil filter. The engine and exhaust components will be warm during the actual work, so note how they are situated to avoid touching them when working under the vehicle.

6   Warm the engine to normal operating

temperature. If the new oil or any tools are needed, use this warm-up time to gather everything necessary for the job. The correct type of oil for your application can be found in Recommended lubricants and fluids at the beginning of this Chapter.

7   With the engine oil warm (warm engine oil will drain better and more built-up sludge will be removed with it), raise and support the vehicle. Make sure it's safely supported!

8   Move all necessary tools, rags and newspapers under the vehicle. Set the drain pan under the drain plug. Keep in mind that the oil will initially flow from the pan with some force; position the pan accordingly.

9   Being careful not to touch any of the hot exhaust components, use a wrench to remove the drain plug near the bottom of the oil pan **(see illustration)**. Depending on how hot the oil is, you may want to wear gloves while unscrewing the plug the final few turns.

10   Allow the old oil to drain into the pan. It may be necessary to move the pan as the oil flow slows to a trickle.

11   After all the oil has drained, wipe off the

drain plug with a clean rag. Small metal particles may cling to the plug and would immediately contaminate the new oil.

12   Clean the area around the drain plug opening and reinstall the plug. Tighten the plug securely with the wrench. If a torque wrench is available, use it to tighten the plug.

13   Move the drain pan into position under the oil filter.

14   Use the filter wrench to loosen the oil filter **(see illustration)**. Chain or metal band filter wrenches may distort the filter canister, but it doesn't matter since the filter will be discarded anyway.

15   Completely unscrew the old filter. Be careful; it's full of oil. Empty the oil inside the filter into the drain pan.

16   Compare the old filter with the new one to make sure they're the same type **(see illustration)**.

17   Use a clean rag to remove all oil, dirt and sludge from the area where the oil filter mounts to the engine. Check the old filter to make sure the rubber gasket isn't stuck to the engine. If the gasket is stuck to the engine (use a flashlight if necessary), remove it.

13.1 Materials required for chassis and body lubrication

1   **Engine oil** - Light engine oil in a can like this can be used for door and hood hinges
2   **Graphite spray** - Used to lubricate lock cylinders
3   **Grease** - Grease, in a variety of types and weights, is available for use in a grease gun. Check the Specifications for your requirements
4   **Grease gun** - A common grease gun, shown here with a detachable hose and nozzle, is needed for chassis lubrication. After use, clean it thoroughly!

18   Apply a light coat of clean oil to the rubber gasket on the new oil filter **(see illustration)**.
19   Attach the new filter to the engine, following the tightening directions printed on the filter canister or packing box. Most filter manufacturers recommend against using a filter wrench due to the possibility of overtightening and damage to the seal.
20   Remove all tools, rags, etc. from under the vehicle, being careful not to spill the oil in the drain pan, then lower the vehicle.
21   Move to the engine compartment and locate the oil filler cap.
22   If an oil can spout is used, push the spout into the top of the oil can and pour the fresh oil through the filler opening. A funnel may also be used.
23   On four-cylinder and V6 engines, pour four quarts of fresh oil into the engine. On inline six-cylinder engines, pour in five quarts. Wait a few minutes to allow the oil to drain into the pan, then check the level on the oil dipstick (see Section 4 if necessary). If the oil level is above the ADD mark, start the engine and allow the new oil to circulate.
24   Run the engine for only about a minute checking the pressure gauge or indicator light to make sure normal oil pressure is achieved. Shut off the engine. Immediately look under the vehicle and check for leaks at the oil pan drain plug and around the oil filter.

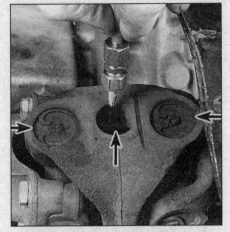

13.9 In addition to the conventional universal joints at each end of the driveshaft, 4WD models have a Constant Velocity (CV) joint which requires a needle-like adapter on the grease gun - the arrows show the locations of the CV joint grease fittings

If either is leaking, tighten with a bit more force.
25   With the new oil circulated and the filter now completely full, recheck the level on the dipstick and add more oil as necessary.
26   During the first few trips after an oil change, make it a point to check frequently for leaks and proper oil level.
27   The old oil drained from the engine cannot be reused in its present state and should be disposed of. Oil reclamation centers, auto repair shops and gas stations will normally accept the oil, which can be refined and used again. After the oil has cooled it can be drained into a suitable container (capped plastic jugs, topped bottles, milk cartons, etc.) for transport to one of these disposal sites.

13 Chassis lubrication

Refer to illustrations 13.1, 13.9, 13.10, 13.11 and 13.14
1   Refer to Recommended lubricants and fluids at the front of this Chapter to obtain the necessary grease, etc. You will also need a grease gun **(see illustration)**. Occasionally plugs will be installed rather than grease fittings. If so, grease fittings will have to be purchased and installed.
2   Look under the vehicle for grease fittings or plugs on the steering, suspension, and driveline components. They are normally found on the balljoints, tie-rod ends and universal joints. If there are plugs, remove them and buy grease fittings, which will thread into the component. A dealer or auto parts store will be able to supply the correct fittings. Straight, as well as angled, fittings are available.
3   For easier access under the vehicle,

13.10 The slip joint grease fitting is located on the collar - pump grease into it until it comes out of the slip joint seal

raise it with a jack and place jackstands under the frame. Make sure it is safely supported by the stands. If the wheels are to be removed at this interval for tire rotation or brake inspection, loosen the lug nuts slightly while the vehicle is still on the ground.
4   Before beginning, force a little grease out of the nozzle to remove any dirt from the end of the gun. Wipe the nozzle clean with a rag.
5   With the grease gun and plenty of clean rags, crawl under the vehicle and begin lubricating the components.
6   Wipe the balljoint grease fitting nipple clean and push the nozzle firmly over it. Squeeze the trigger on the grease gun to force grease into the component. The balljoints should be lubricated until the rubber seal is firm to the touch. Do not pump too much grease into the fittings as it could rupture the seal. For all other suspension and steering components, continue pumping grease into the fitting until it oozes out of the joint between the two components. If it escapes around the grease gun nozzle, the nipple is clogged or the nozzle is not completely seated on the fitting. Resecure the gun nozzle to the fitting and try again. If necessary, replace the fitting with a new one.
7   Wipe the excess grease from the components and the grease fitting. Repeat the procedure for the remaining fittings.
8   On models where the manual transmission shift linkage is accessible, lubricate the shift linkage with a little multi-purpose grease.
9   On 4WD models, lubricate the front driveshaft Constant Velocity (CV) joint, located at the transfer case end, using a special needle-like adapter on the grease gun **(see illustration)**. Lubricate the transfer case shift mechanism contact surfaces with clean engine oil.
10   If equipped, lubricate the driveshaft slip joints by pumping grease into the fitting until it can be seen coming out of the slip joint seal **(see illustration)**.

1

11   Lubricate conventional universal joints until grease can be seen coming out of the contact points **(see illustration)**.

12   While you are under the vehicle, clean and lubricate the parking brake cable along with the cable guides and levers. This can be done by smearing some chassis grease onto the cable and its related parts with your fingers.

13   The manual steering gear seldom requires the addition of lubricant, but if there is obvious leakage of grease at the seals, remove the plug or cover and check the lubricant level. If the level is low, add the specified lubricant.

14   Lubricate the contact points on the steering knuckle and adjustment bolt **(see illustration)**.

15   Open the hood and smear a little chassis grease on the hood latch mechanism. Have an assistant pull the hood release lever from inside the vehicle as you lubricate the cable at the latch.

16   Lubricate all the hinges (door, hood, etc.) with engine oil to keep them in proper working order.

17   The key lock cylinders can be lubricated with spray-on graphite or silicone lubricant, which is available at auto parts stores.

18   Lubricate the door weatherstripping with silicone spray. This will reduce chafing and retard wear.

14   Suspension and steering check

Refer to illustration 14.4

1   Indications of a fault in these systems are excessive play in the steering wheel before the front wheels react, excessive sway around corners, body movement over rough roads, noise from the axle or wheel areas or binding at some point as the steering wheel is turned.

2   Raise the front of the vehicle periodically and visually check the suspension and steering components for wear. Because of the work to be done, make sure the vehicle cannot fall from the stands.

3   Check the wheel bearings. Do this by spinning the front wheels. Listen for any

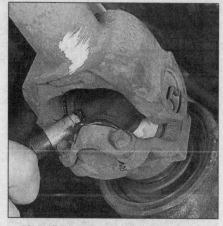

13.11 Pump grease into conventional universal joints until it can be seen coming out of the contact surfaces

abnormal noises and watch to make sure the wheel spins true (doesn't wobble). Grab the top and bottom of the tire and pull in-and-out on it. Notice any movement which would indicate a loose wheel bearing assembly. If the bearings are suspect, refer to Section 38 (for 2WD vehicles) and Chapter 10 for more information.

4   The front axle Constant Velocity (CV) joints on some 4WD models are protected by rubber boots. Check the boots for cuts, wear and signs of leaking grease **(see illustration)**. The boots must be replaced if they are damaged, otherwise the CV joint will be contaminated and eventually fail (see Chapter 8).

5   From under the vehicle, check for loose bolts, broken or disconnected parts and deteriorated rubber bushings on all suspension and steering components. Look for grease or fluid leaking from the steering assembly. Check the power steering hoses and connections for leaks.

6   Have an assistant turn the steering wheel from side-to-side and check the steering components for free movement, chafing and binding. If the steering doesn't react simultaneously with the movement of the steering wheel, try to determine where the slack is located.

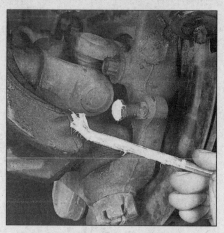

13.14 Use white lithium base grease to lubricate the contact points on the steering knuckle and adjustment bolt

15   Exhaust system check

1   With the engine cold (at least three hours after the vehicle has been driven), check the complete exhaust system from the manifold to the end of the tailpipe. Be careful around the catalytic converter, which may be hot even after three hours. The inspection should be done with the vehicle on a hoist to permit unrestricted access. If a hoist isn't available, raise the vehicle and support it securely on jackstands.

2   Check the exhaust pipes and connections for signs of leakage and/or corrosion indicating a potential failure. Make sure that all brackets and hanger are in good condition and tight.

3   Inspect the underside of the body for holes, corrosion, open seams, etc. which may allow exhaust gases to enter the passenger compartment. Seal all body openings with silicone or body putty.

4   Rattles and other noises can often be traced to the exhaust system, especially the hangers, mounts and heat shields. Try to move the pipes, mufflers and catalytic converter. If the components can come in contact with the body or suspension parts,

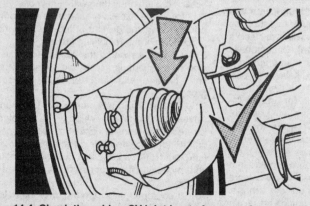

14.4 Check the rubber CV joint boots for wear, damage and leaking grease

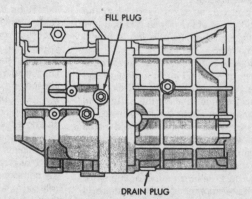

16.1a AX 4/5 manual transmission drain and fill plug location. Fill plug is located on the driver's side

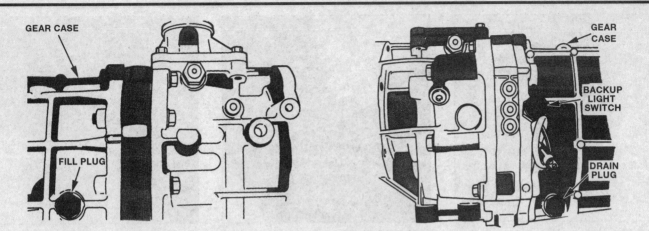

16.1b AX15 transmission check/fill and drain plug locations. Check/fill plug is on the driver's side. Drain plug is located on the passenger's side

secure the exhaust system with new brackets and hangers.

16   Manual transmission lubricant level check

Refer to illustrations 16.1a and 16.1b

1   The manual transmission has a fill plug which must be removed to check the lubricant level **(see illustrations)**. If the vehicle is raised to gain access to the plug, be sure to support it safely on jackstands - DO NOT crawl under a vehicle which is supported only by a jack!

2   Remove the fill plug from the transmission and use your little finger to reach inside the housing to feel the oil level. The level should be at or near the bottom of the plug hole.

3   If it isn't, add the recommended oil through the plug hole with a syringe or squeeze bottle.

4   Install and tighten the plug and check for leaks after the first few miles of driving.

17   Differential lubricant level check

Refer to illustration 17.2

Note: *4WD vehicles have two differentials, one in the center of each axle. 2WD vehicles have one differential, in the center of the rear axle. On 4WD vehicles, be sure to check the lubricant level in both differentials.*

1   The differential has a check/fill plug which must be removed to check the lubricant level. If the vehicle must be raised to gain access to the plug, be sure to support it safely on jackstands, DO NOT crawl under the vehicle when it's supported only by the jack.

2   Remove the oil check/fill plug from the back of the rear differential or the front of the front differential **(see illustration)**.

3   The oil level should be at the bottom of the plug opening. If it isn't, use a syringe to add the specified lubricant until it just starts to run out of the opening. On some models a tag is located in the area of the plug which gives information regarding lubricant type, particularly on models equipped with a lim-

ited slip differential.

4   Install the plug and tighten it securely.

18   Transfer case lubricant level check (4WD models)

Refer to illustration 18.1

1   The transfer case lubricant level is checked by removing the upper plug located in the side of the case **(see illustration)**.

2   After removing the plug, reach inside the hole. The lubricant level should be just at the bottom of the hole. If not, add the appropriate lubricant through the opening.

19   Tire rotation

Refer to illustration 19.2

1   The tires should be rotated at the specified intervals and whenever uneven wear is noticed.

2   Refer to the accompanying illustration for the preferred tire rotation pattern.

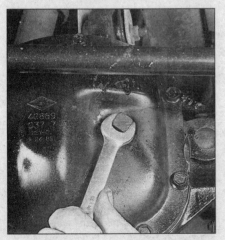

17.2 Use the correct size open-end wrench to remove the differential check/fill plug

18.1 The transfer case drain and fill plugs (arrows) are located on the front face of the housing - the upper one is the check/fill plug and the lower one is the drain plug

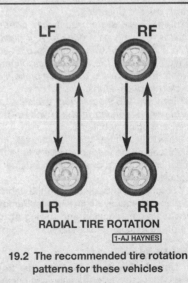

19.2 The recommended tire rotation patterns for these vehicles

20.6 You will find an inspection hole like this in each caliper - placing a steel ruler across the window should enable you to determine the thickness of remaining pad material for both inner and outer pads

20.12 The lining thickness of the rear brake shoe is measured from the outer surface of the lining to the metal shoe

20.14 Peel the wheel cylinder boot back carefully and check for leaking fluid - any leakage indicates the cylinder must be replaced or rebuilt

3   Refer to the information in Jacking and towing at the front of this manual for the proper procedures to follow when raising the vehicle and changing a tire. If the brakes are to be checked, don't apply the parking brake as stated. Make sure the tires are blocked to prevent the vehicle from rolling as it's raised.

4   Preferably, the entire vehicle should be raised at the same time. This can be done on a hoist or by jacking up each corner and then lowering the vehicle onto jackstands placed under the frame rails. Always use four jackstands and make sure the vehicle is safely supported.

5   After rotation, check and adjust the tire pressures as necessary and be sure to check the lug nut tightness.

6   For additional information on the wheels and tires, refer to Chapter 10.

20   Brake check

Refer to illustrations 20.6, 20.12 and 20.14
Note: *For detailed photographs of the brake system, refer to Chapter 9.*
Warning: *Brake system dust may contain asbestos, which is hazardous to your health. DO NOT blow it out with compressed air and DO NOT inhale it. DO NOT use gasoline or solvents to remove the dust. Use brake system cleaner or denatured alcohol only.*
Caution: *On ABS-equipped vehicles, be careful when working in the vicinity of the brake sensing components.*

1   In addition to the specified intervals, the brakes should be inspected every time the wheels are removed or whenever a defect is suspected.

2   To check the brakes, raise the vehicle and place it securely on jackstands. Remove the wheels (see Jacking and towing at the front of the manual, if necessary).

Disc brakes

3   Disc brakes are used on the front wheels. Extensive rotor damage can occur if

the pads are not replaced when needed.

4   These vehicles are equipped with a wear sensor attached to the inner pad. This is a small, bent piece of metal which is visible from the inner side of the brake caliper. When the pad wears to the specified limit, the metal sensor rubs against the rotor and makes a squealing sound.

5   The disc brake calipers, which contain the pads, are visible with the wheels removed. There is an outer pad and an inner pad in each caliper. All pads should be inspected.

6   Each caliper has an inspection hole to inspect the pads. Check the thickness of the pad lining by looking into the caliper at each end and down through the inspection hole at the top of the housing **(see illustration)**. If the wear sensor is very close to the rotor or the pad material has worn to about 1/8-inch or less, the pads should be replaced.

7   If you're unsure about the exact thickness of the remaining lining material, remove the pads for further inspection or replacement (refer to Chapter 9).

8   Before installing the wheels, check for leakage and/or damage (cracks, splitting, etc.) around the brake hose connections. Replace the hose or fittings as necessary, referring to Chapter 9.

9   Check the condition of the rotor. Look for score marks, deep scratches and burned spots. If any of these conditions exist, the hub/rotor assembly should be removed for servicing (see Section 38 for 2WD vehicles or Chapter 9 for 4WD vehicles).

Drum brakes

10   On rear brakes, remove the drum by pulling it off the axle and brake assembly (see Chapter 9).

11   With the drum removed, do not touch any brake dust (see the Warning at the beginning of this Section).

12   Note the thickness of the lining material on both the front and rear brake shoes. If the material has worn away to within 1/16-inch of the recessed rivets or metal backing, the shoes should be replaced **(see illustration)**. The shoes should also be replaced if they're

cracked, glazed (shiny surface) or contaminated with brake fluid.

13   Make sure that all the brake assembly springs are connected and in good condition.

14   Check the brake components for any signs of fluid leakage. With your finger or a small screwdriver, carefully pry back the rubber cups on the wheel cylinders located at the top of the brake shoes **(see illustration)**. Any leakage is an indication that the wheel cylinders should be overhauled immediately (Chapter 9). Also check brake hoses and connections for signs of leakage.

15   Wipe the inside of the drum with a clean rag and brake cleaner or denatured alcohol. Again, be careful not to breath the dangerous asbestos dust.

16   Check the inside of the drum for cracks, score marks, deep scratches and hard spots, which will appear as small discolorations. If these imperfections cannot be removed with fine emery cloth, the drum must be taken to a machine shop equipped to turn the drums.

17   If after the inspection process all parts are in good working condition, reinstall the brake drum (see Chapter 9).

18   Install the wheels and lower the vehicle.

Parking brake

19   The parking brake operates from a foot pedal or hand lever and locks the rear brake system. The easiest, and perhaps most obvious method of periodically checking the operation of the parking brake assembly is to park the vehicle on a steep hill with the parking brake set and the transmission in Neutral. If the parking brake cannot prevent the vehicle from rolling, it's in need of adjustment (see Chapter 9).

21   Fuel system check

Warning: *Take certain precautions when inspecting or servicing the fuel system components. Work in a well ventilated area and don't allow open flames (cigarettes, appliance pilot lights, etc.) in the work area. Mop up spills immediately and don't store fuel soaked* *rags where they could ignite. On fuel injection equipped models the fuel system is under pressure. No components should be disconnected until the pressure has been relieved (see Chapter 4).*

1   On all models, the fuel tank is located under the rear of the vehicle, covered by a shield.

2   The fuel system is most easily checked with the vehicle raised on a hoist so the components underneath the vehicle are readily visible and accessible.

3   If the smell of gasoline is noticed while driving or after the vehicle has been in the sun, the system should be thoroughly inspected immediately.

4   Remove the gas tank cap and check it for damage, corrosion and an unbroken sealing imprint on the gasket. Replace the cap with a new one if necessary.

5   With the vehicle raised, check the gas tank and filler neck for punctures, cracks and other damage. The connection between the filler neck and the tank is especially critical. Sometimes a rubber filler neck will leak due to loose clamps or deteriorated rubber; these are problems a home mechanic can usually rectify. **Warning:** *Do not, under any circumstances, try to repair a fuel tank yourself (except rubber components). A welding torch or any open flame can easily cause the fuel vapors to explode if the proper precautions are not taken!*

6   Carefully check all rubber hoses and metal lines attached to the fuel tank. Look for loose connections, deteriorated hoses, crimped lines and other damage. Follow the lines to the front of the vehicle, carefully inspecting them all the way. Repair or replace damaged sections as necessary.

7   If a fuel odor is still evident after the inspection, refer to Section 41.

22   Carburetor choke check

1   The choke operates only when the engine is cold, so this check should be performed before the engine has been started for the day.

2   Remove the top plate of the air cleaner assembly. It's usually held in place by a wing nut at the center. If any vacuum hoses must be disconnected, make sure you tag the hoses for reinstallation in their original positions. Place the top plate and wing nut aside, out of the way of moving engine components.

3   Look at the center of the air cleaner housing. You will notice a flat plate at the carburetor opening. This is the choke plate.

4   Press the accelerator pedal to the floor. The plate should close completely. Start the engine while you watch the choke plate. Don't position your face near the carburetor, as the engine could backfire, causing serious burns. When the engine starts, the choke plate should open slightly.

5   Allow the engine to continue running at an idle speed. As the engine warms up to operating temperature, the plate should slowly open, allowing more air to enter through the top of the carburetor.

6   After a few minutes, the choke plate should be fully open to the vertical position. Tap the accelerator to make sure the fast idle cam disengages.

7   You'll notice that the engine speed corresponds to the plate opening. With the plate fully closed, the engine should run at a fast idle speed. As the plate opens and the throttle is moved to disengage the fast idle cam, the engine speed will decrease.

8   Refer to Chapter 4 for specific information on adjusting and servicing the choke components.

23   Carburetor/throttle body mounting nut torque check

1   The carburetor or Throttle Body Injection (TBI) unit is attached to the top of the intake manifold by several bolts or nuts. These fasteners can sometimes work loose from vibration and temperature changes during normal engine operation and cause a vacuum leak.

2   If you suspect that a vacuum leak exists at the bottom of the carburetor or throttle body, obtain a length of hose. Start the engine and place one end of the hose next to your ear as you probe around the base with the other end. You will hear a hissing sound if a leak exists (be careful of hot or moving engine components when performing this check).

3   Remove the air cleaner assembly, tagging each hose to be disconnected with a piece of numbered tape to make reassembly easier.

4   Locate the mounting nuts or bolts at the base of the carburetor or throttle body. Decide what special tools or adapters will be necessary, if any, to tighten the fasteners.

5   Tighten the nuts to the specified torque. Don't overtighten them, as the threads could strip.

6   If, after the nuts or bolts are properly tightened, a vacuum leak still exists, the carburetor or throttle body must be removed and a new gasket installed. See Chapter 4 for more information.

7   After tightening the fasteners, reinstall the air cleaner and return all hoses to their original positions.

24   Throttle linkage inspection

1   Inspect the throttle linkage for damage and missing parts and for binding and interference when the accelerator pedal is depressed.

2   Lubricate the various linkage pivot points with engine oil.

25   Thermostatic air cleaner check

Refer to illustrations 25.3a and 25.3b

1   Some engines are equipped with a thermostatically controlled air cleaner which draws air to the carburetor from different locations, depending on engine temperature.

2   This is a visual check. If access is limited, a small mirror may have to be used.

3   Locate the air valve inside the air cleaner assembly. It's inside the long snorkel of the air cleaner housing **(see illustrations)**.

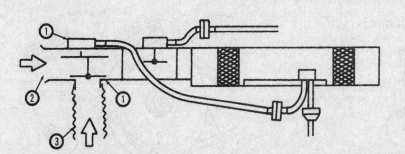
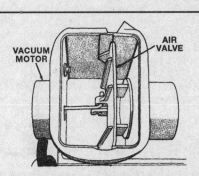

25.3a Thermostatic air cleaner for V6 engines and some four-cylinder engines (air valve shown open) - when the engine is cold, the air valve (1) should close off the snorkel end (2), allowing warm air to flow through the flexible duct (3), which is attached to the exhaust manifold and heat stove passage - when the engine is at normal operating temperature, the air valve should be open, allowing cool air to flow through the snorkel end (2)

25.3b On inline six-cylinder engines, the thermostatic air cleaner is located towards the front of the driver's side of the engine compartment - you must put a mirror in front of the radiator support to view the air valve

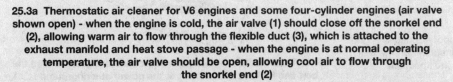

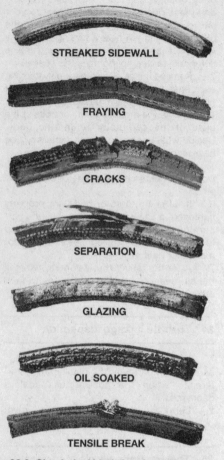

26.3 Check the V-belt for signs of wear like these - if it looks worn, replace it

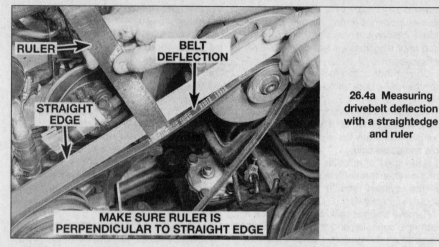

26.4a Measuring drivebelt deflection with a straightedge and ruler

4   If there is a flexible air duct attached to the end of the snorkel, leading to an area behind the grille, disconnect it at the snorkel. This will enable you to look through the end of the snorkel and see the air valve inside.

5   The check should be done when the engine is cold. Start the engine and look through the snorkel at the air valve, which should move to a closed position. With the valve closed, air cannot enter through the end of the snorkel, but instead enters the air cleaner through the flexible duct attached to the exhaust manifold and the heat stove passage.

6   As the engine warms up to operating temperature, the air valve should open to allow air through the snorkel end. Depending on outside temperature, this may take 10-to-15 minutes. To speed up this check, you can reconnect the snorkel air duct, drive the vehicle until it reaches normal operating temperature, then check to see if the air valve is completely open.

7   If the thermostatically controlled air cleaner isn't operating properly, see Chapter 6 for more information.

26   Drivebelt check, adjustment and replacement

Refer to illustrations 26.3, 26.4a, 26.4b, 26.5a, 26.5b, 26.12a and 26.12b

1   The drivebelts, or V-belts as they are often called, are located at the front of the engine and play an important role in the overall operation of the engine and accessories.

Due to their function and material makeup, the belts are prone to failure after a period of time and should be inspected and adjusted periodically to prevent major engine damage.

2   The number of belts used on a particular vehicle depends on the accessories installed. Drivebelts are used to turn the alternator, power steering pump, water pump and air conditioning compressor. Depending on the pulley arrangement, more than one of these components may be driven by a single belt. Later models are equipped with one serpentine drivebelt that runs all engine accessories.

3   With the engine off, locate the drivebelts at the front of the engine. Using your fingers (and a flashlight, if necessary), move along the belts checking for cracks and separation of the belt plies. Also check for fraying and glazing, which gives the belt a shiny appearance **(see illustration)**. Both sides of each belt should be inspected, which means you will have to twist the belt to check the underside. Check the pulleys for nicks, cracks, distortion and corrosion.

4   The tension of each V-belt is checked by pushing on it at a distance halfway between the pulleys. Push firmly and see how much the belt moves (deflects) **(see illustration)**. A rule of thumb is that if the distance

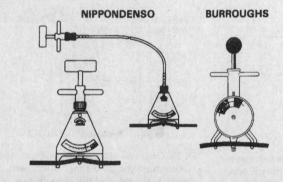

26.4b If you are able to borrow either a Nippondenso or Burroughs belt tension gauge, this is how it's installed on the belt - compare the reading on the scale with the specified drivebelt tension

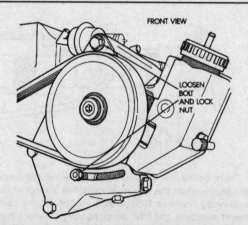

26.5a To adjust the tension on a serpentine drivebelt, loosen the bolt and locknut on the front of the power steering pump . . .

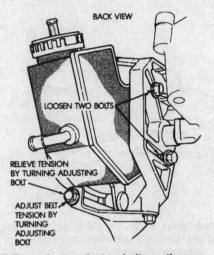

26.5b ... loosen the two bolts on the rear of the power steering pump and turn the adjusting bolt

from pulley center-to-pulley center is between 7 and 11-inches, the belt should deflect 1/4-inch. If the belt travels between pulleys spaced 12-to-16 inches apart, the belt should deflect 1/2-inch. The tension on serpentine belts can only be checked using a belt tension gauge, available at auto parts stores **(see illustration)**.

5   If adjustment is needed, either to make the belt tighter or looser, it's done by moving the belt-driven accessory on the bracket. On later models with a serpentine belt, belt tension is adjusted at the power steering pump (or idler pulley if not equipped with power steering) **(see illustration)**.

6   For each component there will be an adjusting bolt and a pivot bolt. Both bolts must be loosened slightly to enable you to move the component.

7   After the two bolts have been loosened, move the component away from the engine to tighten the belt or toward the engine to loosen the belt. Hold the accessory in posi-

tion and check the belt tension. If it's correct, tighten the two bolts until just snug, then recheck the tension. If the tension is all right, tighten the bolts.

8   It will often be necessary to use some sort of pry bar to move the accessory while the belt is adjusted. If this must be done to gain the proper leverage, be very careful not to damage the component being moved or the part being pried against.

9   To replace a belt, follow the above procedures for drivebelt adjustment, but slip the belt off the pulleys and remove it. Since belts tend to wear out more or less at the same time, it's a good idea to replace all of them at the same time. Mark each belt and the corresponding pulley grooves so the replacement belts can be installed properly.

10   Take the old belts with you when purchasing new ones in order to make a direct comparison for length, width and design.

11   Adjust the belts as described earlier in this Section.

12   When replacing a serpentine drivebelt (used on later models), make sure the new belt is routed correctly or the water pump could turn backwards, causing overheating **(see illustrations)**. Also, the belt must completely engage the grooves in the pulleys.

27   Seatbelt check

1   Check the seat belts, buckles, latch plates and guide loops for any obvious damage or signs of wear.

2   Make sure the seatbelt reminder light comes on when the key is turned on.

3   The seat belts are designed to lock up during a sudden stop or impact, yet allow free movement during normal driving. The retractors should hold the belt against your chest while driving and rewind the belt when the buckle is unlatched.

4   If any of the above checks reveal problems with the seatbelt system, replace parts as necessary.

28   Neutral start switch check

Warning: *During the following checks there is a chance that the vehicle could lunge forward, possibly causing damage or injuries. Allow plenty of room around the vehicle, apply the parking brake firmly and hold down the regular brake pedal during the checks.*

1   Automatic transmission equipped models have a Neutral start switch which prevents the engine from starting unless the shift lever is in Neutral or Park.

2   Try to start the vehicle in each gear. The engine should crank only in Park or Neutral.

3   Make sure the steering column lock allows the key to go into the Lock position only when the shift lever is in Park.

4   The ignition key should come out only in the Lock position.

5   Refer to Chapter 7, Part B for further information on the Neutral start switch.

29   Spare tire and jack check

1   Check the spare tire to make sure it's securely fastened so it cannot come loose when the vehicle is in motion.

2   Make sure the jack and components are secured in place.

30   Idle speed check and adjustment (1984 through 1986 carburetor-equipped models)

Note: *Idle speed check and adjustment are not routine maintenance procedures on vehicles other than those identified in the heading above.*
Refer to illustrations 30.7, 30.11a and 30.11b

1   Engine idle speed is the speed at which the engine operates when no throttle pedal pressure is applied. The idle speed is critical to the performance of the engine itself, as

1

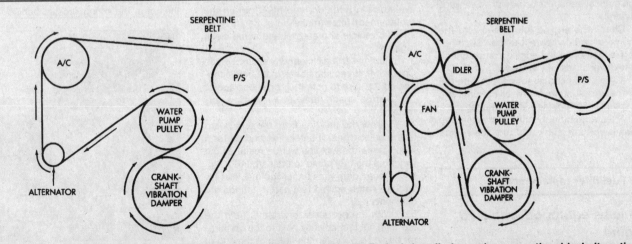

26.12a Typical four-cylinder engine serpentine drivebelt routing   **26.12b Typical six-cylinder engine serpentine drivebelt routing**

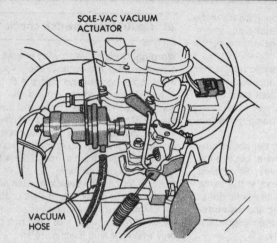

SOLE-VAC VACUUM ACTUATOR

VACUUM HOSE

30.7 Before checking the idle speed on four-cylinder engines, disconnect the vacuum hose from the sole-vac vacuum actuator

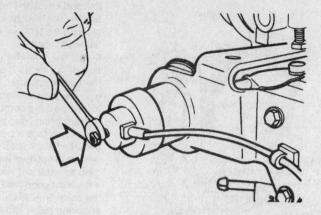

30.11a On four-cylinder engines, adjust the idle speed by turning the 1/4-inch nut (arrow) on the end of the sole-vac vacuum actuator

well as many engine sub-systems.

2   A hand-held tachometer must be used when adjusting idle speed to get an accurate reading. The exact hook-up for these meters varies with the manufacturer, so follow the particular directions included.

3   Set the parking brake and block the wheels. Be sure the transmission is in Neutral (manual transmission) or Park (automatic transmission).

4   Turn off the air conditioner (if equipped), the headlights and any other accessories during this procedure.

5   Start the engine and allow it to reach normal operating temperature.

6   Shut off the engine.

7   On four-cylinder models, disconnect and plug the vacuum hose on the sole-vac vacuum actuator **(see illustration)**.

8   On V6 models, disconnect and plug the EGR vacuum hose at the EGR valve and the purge control vacuum hose on the canister (see Sections 41 and 42 to locate the EGR valve and canister)

9   On automatic transmission equipped vehicles, have an assistant shift to Drive while keeping the brake pedal firmly depressed. Place manual transmission equipped vehicles in Neutral.

10   Check the engine idle speed with the tachometer and compare it to the VECI label, which is located in the engine compartment, on the driver's side of the firewall.

11   If the idle speed is not correct, turn the idle speed adjusting screw or nut until the idle speed is correct **(see illustrations)**.

12   After adjustment, shift the automatic transmission into Park and turn the engine off.

31   Fuel filter replacement

Vehicles equipped with a V6 engine

Refer to illustrations 31.6 and 31.8

1   On these models, the fuel filter is

30.11b On V6 engines, turn the idle speed screw (1) to adjust the idle speed - on V6 engines equipped with air conditioning, unplug the electrical lead from the air conditioning compressor, turn the air conditioning on, then open the throttle slightly to allow the idle speed adjustment screw on the solenoid (2) to extend. Turn the screw to adjust the idle speed (DO NOT touch 3 and 4)

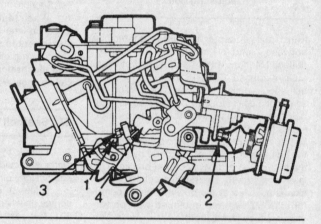

located inside the fuel inlet nut at the carburetor. It's made of either pleated paper or porous bronze and cannot be cleaned or reused.

2   The job should be done with the engine cold (after sitting at least three hours). The necessary tools include open-end wrenches to fit the fuel line nuts. Flare nut wrenches (which wrap around the nut) should be used if available. In addition, you must obtain the replacement filter (make sure it's for your specific vehicle and engine) and some clean rags.

3   Remove the air cleaner assembly (see Chapter 4). If vacuum hoses must be disconnected, be sure to note their positions and/or tag them to ensure they are reinstalled correctly.

4   Follow the fuel line from the fuel pump to the point where it enters the carburetor. I most cases the fuel line will be metal all the way from the fuel pump to the carburetor.

5   Place some rags under the fuel inlet fittings to catch spilled fuel as the fittings are disconnected.

6   With the proper size wrench, hold the fuel inlet nut immediately next to the carburetor body. Now loosen the fitting at the end of the metal fuel line with a flare-nut wrench (if available). Make sure the fuel inlet nut next to

the carburetor is held securely while the fuel line is disconnected **(see illustration)**.

7   After the fuel line is disconnected, move it aside for better access to the inlet nut. Don't crimp the fuel line.

8   Unscrew the fuel inlet nut, which was previously held steady. As this fitting is drawn away from the carburetor body, be careful not to lose the thin washer-type gasket on the nut or the spring, located behind the fuel

31.6 Two wrenches are required to loosen the carburetor fuel inlet line

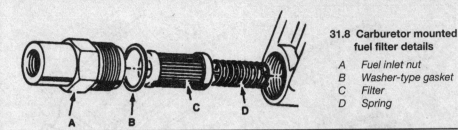

31.8 Carburetor mounted fuel filter details

A   Fuel inlet nut
B   Washer-type gasket
C   Filter
D   Spring

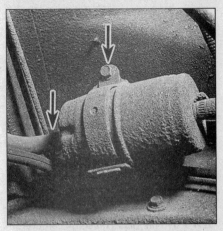

31.15 To remove the inline fuel filter, unscrew the fuel hose clamps and pull off the hoses, then remove the securing strap bolt (arrows)

filter. Also pay close attention to how the filter is installed **(see illustration)**.

9   Compare the old filter with the new one to make sure they're the same length and design.

10   Reinstall the spring in the carburetor body.

11   Place the filter in position (a gasket is usually supplied with the new filter) and tighten the nut. Make sure it's not cross-threaded. Tighten it securely, but be careful not to overtighten it as the threads can strip easily, causing fuel leaks. Reconnect the fuel line to the fuel inlet nut, being careful not to cross-thread the nut. Use a back-up wrench on the fuel inlet nut while tightening the fuel line fitting.

12   Start the engine and check carefully for leaks. If the fuel line fitting leaks, disconnect it and check for stripped or damaged threads. If the fuel line fitting has stripped threads, remove the entire line and have a repair shop install a new fitting. If the threads look all right, purchase some thread sealing tape and wrap the threads with it. Inlet nut repair kits are available at most auto parts stores to overcome leaking at the fuel inlet nut.

Vehicles equipped with a four- or inline six-cylinder engine (1996 and earlier)

Note: *There is no specified maintenance interval for the fuel filter on 1997 and later models. The fuel filter is part of the fuel pump/sending unit assembly inside the fuel tank.*

Refer to illustrations 31.15 and 31.16

13   These engines employ an in-line fuel filter. The filter is located on the left side frame rail, near the fuel tank. On six-cylinder port fuel injected engines, the system is under pressure even when the engine is off. **Warning:** *On inline six-cylinder engines, the system must be depressurized (see Chapter 4) before any work is performed*.

14   With the engine cold, place a container, newspapers or rags under the fuel filter.

15   Disconnect the fuel hoses and detach the filter from the frame **(see illustration)**.

16   Install the new filter by reversing the removal procedure. Make sure the arrow or the word -OUT" on the filter points toward the engine, not the fuel tank **(see illustration)**. Tighten the screw clamps securely, but not to the point where the rubber hose is badly distorted.

32   Air filter and PCV filter replacement

Refer to illustrations 32.2a, 32.2b, 32.4 and 32.6

1   At the specified intervals, the air filter and (if equipped) PCV filter should be replaced with new ones. The engine air cleaner also supplies filtered air to the PCV system.

2   The filter on carburetor-equipped models is located on top of the carburetor and is replaced by unscrewing the wing nut from the top of the filter housing, disconnecting the clips and lifting off the cover **(see illustration)**. On most fuel-injected models, the filter is located in a housing in the engine compartment. It can be replaced after disengaging

31.16 Make sure the word OUT on the fuel filter faces the engine

the clips holding the top plate in place **(see illustration)**.

3   While the top plate is off, be careful not to drop anything into the carburetor, TBI or air cleaner assembly.

4   Lift the air filter element out **(see illustration)** and wipe out the inside of the air cleaner housing with a clean rag.

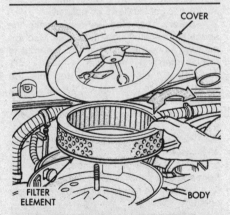

32.2a On carburetor-equipped models, remove the air cleaner wing nut, disconnect the clips, lift the cover off and remove the filter element

32.2b On fuel-injected models, clips on the sides hold the top plate of the air cleaner to the housing - they can be pried loose with a screwdriver

32.4 Hold the top plate up and lift the filter element out of the housing (fuel-injected model shown)

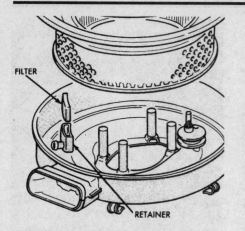

32.6 PCV filter (used on some carburetor-equipped models)

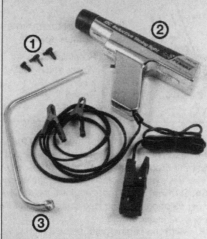

33.1 Tools needed to check and adjust ignition timing

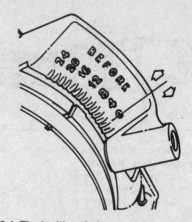

33.4 The ignition timing marks are located at the front of the engine (four-cylinder shown)

5   Place the new filter in the air cleaner housing. Make sure it seats properly in the bottom of the housing.

6   The PCV filter is also located inside the air cleaner housing on some models **(see illustration)**. Remove the cover and air filter as previously described, then locate the PCV filter on the inside of the housing.

7   Remove the old filter.

8   Install the new PCV filter and the new air filter.

9   Install the cover and any hoses which were disconnected.

33 Ignition timing check and adjustment (1984 through 1986 carburetor-equipped models)

Refer to illustrations 33.1 and 33.4
Note: *Ignition timing check and adjustment are not required on vehicles other than those identified in the heading above. If the information in this Section differs from the Vehicle Emission Control Information label (located on the driver's side of the firewall in the engine compartment of your vehicle), the label should be considered correct.*

1   The engine must be at normal operating temperature and the air conditioner must be Off. Make sure the idle speed is correct (see Section 30). Some special tools will be required for this procedure **(see illustration)**.

2   Apply the parking brake and block the wheels to prevent movement of the vehicle. The transmission must be in Park (automatic) or Neutral (manual).

3   On four-cylinder engines, disconnect and plug the hose connected to the vacuum advance unit on the distributor and unplug the three wire connector to the vacuum input switches.

4   Locate the timing marks at the front of the engine (they should be visible from above after the hood is opened) **(see illustration)**. The crankshaft pulley or vibration damper has a groove in it and a scale with notches and numbers is either molded into or

1   *Vacuum plugs - Vacuum hoses will, in most cases, have to be disconnected and plugged. Molded plugs in various shapes and sizes are available for this*

2   *Inductive pick-up timing light - Flashes a bright concentrated beam of light when the number one spark plug fires. Connect the leads according to the instructions supplied with the light*

3   *Distributor wrench - On some models, the hold-down bolt for the distributor is difficult to reach and turn with conventional wrenches or sockets. A special wrench like this must be used*

attached to the engine's timing cover. Clean the scale with solvent so the numbers are visible.

5   Use chalk or white paint to mark the groove in the pulley/vibration damper.

6   Highlight the notch or point on the scale that corresponds to the ignition timing specification on the Vehicle Emission Control Information label.

7   Hook up the timing light, following the manufacturer's instructions (an inductive pick-up timing light is preferred). Generally, the power leads are attached to the battery terminals and the pick-up lead is attached to the number one spark plug wire. On four and six-cylinder engines, the number one spark plug is the very front one. On the V6 engine, the number one spark plug is the front one on the passenger side of the engine. **Caution:** *If an inductive pick-up timing light isn't available, don't puncture the spark plug wire to attach the timing light pick-up lead. Instead, use an adapter between the spark plug and plug wire. If the insulation on the plug wire is damaged, the secondary voltage will jump to ground at the damaged point and the engine will misfire.*

8   Make sure the timing light wires are routed away from the drivebelts and fan, then start the engine.

9   Allow the idle speed to stabilize, then point the flashing timing light at the timing

marks, be very careful of moving engine components!

10   The mark on the pulley/vibration damper will appear stationary.

If it's aligned with the specified point on the scale, the ignition timing is correct.

11   If the marks aren't aligned, adjustment is required. Loosen the distributor hold-down bolt and turn the distributor very slowly until the marks are aligned. Since access to the bolt is tight, a special distributor wrench may be needed.

12   Tighten the bolt and recheck the timing.

13   Turn off the engine and remove the timing light (and adapter, if used).

14   Reconnect any components which were disconnected.

34 Automatic transmission fluid and filter change

Refer to illustrations 34.10a and 34.10b

1   At the specified time intervals, the transmission fluid should be drained and replaced. Since the fluid will remain hot long after driving, perform this procedure only after the engine has cooled down completely.

2   Before beginning work, purchase the specified transmission fluid (see Recommended lubricants and fluids at the front of this Chapter) and a new filter.

3   Other tools necessary for this job include jackstands to support the vehicle in a raised position, a drain pan capable of holding at least eight pints, newspapers and clean rags.

4   Raise the vehicle and support it securely on jackstands.

5   With a drain pan in place, remove the front and side pan mounting bolts.

6   Loosen the rear pan bolts approximately one turn.

7   If the pan does not loosen and fluid does not begin to drain, carefully pry the transmission pan loose with a putty knife.

8   Remove the remaining bolts, pan and

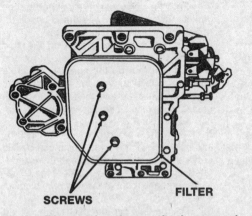

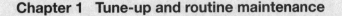

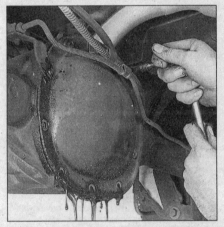

34.10a Filter bolt locations on the three-speed automatic transmission

34.10b Filter bolt locations on the four-speed automatic transmission

gasket. Carefully clean the gasket surface of the transmission to remove all traces of the old gasket and sealant.

9   Drain the fluid from the transmission pan, clean it with solvent and dry it with compressed air.

10   Remove the fluid filter from the transmission valve body **(see illustrations)**.

11   Install a new filter.

12   Make sure the gasket surface on the transmission pan is clean, then install a new gasket. Put the pan in place against the transmission and, working around the pan, tighten each bolt a little at a time until the final torque figure is reached.

13   Lower the vehicle and add the specified amount of automatic transmission fluid through the filler tube (Section 6).

14   With the transmission in Park and the parking brake set, run the engine at a fast idle, but don't race it.

15   Move the gear selector through each range and back to Neutral. Check the fluid level.

16   Check under the vehicle for leaks during the first few trips.

35   Manual transmission lubricant change

1   Raise the vehicle and support it securely on jackstands.

2   Move a drain pan, rags, newspapers and wrenches under the transmission.

3   Remove the transmission drain plug at the bottom of the case **(see illustration 16.1)** and allow the oil to drain into the pan.

4   After the oil has drained completely, reinstall the plug and tighten it securely.

5   Remove the fill plug from the side of the transmission case. Using a hand pump, syringe or fuller, fill the transmission with the correct amount of the specified lubricant. Reinstall the fill plug and tighten it securely.

6   Lower the vehicle for a short distance then check the drain and fill plugs for leakage.

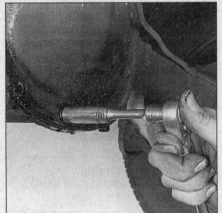

36.6a Remove the bolts from the lower edge of the cover . . .

7   Drive the vehicle for a short distance, then check the drain and fill plugs for leakage.

36   Differential lubricant change

Refer to illustrations 36.6a, 36.6b, 36.6c and 36.8

1   Some differentials can be drained by removing the drain plug, while on others it's necessary to remove the cover plate on the differential housing. As an alternative, a hand suction pump can be used to remove the differential lubricant through the filler hole. If the gasket is leaking or there is no drain plug and a suction pump isn't available, be sure to obtain a new gasket at the same time the gear lubricant is purchased.

2   Raise the vehicle and support it securely on jackstands. Move a drain pan, rags, newspapers and wrenches under the vehicle.

3   Remove the fill plug from the differential.

4   If equipped with a drain plug, remove the plug and allow the differential lubricant to drain completely. After the lubricant has drained, install the plug and tighten it securely.

5   If a suction pump is being used, insert the flexible hose. Work the hose down to the

36.6b . . . then loosen the top bolts and let the lubricant drain out

bottom of the differential housing and pump the lubricant out.

6   If the differential is being drained by removing the cover, remove the bolts on the lower half of the cover **(see illustration)**. Loosen the bolts on the upper half and use them to keep the cover loosely attached **(see illustration)**. Allow the lubricant to drain into the pan, then completely remove the cover **(see illustration)**.

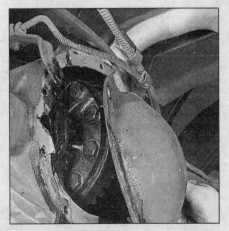

36.6c After the lubricant has completely drained, remove the cover

36.8 Carefully scrape the old gasket or sealant material off to ensure a clean surface for the new gasket to seal against

7   Using a lint-free rag, clean the inside of the cover and the accessible areas of the differential housing. As this is done, check for chipped gears and metal particles in the lubricant, indicating that the differential should be more thoroughly inspected and/or repaired.

8   Thoroughly clean the gasket mating surfaces of the differential housing and the cover plate. Use a gasket scraper or putty knife to remove all traces of the old gasket or sealant **(see illustration)**.

9   Apply a thin layer of RTV sealant to the cover flange and then press a new gasket into position on the cover. Make sure the bolt holes align properly.

10   Place the cover on the differential housing and install the bolts. Tighten the bolts securely.

11   On all models, use a hand pump, syringe or funnel to fill the differential housing with the specified lubricant until it's level with the bottom of the plug hole.

12   Install the filler plug and tighten it securely.

37   Transfer case lubricant change (4WD models)

1   Drive the vehicle for at least 15 minutes in stop and go traffic to warm the lubricant in the case. Perform this warm-up procedure in 4WD. Use all gears, including Reverse, to ensure the lubricant is sufficiently warm to drain completely.

2   Raise the vehicle and support it securely on jackstands.

3   Remove the filler plug from the case **(see illustration 18.1)**.

4   Remove the drain plug from the lower part of the case and allow the old lubricant to drain completely.

5   Carefully clean and install the drain plug after the case is completely drained. Tighten the plug to the specified torque.

6   Fill the case with the specified lubricant until it is level with the lower edge of the filler hole.

7   Install the filler plug and tighten it securely.

8   Drive the vehicle for a short distance and recheck the lubricant level. In some instances a small amount of additional lubricant will have to be added.

38   Front wheel bearing check, repack and adjustment (2WD models)

Refer to illustrations 38.1, 38.6, 38.7, 38.8, 38.11 and 38.15

1   In most cases the front wheel bearings will not need servicing until the brake pads are changed. However, the bearings should be checked whenever the front of the vehicle is raised for any reason. Several items, including a torque wrench and special grease, are required for this procedure **(see illustration)**.

2   With the vehicle securely supported on jackstands, spin each wheel and check for noise, rolling resistance and free play.

3   Grasp the top of each tire with one hand and the bottom with the other. Move the wheel in-and-out on the spindle. If there's any noticeable movement, the bearings should be checked and then repacked with grease or replaced if necessary.

4   Remove the wheel.

5   Fabricate a wood block (1-1/16 inch by 1/2-inch by 2-inches long) which can be slid between the brake pads to keep them separated. Remove the brake caliper (see Chapter 9) and hang it out of the way on a piece of wire.

6   Pry the dust cap out of the hub using a screwdriver or hammer and chisel **(see illustration)**.

7   Straighten the bent ends of the cotter pin, then pull the cotter pin out of the nut retainer **(see illustration)**. Discard the cotter pin and use a new one during reassembly.

8   Remove the nut retainer, adjustment nut and thrust washer from the end of the spindle **(see illustration)**.

9   Pull the rotor/hub assembly out slightly, then push it back into its original position. This should force the outer bearing off the spindle enough so it can be removed.

10   Pull the rotor/hub assembly off the spindle.

11   Use a screwdriver to pry the grease seal out of the rear of the hub **(see illustration)**. As this is done, note how the seal is installed.

12   Remove the inner wheel bearing from the hub.

13   Use solvent to remove all traces of the old grease from the bearings, hub and spindle. A small brush may prove helpful; however make sure no bristles from the brush embed themselves inside the bearing rollers. Allow the parts to air dry.

14   Carefully inspect the bearings for cracks, heat discoloration, worn rollers, etc. Check the bearing races inside the hub for

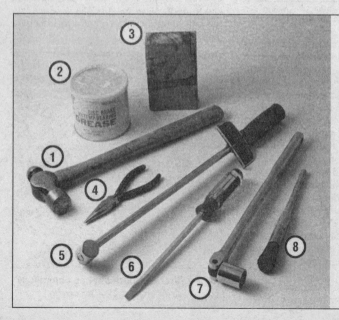

38.1 Tools and material required for front wheel bearing maintenance

1   *Hammer - A common hammer will do just fine*

2   *Grease - High-temperature grease which is formulated specially for front wheel bearings should be used*

3   *Wood block - If you have a scrap piece of 2x4, it can be used to drive the new seal into the hub*

4   *Needle-nose pliers - Used to straighten and remove the cotter pin in the spindle*

5   *Torque wrench - This is very important in this procedure; if the bearing is too tight, the wheel won't turn freely - if it is too loose, the wheel will 'wobble' on the spindle. Either way, it could mean extensive damage*

6   *Screwdriver - Used to remove the seal from the hub (a long screwdriver would be preferred)*

7   *Socket/breaker bar - Needed to loosen the nut on the spindle if it is extremely tight*

8   *Brush - Together with some clean solvent, this will be used to remove old grease from the hub and spindle*

38.6 Dislodge the dust cap by working around the outer edge with a screwdriver or hammer and chisel

38.7 Remove the cotter pin and discard it - use a new one when the hub is reinstalled

wear and damage. If the bearing races are defective, the rotor/hub assemblies should be taken to a machine shop with the facilities to remove the old races and press new ones in. Note that the bearings and races come as matched sets and old bearings should never be installed on new races.

15   Use high-temperature front wheel bearing grease to pack the bearings. Work the grease completely into the bearings, forcing it between the rollers, cone and cage from the back side **(see illustration)**.

16   Apply a thin coat of grease to the spindle at the outer bearing seat, inner bearing seat, shoulder and seal seat.

17   Put a small quantity of grease inboard of each bearing race inside the hub. Using your finger, form a dam at these points to provide extra grease availability and to keep thinned grease from flowing out of the bearing.

18   Place the grease-packed inner bearing into the rear of the hub and put a little more grease outboard of the bearing.

19   Place a new seal over the inner bearing and tap the seal evenly into place with a hammer and block of wood until it's flush with the hub.

20   Carefully place the rotor/hub assembly onto the spindle and push the grease-packed outer bearing into position.

21   Install the thrust washer and adjustment nut. Tighten the nut only slightly (no more than 12 ft-lbs of torque).

22   Spin the hub in a forward direction to seat the bearings and remove any grease or burrs which could cause excessive bearing play later.

23   Check to see that the tightness of the adjustment nut is still approximately 12 ft-lbs.

24   Loosen the adjustment nut until it's just loose, no more.

25   Using your hand (not a wrench of any kind), tighten the nut until it's snug. Install the nut retainer and a new cotter pin through the hole in the spindle and nut retainer. If the nut slots don't line up, loosen the nut slightly until they do. From the hand-tight position, the nut should not be loosened more than one-half

flat to install the cotter pin.

26   Bend the ends of the cotter pin until they're flat against the nut. Cut off any extra length which could interfere with the dust cap.

27   Install the dust cap, tapping it into place with a hammer.

28   Place the brake caliper near the rotor

and carefully remove the wood spacer. Install the caliper (see Chapter 9).

29   Install the tire/wheel assembly on the hub and tighten the lug nuts.

30   Grasp the top and bottom of the tire and check the bearings in the manner described earlier in this Section.

31   Lower the vehicle.

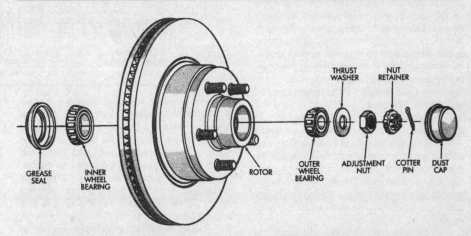

38.8 Front wheel hub and bearing components (2WD models) - exploded view

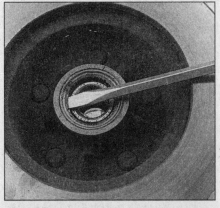

38.11 Use a screwdriver to pry the grease seal from the rear of the hub

38.15 Work the grease into each bearing from the large diameter side until grease oozes out the small-diameter side

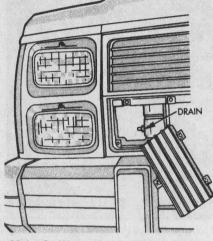

39.4a On Wagoneer models, remove the park/turn signal light for access to the radiator drain

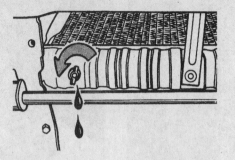

39.4b On models other than Wagoneer, first check the bottom of the radiator for the drain fitting

39.4c If the radiator drain on non-Wagoneer models isn't located on the bottom of the radiator, you'll have to remove the grille (Chapter 11) for access to the drain (arrow)

39 Cooling system servicing (draining, flushing and refilling)

Refer to illustrations 39.4a, 39.4b, 39.4c and 39.5

Warning: *Antifreeze is a corrosive and poisonous solution, so be careful not to spill any of the coolant mixture on the vehicle's paint or your skin. If this happens, rinse immediately with plenty of clean water. Consult local authorities regarding proper disposal procedures for antifreeze before draining the cooling system. In many areas, reclamation centers have been established to collect used oil and coolant mixtures.*

1   Periodically, the cooling system should be drained, flushed and refilled to replenish the antifreeze mixture and prevent formation of rust and corrosion, which can impair the performance of the cooling system and cause engine damage. When the cooling system is serviced, all hoses and the radiator cap should be checked and replaced if necessary.

2   Apply the parking brake and block the wheels. If the vehicle has just been driven, wait several hours to allow the engine to cool down before beginning this procedure.

3   Once the engine is completely cool, remove the radiator cap (four cylinder and V6 engines) or coolant pressure bottle cap (inline six-cylinder engines).

4   Move large container under the radiator drain to catch the coolant. Attach a hose to the drain fitting to direct the coolant into the container, then open the drain fitting **(see illustrations)** by turning it counterclockwise (a pair of pliers may be required to turn it).

5   After the coolant stops flowing out of the radiator, move the container under the engine block drain plug(s)**(see illustration)**. Remove the plug(s) and allow the coolant in the block to drain.

6   While the coolant is draining, check the condition of the radiator hoses, heater hoses and clamps (refer to Section 9 if necessary).

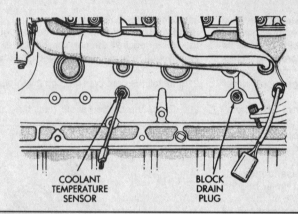

COOLANT TEMPERATURE SENSOR

BLOCK DRAIN PLUG

39.5 On four and inline six-cylinder engines, the block drain plug is located on the driver's side of the engine block, below the exhaust manifold; V6 engines have two block drain plugs at the lower rear of each side of the engine block (inline six-cylinder engine shown)

7   Replace any damaged clamps or hoses.

8   Once the system is completely drained, flush the radiator with fresh water from a garden hose until it runs clear at the drain. On inline six-cylinder engines, add the slush water through the coolant pressure bottle. The flushing action of the water will remove sediments from the radiator but will not remove rust and scale from the engine and cooling tube surfaces.

9   These deposits can be removed with a chemical cleaner. Follow the procedure outlined in the manufacturer's instructions. If the radiator is severely corroded, damaged or leaking, it should be removed (Chapter 3) and taken to a radiator repair shop.

10   Remove the overflow hose from the coolant recovery reservoir (four-cylinder and V6 engines). Drain the reservoir and flush it with clean water, then reconnect the hose.

11   Close and tighten the radiator drain. Install and tighten the block drain plugs.

12   Place the heater temperature control in the maximum heat position.

13   On four-cylinder and V6 engines, slowly add new coolant (a 50/50 mixture of water and antifreeze) to the radiator until it's full. Add coolant to the reservoir up to the lower mark. On inline six-cylinder engines, add coolant to the pressure bottle until the level is at the top of the post (see Section 9).

14   Leave the radiator or pressure bottle

cap off and run the engine in a well-ventilated area until the thermostat opens (coolant will begin flowing through the radiator and the upper radiator hose will become hot).

15   Turn the engine off and let it cool. Add more coolant mixture to bring the level back up to the lip on the radiator filler neck or top of the post (inline six-cylinder engine).

16   Squeeze the upper radiator hose to expel air, then add more coolant mixture if necessary. Replace the radiator cap.

17   Start the engine, allow it to reach normal operating temperature and check for leaks.

40 Positive Crankcase Ventilation (PCV) valve or Crankcase Ventilation (CCV) hose check, cleaning and replacement

Refer to illustration 40.11

Check

1   The PCV valve or CCV hose is located in the rocker arm cover.

2   With the engine idling at normal operating temperature, pull the valve (with hose attached) or CCV hose from the rubber grommet in the cover.

3   Place your finger over the valve or hose opening. If there's no vacuum, check for a

40.11 A paper clip can be used to clean the CCV system orifice

41.2 The charcoal canister (arrow) is located on the firewall in the engine compartment - check the hoses connected to it for damage

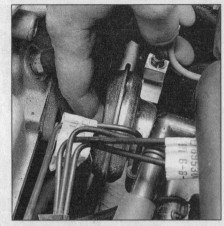

42.2 Move the EGR valve diaphragm in and out to make sure it isn't stuck

plugged hose, manifold port, or valve. Replace any plugged or deteriorated hoses.

4   Turn off the engine and shake the PCV valve, listening for a rattle. If the valve doesn't rattle, replace it with a new one.

PCV valve replacement

5   To replace the valve, pull it from the end of the hose, noting its installed position and direction.

6   When purchasing a replacement PCV valve, make sure it's for your particular vehicle and engine size. Compare the old valve with the new one to make sure they're the same.

7   Push the valve into the end of the hose until it's seated.

8   Inspect the rubber grommet for damage and replace it with a new one if necessary.

9   Push the PCV valve and hose securely into position.

CCV hose and orifice cleaning

10   Later models are equipped with a CCV system which performs the same function as the PCV system but uses a rubber fitting with a molded-in orifice which is pressed into a hole in the rocker arm cover. The fitting is connected to the intake manifold by a plastic hose.

11   If there is no vacuum at the end of the

hose (Step 3), turn off the engine, remove the fitting and clean the hose with solvent, clean the fitting orifice **(see illustration)** if it's plugged. If the fitting or hose are cracked or deteriorated, replace them with new ones.

12   Install the fitting and hose securely in the rocker arm cover.

41   Evaporative emissions control system check

Refer to illustration 41.2

1   The function of the evaporative emissions control system is to draw fuel vapors from the gas tank and fuel system, store them in a charcoal canister and route them to the intake manifold during normal engine operation.

2   The most common symptom of a fault in the evaporative emissions system is a strong fuel odor in the engine compartment. If a fuel odor is detected, inspect the charcoal canister, located in the engine compartment **(see illustration)**. Check the canister and all hoses for damage and deterioration.

3   The evaporative emissions control system is explained in more detail in Chapter 6.

42   Exhaust Gas Recirculation (EGR) system check

Refer to illustration 42.2

1   The EGR valve is usually located on the intake manifold, adjacent to the carburetor or TBI unit. Most of the time when a problem develops in this emissions system, it's due to a stuck or corroded EGR valve.

2   With the engine cold to prevent burns, push on the EGR valve diaphragm. Using moderate pressure, you should be able to press the diaphragm in-and-out within the housing **(see illustration)**.

3   If the diaphragm doesn't move or moves only with much effort, replace the EGR valve with a new one. If in doubt about the condition of the valve, compare the free movement of the valve with a new one.

4   Refer to Chapter 6 for more information on the EGR system.

43   Spark plug replacement

Refer to illustrations 43.2, 43.5a, 43.5b, 43.6 and 43.10

1   Open the hood.

2   In most cases, the tools necessary for

43.2 Tools required for changing spark plugs

1   **Spark plug socket** - This will have special padding inside to protect the spark plug porcelain insulator

2   **Torque wrench** - Although not mandatory, use of this tool is the best way to ensure that the plugs are tightened properly

3   **Ratchet** - Standard hand tool to fit the plug socket

4   **Extension** - Depending on model and accessories, you may need special extensions and universal joints to reach one or more of the plugs

5   **Spark plug gap gauge** - This gauge for checking the gap comes in a variety of styles. Make sure the gap for your engine is included

43.5a Spark plug manufacturers recommend using a wire-type gauge when checking the gap - if the wire does not slide between the electrodes with a slight drag, adjustment is required

43.5b To adjust the gap, bend the side electrode only, as indicated by the arrows, and be very careful not to crack or chip the porcelain insulator surrounding the center electrode

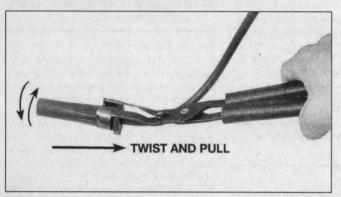

TWIST AND PULL

43.6 When removing the spark plug wires, pull only on the boot and use a twisting/pulling motion

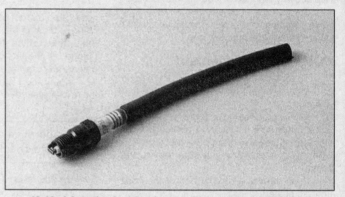

43.10 A length of rubber hose will save time and prevent damaged threads when installing the spark plugs

spark plug replacement include a spark plug socket which fits onto a ratchet (spark plug sockets are padded inside to prevent damage to the porcelain insulators on the new plugs), various extensions and a gap gauge to check and adjust the gaps on the new plugs **(see illustration)**. A special plug wire removal tool is available for separating the wire boots from the spark plugs, but it isn't absolutely necessary. A torque wrench should be used to tighten the new plugs.

3   The best approach when replacing the spark plugs is to purchase the new ones in advance, adjust them to the proper gap and replace them one at a time. When buying the new spark plugs, be sure to obtain the correct plug type for your particular engine. This information can be found on the Vehicle Emission Control Information label located in the engine compartment on the driver's side of the firewall and in the factory owner's manual. If differences exist between the plug specified on the emissions label and in the owner's manual, assume that the emissions label is correct.

4   Allow the engine to cool completely before attempting to remove any of the plugs. While you're waiting for the engine to cool, check the new plugs for defects and adjust the gaps.

5   The gap is checked by inserting the proper thickness gauge between the electrodes at the tip of the plug **(see illustration)**. The gap between the electrodes should be the same as the one specified on the Vehicle Emissions Control Information label. The wire should just slide between the electrodes with a slight amount of drag. If the gap is incorrect, use the adjuster on the gauge body to bend the curved side electrode slightly until the proper gap is obtained **(see illustration)**. If the side electrode is not exactly over the center electrode, bend it with the adjuster until it is. Check for cracks in the porcelain insulator (if any are found, the plug should not be used).

6   With the engine cool, remove the spark plug wire from one spark plug. Pull only on the boot at the end of the wire, do not pull on the wire. A plug wire removal tool should be used if available **(see illustration)**.

7   If compressed air is available, use it to blow any dirt or foreign material away from the spark plug hole. A common bicycle pump will also work. The idea here is to eliminate the possibility of debris falling into the cylinder as the spark plug is removed.

8   Place the spark plug socket over the plug and remove it from the engine by turning it in a counterclockwise direction.

9   Compare the spark plug to those shown in the color photos on the inside back cover to get an indication of the general running condition of the engine.

10   Thread one of the new plugs into the hole until you can no longer turn it with your fingers, then tighten it with a torque wrench (if available) or the ratchet. It is a good idea to slip a short length of rubber hose over the end of the plug to use as a tool to thread it into place **(see illustration)**. The hose will grip the plug well enough to turn it, but will start to slip if the plug begins to cross-thread in the hole, this will prevent damaged threads and the accompanying repair costs.

11   Before pushing the spark plug wire onto the end of the plug, inspect it following the procedures outlined in Section 44.

12   Attach the plug wire to the new spark plug, again using a twisting motion on the boot until it's seated on the spark plug.

13   Repeat the procedure for the remaining spark plugs, replacing them one at a time to prevent mixing up the spark plug wires.

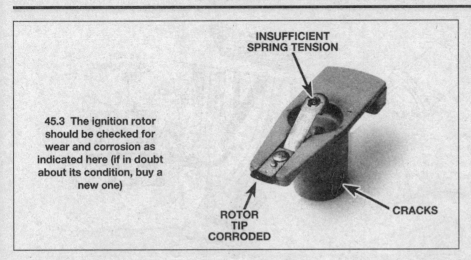

45.3 The ignition rotor should be checked for wear and corrosion as indicated here (if in doubt about its condition, buy a new one)

INSUFFICIENT SPRING TENSION

ROTOR TIP CORRODED

CRACKS

Check

1   To gain access to the distributor cap, especially on a V6 engine, it may be necessary to remove the air cleaner assembly (see Chapter 4).

2   Loosen the distributor cap mounting screws (note that the screws have a shoulder so they don't come completely out). On some models, the cap is held in place with latches that look like screws, to release them, push down with a screwdriver and turn them about 1/2-turn. Pull up on the cap, with the wires attached, to separate it from the distributor, then position it to one side.

3   The rotor is now visible on the end of the distributor shaft. Check it carefully for cracks an carbon tracks. Make sure the center terminal spring tension is adequate and look for corrosion and wear on the rotor tip (see illustration). If in doubt about its condition, replace it with a new one.

4   If replacement is required, detach the rotor from the shaft and install a new one. On some models, the rotor is press fit on the shaft and can be pried or pulled off. On other models, the rotor is attached to the distributor shaft with two screws.

5   The rotor is indexed to the shaft so it can only be installed one way. Press fit rotors have an internal key that must line up with a slot in the end of the shaft (or vice versa). Rotors held in place with screws have one square and one round peg on the underside that must fit into holes with the same shape. Apply a very thin coat of silicone dielectric compound to the rotor blade, if a new rotor is being installed.

6   Check the distributor cap for carbon tracks, cracks and other damage. Closely examine the terminals on the inside of the cap for excessive corrosion and damage (see illustration). Slight deposits are normal. Again, if in doubt about the condition of the cap, replace it with a new one. Be sure to apply a small dab of silicone lubricant to each terminal before installing the cap. Also, make sure the carbon brush (center terminal) is correctly installed in the cap, a wide gap

44 Spark plug wire check and replacement

1   The spark plug wires should be checked at the recommended intervals and whenever new spark plugs are installed in the engine.

2   The wires should be inspected one at a time to prevent mixing up the order, which is essential for proper engine operation.

3   Disconnect the plug wire from one spark plug. To do this, grab the rubber boot, twist slightly and pull the wire free. Do not pull on the wire itself, only on the rubber boot (see illustration 43.6).

4   Check inside the boot for corrosion, which will look like a white crusty powder. Push the wire and boot back onto the end of the spark plug. It should be a tight fit on the plug. If it isn't, remove the wire and use a pair of pliers to carefully crimp the metal connector inside the boot until it fits securely on the end of the spark plug.

5   Using a clean rag, wipe the entire length of the wire to remove any built-up dirt and grease. Once the wire is clean, check for holes, burned areas, cracks and other damage. Don't bend the wire excessively or the conductor inside might break.

6   Disconnect the wire from the distributor cap. A retaining ring at the top of the distributor may have to be removed to free the wires. Again, pull only on the rubber boot. Check for corrosion and a tight fit in the same manner as the spark plug end. Reattach the wire to the distributor cap.

7   Check the remaining spark plug wires one at a time, making sure they are securely fastened at the distributor and the spark plug when the check is complete.

8   If new spark plug wires are required, purchase a new set for your specific engine model. Wire sets are available pre-cut, with the rubber boots already installed. Remove and replace the wires one at a time to avoid mix-ups in the firing order. The wire routing is extremely important, so be sure to note exactly how each wire is situated before removing it.

45 Distributor cap and rotor check and replacement

Refer to illustrations 45.3 and 45.6
Note: *It's common practice to install a new distributor cap and rotor whenever new spark plug wires are installed.*

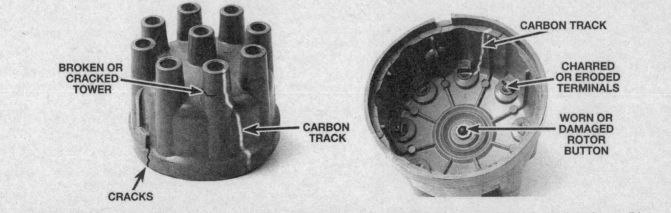

BROKEN OR CRACKED TOWER

CARBON TRACK

CRACKS

CARBON TRACK

CHARRED OR ERODED TERMINALS

WORN OR DAMAGED ROTOR BUTTON

45.6 Shown here are some of the common defects to look for when inspecting the distributor cap (if in doubt about its condition, install a new one)

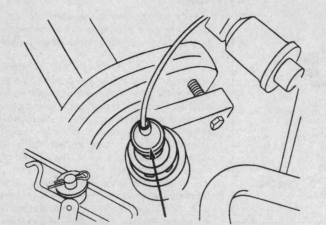

46.3 The oxygen sensor (arrow) is screwed into the exhaust pipe, just below where the pipe and exhaust manifold join

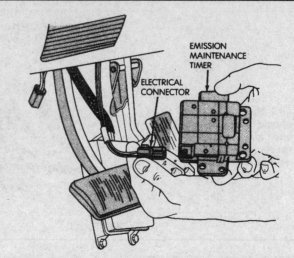

46.8 Emissions maintenance timer installation details

between the brush and rotor will result in rotor burn-through and/or damage to the distributor cap.

Replacement

7   Separate the cap from the distributor and transfer the spark plug wires, one at a time, to the new cap. Be very careful not to mix up the wires!

8   Reattach the cap to the distributor, then tighten the screws or reposition the latches to hold it in place.

46   Oxygen sensor and emission maintenance timer replacement (1988 and later 49-state models)

Refer to illustrations 46.3 and 46.8

Note: *Special care must be taken when handling the oxygen sensor- it's very sensitive:*

a) *The oxygen sensor has a permanently attached pigtail and connector, which should not be removed. Damage or removal of the pigtail or connector can adversely affect sensor operation.*

b) *Grease, dirt and other contaminants should be kept away from the electrical connector and the louvered end of the sensor.*

c) *Do not use cleaning solvents of any kind on the oxygen sensor.*

d) *Do not drop or roughly handle the sensor.*

1   The sensor is located in the exhaust manifold or exhaust pipe and is accessible from under the vehicle.

2   Since the oxygen sensor may be difficult to remove with the engine cold, begin by operating the engine until it has warmed to at least 120-degrees F (48-degrees C).

3   Disconnect the electrical wire from the oxygen sensor and carefully unscrew the oxygen sensor from the exhaust pipe **(see illustration)**. Be advised that excessive force may damage the threads. Inspect the oxygen sensor for damage.

4   A special anti-seize compound must be used on the threads of the oxygen sensor to aid in future removal. New or service sensors will have this compound already applied, but if for any reason an oxygen sensor is removed and then reinstalled, the threads

must be coated before reinstallation.

5   Install the sensor and tighten it securely.

6   Connect the electrical wire.

7   On these models an emission maintenance timer (mounted on the dash panel, to the right of the steering column) activates the emissions maintenance indicator light on the instrument panel when the oxygen sensor is scheduled for replacement (approximately 82,500 miles). The timer cannot be reset and must be replaced or disconnected to turn out the indicator light.

8   To replace the timer, remove the cruise control module (if equipped). Unplug the electrical connector, remove the screws and lower the timer from the instrument panel **(see illustration)**. Installation is the reverse of removal.

9   On 1991 and later models, a special tool (Chrysler DRB-II or equivalent) is required to reset the emissions maintenance indicator light. After replacing the oxygen sensor and performing the required emissions maintenance, take the vehicle to a dealership service department or other properly equipped repair facility and have the light reset.

Chapter 2 Part A
Four-cylinder engine

Contents

Specifications

General

Displacement	150 cu. in. (2.5 liters)
Cylinder numbers (front-to-rear)	1-2-3-4
Firing order	1-3-4-2
Camshaft	
Lobe lift (intake and exhaust)	0.265 in
End play	None
Journal-to-bearing (oil) clearance	0.001 to 0.003 in
Journal out-of-round limit	0.002 in
Fuel pump eccentric diameter	1.615 to 1.625 in
Journal diameter (journals numbered from front-to-rear of engine)	
No. 1	2.029 to 2.030 in
No. 2	2.019 to 2.020 in
No. 3	2.009 to 2.010 in
No. 4	1.999 to 2.000 in

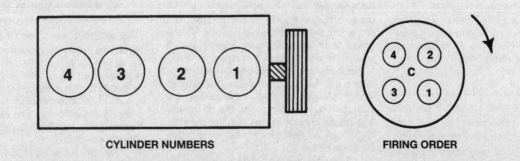

CYLINDER NUMBERS   FIRING ORDER

Torque specifications

	Ft–lbs (unless otherwise indicated)
Camshaft sprocket bolt	80
Crankshaft pulley-to-vibration damper bolts	20
Cylinder head bolts	
1984 through 1988	
Step 1	25
Step 2	50
Step 3	
Bolt 8 only	75
All other bolts	85
1989 on	
Step 1	22
Step 2	45
Step 3	
Bolt 8 only	100
All other bolts	110
Driveplate bolts	
Step 1	40
Step 2	Turn an additional 60°
Exhaust manifold nuts	23
Flywheel bolts	
Step 1	50
Step 2	Turn an additional 60°
Intake manifold-to-cylinder head bolts	23
Oil pan mounting bolts	
1/4 x 20	7
5/16 x 18	11
Oil pump bolts	
Short	10
Long	17
Rocker arm capscrews	19
Rocker arm cover-to-cylinder head bolts	
With RTV	55 in-lbs
With pre-cured permanent gasket	44 in-lbs
Tensioner bracket-to-block bolts	14
Timing chain cover-to-block	
Bolts	5
Nuts	16
Vibration damper bolt (lubricated)	80

1   General information

This Part of Chapter 2 is devoted to in-vehicle repair procedures for the 2.5 liter four-cylinder engine. Information concerning engine removal and installation, as well as engine block and cylinder head overhaul, is in Part D of this Chapter.

The following repair procedures are based on the assumption that the engine is installed in the vehicle. If the engine has been removed from the vehicle and mounted on a stand, many of the steps included in this Part of Chapter 2 will not apply.

The Specifications included in this Part of Chapter 2 apply only to the engine and procedures in this Part. The Specifications necessary for rebuilding the block and cylinder head are found in Part D.

2   Repair operations possible with the engine in the vehicle

Many major repair operations can be accomplished without removing the engine from the vehicle.

Clean the engine compartment and the exterior of the engine with some type of pressure washer before any work is done. A clean engine will make the job easier and will help keep dirt out of the internal areas of the engine.

Depending on the components involved, remove the engine cover and, if necessary, the hood to improve access to the engine as repairs are performed (refer to Chapter 11 if necessary).

If vacuum, exhaust, oil or coolant leaks develop, indicating a need for gasket or seal replacement, the repairs can generally be made with the engine in the vehicle. The intake and exhaust manifold gaskets, oil pan gasket and cylinder head gasket are all accessible with the engine in place.

Exterior engine components such as the intake and exhaust manifolds, the oil pan (and the oil pump), the water pump, the starter motor, the alternator, the distributor and the carburetor or fuel injection components can be removed for repair with the engine in place.

Since the cylinder head can be removed without pulling the engine, valve component servicing can also be accomplished with the engine in the vehicle.

In extreme cases caused by a lack of necessary equipment, repair or replacement of piston rings, pistons, connecting rods and rod bearings is possible with the engine in the vehicle. However, this practice is not recommended because of the cleaning and preparation work that must be done to the components involved.

3   Top Dead Center (TDC) for number one piston - locating

Refer to illustrations 3.6 and 3.8
Note: *The following procedure is based on the assumption that the distributor is correctly installed. If you are trying to locate TDC to install the distributor correctly, piston position must be determined by feeling for compression at the number one spark plug hole, then aligning the ignition timing marks as described in step 8.*

1   Top Dead Center (TDC) is the highest point in the cylinder that each piston reaches

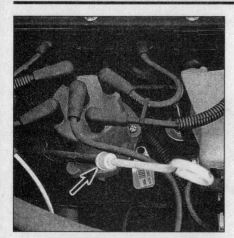

3.6 Make a mark on the aluminum distributor body directly below the number one spark plug wire terminal on the distributor cap (arrow)

3.8 Turn the crankshaft clockwise until the notch aligns with the O

4.3 Typical rocker arm cover - before removing, disconnect the CCV vacuum and fresh air inlet hoses (arrows)

2A

as it travels up-and-down when the crankshaft turns. Each piston reaches TDC on the compression stroke and again on the exhaust stroke, but TDC generally refers to piston position on the compression stroke.

2   Positioning the piston(s) at TDC is an essential part of many procedures such as rocker arm removal, camshaft and timing chain/sprocket removal and distributor removal.

3   Before beginning this procedure, be sure to place the transmission in Neutral and apply the parking brake or block the rear wheels. Also, remove the spark plugs (see Chapter 1) and disable the ignition system using one of the following methods:

a) *On ignition systems with the ignition coil mounted separately from the distributor, detach the coil wire from the center terminal of the distributor cap and ground it on the block with a jumper wire.*

b) *On ignition systems with the ignition coil in the distributor cap (no center terminal in the distributor cap), unplug the BAT electrical connector from the coil (see Chapter 5).*

4   In order to bring any piston to TDC, the crankshaft must be turned using one of the methods outlined below. When looking at the front of the engine, normal crankshaft rotation is clockwise.

a) *The preferred method is to turn the crankshaft with a socket and ratchet attached to the bolt threaded into the front of the crankshaft.*

b) *A remote starter switch, which may save some time, can also be used. Follow the instructions included with the switch. Once the piston is close to TDC, use a socket and ratchet, as described in the previous paragraph.*

c) *If an assistant is available to turn the ignition switch to the Start position in short bursts, you can get the piston close to TDC without a remote starter switch. Make sure your assistant is out*

of the vehicle, away from the ignition switch, then use a socket and ratchet (as described in Paragraph a) to complete the procedure.

5   Note the position of the terminal for the number one spark plug wire on the distributor cap. If the terminal isn't marked, follow the plug wire from the number one cylinder spark plug to the cap.

6   Use a felt-tip pen or chalk to make a mark on the distributor body directly under the terminal **(see illustration)**.

7   Detach the cap from the distributor and set it aside (see Chapter 1 if necessary).

8   Turn the crankshaft (see Paragraph 3 above) until the notch in the crankshaft pulley is aligned with the 0 on the timing plate (located at the front of the engine) **(see illustration)**.

9   Look at the distributor rotor - it should be pointing directly at the mark you made on the distributor body. If it is, go to Step 12.

10   If the rotor is 180° off, the number one piston is at TDC on the exhaust stroke. Go to Step 11.

11   To get the piston to TDC on the compression stroke, turn the crankshaft one complete turn (360°) clockwise. The rotor should now be pointing at the mark on the distributor. When the rotor is pointing at the number one spark plug wire terminal in the distributor cap and the ignition timing marks are aligned, the number one piston is at TDC on the compression stroke.

12   After the number one piston has been positioned at TDC on the compression stroke, TDC for any of the remaining pistons can be located by turning the crankshaft and following the firing order. Mark the remaining spark plug wire terminal locations on the distributor body just like you did for the number one terminal, then number the marks to correspond with the cylinder numbers. As you turn the crankshaft, the rotor will also turn. When it's pointing directly at one of the marks on the distributor, the piston for that particular cylinder is at TDC on the compression stroke.

4   Rocker arm cover - removal and installation

Refer to illustrations 4.3, 4.5, 4.6 and 4.8

1   Disconnect the negative cable from the battery.

2   Remove the air cleaner (see Chapter 4).

3   Label and then remove all hoses and/or wires necessary to provide clearance for rocker arm cover removal **(see illustration)**.

4   On carburetor equipped models, disconnect the fuel pipe at the fuel pump and swivel the pipe to allow removal of the rocker arm cover. Cap the open fuel fittings (see Chapter 4).

5   Remove the rocker arm cover retaining bolts and lift off the cover. On 1984 through 1986 models, the cover may stick. Detach the cover by breaking the seal with a putty knife or razor blade. Locations for prying have been provided **(see illustration)**. **Caution:** *To avoid damaging the cover, do not pry up until the seal has been broken.*

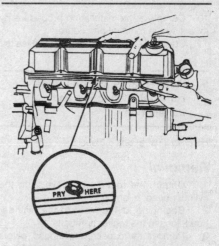

4.5 On 1984 through 1986 models, carefully pry the cover loose after breaking the seal with a putty knife or razor blade

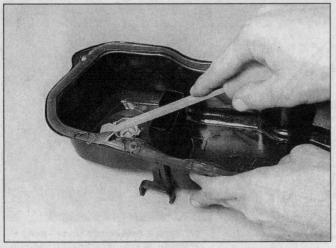

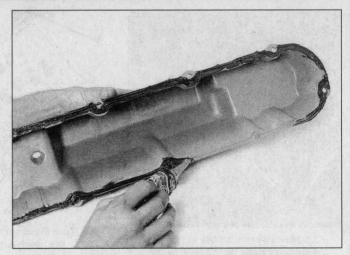

4.6 Remove the old sealant from the rocker arm cover flange and the cylinder head with a gasket scraper, then clean the mating surfaces with lacquer thinner or acetone

4.8 Make sure the sealant is applied to the INSIDE of the bolt holes or oil will leak out around the bolt threads

6   Prior to installation, remove all traces of dirt, oil and old gasket material from the cover and cylinder head with a scraper **(see illustration)**. Clean the mating surfaces with lacquer thinner or acetone and a clean rag.

7   Inspect the mating surface on the cover for damage and warpage. Correct or replace as necessary.

8   On models which use RTV, apply a continuous 1/8-inch (3mm) bead of sealant to the cover flange. Be sure the sealant is applied to the inside of the bolt holes **(see illustration)**. **Note:** *On models with plastic rocker arm covers, RTV or a gasket may be used. On models equipped with an aluminum rocker arm cover, RTV must be used. Later models use a pre-cured reuseable gasket - install these without sealant.*

9   Place the rocker arm cover on the cylinder head while the sealant (if used) is still wet and install the mounting bolts. Tighten the bolts a little at a time until the specified torque is reached.

10   Complete the installation procedure by reversing the removal procedure.

11   Start the engine and check for oil leaks.

5   Rocker arms and pushrods - removal, inspection and installation

Refer to illustrations 5.3 and 5.4

Removal

1   Detach the rocker arm cove from the cylinder head, referring to Section 4.

2   Beginning at the front of the cylinder head, loosen the rocker arm bolts.

3   Remove the capscrews, bridges, pivots and rocker arms **(see illustration)**. Store them in marked containers (they must be reinstalled in their original locations).

4   Remove the pushrods and store them

separately to make sure they don't get mixed up during installation **(see illustration)**.

Inspection

5   Check each rocker arm for wear, cracks and other damage, especially where the pushrods and valve stems contact the rocker arm faces.

6   Make sure the hole at the pushrod end of each rocker arm is open.

7   Check each rocker arm pivot area for wear, cracks and galling.

If the rocker arms are worn or damaged, replace them with new ones and use new pivots as well.

8   Inspect the pushrods for cracks and excessive wear at the ends. Roll each pushrod across a piece of plate glass to see if it's bent (if it wobbles, it's bent).

Installation

9   Lubricate the lower ends of the pushrods with clean engine oil or moly-base grease and install them in their original locations. Make sure each pushrod seats completely in the lifter socket.

10   Apply moly-base grease to the ends of the valve stems and the upper ends of the pushrods before positioning the rocker arms and installing the capscrews.

11   Set the rocker arms in place, then install the pivots, bridges and capscrews. Apply moly-base grease to the pivots to prevent damage to the mating surfaces before engine oil pressure builds up. Tighten the bolts to the specified torque.

12   Reinstall the rocker arm cover and run the engine. Check for oil leaks and unusual valve train noises.

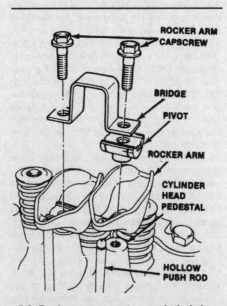

5.3 Rocker arm mounts - exploded view

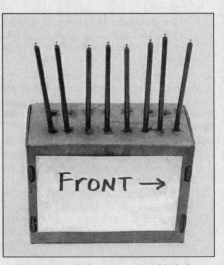

5.4 If more than one pushrod is being removed, store them in a perforated cardboard box to prevent mixups during installation - note the label indicating the front of the engine

6.4 This is what the air hose adapter that threads into the spark plug hole looks like - they're commonly available from auto parts stores

6.9 Once the spring is depressed, the keepers can be removed with a small magnet or needle-nose pliers (a magnet is preferred to prevent dropping the keepers)

6.17 Apply a small dab of grease to the keepers before installation - it will hold them in place on the valve stem as the spring is released

2A

6   Valve spring, retainer and seals - replacement

Refer to illustrations 6.4, 6.9 and 6.17

Note: *Broken valve springs and defective valve stem seals can be replaced without removing the cylinder heads. Two special tools and a compressed air source are normally required to perform this operation, so read through this Section carefully and rent or buy the tools before beginning the job. If compressed air isn't available, a length of nylon rope can be used to keep the valves from falling into the cylinder during this procedure.*

1   Remove the rocker arm cover referring to Section 4.

2   Remove the spark plug from the cylinder which has the defective component. If all of the valve stem seals are being replaced, all of the spark plugs should be removed.

3   Turn the crankshaft until the piston in the affected cylinder is at top dead center on the compression stroke (refer to Section 3 for instructions). If you're replacing all of the valve stem seals, begin with cylinder number one and work on the valves for one cylinder at a time. Move from cylinder-to-cylinder following the firing order sequence (see the Specifications).

4   Thread an adapter into the spark plug hole **(see illustration)** and connect an air hose from a compressed air source to it. Most auto parts stores can supply the air hose adapter. **Note:** *Many cylinder compression gauges utilize a screw-in fitting that may work with your air hose quick-disconnect fitting.*

5   Remove the rocker arm and pivot for the valve with the defective part and pull out the pushrod. If all of the valve stem seals are being replaced, all of the rocker arms and pushrods should be removed (refer to Section 5).

6   Apply compressed air to the cylinder. **Warning:** *The piston may be forced down by compressed air, causing the crankshaft to turn suddenly. If the wrench used when positioning the number one piston at TDC is still attached to the bolt in the crankshaft nose, it*

could cause damage or injury when the crankshaft moves.

7   The valves should be held in place by the air pressure. If the valve faces or seats are in poor condition, leaks may prevent air pressure from retaining the valves - refer to the alternative procedure below.

8   If you don't have access to compressed air, an alternative method can be used. Position the piston at a point just before TDC on the compression stroke, then feed a long piece of nylon rope through the spark plug hole until it fills the combustion chamber. Be sure to leave the end of the rope hanging out of the engine so it can be removed easily. Use a large ratchet and socket to rotate the crankshaft in the normal direction of rotation until slight resistance is felt.

9   Stuff shop rags into the cylinder head holes above and below the valves to prevent parts and tools from falling into the engine, then use a valve spring compressor to compress the spring. Remove the keepers with small needle-nose pliers or a magnet **(see illustration)**. **Note:** *A couple of different types of tools are available for compressing the valve springs with the head in place. One type grips the lower spring coils and presses on the retainer as the knob is turned, while the other type, shown here, utilizes the rocker arm capscrew for leverage. Both types work very well, although the lever type is usually less expensive.*

10   Remove the spring retainer, oil shield and valve spring, then remove the guide seal. **Note:** *If air pressure fails to hold the valve in the closed position during this operation, the valve face and/or seat is probably damaged. If so, the cylinder head will have to be removed for additional repair operations.*

11   Wrap a rubber band or tape around the top of the valve stem so the valve won't fall into the combustion chamber, then release the air pressure. **Note:** *If a rope was used instead of air pressure, turn the crankshaft slightly in the direction opposite normal rotation.*

12   Inspect the valve stem for damage. Rotate the valve in the guide and check the

end for eccentric movement, which would indicate that the valve is bent.

13   Move the valve up-and-down in the guide and make sure it doesn't bind. If the valve stem binds, either the valve is bent or the guide is damaged. In either case, the head will have to be removed for repair.

14   Reapply air pressure to the cylinder to retain the valve in the closed position, then remove the tape or rubber band from the valve stem. If a rope was used instead of air pressure, rotate the crankshaft in the normal direction of rotation until slight resistance is felt.

15   Lubricate the valve stem with engine oil and install a new guide seal.

16   Install the spring and shield in position over the valve.

17   Install the valve spring retainer. Compress the valve spring and carefully position the keepers in the groove. Apply a small dab of grease to the inside of each keeper to hold it in place **(see illustration)**.

18   Remove the pressure from the spring tool and make sure the keepers are seated.

19   Disconnect the air hose and remove the adapter from the spark plug hole. If a rope was used in place of air pressure, pull it out of the cylinder.

20   Refer to Section 5 and install the rocker arm(s) and pushrod(s).

21   Install the spark plug(s) and hook up the wire(s).

22   Refer to Section 4 and install the rocker arm cover.

23   Start and run the engine, then check for oil leaks and unusual sounds coming from the rocker arm cover area.

7   Intake manifold - removal and installation

Refer to illustrations 7.8, 7.9 and 7.12

1   Disconnect the negative cable from the battery.

2   Drain the cooling system (see Chapter 1).

3   Remove the carburetor or throttle body (see Chapter 4).

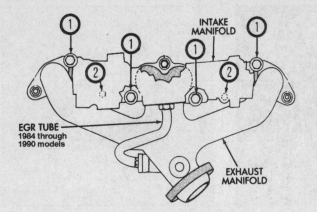

7.8 Remove the bolts (1), pull the manifold away from the engine slightly to disengage it from the dowel pins (2), then lift the manifold from the engine

7.9 Remove the old intake manifold gasket with a scraper- don't leave any material on the mating surfaces

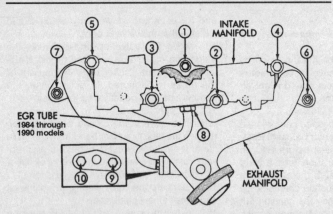

7.12 Intake and exhaust manifold bolt tightening sequence

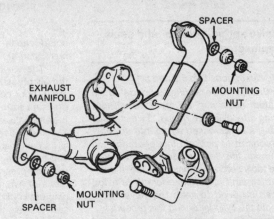

8.5 Typical exhaust manifold mounting details

4   Label and then disconnect any wiring, hoses and control cables still connected to the intake manifold.

5   Unbolt the power steering pump (if equipped), and set it aside without disconnecting the hoses (see Chapter 10).

6   Disconnect the throttle valve (TV) linkage, if equipped with an automatic transmission (see Chapter 7).

7   On 1984 through 1990 models, disconnect the EGR tube from the intake manifold (see Chapter 6).

8   Remove the intake manifold bolts (see illustration). Pull the manifold away from the engine slightly to disengage it from the dowel pins in the cylinder head, then lift the manifold from the engine. If the manifold sticks to the engine after all the bolts are removed, tap it with a soft face hammer or a block of wood and a hammer while supporting the manifold.

9   Thoroughly clean the mating surfaces, removing all traces of gasket material (see illustration).

10   If the manifold is being replaced, transfer all fittings to the new one.

11   Position the replacement gasket on the cylinder head and install the manifold.

12   Install the intake manifold bolts and tighten them (along with the exhaust manifold nuts and EGR tube nut and bolts, if applica-

ble) in several stages, following the sequence shown to the specified torque. (see illustration).

13   Reinstall the remaining parts in the reverse order of removal.

14   Run the engine and check for vacuum leaks and proper operation.

8   Exhaust manifold - removal and installation

Refer to illustration 8.5

Warning: Allow the engine to cool to room temperature before following this procedure.

1   Remove the intake manifold (see Section 7).

2   On 1984 through 1990 models, disconnect the EGR tube (see Chapter 6).

3   Remove the two nuts and bolts that secure the exhaust pipe to the exhaust manifold. It may be necessary to apply penetrating oil to the threads.

4   Unplug the oxygen sensor wire connector (see Chapter 1, if necessary).

5   Remove the mounting nuts and spacers (see illustration) and detach the exhaust manifold from the engine.

6   Clean the mating surfaces and, if the manifold is being replaced, transfer the oxy-

gen sensor to the new manifold.

7   Reinstall the exhaust manifold and finger tighten the nuts on the end studs.

8   Reinstall the intake manifold (see Section 7) and tighten all mounting fastener to the specified torque, following the sequence shown in illustration 7.12. Reinstall the remaining components in the reverse order of removal.

9   Run the engine and check for exhaust leaks.

9   Cylinder head - removal and installation

Refer to illustrations 9.9 and 9.15

Warning: Allow the engine to cool to room temperature before following this procedure.

1   Remove the rocker arms and pushrods (see Section 5).

2   Remove the intake and exhaust manifolds (see Sections 7 and 8).

3   Remove the drivebelt(s) as described in Chapter 1.

4   Unbolt the power steering pump (if equipped) and set it aside without disconnecting the hoses.

5   On air conditioned models, remove the bolt that secures the air conditioning com-

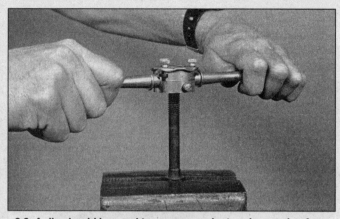

9.9 A die should be used to remove sealant and corrosion from the head bolt threads

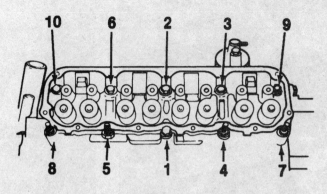

9.15a Cylinder head bolt/nut tightening sequence (1984 through 1989 models)

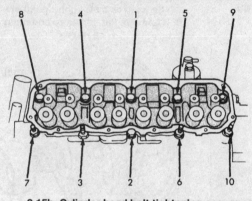

9.15b Cylinder head bolt tightening sequence (1990 and later models)

10.5 Removing lifters with special tool

pressor/alternator bracket to the cylinder head, then unbolt the compressor/alternator bracket from the engine. Set the compressor aside without disconnecting the hoses.

6   Label, then disconnect the wire from the coolant temperature sending unit on the cylinder head.

7   Label the spark plug wires and remove the spark plugs.

8   Remove the ten bolts that secure the cylinder head and lift it off the engine. If the head is stuck to the engine block, it may be necessary to tap it with a soft-face hammer or a block of wood and a hammer to break the seal.

9   Stuff clean shop towels into the cylinders. Thoroughly clean the gasket surfaces, removing all traces of gasket material. Run an appropriate sized tap into the bolt holes in the cylinder head and run a die over the bolt threads **(see illustration)**. Ensure all bolt holes are clean and dry.

10   Inspect the cylinder head for cracks and check it for warpage. Refer to Chapter 2, Part D, for cylinder head servicing procedures.

11   1984 through 1987 four-cylinder engines use different head gaskets than 1988 and later models. The two types of gaskets are NOT interchangeable and require different tightening torques.

1984 through 1987 models

12   These engines use stamped steel gaskets. Apply an even coat of a copper coat type gasket sealant to both sides of the new gasket.

1988 and later models

13   These engines use a composition gasket. Install it dry (without any sealing compound).

All models

14   Install the new gasket with the word Top on the cylinder head side. Place the cylinder head on the engine.

15   Coat the threads of the stud bolt (no. 8 in the tightening sequence) with Teflon pipe sealant. Install the bolts and tighten them in the sequence shown **(see illustration)**. Tighten them in the steps and to the torque listed in the specifications. **Caution:** *During the final tightening step, bolt no. 8 is tightened to a lower torque than the other bolts.*

16   Install the remaining components in the reverse order of removal.

17   Change the oil and filter (see Chapter 1).

18   Refill the cooling system and run the engine, checking for leaks and proper operation.

10   Hydraulic lifters - removal, inspection and installation

Removal

Refer to illustrations 10.5 and 10.6

1   A noisy valve lifter can be isolated when the engine is idling. Place a length of hose or tubing on the rocker arm cover near the position of each valve while listening at the other end. Or remove the rocker arm cover and, with the engine idling, place a finger on each of the valve spring retainers, one at a time. If a valve lifter is defective, it'll be evident from the shock felt at the retainer as the valve opens.

2   The most likely cause of a noisy valve lifter is a piece of dirt trapped between the plunger and the lifter body.

3   Remove the rocker arm cover (see Section 4).

4   Remove both rocker arms and both pushrods at the cylinder with the noisy lifter (see Section 5).

5   Remove the lifters through the pushrod openings in the cylinder head. A special removal tool is available **(see illustration)**, but isn't always necessary. On newer engines without a lot of varnish buildup, lifters can often be removed with a magnet attached to a long handle.

10.6 If you're removing more than one lifter, keep them in order in a clearly labeled box

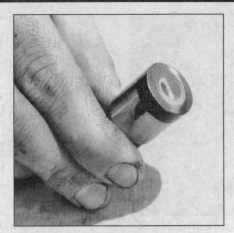

10.8a If the bottom (foot) of any lifter is worn concave (shown here), scratched or galled, replace the entire set with new lifters

10.8b The foot of each lifter should be slightly convex - the side of another lifter can be used as a straightedge to check it; if it appears flat, it's worn and must not be reused

10.8c Check the pushrod seat (arrow) in the top of each lifter for wear

10.8d If the lifters are pitted or rough, they shouldn't be reused

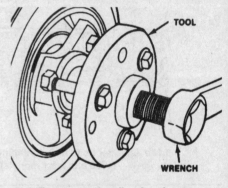

11.6 To remove the vibration damper, use a wheel puller which attaches to the bolt holes

6   Store the lifters in a clearly labeled box to insure their reinstallation in the same lifter bores **(see illustration)**.

Inspection

Refer to illustrations 10.8a, 10.8b, 10.8c and 10.8d

7   Clean the lifters with solvent and dry them thoroughly. Do this one lifter at a time to avoid mixing them up.
8   Check each lifter wall, pushrod seat and foot for scuffing, score marks and uneven wear. Each lifter foot (the surface that rides on the cam lobe) must be slightly convex, although this can be difficult to determine by eye. If the base of the lifter is concave or rough **(see illustrations)**, the lifters and camshaft must be replaced. If the lifter walls are damaged or worn (which isn't very likely), inspect the lifter bore in the engine block as well. If the pushrod seats **(see illustration)** are worn, check the pushrod ends.
9   If new lifters are being installed, a new camshaft must also be installed. If a new camshaft is installed, then use new lifters as well. Never install used lifters unless the original camshaft is used and the lifters can be installed in their original locations!

Installation

10   The used lifters must be installed in their original bores. Coat them with moly-base grease or engine assembly lube.
11   Lubricate the bearing surfaces of the lifter bores with engine oil.
12   Install the lifter(s) in the lifter bore(s).
13   Install the pushrods and rocker arms (see Section 5). **Caution:** *Make sure that each pair of lifters is on the base circle of the camshaft; that is, with both valves closed, before tightening the rocker arm bolts.*
14   Tighten the rocker arm capscrews to the specified torque.
15   Install the rocker arm cover (Section 4).

11   Vibration damper - removal and installation

Refer to illustration 11.6
1   Remove the cable from the negative battery terminal.
2   Remove the drivebelts (Chapter 1). Tag each belt as it's removed to simplify reinstallation. If the vehicle is equipped with a fan shroud, unscrew the mounting bolts and

position the shroud out of the way.
3   Raise the vehicle and place it securely on jackstands.
4   If the vehicle is equipped with a manual transmission, apply the parking brake and put the transmission in gear to prevent the crankshaft from turning, then remove the crankshaft pulley bolts. If your vehicle is equipped with an automatic transmission, it may be necessary to remove the starter motor (Chapter 5) and immobilize the starter ring gear with a large screwdriver while an assistant loosens the pulley bolts. **Note:** *On vehicles with serpentine belts, the pulley and vibration damper are combined as one piece - there are no pulley bolts to remove.*
5   To loosen the vibration damper retaining bolt, install a bolt in one of the pulley bolt holes. Attach a breaker bar, extension and socket to the vibration damper retaining bolt and immobilize the crankshaft by wedging a large screwdriver between the bolt and the socket. Remove the retaining bolt.
6   Remove the vibration damper. Use a puller if necessary **(see illustration)**.
7   Refer to Section 12 for the front oil seal replacement procedure.
8   Apply a thin layer of moly-base grease

12.2 The front crankshaft seal can be removed with a seal removal tool or a large screwdriver (V6 engine shown - four-cylinder similar)

12.6 Once the timing chain cover is removed, place it on a flat surface and gently pry the old seal out with a large screwdriver

12.8a Clean the bore, then apply a small amount of oil to the outer edge of the new seal and drive it squarely into the opening with a large socket . . .

to the seal contact surface of the vibration damper.

9   Slide the vibration damper onto the crankshaft. Note that the slot in the hub must be aligned with the Woodruff key in the end of the crankshaft. Once the key is aligned with the slot, tap the damper onto the crankshaft with a soft-face hammer. The retaining bolt can also be used to press the damper into position.

10   Tighten the vibration damper-to-crankshaft bolt to the specified torque.

11   Install the crankshaft pulley (if equipped) on the hub and tighten the bolts to the specified torque. Use Loctite on the bolt threads.

12   Install the drivebelt(s) (Chapter 1) and replace the fan shroud (if equipped).

12   Front crankshaft oil seal - replacement

Note: *The front crankshaft oil seal can be replaced with the timing chain cover in place. However, due to the limited amount of room available, you may conclude that the procedure would be easier if the cover were removed from the engine first. If so, refer to Section 13 for the cover removal and installation procedure.*

Timing chain cover in place
Refer to illustration 12.2

1   Disconnect the negative battery cable from the battery, then remove the vibration damper (see Section 11).

2   Note how the seal is installed - the new one must face the same direction! Carefully pry the oil seal out of the cover with a seal puller or a large screwdriver **(see illustration)**. Be very careful not to distort the cover or scratch the crankshaft!

3   Apply clean engine oil or multi-purpose grease to the outer edge of the new seal, then install it in the cover with the lip (open end) facing in. Drive the seal into place with a large socket and a hammer (if a large socket isn't available, a piece of pipe will also work).

12.8b . . . or a block of wood and a hammer - don't damage the seal in the process!

Make sure the seal enters the bore squarely and stop when the front face is flush with the cover.

4   Install the vibration damper (see Section 11).

Timing chain cover removed
Refer to illustrations 12.6, 12.8a and 12.8b

5   Remove the timing chain cover as described in Section 13.

6   Using a large screwdriver, pry the old seal out of the cover **(see illustration)**. Be careful not to distort the cover or scratch the wall of the seal bore. If the engine has accumulated a lot of miles, apply penetrating oil to the seal-to-cover joint and allow it to soak in before attempting to remove the seal.

7   Clean the bore to remove any old seal material and corrosion. Support the cover on a block of wood and position the new seal in the bore with the lip (open end) of the seal facing in. A small amount of oil applied to the outer edge of the new seal will make installation easier- don't overdo it.

8   Drive the seal into the bore with a large socket and hammer until it's completely seated **(see illustration)**. Select a socket

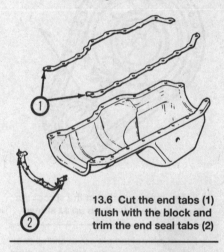

13.6 Cut the end tabs (1) flush with the block and trim the end seal tabs (2)

that's the same outside diameter as the seal. A section of pipe or even a block of wood can be used if a socket isn't available **(see illustration)**.

9   Reinstall the timing chain cover.

13   Timing chain cover - removal and installation

Refer to illustrations 13.6, 13.7 and 13.13

1   Remove the vibration damper (see Section 11).

2   Remove the fan and hub assembly (see Chapter 3).

3   Remove the air conditioning compressor (if equipped) and the alternator bracket assembly from the cylinder head and set it aside.

4   Remove the oil pan-to-timing chain cover bolts and the timing chain cover-to-block bolts.

5   Separate the timing chain cover from the engine. Avoid damaging the sealing surfaces; do not force tools between the cover and block.

6   Cut off the oil pan gasket end tabs flush with the front face of the cylinder block and trim off the end seal tabs **(see illustration)**.

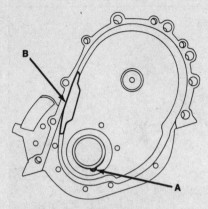

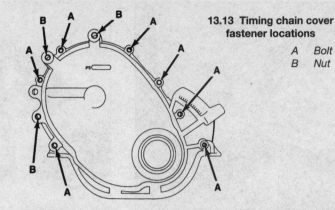

13.7 Inside view of timing chain cover

A   Cutout for driving out oil seal
B   Timing chain guide

13.13 Timing chain cover fastener locations

A   Bolt
B   Nut

14.3 The timing marks on the sprockets (arrows) should be lined up as shown

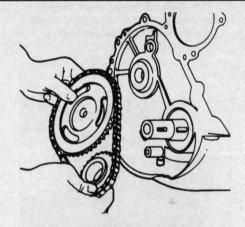

14.5 The timing chain and sprockets are removed and installed as an assembly

14.7a The timing chain tensioner is located below the timing chain

7   Thoroughly clean the cover and all sealing surfaces, removing any traces of gasket material. Drive the old oil seal out from the rear of the timing chain cover and replace it with a new one (see Section 12). Also replace the chain guide, if necessary **(see illustration)**.

8   Apply sealing compound (Perfect Seal or equivalent) to both sides of the new timing cover gasket and position the gasket on the engine block.

9   Trim the end tabs off the new oil pan gaskets to correspond with those cut off the original gasket. Attach the new end tabs to the oil pan with cement.

10   Coat the timing chain cover-to-oil pan seal tab recesses generously with RTV sealant and position the seal onto the timing chain cover. Then apply a film of oil to the seal-to-oil pan contact surface.

11   Position the timing chain cover on the engine block.

12   Install the vibration damper to center the timing chain cover.

13   Install the cover-to-block nuts and bolts and the oil pan-to-cover bolts and tighten them to the specified torque **(see illustration)**.

14   Reinstall the remaining parts in the reverse order of removal.

15   Run the engine and check for oil leaks.

14   Timing chain and sprockets - removal, inspection and installation

Refer to illustrations 14.3, 14.5, 14.7a, 14.7b, 14.10a and 14.10b

1   Set the number one piston at Top Dead Center (see Section 3).

2   Remove the timing chain cover (see Section 13).

3   Reinstall the vibration damper bolt and rotate the crankshaft until the zero timing mark on the crankshaft sprocket is lined up with the timing mark on the camshaft sprocket **(see illustration)**.

4   Slide the oil slinger off the crankshaft.

5   Remove the camshaft retaining bolt and slip both sprockets and the chain off as an assembly **(see illustration)**.

6   Clean the components and inspect for wear and damage. Excessive chain slack and teeth that are deformed, chipped, pitted or discolored call for replacement. Always replace the sprockets and chain as a set. Inspect the tensioner for excessive wear and replace it, if necessary. **Note:** *The oil pan must be removed for timing chain tensioner replacement (see Section 17).*

7   Turn the tensioner lever to the unlock position. Pull the tensioner block toward the tensioner lever to compress the spring. Hold

the block and turn the tensioner lever to the lock position **(see illustrations)**.

8   Install the crankshaft/camshaft sprockets and timing chain. Ensure the marks on the sprockets are still properly aligned **(see illustration 14.3)**.

9   Install the camshaft sprocket retaining bolt and washer and tighten to the specified torque.

10   To verify correct installation of the timing chain, turn the crankshaft to place the camshaft sprocket timing mark at approximately the one o'clock position (1984 through 1991 models) or 3 o'clock (1992 and later models). Count the number of chain pins between the timing marks of both sprockets. There must be 20 pins **(see illustrations)**.

11   Release the tensioner by turning the lever to the unlock position. Be sure the tensioner is released before installing the timing cover.

12   Install the oil slinger and the remaining parts in the reverse order of removal. Refer to the appropriate sections for instructions.

15   Camshaft and bearings - removal, inspection and installation

Camshaft lobe lift check

Refer to illustration 15.3

1   To determine the extent of cam lobe

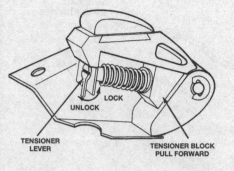

14.7b The timing chain tensioner is locked or unlocked with the tensioner lever

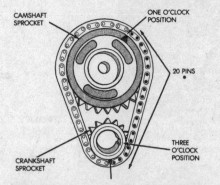

14.10a With the camshaft sprocket timing mark at one o'clock, count the chain pins between the two timing marks - there must be 20 pins (1984 through 1991 models)

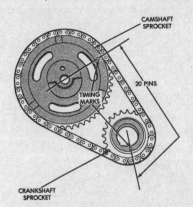

14.10b On 1992 and later models, rotate the cam sprocket timing mark to the 3 o'clock position and verify that there are 20 pins between the marks

2A

15.3 When checking the camshaft lobe lift, the dial indicator plunger must be positioned directly above the pushrod

wear, the lobe lift should be checked prior to camshaft removal. Refer to Section 4 and remove the rocker arm cover.

2   Position the number one piston at TDC on the compression stroke (see Section 3).
3   Beginning with the number one cylinder valves, mount a dial indicator on the engine and position the plunger against the top surface of the first rocker arm. The plunger should be directly above and in line with the pushrod **(see illustration)**.
4   Zero the dial indicator, then very slowly turn the crankshaft in the normal direction of rotation until the indicator needle stops and begins to move in the opposite direction. The point at which it stops indicates maximum cam lobe lift.
5   Record this figure for future reference, then reposition the piston at TDC on the compression stroke.
6   Move the dial indicator to the remaining number one cylinder rocker arm and repeat the check. Be sure to record the results for each valve.
7   Repeat the check for the remaining valves. Since each piston must be at TDC on the compression stroke for this procedure, work from cylinder-to-cylinder following the firing order sequence.

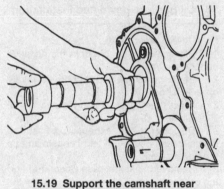

15.19 Support the camshaft near the block

8   After the check is complete, compare the results to the specifications. If camshaft lobe lift is less than specified, cam lobe wear has occurred and a new camshaft should be installed.

Removal

Refer to illustration 15.19

9   Set the number one piston at Top Dead Center (see Section 3).
10   Disconnect the negative cable from the battery.
11   Remove the radiator (see Chapter 3).
12   On models equipped with air conditioning, unbolt the air conditioning compressor and set it aside without disconnecting the refrigerant lines.
13   On carburetor equipped models, remove the fuel pump (see Chapter 4).
14   Remove the distributor (see Chapter 5).
15   If not removed already, detach the rocker arm cover (see Section 4).
16   Remove the rocker arms and pushrods (see Section 5).
17   Remove the hydraulic lifters (see Section 10).
18   Remove the timing chain and sprockets (see Section 14). **Note:** *If the camshaft appears to have been rubbing against the timing chain cover, examine the oil pressure relief holes in the rear cam journal and ensure they are free of debris.*

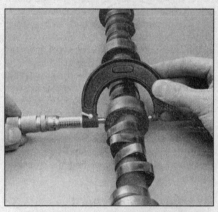

15.21 The camshaft bearing journal diameters are checked to pinpoint excessive wear and out-of-round conditions

19   Install a bolt in the end of the camshaft to use as a handle. Carefully slide the camshaft out of the block. **Caution:** *To avoid damage to the camshaft bearings as the lobes pass over them, support the camshaft near the block as it is withdrawn* **(see illustration)**.

Inspection

Refer to illustration 15.21

20   After the camshaft has been removed from the engine, cleaned with solvent and dried, inspect the bearing journals for uneven wear, pitting and evidence of seizure. If the journals are damaged, the bearing inserts in the block are probably damaged also. Both the camshaft and bearings will have to be replaced.
21   If the bearing journals are in good condition, measure them with a micrometer and record the measurements **(see illustration)**. Measure each journal at several locations around its circumference. If you get different measurements at different locations, the journal is out of round.
22   Check the inside diameter of each camshaft bearing with a telescoping gauge and measure the gauge with a micrometer. Subtract each cam journal diameter from the

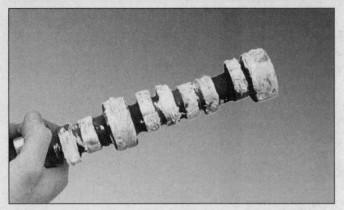

15.27 Be sure to apply moly-base grease or engine assembly lube to the cam lobes and bearing journals before installing the camshaft

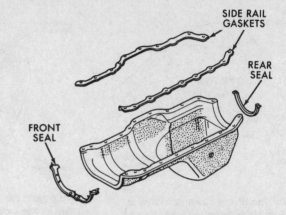

16.8 Oil pan components - exploded view

corresponding camshaft bearing inside diameter to obtain the bearing oil clearance.

23   Compare the clearance for each bearing to the specifications. If it is excessive, for any of the bearings, have new bearings installed by an automotive machine shop.

24   Inspect the distributor drive gear for wear. Replace the camshaft if the gear is worn.

25   Inspect the camshaft lobes (including the fuel pump lobe on carburetor equipped models) for heat discoloration, score marks, chipped areas, pitting and uneven wear. If the lobes are in good condition and if the lobe lift measurements are as specified, the camshaft can be reused.

Bearing replacement

26   Camshaft bearing replacement requires special tools and expertise that place it outside the scope of the home mechanic. Take the engine block (see Part D of this Chapter) to an automotive machine shop to ensure the job is done correctly.

Installation

Refer to illustration 15.27

27   Lubricate the camshaft bearing journals and lobes with moly-base grease or engine assembly lube **(see illustration)**.

28   Slide the camshaft into the engine. Support the cam near the block and be careful not to scrape or nick the bearings.

29   Temporarily place the camshaft sprocket onto the camshaft and turn the camshaft until the timing mark is aligned with the centerline of the crankshaft (see Section 14). Remove the sprocket.

30   Install the timing chain and sprockets and the remaining components in the reverse order of removal. Refer to the appropriate Sections for installation instructions. **Note:** *If the original cam and lifters are being reinstalled, be sure to install the lifters in their original locations. If a new camshaft was installed, be sure to install new lifters also.*

31   Add coolant and change the oil and filter (see Chapter 1).

32   Start the engine and check the ignition timing. Check for leaks and unusual noises.

16   Oil pan - removal and installation

Refer to illustration 16.8

1   Disconnect the cable from the negative battery terminal.

2   Raise the vehicle and support it securely on jackstands.

3   Drain the engine oil and remove the oil filter (Chapter 1).

4   Disconnect the exhaust pipe at the manifold (see Section 8) and hangers and tie the system aside.

5   Remove the starter (see Chapter 5) and the bellhousing dust cover.

6   Remove the bolts and detach the oil pan. Don't pry between the block and pan or damage to the sealing surfaces may result and oil leaks could develop. If the pan is stuck, dislodge it with a soft-face hammer or a block of wood and a hammer.

7   Use a scraper to remove all traces of sealant from the pan and block, then clean the mating surfaces with lacquer thinner or acetone.

8   Using gasket adhesive, position new oil pan seals and gaskets on the engine **(see illustration)**.

9   Install the oil pan and tighten the mounting bolts to the specified torque. Note that the 1/4-inch diameter and 5/16-inch diameter bolts have different torques. Start at the center of the pan and work out toward the ends in a spiral pattern.

10   Install the bellhousing dust cover and the starter, then reconnect the exhaust pipe to the manifold and hanger brackets.

11   Lower the vehicle.

12   Install a new filter and add oil to the engine.

13   Reconnect the negative battery cable.

14   Start the engine and check for leaks.

17   Oil pump - removal and installation

Refer to illustration 17.2

1   Remove the oil pan (see Section 16).

2   Remove the two oil pump attaching

bolts from the engine block **(see illustration)**.

3   Detach the oil pump and strainer assembly from the block.

4   If the pump is defective, replace it with a new one. If the engine is completely being overhauled, install a new oil pump - don't reuse the original or attempt to rebuild it. **Note:** *It is a good idea to prime the new oil pump with clean engine oil before installing it.*

5   On four-cylinder engines, to install the oil pump, turn the shaft so the gear tang mates with the slot on the lower end of the distributor shaft. The oil pump should slide easily into place. If it doesn't, pull it off and turn the tang until it's aligned with the lower end of the shaft.

6   On six-cylinder engines, to install the oil pump, turn the shaft so the gear tang mates with the slot on the lower end of the distributor shaft on 1987 through 1999 models, or the camshaft position sensor drive shaft on 2000 models. The oil pump should slide easily into place. If it doesn't, pull it off and turn the tang until it's aligned with the lower end of the shaft.

7   Install the pump attaching bolts. Tighten them to the specified torque.

8   Reinstall the oil pan (see Section 16).

9   Add oil, run the engine and check for leaks.

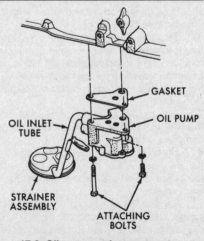

17.2 Oil pump and components

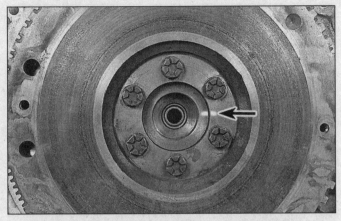

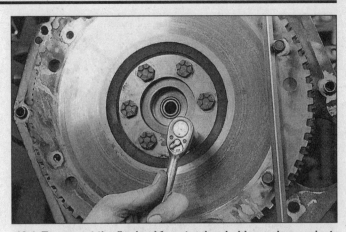

18.3 Before removing the flywheel, index it to the crankshaft (arrow)

18.4 To prevent the flywheel from turning, hold a pry bar against two bolts or wedge a large screwdriver into the flywheel ring gear

2A

18 Flywheel/driveplate - removal and installation

Refer to illustrations 18.3, 18.4, 18.8 and 18.10

1   Raise the vehicle and support it securely on jackstands, then refer to Chapter 7 and remove the transmission. If it's leaking, now would be a very good time to replace the front pump seal/O-ring (automatic transmission only).

2   Remove the pressure plate and clutch disc (see Chapter 8) (manual transmission equipped vehicles). Now is a good time to check/replace the clutch components and pilot bearing.

3   Use paint or a center punch to make alignment marks on the flywheel/driveplate and crankshaft to ensure correct alignment during reinstallation (see illustration).

4   Remove the bolts that secure the flywheel/driveplate to the crankshaft (see illustration). If the crankshaft turns, hold the flywheel with a pry bar or wedge a screwdriver into the ring gear teeth to jam the flywheel.

5   Remove the flywheel/driveplate from the crankshaft. Since the flywheel is fairly heavy, be sure to support it while removing the last bolt.

6   Clean the flywheel to remove grease and oil. Inspect the surface for cracks, rivet grooves, burned areas and score marks. Light scoring can be removed with emery cloth. Check for cracked and broken ring gear teeth or a loose ring gear. Lay the flywheel on a flat surface and use a straightedge to check for warpage.

7   Clean and inspect the mating surfaces of the flywheel/driveplate and the crankshaft. If the crankshaft rear seal is leaking, replace it before reinstalling the flywheel/driveplate.

8   On models with a four-cylinder engine and an automatic transmission, the trigger wheel portion of the driveplate assembly provides the timing signal for the fuel and ignition systems. If the trigger wheel becomes damaged when removing/installing the engine, transmission or torque convertor,

ignition performance will be affected. The general result is either rough engine operation and backfire, or a no-start condition. Check for suspected trigger wheel damage as follows:

a) *Check the trigger wheel radial runout on the inner surface (A) with a dial indicator (see illustration). Maximum allowable runout is 0.016-inch (0.40 mm). Replace the driveplate assembly if runout exceeds this figure.*

b) *Inspect the timing slot separators (B) in the trigger wheel. Replace the driveplate assembly if the separators are dented, distorted, or cracked. Do not attempt to repair the wheel as the results are usually not satisfactory.*

9   Position the flywheel/driveplate against the crankshaft. Be sure to align the marks made during removal. Note that some engines have an alignment dowel or staggered bolt holes to ensure correct installation. Before installing the bolts, apply thread locking compound to the threads.

10   Wedge a screwdriver into the ring gear teeth to keep the flywheel/driveplate from turning as you tighten the bolts to the specified torque. After the initial torque has been reached, turn each bolt an additional 60° as outlined in the specifications. **Note:** *A 3/4-inch socket should be marked every 60° on the outside* **(see illustration).** *After reaching the specified torque, make a reference mark below one of the marks on the socket. Turn the bolt until the next 60 degrees mark is reached.*

11   The remainder of installation is the reverse of the removal procedure.

19 Rear main oil seal - replacement

Refer to illustrations 19.5, 19.8a and 19.8b

1   The rear main bearing oil seal can be replaced without removing the oil pan or crankshaft.

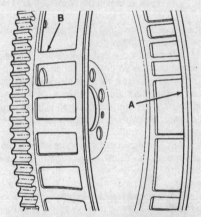

18.8 The driveplate on a four-cylinder engine with an automatic transmission has a trigger wheel - check the inner surface (A) for excessive runout with a dial indicator - inspect the timing slot separators (B) for dents, distortion and cracks

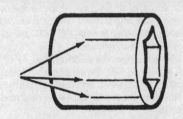

18.10 Make marks every 60 degrees on the socket to indicate how much you have turned the bolt

2   Remove the transmission (see Chapter 7).

3   If equipped with a manual transmission, remove the pressure plate and clutch disc (see Chapter 8).

4   Remove the flywheel or driveplate (see Section 18).

5   Using a seal removal tool or a large screwdriver, carefully pry the seal out of the

19.5 Carefully pry the oil seal out with a removal tool or a screwdriver - don't nick or scratch the crankshaft or the new seal will be damaged and leaks will develop

19.8a If the special tool isn't available, tap around the outer edge of the new seal with a hammer and punch to seat it squarely in the bore

block **(see illustration)**. Don't scratch or nick the crankshaft in the process.

6   Clean the bore in the block and the seal contact surface on the crankshaft. Check the crankshaft surface for scratches and nicks that could damage the new seal lip and cause oil leaks. If the crankshaft is damaged, the only alternative is a new or different crankshaft.

7   Apply a light coat of engine oil or multi-purpose grease to the outer edge of the new seal. Lubricate the seal lip with moly-base grease.

8   Press the new seal into place with a special tool, available at most auto parts stores. The seal lip must face toward the front of the engine. If the special tool isn't available, carefully work the seal lip over the end of the crankshaft and tap the seal in with a hammer and punch until it's seated in the bore **(see illustration)**. **Note:** *A new rear main oil seal with improved sealing characteristics is installed on engines built since December 8, 1986. A special installation tool, available at most auto parts stores, comes with a removable shim* **(see illustration)**. *This shim is only used with the old style seal. DO NOT use this shim with the new style seal.*

9   Install the flywheel or driveplate.

10   If equipped with a manual transmission, reinstall the clutch disc and pressure plate.

11   Reinstall the transmission as described in Chapter 7.

20   Engine mounts - check and replacement

Refer to illustration 20.8

1   Engine mounts seldom require attention, but broken or deteriorated mounts should be replaced immediately or the added strain placed on the driveline components may cause damage or wear.

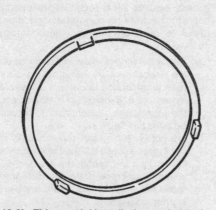

19.8b This special installation tool shim is ONLY used when installing the old-style seal - we recommend using the new style seal

Check

2   During the check, the engine must be raised slightly to remove the weight from the mounts.

3   Raise the vehicle and support it securely on jackstands, then position a jack under the engine oil pan. Place a large block of wood between the jack head and the oil pan, then carefully raise the engine just enough to take the weight off the mounts. **Warning:** *DO NOT place any part of your body under the engine when it's supported only by a jack!*

4   Check the mounts to see if the rubber is cracked, hardened or separated from the metal plates. Sometimes the rubber will split right down the center.

5   Check for relative movement between the mount plates and the engine or frame (use a large screwdriver or pry bar to attempt to move the mounts). If movement is noted, lower the engine and tighten the mount fasteners.

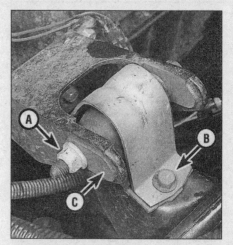

20.8 Remove the through-bolt and nut (A) and mount bolt (B), then remove the mount nut (not visible in this photo) from beneath the frame bracket (C)

6   Rubber preservative should be applied to the mounts to slow deterioration.

Replacement

7   Disconnect the negative battery cable from the battery, then raise the vehicle and support it securely on jackstands (if not already done).

8   Loosen the nut on the through bolt and remove the bolt and nut that secure the mount to the frame bracket **(see illustration)**.

9   Raise the engine slightly with a jack or hoist (make sure the fan doesn't hit the radiator or shroud). Remove the through bolt and nut and detach the mount.

10   Installation is the reverse of removal. Use thread locking compound on the mount bolts and be sure to tighten them securely.

Chapter 2 Part B
V6 engine

Contents

2B

Specifications

General

Displacement	171 cu. in. (2.8 liters)
Cylinder numbers	
Left bank (driver's side)	2-4-6
Right bank	1-3-5
Firing order	1-2-3-4-5-6

Camshaft

Lobe lift	
Intake	0.2311 in
Exhaust	0.2625 in
Journal diameter	1.867 to 1.869 in
Journal-to-bearing (oil) clearance	0.0010 to 0.0039 in

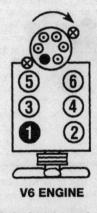

V6 ENGINE

Cylinder location and distributor rotation

The blackened terminal shown on the distributor cap indicates the Number One spark plug wire position

Torque specifications

	Ft-lbs (unless otherwise indicated)
Camshaft sprocket bolt	15 to 20
Camshaft cover (rear) bolts	6 to 9
Crankshaft pulley-to-vibration damper bolts	20 to 30
Cylinder head bolts	70
Exhaust manifold bolts	25
Flywheel/driveplate bolts	45 to 55
Intake manifold-to-cylinder head bolts	23
Oil pan mounting bolts	
6 x 1.0 mm	6 to 9
8 x 1.25 mm	14 to 22
Oil pump-to-rear main bearing cap bolt	26 to 35
Rear main bearing cap bolts	70
Rocker arm stud	43 to 49
Rocker arm cover-to-cylinder head bolts	
With gasket	43 to 66 in-lbs
RTV only	8
Tensioner bracket-to-engine block bolts	14
Timing chain cover bolts	
8 x 1.25 mm	13 to 18
10 x 1.5 mm	20 to 30
Timing chain tensioner bolts	13 to 18
Vibration damper center bolt	66 to 84

3.1 V6 engine timing marks are located at the lower front of the engine

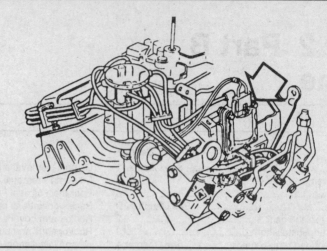

4.5 Remove the ignition coil (arrow) for access to the right rocker arm cover

1   General information

This Part of Chapter 2 is devoted to in-vehicle repair procedures for the V6 engine. All information concerning engine removal and installation and engine block and cylinder head overhaul can be found in Part of this Chapter.

The following repair procedures are based on the assumption that the engine is installed in the vehicle. If the engine has been removed from the vehicle and mounted on a stand, many of the steps outlined in this Part of Chapter 2 will not apply.

The specifications included in this Part of Chapter 2 apply only to the procedures contained in this Part. Part D of Chapter 2 contains the specifications necessary for cylinder head and engine block rebuilding.

2   Repair operations possible with the engine in the vehicle

Many major repair operations can be accomplished without removing the engine from the vehicle.

Clean the engine compartment and the exterior of the engine with some type of degreaser before any work is done. It will make the job easier and help keep dirt out of the internal areas of the engine.

Depending on the components involved, it may be helpful to remove the hood to improve access to the engine as repairs are performed (refer to Chapter 11 if necessary). Cover the fenders to prevent damage to the paint. Special pads are available, but an old bedspread or blanket will also work.

If vacuum, exhaust, oil or coolant leaks develop, indicating a need for gasket or seal replacement, the repairs can generally be made with the engine in the vehicle. The intake and exhaust manifold gaskets, timing

cover gasket, oil pan gasket, crankshaft oil seals and cylinder head gasket are all accessible with the engine in place.

Exterior engine components, such as the intake and exhaust manifolds, the oil pan (and the oil pump), the water pump, the starter motor, the alternator, the distributor and the fuel system components can be removed for repair with the engine in place.

Since the cylinder heads can be removed without pulling the engine, valve component servicing can also be accomplished with the engine in the vehicle. Replacement of the timing chain, sprockets and camshaft is also possible with the engine in the vehicle.

In extreme cases caused by a lack of necessary equipment, repair or replacement of piston rings, pistons, connecting rods and rod bearings is possible with the engine in the vehicle. However, this practice is not recommended because of the cleaning and preparation work that must be done to the components involved.

3   Top Dead Center (TDC) for number one piston - locating

Refer to illustration 3.1

See Chapter 2, Part A, Section 3 for this procedure. The timing mark tab is attached to the timing chain cover **(see illustration)**. Be sure to use the firing order in the specifications in this Part of Chapter 2 for the V6 engine.

4   Rocker arm covers - removal and installation

Refer to illustrations 4.5, 4.11 and 4.12

1   Disconnect the negative cable from the battery.
2   Remove the air cleaner assembly (see Chapter 4).
3   Label and then disconnect the wires and hoses which would interfere with removal of the rocker arm cover(s).

4   Label and then detach the spark plug wires and unclip the wire retainers from the studs.

Right side
5   Remove the air injection diverter valve and the ignition coil **(see illustration)**.
6   Disconnect the rubber hose from the air injection manifold (the air injection manifold is the group of metal tubes which are screwed into the exhaust manifold).
7   Disconnect the carburetor controls and remove them from the bracket.

Left side
8   Disconnect the pipe bracket.

Both sides
9   Remove the six rocker arm cover attaching bolts and detach the cover. Note that some of the bolts have posts attached to the ends- you'll have to use a deep socket to remove these. If the cover is stuck, use a soft-face hammer or a block of wood and a hammer to dislodge it. If the cover still will not come loose, pry on it carefully with a putty knife, but do not distort the sealing flange surface.
10   Clean all dirt, oil and old gasket material from the sealing surfaces with a scraper, solvent and clean rags.
11   On models that use RTV sealant in place of a gasket, apply a 3mm (1/8-inch) bead of RTV to the flange of the cover. Be sure to apply sealant on the inside of the bolt holes or oil will leak past the bolt threads **(see illustration)**. **Caution:** *When applying RTV sealant, keep sealant out of the bolt holes as this could result in damage to the head casting.*
12   On models that use gaskets, place a small amount of RTV sealant on the seam area where the cylinder head and intake manifold meet **(see illustration)** before installing the gasket.
13   Install the cover(s) while the RTV is still wet. Tighten the bolts a little at a time to the specified torque.
14   Reinstall the remaining parts in the

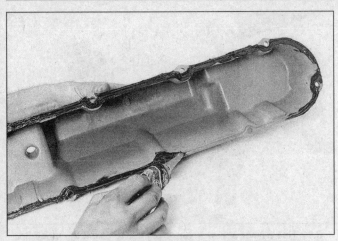

4.11 Make sure the sealant is applied to the inside of the bolt holes or oil will leak out around the bolt threads

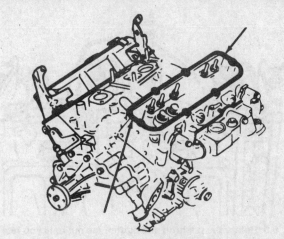

4.12 On models that use gaskets, apply RTV sealant to the joints between the head and intake manifold (arrows) prior to assembly

2B

5.4 When removing the pushrods, be sure to store them separately to ensure reinstallation in their original positions

5.10 The ends of the pushrods and the valve stems should be lubricated with moly-base grease prior to installation of the rocker arms

5.11 Moly-base grease applied to the pivot balls will ensure adequate lubrication until oil pressure builds up when the engine is started

reverse order of removal.

15   Run the engine and check for oil leaks.

---

5   Rocker arms and pushrods - removal, inspection and installation

Removal

Refer to illustration 5.4

1   Refer to Section 3 and detach the rocker arm cover(s) from the cylinder head(s).

2   Beginning at the front of one cylinder head, loosen and remove the rocker arm stud nuts. Store them separately in marked containers to ensure that they will be reinstalled in their original locations. **Note:** *If the pushrods are the only items being removed, loosen each nut just enough to allow the rocker arms to be rotated to the side so the pushrods can be lifted out.*

3   Lift off the rocker arms and pivot balls and store them in the marked containers with the nuts (they must be reinstalled in their original locations).

4   Remove the pushrods and store them separately to make sure they don't get mixed up during installation **(see illustration)**.

Inspection

5   Check each rocker arm for wear, cracks and other damage, especially where the pushrods and valve stems contact the rocker arm faces.

6   Make sure the hole at the pushrod end of each rocker arm is open.

7   Check each rocker arm pivot area for wear, cracks and galling. If the rocker arms are worn or damaged, replace them with new ones and use new pivot balls as well.

8   Inspect the pushrods for cracks and excessive wear at the ends. Roll each pushrod across a piece of plate glass to see if it's bent (if it wobbles, it's bent).

Installation

Refer to illustrations 5.10 and 5.11

9   Lubricate the lower end of each pushrod with clean engine oil or moly-base grease and install them in their original locations.

Make sure each pushrod seats completely in the lifter.

10   Apply moly-base grease to the ends of the valve stems and the upper ends of the pushrods before positioning the rocker arms over the studs **(see illustration)**.

11   Set the rocker arms in place, then install the pivot balls and nuts. Apply moly-base grease to the pivot balls to prevent damage to the mating surfaces before engine oil pressure builds up **(see illustration)**. Be sure to install each nut with the flat side against the pivot ball.

12   Adjust the valve lash (see Section 6).

13   Reinstall the rocker arm covers as described in Section 4.

---

6   Valve lash - adjustment

Refer to illustration 6.5

1   Disconnect the cable from the negative battery terminal.

2   If the rocker arm covers are still on the engine, refer to Section 4 and remove them.

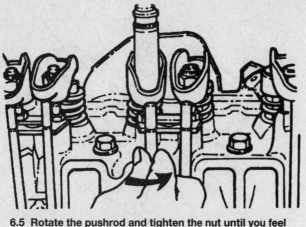

6.5 Rotate the pushrod and tighten the nut until you feel resistance to rotation

8.11 Use a large screwdriver or pry bar to break the manifold gasket seal

3   If the valve train components have been serviced just prior to this procedure, make sure that the components are completely reassembled.

4   Rotate the crankshaft until the number one piston is at top dead center (TDC) on the compression stroke (see Section 3). To make sure you do not mix up the TDC positions of the number one and four pistons, place your fingers on the number one rocker arms as the timing marks line up at the crankshaft pulley. If the rocker arms are not moving, the number one piston is at TDC. If they move as the timing marks line up, the number four piston is at TDC.

5   Back off the rocker arm nut until play is felt at the pushrod, then turn it back in until all play is removed. This can be determined by rotating the pushrod while tightening the nut. Just when drag is felt at the pushrod, all lash has been removed **(see illustration)**. Now turn the nut an additional 1-1/2 turns.

6   Adjust the number one, five and six cylinder intake valves and the number one, two and three cylinder exhaust valves, with

the crankshaft in this position, using the method just described.

7   Rotate the crankshaft until the number four piston is at TDC on the compression stroke and adjust the number two, three and four cylinder intake valves and the number four, five and six cylinder exhaust valves.

8   Refer to Section 4 and install the rocker arm covers.

7   Valve spring, retainer and seals - replacement

Refer to Chapter 2, Part A, Section 6 for this procedure. Remove the rocker arm covers, rocker arms and pushrods; adjust the valve lash following the procedures in this Part.

8   Intake manifold - removal and installation

Refer to illustrations 8.11, 8.13, 8.17, 8.20 and 8.25

1   Disconnect the negative cable from the battery.

2   Drain the coolant from the radiator (See Chapter 1).

3   Remove the carburetor (See Chapter 4).

4   Label and disconnect all wires and hoses connecting the intake manifold to the vehicle.

5   Unbolt the air conditioning compressor and set it aside without disconnecting the hoses (See Chapter 3).

6   Label and then disconnect the spark

plug wires from the plugs, then remove the distributor cap (See Chapter 1).

7   Remove the distributor (See Chapter 5).

8   Remove the EGR and diverter valves (See Chapter 6).

9   Remove the rocker arm covers (See Section 4).

10   Disconnect the upper radiator and heater hoses.

11   Unbolt the intake manifold and lift it off the engine. If it is stuck, carefully pry against a protrusion of the manifold casting **(see illustration)**. Do not pry between the gasket surfaces.

12   Thoroughly clean all sealing surfaces, removing all traces of oil and old gasket material. Clean the intake manifold bolt holes in the cylinder head by chasing them with a tap. Compressed air can be used to remove the debris after chasing. **Warning:** *Wear eye protection when using compressed air.*

13   Inspect the underside of the rear of the intake manifold to see if a machined groove **(see illustration)** is present. This groove was added to later production models to improve oil sealing.

14   You may add this groove to early models yourself or take the manifold to a dealer or automotive machine shop to have it done.

15   Place tape over the intake manifold ports to keep out metal chips.

16   Scribe marks on the intake manifold rear flange **(see illustration 8.13)** to outline the area to be ground out.

17   Obtain a die grinder or electric drill and a spherical cutter with a 3/16-inch diameter ball **(see illustration)**. **Warning:** *Wear eye protection!*

18   Cut a groove 3/16-inch wide by 1/16 to

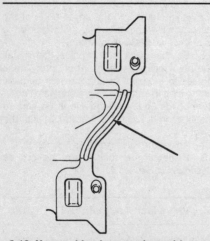

8.13 If a machined groove (arrow) is not present on the underside of the rear of the manifold, scribe marks to outline the area to be ground out

8.17 Use a 3/16-inch diameter cutter to form a groove in the manifold

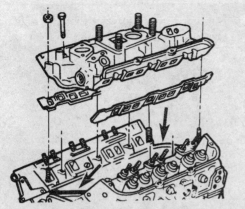

8.20 Apply RTV to the block ridges between the heads (arrows)

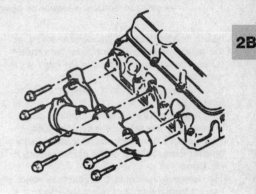

FRONT

8.25 Intake manifold bolt tightening sequence

1/8-inch deep. This provides a groove to hold RTV sealant when the manifold is installed on the engine.

19 Thoroughly clean the intake manifold, removing all chips and tape.

20 Apply a 3/16-inch bead of RTV to each of the ridges between the heads **(see illustration)**.

21 Install new intake manifold gaskets, noting that they are marked Right and Left. Be sure to install them as indicated.

22 Hold the gaskets in place by extending the RTV sealant bead up 1/4-inch onto the gasket ends.

23 The intake manifold gaskets will have to be cut to position the tops behind the pushrods. Cut only those areas necessary to clear the pushrods.

24 Install the intake manifold on the engine and hand tighten the bolts. Ensure the areas between the block ridges and intake manifold are completely sealed.

25 Working in the sequence shown **(see illustration)**, tighten the manifold bolts, in several steps, to the specified torque.

26 Reinstall the remaining parts in the reverse order of removal.

27 Run the engine, adjust the ignition timing and check for oil and vacuum leaks.

9 Exhaust manifolds - removal and installation

Refer to illustration 9.7
Warning: *Allow the engine to cool to room temperature before performing this procedure.*

Right side

1 Remove the cable from the negative battery terminal.

2 Raise the front of the vehicle and support it securely on jackstands. Block the rear wheels to keep the vehicle from rolling.

3 Remove the bolts attaching the exhaust pipe to the exhaust manifold, then separate the pipe from the manifold.

4 Remove the jackstands and lower the vehicle.

5 Disconnect the air injection hose from the air injection manifold (the air injection manifold is a group of metal tubes screwed into the exhaust manifold).

6 Disconnect the spark plug wires from the spark plugs, labeling them as they are disconnected to simplify installation.

7 Remove the exhaust manifold mounting bolts **(see illustration)** and separate the manifold from the engine.

8 Installation is the reverse of the removal procedure. Before installing the manifold, be sure to thoroughly clean the mating surfaces on the manifold and cylinder head. Tighten the manifold mounting bolts to the specified torque.

Left side

9 Disconnect the cable from the negative battery terminal.

10 Raise the front of the vehicle and place it securely on jackstands. Block the rear wheels to keep the vehicle from rolling.

11 Remove the bolts attaching the exhaust pipe to the manifold, then disconnect the pipe from the manifold.

12 Remove the four manifold mounting bolts accessible at the rear of the manifold.

13 Remove the jackstands and lower the vehicle.

14 Remove the air cleaner assembly (see Chapter 4), labeling all hoses.

15 Disconnect the hoses leading to the air injection valve.

16 Disconnect and label any wire that will interfere with the removal of the manifold.

17 Remove the power steering pump bracket from the cylinder head. Loosen the pump adjusting bracket bolt and remove the pump drivebelt from the pulley first. After removing the bracket from the cylinder head, place the steering pump assembly aside, out of the way. Do not disconnect any hoses and be sure to keep the top of the pump up so that no fluid spills.

18 Remove the remaining manifold bolts and separate the manifold and heat shield from the engine.

19 Installation is the reverse of the removal procedure. Be sure to thoroughly clean the cylinder head and manifold surfaces before

9.7 Right-side exhaust manifold - exploded view (left side similar)

installing the manifold. Tighten the bolts to the specified torque.

10 Cylinder heads - removal and installation

Refer to illustrations 10.13 and 10.14
Warning: *Allow the engine to cool to room temperature before performing this procedure.*

Left side head

1 Remove the intake manifold (refer to Section 8).

2 Raise the vehicle and place it securely on jackstands.

3 Locate the engine block drain plugs to the rear of the engine mounts (the plug on the left side is just above the oil filter). Remove the plugs and drain the block.

4 Disconnect the exhaust pipe from the exhaust manifold.

5 Unbolt and remove the oil dipstick tube assembly from the side of the engine.

6 Remove the jackstands and lower the vehicle.

7 **Note:** *Steps 8 through 11 are to be followed if the head is to be replaced with a new one. These Steps may be performed either*

10.13 Tightening sequence for V6 engine cylinder head bolts - loosen the bolts in a sequence opposite to this

10.14 Be careful not to damage the cylinder head sealing surface when breaking the head loose with a pry bar or large screwdriver

before or after the head has been removed. In the accompanying illustrations, the procedures were performed before the head was removed.

8   Remove the exhaust manifold (refer to Section 9).

9   Remove the power steering pump bracket from the side of the cylinder head.

10   Remove the air-conditioner compressor bracket from the front of the cylinder head, if equipped.

11   Remove the coolant temperature sending unit from the front of the cylinder head.

12   Loosen the rocker arm nuts enough to allow removal of the pushrods, then remove the pushrods (see Section 5).

13   Loosen the head bolts in a sequence opposite to the one used for tightening them **(see illustration)**.

14   Remove the cylinder head. To break the gasket seal, use a long screwdriver or pry bar under the cast "ears" of the cylinder head. Be sure not to damage the cylinder head sealing surface **(see illustration)**.

15   If a new cylinder head is being installed, attach the components previously removed from the old head. Before installing the cylinder head, the gasket surfaces of both the head and the engine block must be clean and free of nicks and scratches. Also, the threads in the block and on the head bolts must be completely clean, as any dirt remaining in the threads will affect bolt torque. You can clean the threads in the block with a tap and the threads on the bolts with a die.

16   Place the gasket in position over the locating dowels, with the mark This Side Up visible.

17   Position the cylinder head over the gasket.

18   Coat the cylinder head bolts with RTV sealant and install the bolts.

19   Tighten the bolts in the proper sequence **(see illustration 10.13)** to the specified torque. Work up to the final torque in three steps.

20   Install the pushrods, making sure the lower ends are in the lifter seats. Place the

rocker arm ends over the pushrods and loosely install the rocker arm nuts.

21   The remaining installation steps are the reverse of those for removal. Before installing the rocker arm covers, adjust the valve lash (refer to Section 6).

Right side head

22   Remove the intake manifold (refer to Section 8).

23   Raise the vehicle and place it securely on jackstands.

24   Locate the engine block drain plugs to the rear of the motor mounts (the plug on the left side is just above the oil filter), remove the plugs and drain the block.

25   Remove the two nuts and disconnect the exhaust pipe from the exhaust manifold.

26   Remove the jackstands and lower the vehicle.

27   Remove the alternator from the alternator bracket, then remove the bracket from the head.

28   Remove the lifting "eye" from the rear of the head (necessary only if the head is to be replaced with a new one).

29   Remove the exhaust manifold (if the head is to be replaced with a new one) as described in Section 9.

30   Loosen the rocker arm nuts sufficiently to allow removal of the pushrods, then remove the pushrods (see Section 5).

31   Loosen the head bolts in a sequence opposite to the one used for tightening them **(see illustration 10.13)**.

32   Remove the cylinder head. To break the gasket seal, insert a bar into one of the exhaust ports, then carefully lift on the tool.

33   To install the head, refer to Steps 15 through 21.

11   Vibration damper - removal and installation

Refer to illustrations 11.5 and 11.8

1   Remove the bolts and separate the radiator shroud from the radiator.

2   Remove the cooling fan mounting bolts and lift off the cooling fan and the radiator shroud.

3   Loosen the accessory drivebelt adjusting bolts as necessary, then remove the drivebelts, tagging each one as it is removed to simplify reinstallation.

4   Remove the bolts from the crankshaft pulley (a screwdriver can be used to lock the starter ring gear on the flywheel so the crankshaft won't rotate), then remove the pulley. Remove the damper center bolt.

5   Attach a puller to the damper. Draw the damper off the crankshaft, being careful not

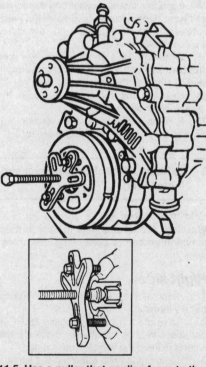

11.5 Use a puller that applies force to the center of the hub - be sure the center puller bolt doesn't damage the crankshaft threads

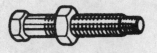

11.8 A special tool is recommended to install the vibration damper

12.3 A special tool is available for crankshaft front seal installation

12.7 Driving the seal out of the front cover

to drop it as it breaks free. A common gear puller should not be used to draw the damper off, as it may separate the outer portion of the damper from the hub. Use only a puller which bolts to the hub **(see illustration)**.

6   Before installing the damper, coat the front cover sealing surface on the damper with moly-base grease.

7   Place the damper in position over the key on the crankshaft. Make sure the damper keyway lines up with the key.

8   Using a damper installation tool, available at most auto parts stores, push the damper onto the crankshaft. The special tool **(see illustration)** distributes the pressure evenly around the hub.

9   Remove the installation tool and install the damper retaining bolt. Tighten the bolt to the specified torque.

10   To install the remaining components, reverse the removal procedure.

11   Adjust the drivebelts (refer to Chapter 1).

12   Crankshaft front oil seal - replacement

Refer to illustrations 12.3, 12.7 and 12.9

With front cover installed on engine

1   With the vibration damper removed (see Section 11), pry the old seal out of the crankcase front cover with a large screwdriver. Be very careful not to damage the surface of the crankshaft.

2   Place the new seal in position with the open end of the seal (seal lip) toward the inside of the cover.

3   Drive the seal into the cover until it is seated. A special tool **(see illustration)** is available from most auto parts stores for this purpose. This tool is designed to exert even pressure around the entire circumference of the seal as it is hammered into place. A section of large-diameter pipe or a large socket could also be used. Be careful not to distort the front cover.

4   Reinstall the remaining parts in the reverse order of removal.

With front cover removed from engine

5   This method is preferred, as the cover can be supported while the old seal is

removed and the new one is installed.

6   Remove the timing chain cover (refer to Section 13).

7   Using a large screwdriver, pry the old seal out of the bore in the front of the cover. Alternatively, support the cover and drive the seal out from the rear **(see illustration)**. Be careful not to damage the cover.

8   With the front of the cover facing up, place the new seal in position with the open end of the seal toward the inside of the cover.

9   Using a block of wood and hammer, drive the new seal into the cover until it is completely seated **(see illustration)**.

10   Install the timing chain cover by reversing the removal procedure in Section 13.

13   Timing chain cover - removal and installation

Refer to illustrations 13.6a, 13.6b and 13.8

1   Remove the water pump as described in Chapter 3.

2   If equipped with air conditioning, remove the compressor from the mounting bracket and secure it out of the way. Do not disconnect any of the air conditioning system hoses without having the system depressurized by a dealer service department or air conditioning technician.

3   Remove the compressor mounting bracket.

12.9 Installing the new oil seal with a wood block and a hammer

4   Remove the vibration damper as described in Section 11.

5   Disconnect the lower radiator hose at the front cover.

6   Remove the timing chain cover mounting bolts and separate the cover from the

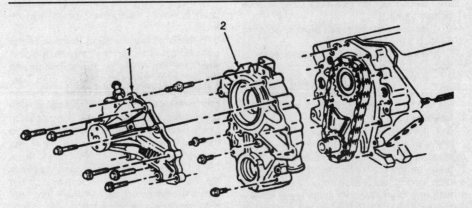

13.6a After the water pump (1) is removed, two bolts and two studs secure the timing chain cover (2) to the engine block . . .

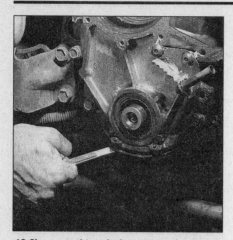

13.6b . . . and two bolts secure the timing chain cover to the front lip of the oil pan

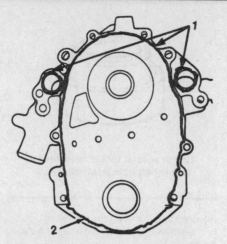

13.8 Apply two types of sealant as shown

1   *3/32-inch bead of anaerobic sealant*
2   *1/8-inch bead of RTV sealant*

14.8 The camshaft and crankshaft timing marks (arrows) should be in exact alignment before removing the timing sprockets and chain

engine **(see illustrations)**.

7   Clean all oil, dirt and old gasket material from the sealing surfaces of the cover and engine block. Replace the oil seal as described in Section 12.

8   Apply a continuous 3/32-inch (2 mm) bead of anaerobic sealant (Loctite 515 or equivalent) to both mating surfaces of the cover (except the mating surface where the cover engages the oil pan lip). Apply RTV-type sealant to the cover-to-oil pan area. Also apply anaerobic sealant to the areas surrounding the coolant passages **(see illustration)**.

9   Place the timing chain cover in position on the engine block and install the mounting bolts. Tighten the bolts to the specified torque.

10   The remaining installation procedures are the reverse of removal.

14   Timing chain and sprockets - inspection, removal and installation

Refer to illustrations 14.8, 14.9, 14.10 and 14.12

1   Disconnect the cable from the negative battery terminal.

2   Remove the vibration damper (refer to Section 11).

3   Remove the timing chain cover (refer to Section 13).

4   Before removing the chain and sprockets, visually inspect the teeth on the sprockets for signs of wear and check the chain for looseness.

5   If either or both sprockets show any signs of wear (edges on the teeth of the camshaft sprocket rounded, bright or blue areas on the teeth of either sprocket, chipping, pitting, etc.), they should be replaced with new ones. Wear in these areas is very common. Failure to replace a worn timing chain and sprockets may result in erratic engine performance, loss of power and low-

14.9 A screwdriver will hold the camshaft sprocket in place while loosening the mounting bolts

ered gas mileage.

6   If any one component (timing chain or either sprocket) requires replacement, the other two components should be replaced as well.

7   If it is determined that the timing components require replacement, proceed as follows.

8   Reinstall the vibration damper center bolt and use it to turn the crankshaft clockwise until the marks on the camshaft and crankshaft are in exact alignment **(see illustration)**. At this point the number one and four pistons will be at top dead center with the number four piston in the firing position (verify by checking the position of the rotor in the distributor, which should point to the number four spark plug wire terminal). **Note:** *Do not attempt to remove either sprocket or the timing chain until this is done and do not turn the crankshaft or camshaft after the sprockets/chain are removed.*

9   Remove the three camshaft sprocket retaining bolts **(see illustration)** and lift the camshaft sprocket and timing chain off the

14.10 A special puller will be needed to remove the crankshaft sprocket

front of the engine. It may be necessary to tap the sprocket with a soft-faced hammer to dislodge it.

10   If it is necessary to remove the crankshaft sprocket, it can be withdrawn from the crankshaft with a special puller **(see illustration)**.

11   Push the crankshaft sprocket onto the crankshaft using a bolt and washer from the puller set.

12   Lubricate the thrust (rear) surface of the camshaft sprocket with moly-base grease or engine assembly lube **(see illustration)**. Install the timing chain over the camshaft sprocket with slack in the chain hanging down over the crankshaft sprocket.

13   With the timing marks aligned, slip the chain over the crankshaft sprocket and then draw the camshaft sprocket into place with the three retaining bolts. Do not hammer or attempt to drive the camshaft sprocket into place, as it could dislodge the welch plug at the rear of the engine.

14   With the chain and both sprockets in place, check again to ensure that the timing marks on the two sprockets are properly

14.12 Lubricating the thrust surface of the camshaft sprocket

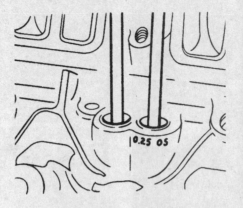

15.7a If the engine is factory equipped with oversize lifters, the lifter boss will be marked with a dab of white paint and will have 0.25 (mm) OS stamped on it

15.7b As the lifters are removed from the engine block, they should be stored separately to ensure reinstallation in their original positions

aligned. If not, remove the camshaft sprocket and move it until the marks align.

15 Lubricate the chain with engine oil and install the remaining components in the reverse order of removal.

15 Valve lifters - removal, inspection and installation

Refer to illustrations 15.7a and 15.7b

1 A noisy valve lifter can be isolated when the engine is idling. Hold a mechanic's stethoscope or a length of hose near the position of each valve while listening at the other end. Another method is to remove the rocker arm cover and, with the engine idling, place a finger on each of the valve spring retainers, one at a time. If a valve lifter is defective, it will be evident from the shock felt at the retainer as the valve seats.

2 The most likely causes of noisy valve lifters are dirt trapped between the plunger and the lifter body or lack of oil flow, viscosity or pressure. Before condemning the lifters, we recommend checking the oil for fuel contamination, correct level, cleanliness and correct viscosity.

Removal

3 Remove the rocker arm cover(s) as described in Section 4.

4 Remove the intake manifold as described in Section 8.

5 Remove the rocker arms and pushrods (Section 5).

6 There are several ways to extract the lifters from the bores. A special tool designed to grip and remove lifters is manufactured by many tool companies and is widely available, but it may not be required in every case. On newer engines without a lot of varnish buildup, the lifters can often be removed with a small magnet or even with your fingers. A machinist's scribe with a bent end can be used to pull the lifters out by positioning the point under the retainer ring in the top of each lifter. **Caution:** *Don't use pliers to remove the*

lifters unless you intend to replace them with new ones (along with the camshaft). The pliers may damage the precision machined and hardened lifters, rendering them useless.

7 Before removing the lifters, arrange to store them in a clearly labeled box to ensure that they're reinstalled in their original locations. **Note:** *Some engines may have both standard and 0.25 mm (0.010-inch) oversize valve lifters installed at the factory. These are specially marked* **(see illustration)**. *Remove the lifters and store them where they won't get dirty* **(see illustration)**.

Inspection and installation

8 Parts for valve lifters are not available separately. The work required to remove them from the engine again if cleaning is unsuccessful outweighs any potential savings from repairing them. Refer to Chapter 2, Part A, for lifter and camshaft inspection procedures. If the lifters are worn, they must be replaced with new ones and the camshaft must be replaced as well - never install used lifters with a new camshaft or new lifters with a used camshaft.

9 When reinstalling used lifters, make sure they're replaced in their original bores. Soak new lifters in oil to remove trapped air. Coat all lifters with moly-base grease or engine assembly lube prior to installation.

10 The remaining installation steps are the reverse of removal.

11 Run the engine and check for oil leaks.

16 Camshaft and bearings - removal, inspection and installation

Refer to illustrations 16.10, 16.11 and 16.13

Note: *Before removing the camshaft, refer to Chapter 2, Part A (Section 15), and measure the lobe lift.*

1 Remove the cable from the negative battery terminal.

2 Drain the oil from the crankcase (refer to Chapter 1).

3 Drain the coolant from the radiator (refer to Chapter 1).

4 Remove the radiator (refer to Chapter 3).

5 If equipped with air conditioning, remove the condenser (refer to Chapter 3). **Caution:** *The air conditioning system must be discharged by an air conditioning technician before the condenser can be removed. Under no circumstances should this be attempted by the home mechanic, as personal injury may result.*

6 Remove the valve lifters (refer to Section 15).

7 Remove the timing chain cover (refer to Section 13).

8 Remove the fuel pump and pushrod (refer to Chapter 4).

9 Remove the timing chain and camshaft sprocket (refer to Section 14).

10 Install two long bolts in the camshaft bolt holes to be used as a handle to pull on and support the camshaft **(see illustration)**.

16.10 Long bolts can be threaded into the camshaft bolt holes to provide a handle for removal and installation of the camshaft - support the cam near the block as it's withdrawn

16.11 A length of wire with a hook on it can be used to support the camshaft as you guide it out of the block

16.13 Be sure to apply moly-base grease or engine assembly lube to the cam lobes and bearing journals before installing the camshaft

11   Carefully draw the camshaft out of the engine block. Do this very slowly to avoid damage to the camshaft bearings as the lobes pass over the bearing surfaces. Always support the camshaft with one hand on the camshaft near the engine block and the other holding a wire to support the camshaft from above **(see illustration)**.

12   Refer to Chapter 2, Part A, for camshaft and bearing inspection procedures.

13   Prior to installing the camshaft, coat each of the lobes and journals with engine assembly lube or moly-base grease **(see illustration)**.

14   Slide the camshaft into the engine block, again taking extra care not to damage the bearings.

15   Install the camshaft sprocket and timing chain as described in Section 14.

16   Install the remaining components in the reverse order of removal by referring to the appropriate Chapter or Section.

17   Adjust the valve lash (refer to Section 6).

18   Have the air conditioning system (if equipped) evacuated and recharged.

17   Oil pan - removal and installation

Refer to illustrations 17.9, 17.11 and 17.13

Note: *The following procedure is based on the assumption that the engine is in place in the vehicle. If it's been removed, merely unbolt the oil pan and detach it from the block.*

Removal

1   Disconnect the negative battery cable from the battery, then refer to Chapter 1 and drain the oil.

2   Refer to Chapter 5 and remove the starter motor.

3   Separate the exhaust pipes from the manifolds.

4   Remove the oil pan mounting bolts.

5   Carefully separate the pan from the block. Don't pry between the block and pan or damage to the sealing surfaces may result and oil leaks could develop. You may have to turn the crankshaft slightly to maneuver the front of the pan past the crankshaft counterweights.

Installation

6   Clean the gasket sealing surfaces with lacquer thinner or acetone. Make sure the bolt holes in the block are clean.

7   Due to oil leakage problems, the factory has provided updated procedures for oil pan sealing.

1984 models

8   Check the oil pan flanges for distortion, particularly around the bolt holes and cor-

ners. If necessary, place the pan on a block of wood and use a hammer to flatten and restore the gasket surface.

9   Apply RTV sealant at the rear cradle corners **(see illustration)** and use a new oil pan gasket kit.

10   Install the oil pan as described in Steps 16 through 18.

1985 and 1986 models

11   Inspect the oil pan cradle corners for deformation using a straightedge across the cradle opening **(see illustration)**.

12   Replace the oil pan if any deformations are found during inspection.

13   These models have stand-offs **(see illustration)** on the oil pan side rails for use with RTV sealant. To use a gasket with these pans, the raised stand-offs must be filed or ground off.

14   Using a new oil pan gasket kit, place the cradle seal into the groove in the rear main bearing cap. Apply a small amount of RTV sealant to the corners of the seal where it contacts the block.

15   Using gasket sealing compound, position the oil pan gasket on the oil pan side rails.

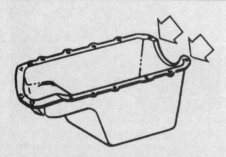

17.9 On 1984 models, apply RTV sealant at the rear cradle corners (arrows)

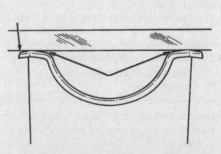

17.11 Using a straightedge, check for distortion at the corners (arrows)

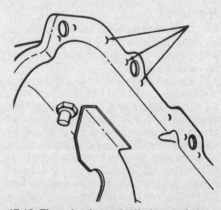

17.13 The raised stand-offs (arrows) must be filed or ground off before a gasket can be used

18.2 Remove the oil pump-to-rear main bearing cap bolt

All models

16 Carefully position the pan against the block and install the bolts finger tight. Tighten the bolts in three steps to the specified torque. Note that there are two bolt sizes which require different torques. Start at the center of the pan and work out toward the ends in a spiral pattern.
17 The remaining steps are the reverse of removal. **Caution:** *Don't forget to refill the engine with oil before starting it (see Chapter 1).*
18 Start the engine and check carefully for oil leaks at the oil pan.

18 Oil pump - removal and installation

Refer to illustration 18.2
1 Remove the oil pan (refer to Section 17).
2 Remove the pump-to-rear main bearing cap bolt **(see illustration)** and separate the pump and extension shaft from the engine.
3 To install the pump, move it into position and align the top end of the hexagonal extension shaft with the hexagonal socket in the

lower end of the distributor drive gear. The distributor drives the oil pump, so it is essential that this alignment is correct.
4 Install the oil pump-to-rear main bearing cap bolt and tighten it to the specified torque.
5 Reinstall the oil pan.

19 Flywheel/driveplate - removal and installation

Refer to Chapter 2, Part A, Section 20 for this procedure. Be sure to use the torque specifications in this Part of Chapter 2 for the V6 engine.

20 Rear main oil seal - replacement (engine in vehicle)

Identifying the seal type

V6 engines covered by this manual may be equipped with four types of rear main oil seals. To avoid unnecessary work, it is important to determine which type of seal your vehicle has before attempting replacement.

Most 1984 models have two-piece neoprene or rope-type seals. You may service these seals without removing the crankshaft. Some 1984 models have been updated with a new design one-piece neoprene seal which fits in the same groove as the old type seal it replaces. This type is less prone to leaking, but the crankshaft must be removed to replace it.

To determine which type seal you have on 1984 models, remove the oil pan and rear main bearing cap. If you have a two-piece oil seal (the lower portion comes off with the cap), follow the procedure outlined here. If you have an updated one-piece seal, the engine and crankshaft must be removed. Follow the procedures outlined in Section 23 of Part D of this Chapter.

On 1985 and later models, check for an oval cast on the left rear surface of the engine

block next to the transmission bellhousing. This signifies the engine has a one-piece seal which may be accessed by removing the flywheel/driveplate.

1984 models with rope-type seals
Refer to illustrations 20.3 and 20.7
Note: *Special tools, as noted in the Steps which follow, are required for this procedure. They are available from your dealer or may in some cases be rented from an auto parts store or tool rental shop.*
1 Although the crankshaft must be removed to install a new seal, the upper portion of the seal can be repaired with the crankshaft in place.
2 Remove the oil pan and oil pump (Sections 17 and 18).
3 Using a packing tool, available at most auto parts stores, drive the old seal gently back into the groove, packing it tight **(see illustration)**. It will pack in to a depth of 1/4 to 3/4-inch.
4 Repeat the procedure on the other end of the seal.
5 Measure the amount that the seal was driven up into the groove on one side and add 1/16-inch. Remove the old seal from the main bearing cap. Use the main bearing cap as a fixture and cut off a piece of the old seal to the predetermined length. Repeat this process for the other side.
6 Install a guide tool, available at most auto parts stores, on the block.
7 Using the packing tool, work the short pieces of the previously cut seal into the guide tool and pack them into the block groove on each side **(see illustration)**. The guide and packing tools have been machined to provide a built-in stop. Use of oil on the seal pieces will ease installation.
8 Remove the guide tool.
9 Install a new seal in the main bearing cap.
10 Apply a thin, even coat of anaerobic-type gasket sealant to the areas of the rear main bearing cap indicated in the illustration in Chapter 2, Part D (Section 23). **Caution:**

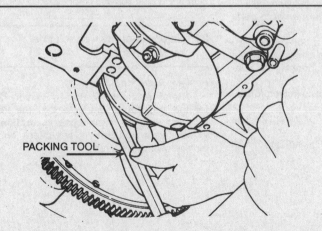

20.3 Using a packing tool to pack the old rear main oil seal into the groove - rope-type seal engines

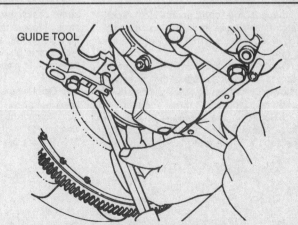

20.7 Using guide tools to install short pieces of seal cut from the old main seal in the upper seal groove - rope type seal engines

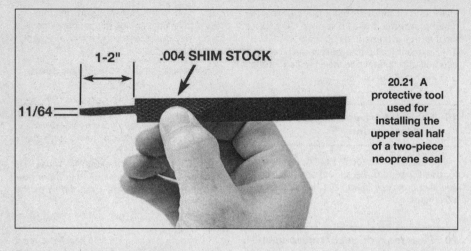

20.21 A protective tool used for installing the upper seal half of a two-piece neoprene seal

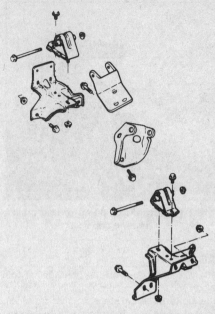

21.1 V6 engine mounts - exploded view

Do not get any sealant on the bearing or seal faces.

11   Tighten the rear main bearing cap bolts to the specified torque.

12   Install the oil pump and oil pan.

1984 models with two-piece neoprene seals

Refer to illustration 20.21

13   Always service both halves of the rear main oil seal. While replacement of this seal is much easier with the engine removed from the vehicle, the job can be done with the engine in place.

14   Remove the oil pan and oil pump as described previously in Sections 17 and 18.

15   Remove the rear main bearing cap from the engine.

16   Using a screwdriver, pry the lower half of the oil seal from the bearing cap.

17   To remove the upper half of the seal, use a small hammer and a brass pin punch to roll the seal around the crankshaft journal. Tap one end of the seal with the hammer and punch (be careful not to strike the crankshaft) until the other end of the seal protrudes enough to pull the seal out with pliers.

18   Remove all sealant and foreign material from the main bearing cap. Do not use an abrasive cleaner for this.

19   Inspect the components for nicks, scratches and burrs at all sealing surfaces. Remove any defects with a fine file or deburring tool.

20   Apply a very thin coat of RTV gasket sealant to the outer surface of the upper seal as shown. Do not get any sealant on the seal lips.

21   Included in the purchase of the rear main oil seal should be a small plastic installation tool; if not, a tool may be fashioned from an old feeler gauge blade **(see illustration)**.

22   With the upper half of the seal positioned so that the seal lip faces toward the front of the engine and the small dust lip faces toward the flywheel, install the seal by rolling it around the crankshaft using the installation tool as a "shoehorn" for protection.

23   Apply sealant as described in Step 20 to the other half of the seal and install it in the bearing cap.

24   Apply a 1/32-inch bead of anaerobic sealant to the cap between the rear main oil seal end and oil pan rear seal groove. Be sure to keep the sealant off the rear main oil seal and bearing and out of the drain slot.

25   Just before installing the cap, apply a light coat of moly-base grease or engine assembly lube to the crankshaft surface that will contact the seal.

26   Install the rear main bearing cap and tighten the bolts to the specified torque.

27   Install the oil pump and oil pan.

1985 and later models with one-piece seals

Note: *Special tools, as noted in the Steps which follow, are required for this procedure. They are available from your dealer or may, in some cases, be rented from an auto parts store or tool rental shop.*

28   Beginning in 1985, a 360-degree lip-type seal is utilized, which allows the oil pan to remain in place when performing this procedure.

29   Remove the transmission (refer to Chapter 7).

30   Remove the flywheel or driveplate (see Section 19).

31   Pry out the old seal, taking care not to mar the crankshaft surface. Inspect the crankshaft for scratches, burrs and nicks on the sealing surfaces.

32   A special seal installation tool, available at most auto parts stores, is required to properly seat the seal in the bore without damaging it. Lubricate the seal bore, seal lip and sealing surface on the crankshaft with engine oil. Slide the seal over the mandril on the tool until the dust lip on the seal bottoms squarely against the collar of the tool.

33   Position the dowel pin on the tool in the dowel pin hole in the crankshaft and secure the tool to the crankshaft.

34   Turn the T-handle of the tool until the collar pushes the seal into the bore completely. Make sure the seal is installed squarely.

35   To complete the operation install the flywheel (or driveplate) and transmission, then start the engine and check for leaks.

21   Engine mounts - check and replacement

Refer to illustration 21.1

Refer to Chapter 2, Part A, Section 22 for this procedure, but note that the V6 mounts are slightly different **(see illustration)** in ways that don't affect the check and replacement procedures.

Chapter 2 Part C
Inline six-cylinder engine

Contents

Specifications

General

Displacement	243 cu. in (4.0 liters)
Cylinder numbers (front-to-rear)	1-2-3-4-5-6
Firing order	1-5-3-6-2-4

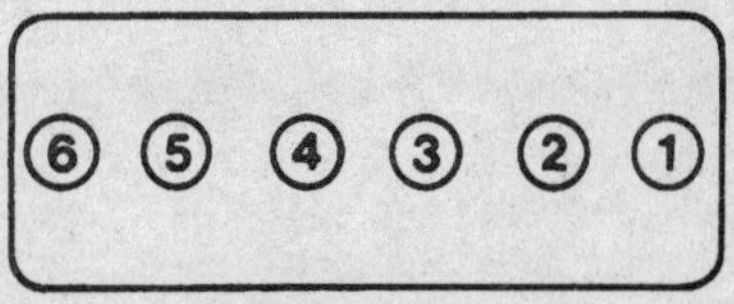

Cylinder numbers

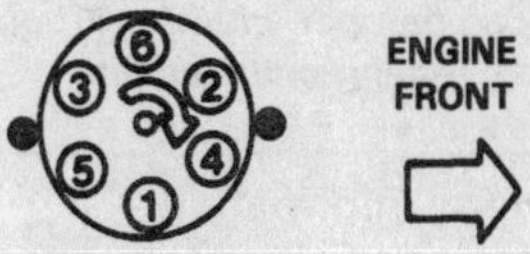

Firing order

ENGINE FRONT

Camshaft

Lobe lift (intake and exhaust)	0.253 in
End play	None
Journal diameter	
No. 1	2.029 to 2.030 in
No. 2	2.019 to 2.020 in
No. 3	2.009 to 2.010 in
No. 4	1.999 to 2.000 in
Journal-to-bearing (oil) clearance	0.001 to 0.003 in

Torque specifications

	Ft-lbs (unless otherwise indicated)
Camshaft sprocket bolt	50
Crankshaft pulley-to-vibration damper bolts	20
Cylinder head bolts	
Step A	22
Step B	45
Step C	
Bolt no. 11	100
All other bolts	110
Flywheel/driveplate bolts	105

Torque specifications (continued)

	Ft-lbs (unless otherwise indicated)
Intake and exhaust manifold retaining bolts and nuts	
1989 and 1990	
Bolt 1	30
Bolts 2 through 11	23
Nuts 12 and 13 (EGR fitting nuts)	30
1991	
Bolts 1 through 5	23
Bolts 6 and 7	17
Bolts 8 through 11	23
1992 and 1993	
Bolts 1 through 5	24
Bolts 6 and 7	17
Bolts 8 through 11	24
1994 on	
Bolts 1 through 5	24
Bolts 6 and 7	23
Bolts 8 through 11	24
Oil pan mounting bolts	
1/4 x 20	7
5/16 x 18	11
Oil pump mounting bolt	
Short	10
Long	17
Rear main bearing cap bolts	80
Rocker arm bolts	19
Rocker arm cover-to-cylinder head bolts	
With RTV	28 in-lbs
With permanent gasket	55 in-lbs
Tensioner bracket-to-block bolts	14
Timing chain cover-to-block	
Bolts	5
Studs	16
Vibration damper center bolt (lubricated)	80

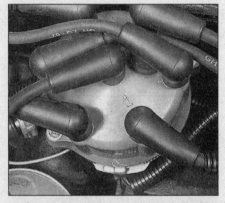

3.1a Align the notch (arrow) on the vibration damper with the "0" on the timing chain cover

3.1b Locate the number one spark plug terminal on the distributor cap, make a mark on the distributor housing, directly under the number one plug terminal . . .

3.1c . . . then remove the distributor cap and verify that the rotor is pointing at the mark (arrow)

1   General information

This Part of Chapter 2 is devoted to in-vehicle repair procedures for the inline six-cylinder engine. All information concerning engine removal and installation and engine block and cylinder head overhaul can be found in Part D of this Chapter.

The following repair procedures are based on the assumption that the engine is installed in the vehicle. If the engine has been removed from the vehicle and mounted on a stand, many of the steps outlined in this Part of Chapter 2 will not apply.

The Specifications included in this Part of Chapter 2 apply only to the procedures contained in this Part. Part D of Chapter 2 contains the Specifications necessary for cylinder head and engine block rebuilding.

2   Repair operations possible with the engine in the vehicle

Many major repair operations can be accomplished without removing the engine from the vehicle.

Clean the engine compartment and the exterior of the engine with some type of pressure washer before any work is done. It will make the job easier and help keep dirt out of the internal areas of the engine.

Remove the hood, if necessary, to improve access to the engine as repairs are performed (refer to Chapter 11 if necessary).

If vacuum, exhaust, oil or coolant leaks develop, indicating a need for gasket or seal replacement, the repairs can generally be made with the engine in the vehicle. The intake and exhaust manifold gaskets, timing cover gasket, oil pan gasket, crankshaft oil seals and cylinder head gaskets are all accessible with the engine in place.

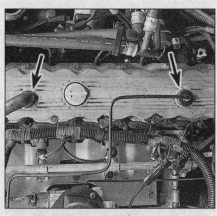

4.1 View from above shows locations of crankcase breather tube and hose (arrows)

Exterior engine components, such as the intake and exhaust manifolds, the oil pan (and the oil pump), the water pump, the starter motor, the alternator, the distributor and the fuel system components can be removed for repair with the engine in place.

Since the cylinder heads can be removed without pulling the engine, valve component servicing can also be accomplished with the engine in the vehicle. Replacement of the camshaft and timing chain and sprockets is also possible with the engine in the vehicle.

In extreme cases caused by a lack of necessary equipment, repair or replacement of piston rings, pistons, connecting rods and rod bearings is possible with the engine in the vehicle. However, this practice is not recommended because of the cleaning and preparation work that must be done to the components involved.

3 Top Dead Center (TDC) for number one piston - locating

Refer to illustrations 3.1a, 3.1b and 3.1c

See Chapter 2, Part A, Section 3 for this procedure, but refer to the illustrations and specifications included in this Section.

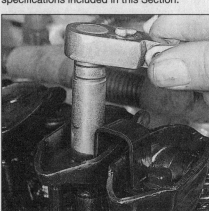

5.2 Go back and forth between the intake and exhaust rocker arms, loosening each bolt 1/4-turn at a time

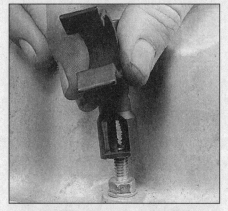

4.3 Pull off the wire loom clips

4 Rocker arm cover - removal and installation

Refer to illustrations 4.1, 4.3 and 4.4

1   Pull the crankcase breather tube and hose off the rocker arm cover **(see illustration)**.
2   Remove the cruise control servo, if equipped.
3   Remove the wire loom clips **(see illustration)**, noting the locations of their studs for reinstallation.
4   Remove the rocker arm cover retaining bolts and lift the cover off. If the cover is stuck, tap on it gently with a soft-face mallet **(see illustration)**. Do not pry on the gasket flange. **Note:** *These covers have a reusable pre-cured RTV gasket that is attached to the cover.*
5   Clean the sealing surfaces, removing any traces of oil with lacquer thinner or acetone and a clean rag.
6   Small cracks in the pre-cured gasket are allowable and can be repaired by applying RTV sealer to the cracked area before the cover is installed.
7   Install the cover and bolts. Tighten the bolts to the specified torque.

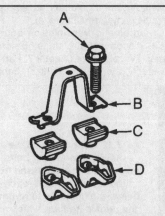

5.3 Rocker arm components - exploded view

A	Bolt	C	Fulcrum
B	Bridge	D	Rocker arm

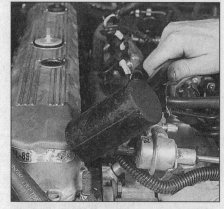

4.4 Use a soft-face mallet to break the cover loose - DO NOT pry between the cover and head

8   Reinstall the crankcase breather hoses and cruise control servo, if equipped.
9   Run the engine and check for oil leaks.

5 Rocker arms and pushrods - removal, inspection and installation

Refer to illustrations 5.2, 5.3 and 5.4

Removal

1   Remove the rocker arm cover (see Section 4).
2   Beginning at the front of the cylinder head, loosen and remove the rocker arm mounting bolts in pairs **(see illustration)**.
3   Remove the rocker arms, bridges and fulcrums **(see illustration)** and store them with their respective mounting bolts. Store each set of rocker arm components separately in a marked plastic bag to ensure they are reinstalled in their original locations. The bridges may be reinstalled in any location.
4   Remove the pushrods and store them separately to make sure they don't get mixed up during installation **(see illustration)**.

5.4 Store the pushrods in a box like this to ensure reinstallation in the same location

7.5 Lift the vacuum connector from its bracket and pull it apart

7.6 Label the connections before detaching them

7.11 Remove the bolts (arrows) and lift the heat shield off

Inspection

5   Check each rocker arm for wear, cracks and other damage, especially where the pushrods and valve stems contact the rocker arm faces. Check the fulcrum seat in each rocker arm and the fulcrum faces. Look for galling, stress cracks and unusual wear patterns. If the rocker arms are worn or damaged, replace them with new ones and install new fulcrums as well.

6   Make sure the oil hole at the pushrod end of each rocker arm is open.

7   Inspect the pushrods for cracks and excessive wear at the ends. Roll each pushrod across a piece of plate glass to see if it's bent (if it wobbles, it's bent).

Installation

8   Lubricate the lower end of each pushrod with clean engine oil or moly-base grease and install it in its original location. Make sure each pushrod seats completely in the lifter socket.

9   Bring the number one piston to top dead center on the compression stroke (Section 3).

10   Apply moly-base grease to the ends of the valve stems and the upper ends of the pushrods before placing the rocker arms in position.

11   Apply moly-base grease to the fulcrums to prevent damage to the mating surfaces before engine oil pressure builds up. Install the rocker arms, fulcrums, bridges and bolts in their original locations. Tighten the bolts to

the specified torque.

12   Install the rocker arm cover (see Section 4).

13   Start the engine, listen for unusual valve train noises and check for oil leaks at the rocker arm cover joint.

---

6   Valve spring, retainer and seals - replacement

See Chapter 2, Part A, Section 6 for this procedure, but be sure to follow the rocker arm cover and rocker arm/pushrod procedures outlined in this Part.

---

7   Intake manifold - removal and installation

Refer to illustrations 7.5, 7.6, 7.11, 7.12, 7.20a, 7.20b and 7.20c

Note: *Since the intake and exhaust manifolds share a common gasket, they must be removed and replaced at the same time.*

1   Disconnect the negative cable from the battery.

2   Remove the air cleaner assembly and the throttle cable (see Chapter 4).

3   On automatic transmission equipped models, disconnect the transmission line pressure (TV) cable (see Chapter 7, Part B).

4   Detach the cruise control cable, if

equipped.

5   Disconnect the vacuum connector on the intake manifold by lifting the connector assembly up and out of the bracket and then pulling it apart **(see illustration)**.

6   Label and then disconnect all vacuum and electrical connectors on the intake manifold **(see illustration)**.

7   Relieve the fuel pressure and then disconnect the fuel supply and return lines from the fuel rail assembly (see Chapter 4). Cap the open ends.

8   Loosen the serpentine drivebelt (see Chapter 1).

9   Remove the power steering pump and bracket from the intake manifold and set it aside without disconnecting the hoses. Be sure to leave the pump in an upright position so fluid won't spill.

10   Remove the fuel rail and injectors (see Chapter 4).

11   Remove the intake manifold heat shield **(see illustration)**.

12   Unbolt the intake manifold, referring to the **accompanying illustration and illustration 7.20**.

13   Remove the EGR tube from the intake manifold (see Chapter 6). Pull the manifold away from the engine slightly to disengage it from the locating dowels in the cylinder head, then lift the manifold out of the engine compartment.

14   Remove the exhaust manifold (see Section 8). This is necessary because the intake and exhaust manifolds share a common gasket.

15   Thoroughly clean the gasket mating surfaces, removing all traces of old gasket material.

16   If the manifold is being replaced, make sure all the fittings, etc. are transferred to the replacement manifold.

17   Position a new gasket on the cylinder head, using the locating dowels to hold it in place. Install the exhaust manifold and hand tighten the nuts.

18   Position the intake manifold loosely on the cylinder head.

19   Install the EGR tube between the manifolds.

7.12 Four of the intake manifold mounting bolts (arrows) can be accessed from the top - you must reach below the manifold to remove the remaining four, which are visible in illustration 7.20

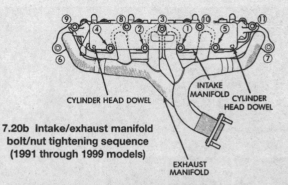

7.20a Intake/exhaust manifold
bolt/nut tightening sequence
(1989 and 1990 models)

7.20b Intake/exhaust manifold
bolt/nut tightening sequence
(1991 through 1999 models)

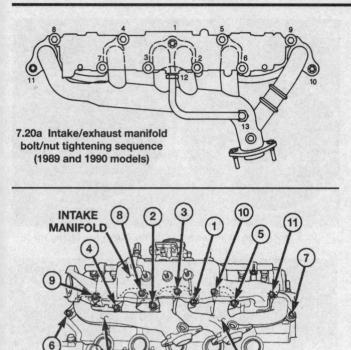

7.20c Intake and exhaust manifold bolt/nut tightening sequence
(2000 models)

9.11 Set the compressor aside with the refrigerant lines still
attached - then remove the upper bracket bolts (arrows)

20   Install the manifold retaining bolts and tighten all fasteners in sequence (see illustrations) to the specified torque. Note that nut number one (1) requires a greater torque than the others.

21   Reinstall the remaining parts in the reverse order of removal. **Caution:** *Before connecting the fuel lines to the fuel rail, replace the O-rings in the quick-connect fuel line couplings (see Chapter 4).*

22   Run the engine and check for fuel, vacuum and exhaust leaks.

8   Exhaust manifold - removal and installation

Warning: *Allow the engine to cool completely before following this procedure.*

1   Remove the intake manifold (see Section 7).

2   Apply penetrating oil to the threads of the exhaust manifold attaching studs and the exhaust pipe-to-manifold attaching bolts.

3   On 1987 through 1999 models, remove the two bolts and nuts that secure the exhaust pipe to the exhaust manifold and detach the pipe from the manifold. On 2000 models, remove the nuts that secure the exhaust pipe assembly to the exhaust manifolds.

4   On 1987 through 1999 models, remove the three nuts that secure the exhaust manifold to the cylinder head and pull the manifold off the engine. On 2000 models, remove the remaining nuts that secure the exhaust

manifolds to the cylinder head and pull both manifolds off the engine.

5   Remove all traces of old gasket material from the mating surfaces.

6   If the manifold gasket was blown out, have the manifold checked for warpage by an automotive machine shop and repaired as necessary.

7   Position a new gasket on the locating dowels and slide the manifold over the studs.

8   Install the attaching nuts on the studs finger tight.

9   Reinstall the intake manifold (see Section 7) and tighten all fasteners, in the sequence shown in **illustration 7.20**, to the specified torque.

10   Reconnect the exhaust pipe, run the engine and check for exhaust leaks.

9   Cylinder head - removal and installation

Refer to illustrations 9.11, 9.15, 9.19 and 9.21

Caution: *Allow the engine to cool completely before following this procedure.*

1   Disconnect the negative cable from the battery.

2   Drain the coolant from the radiator and the engine block (see Chapter 1).

3   Remove the air cleaner assembly (see Chapter 4).

4   Detach the fuel pipe and vacuum advance hose.

5   Remove the rocker arm cover (see Section 4).

6   Remove the rocker arms and pushrods (see Section 5).

7   Unbolt the power steering pump bracket (if equipped) and set the pump aside without disconnecting the hoses. Leave the pump upright so fluid doesn't spill.

8   Remove the intake and exhaust manifolds (see Sections 7 and 8).

Air conditioned models

9   Remove the bracket on the cylinder head that supports the idler pulley for the air conditioning compressor drivebelt (see Chapter 1).

10   Loosen the alternator drivebelt and remove the alternator bracket to-cylinder head mounting bolt.

11   Unplug the wiring and unbolt the air conditioning compressor (see Chapter 3) without disconnecting the refrigerant hoses. Set the compressor aside and remove the upper two bolts from the bracket (see illustration).

All models

12   Label the spark plug wires and remove the distributor cap with the wires attached to it. Remove the spark plugs as described in Chapter 1.

13   Disconnect the wire from the temperature sending unit, which is on the top left rear corner of the cylinder head. Also disconnect the battery ground cable, which is on the right side of the engine.

14   Remove the ignition coil and bracket

9.15 Remove the two bolts (arrows) to disconnect the heater hose bracket

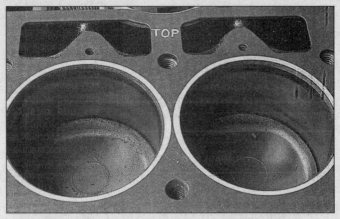

9.19 Install the head gasket with the TOP mark facing up

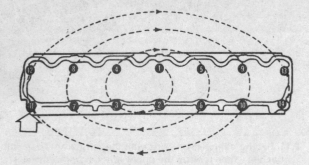

9.21 Cylinder head bolt tightening sequence - be sure to coat the threads of bolt no. 11 (arrow) with Loctite 592 sealant (or equivalent)

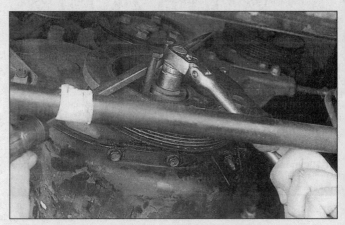

10.6 Install two bolts in the damper and use them to keep the crankshaft from rotating

assembly (see Chapter 5).

15   Remove the heater hose bracket **(see illustration)**.

16   Remove the 14 cylinder head bolts and lift the head off the engine. If the head sticks to the engine, insert a prybar into an exhaust port and pry gently to break the seal.

17   Thoroughly clean the gasket mating surfaces, removing all traces of old gasket material. Stuff shop towels into each cylinder so scraped material doesn't fall in.

18   Inspect the head for cracks and warpage. See Chapter 2, Part D, for cylinder head servicing information. If you are replacing the cylinder head, be sure to transfer all fittings, etc. to the new head.

19   Apply an even coat of Perfect Seal sealing compound (or equivalent) to both sides of the replacement head gasket. Position the new head gasket on the engine block with the word TOP facing up **(see illustration)**.

20   Install the cylinder head on the engine block.

21   Coat the threads of bolt number 11 **(see illustration)** with Loctite 592 sealant (or equivalent) and install the head bolts hand tight. **Note:** *Clean each bolt and mark it with a dab of paint. Replace any bolts which were painted and reused during an earlier servicing operation.*

22   Tighten the cylinder head bolts in sequence **(see illustration 9.21)**, according to the following procedure:

Step A - *tighten all bolts (1 through 14) in sequence to the specified torque for Step A.*

Step B - *tighten all bolts in sequence to the specified torque for Step B.*

Step C - *tighten all bolts except no. 11 to the specified torque for Step C.*

Caution: *In Step C, bolt no. 11 is tightened to a lower torque than the rest of the bolts. Do not overtighten it. Tighten bolt no. 11 to the specified torque for Step C.*

23   Reinstall the remaining components in the reverse order of removal.

24   Add coolant and run the engine, checking for proper operation and coolant and oil leaks.

10 Vibration damper - removal and installation

Refer to illustrations 10.6 and 10.7

1   Disconnect the negative cable from the battery.

2   Raise the front of the vehicle and support it securely on jackstands.

3   Remove the belly pan which is mounted below the front of the engine.

4   Remove the drivebelts (see Chapter 1).

5   Remove the radiator and cooling fan(s)

as described in Chapter 3.

6   Remove the vibration damper retaining bolt and washer. **Note:** *To prevent the crankshaft from rotating, place two 5/16 x 1-1/2-inch long bolts into the damper holes and hold a pry bar between them **(see illustration)**. Rotate the crankshaft until the bar contacts the frame.*

7   Using a vibration damper removal tool **(see illustration)**, pull the damper off the crankshaft.

10.7 Use a vibration damper removal tool such as this one - do not use a gear puller with jaws; it will damage the damper

11.2 Pry the old seal out with a seal removal tool (shown here) or a screwdriver

11.3 Gently drive the new seal into place with a hammer and large socket

12.4 Once the alternator has been removed, remove the bracket retaining nuts (arrows), then unbolt the timing chain cover

2C

8   Clean and inspect the area on the center hub of the damper where the front crankshaft oil seal contacts it. Minor imperfections can be cleaned up with emery cloth. If there is a groove worn in the hub, replace the vibration damper or have a special sleeve installed on the hub to restore the contact surface.
9   Apply clean engine oil to the seal contact surface of the damper hub.
10   Align the key slot of the vibration damper hub with the crankshaft key and tap the damper onto the crankshaft with a soft-face mallet.
11   Install the vibration damper bolt and tighten it to the specified torque.
12   Reinstall the remaining parts in the reverse order of removal.

11   Front crankshaft oil seal - replacement

Refer to illustrations 11.2 and 11.3
1   Remove the vibration damper (Section 10).
2   Carefully pry the oil seal out of the timing chain cover with a seal removal tool or screwdriver **(see illustration)**. Don't scratch the cover bore or damage the crankshaft in the process (if the crankshaft is damaged the new seal will end up leaking).
3   Clean the bore in the cover and coat the outer edge of the new seal with engine oil or multi-purpose grease. Using a socket with an outside diameter slightly smaller than the outside diameter of the seal, carefully drive the new seal into place with a hammer **(see illustration)**. If a socket isn't available, a short section of large diameter pipe will work. Check the seal after installation to be sure that the spring didn't pop out of place.
4   Reinstall the vibration damper.
5   The parts removed to gain access to the damper can now be reinstalled.
6   Run the engine and check for leaks.

12   Timing chain cover - removal and installation

Refer to illustrations 12.4, 12.7 and 12.13
1   Disconnect the negative cable from the battery.
2   Remove the fan, fan shroud, radiator and water pump pulley (see Chapter 3).
3   Remove the vibration damper (see Section 10).
4   Unbolt the alternator and bracket assembly **(see illustration)**.
5   Remove the oil pan-to-timing chain cover bolts and timing chain cover-to-engine block bolts.
6   Separate the timing chain cover from the engine. If necessary, tap on it gently with a soft-face mallet to break the seal. Temporarily stuff a rag into the oil pan opening to prevent entry of debris.
7   Cut off the oil pan side gasket end tabs flush with the front face of the engine block **(see illustration)**. Save the cut off gasket tabs for reference later.
8   Clean the mating surfaces of the timing chain cover, oil pan and engine block, removing all traces of oil and old gasket material.
9   Apply RTV sealer to both sides of the

new timing chain cover-to-engine block gasket and position the gasket on the engine.
10   Using the end tabs you cut off as guides, trim the replacement oil pan side gasket ends to the appropriate sizes. Apply cement to the gasket ends and install them on the exposed portions of the oil pan side rails.
11   Using RTV sealant, generously coat the timing chain cover end tab recesses of the new timing chain cover-to-oil pan seal. Position the seal on the timing chain cover. Apply engine oil to the seal-to-oil pan contact surface.
12   Position the timing chain cover on the engine block.
13   Use the vibration damper to center the timing chain cover **(see illustration)**. Be sure the old oil seal (not the new one) is in place, as it may be damaged.
14   Install the timing chain cover-to-block and oil pan-to-cover bolts and tighten them to the specified torque.
15   Replace the front crankshaft oil seal (see Section 11).
16   With the key inserted in the crankshaft, install the vibration damper as described in Section 10.
17   Reinstall the remaining components in the reverse order of removal.
18   Run the engine and check for oil leaks.

12.7 Cut off the end tabs at both sides where the oil pan and engine block meet

12.13 Use the vibration damper to center the timing chain cover during installation

13.3 Slip the oil slinger off the crankshaft, noting that the cupped side faces away from the engine

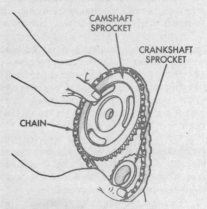

13.9 Remove and install the chain and sprockets as an assembly

13.7 Install the timing chain with the sprocket index dots (arrows) directly opposite each other

13.10a Note that the engine side of the camshaft sprocket has a hole (arrow) . . .

13.8 Remove the thrust pin and spring (arrow), then remove the bolt in the center of the camshaft sprocket - put a large screwdriver through one of the holes in the sprocket to keep it from turning

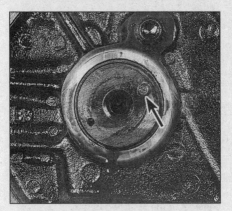

13.10b . . . for the camshaft locating dowel (arrow) - be sure they are aligned properly during installation

13   Timing chain and sprockets - inspection, removal and installation

Refer to illustrations 13.3, 13.7, 13.8, 13.9, 13.10a, 13.10b, 13.11 and 13.13

1   Set the number one piston at Top Dead Center (see Section 3).
2   Remove the timing chain cover (see Section 12).

Inspection

3   Slip the oil slinger off the crankshaft **(see illustration)** and reinstall the vibration damper bolt. Using this bolt, rotate the crankshaft clockwise just enough to take up the slack on one side of the chain.
4   Count the pins on the timing chain. The correct timing chain has 48 pins. A chain with more pins will cause excessive slack.
5   Establish a reference point on the block. Move the slack side of the chain from side-to-side with your fingers and measure the movement. The difference between the two measurements is the deflection.
6   If the deflection exceeds 1/2-inch, replace the timing chain and sprockets.

Removal

7   Align the sprocket timing marks **(see illustration)**.

8   Remove the camshaft thrust pin and spring and the sprocket retaining bolt and washer **(see illustration)**.
9   Pull the crankshaft sprocket, camshaft sprocket and timing chain off as an assembly **(see illustration)**. **Caution:** *Do not turn the crankshaft or camshaft while the timing chain is removed.*

Installation

10   Be sure the crankshaft key is still pointing up. Note the locations of the locating dowel on the camshaft and the corresponding hole in the cam sprocket **(see illustrations)**.
11   Pre-assemble the timing chain, crankshaft sprocket and camshaft sprocket with the timing marks aligned and facing out. Slip the assembly onto the engine in such a way that a line drawn through the timing marks will also pass through the centers of the sprockets **(see illustration)**.
12   Install the camshaft sprocket bolt and tighten it to the specified torque. Reinstall the thrust pin and spring.
13   To verify the correct installation of the timing chain, turn the crankshaft clockwise until the camshaft sprocket timing mark is at the one o'clock position. This positions the crankshaft sprocket timing mark where the adjacent tooth meshes with the chain at the three o'clock position. There must be 15

chain pins between the sprocket timing marks **(see illustration)**.
14   Install the crankshaft oil slinger on the crankshaft with the cupped side facing out.
15   Install the timing chain cover and vibration damper as described in Sections 10 and 12.
16   Reinstall the remaining parts in the reverse order of removal.
17   Run the engine and check for oil leaks and proper operation.

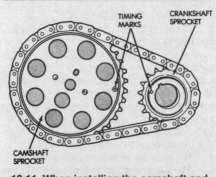

13.11 When installing the camshaft and camshaft sprockets, make sure that the timing marks are aligned as shown

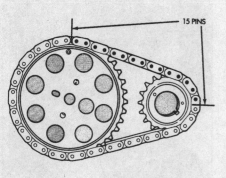

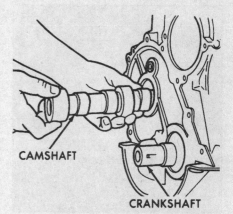

13.13 To verify correct installation, turn the crankshaft clockwise until the timing marks are positioned as shown and count the pins between the marks

14.9 Remove the lifters with a magnetic pick-up tool (shown here) or a special lifter removal tool

14.11 Support the camshaft as you slowly withdraw it from the block

14 Camshaft, bearings and lifters - removal, inspection and installation

Refer to illustrations 14.9 and 14.11

1   The extent of camshaft wear can be determined by measuring the lobe lift. This procedure does not involve removing the camshaft. Refer to Chapter 2, Part A, Section 16 for the lobe lift measuring procedure, but use the specifications provided in this Part of Chapter 2.

2   To remove the lifters only, follow steps 7 through 9, 12, 16 and 17 of this procedure.

3   Remove the radiator, fan and fan shroud (see Chapter 3).

4   On air conditioned models, unbolt the condenser assembly as a charged unit (see Chapter 3) without disconnecting the refrigerant lines. Set the condenser aside. It may be necessary to remove the battery case (see Chapter 5).

5   Remove the distributor (see Chapter 5).

6   Remove the front bumper and/or grille as necessary for the camshaft to be slid out the front of the engine (see Chapter 11).

7   Label and then remove the spark plug wires from the spark plugs.

8   Remove the cylinder head (see Section 9).

9   Remove the valve lifters **(see illustration)** and store them separately so they can be reinstalled in the same bores.

10   Remove the timing chain and sprockets (see Section 13).

11   Carefully pull the camshaft out. Temporarily install the sprocket bolt, if necessary, to use as a handle. Support the cam so the lobes don't nick or gouge the bearings as it's withdrawn **(see illustration)**.

12   See Chapter 2, Part A, Section 11 for the lifter inspection procedure. Camshaft and bearing inspection are covered in Chapter 2, Part A, Section 16. Be sure to use the specifications provided in this Part of Chapter 2.

13   Camshaft bearing replacement requires

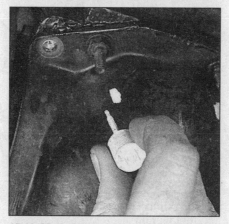

15.10 Mark the locations of the studs with paint to ensure proper reassembly

15.12 The oil pickup tube interferes with pan removal - remove the two attaching bolts and detach the oil pump (arrow) . . .

special tools and expertise that place it outside the scope of the home mechanic. Remove the engine and take the block to an automotive machine shop to ensure the job is done correctly. **Note:** *If the camshaft appears to have been rubbing hard against the timing chain cover, first check the camshaft thrust pin and spring and then examine the oil pressure relief holes in the rear cam journal to make sure they are open.*

14   Lubricate the camshaft journals and lobes with moly-base grease or engine assembly lube.

15   Slide the camshaft into the engine. Support the cam near the block and be careful not to scrape or nick the bearings.

16   The rest of the installation procedure is the reverse of removal.

17   Before starting and running the engine, change the oil and filter (see Chapter 1).

15 Oil pan - removal and installation

Refer to illustrations 15.10, 15.12, 15.13 and 15.16

1   Disconnect the negative cable from the battery.

2   Raise the vehicle and support it securely on jackstands.

3   Remove the belly pan from under the front of the engine.

4   Drain the oil and replace the oil filter (see Chapter 1).

5   Remove the starter motor (see Chapter 5).

6   Remove the bellhousing inspection cover from the front of the transmission.

7   Detach the steering damper from the center steering link (see Chapter 10, Section 16).

8   Support the front axle with a jack and remove the lower shock absorber bolts.

9   Lower the jack and allow the axle to hang free.

10   Mark the locations of the oil pan mounting studs **(see illustration)**.

11   Remove the oil pan mounting bolts/studs and carefully separate the pan from the engine block. If the pan sticks to the block, tap the side of the pan gently with a soft-face mallet.

12   Set the oil pan on the axle and remove the oil pump and pickup tube **(see illustration)**.

15.13 . . . then remove the oil pan from the rear by sliding it out between the axle and bellhousing

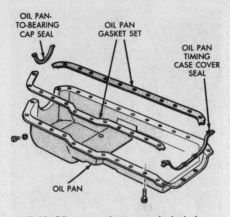

15.16 Oil pan gaskets - exploded view

18.3a Drive one side of the upper seal in . . .

18.3b . . . until the other side protrudes far enough to grasp it with needle-nose pliers and pull it out

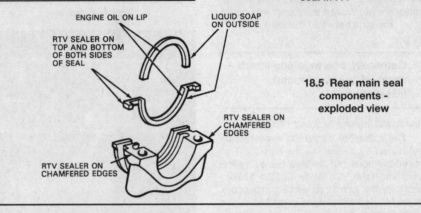

18.5 Rear main seal components - exploded view

13   Remove the oil pan by sliding it out to the rear **(see illustration)**.
14   Thoroughly clean the mating surfaces, removing all traces of oil and old gasket material.
15   Check the oil pan flange for distortion and warpage. Straighten the flange by placing the distorted area on a block of wood and pounding it flat with a hammer.
16   Position new gaskets **(see illustration)** on the pan with gasket spray. Apply a generous amount of RTV sealant at the corners where the gaskets join. Coat the inside curved surface of the replacement rear gasket section (where it contacts the bearing cap) with soap.
17   Slide the oil pan up under the engine and reinstall the oil pump and pickup tube, tightening the oil pump mounting bolts to the specified torque. Install the oil pan and tighten the bolts to the specified torque, working from the center out in several steps.
18   Reinstall the remaining parts in the reverse order of removal.
19   Check that the drain plug is tight and then add the amount of oil specified in Chapter 1.
20   Run the engine and check for oil leaks.

16   Oil pump - removal and installation

See Chapter 2, Part A, Section 17 for this procedure, but be sure to use the torque specifications and **illustration 15.12** in this Part of Chapter 2.

17   Flywheel/driveplate - removal and installation

Refer to Chapter 2, Part A, Section 20 for this procedure, but be sure to use the torque specifications in this Part of Chapter 2.

18   Rear crankshaft oil seal - replacement

Refer to illustrations 18.3a, 18.3b and 18.5
1   Remove the oil pan (see Section 15).
2   Remove the rear main bearing cap and pry the old seal half out of the bearing cap with a small screwdriver.
3   Carefully drive the old upper main seal out with a small brass punch and a hammer until it protrudes sufficiently from the engine block to be gripped with needle-nose pliers and removed **(see illustrations)**. Use great care to avoid damaging the crankshaft.
4   Thoroughly clean the main bearing cap and the rear of the block/crankshaft, removing all traces of oil and old sealer.
5   Coat the lip of the new upper seal with engine oil; coat the outside surface with liquid soap **(see illustration)**.
6   Insert the seal into the groove in the engine block with the lip facing forward.
7   Coat both sides of the lower seal ends with RTV sealer and put liquid soap on the outside of the seal **(see illustration 18.5)**. Do not apply RTV or soap to the seal lip. Press the seal into place in the cap and apply a film of engine oil to the seal lip. **Caution:** *Do not apply sealant to the cylinder block mating surfaces of the rear main bearing cap. Doing this would alter the bearing-to-journal clearance.*
8   Apply RTV sealer to the chamfered edges of the rear main bearing cap **(see illustration 18.5)** and install the cap. Tighten the bolts to the specified torque.
9   Reinstall the remaining components in the reverse order of removal.
10   Add the amount of oil specified in Chapter 1, run the engine and check for oil leaks.

19   Engine mounts - check and replacement

Refer to Chapter 2, Part A, Section 22. The inline six-cylinder engine mounts are slightly different, but this doesn't affect the check and replacement procedures.

Chapter 2 Part D
General engine overhaul procedures

Contents

2D

Specifications

Four-cylinder engine

General

Cylinder compression pressure	155 to 185 psi
Maximum allowable variation between cylinders	30 psi
Oil pressure	
At idle (800 rpm)	25 to 35 psi
Above 1600 rpm	37 to 75 psi

Engine block

Cylinder bore diameter (standard)	3.8751 to 3.8775 in
Maximum allowable taper and out-of-round	0.001 in
Warpage limit	0.002 in per 6 inches
Valve lifter bore diameter	0.9055 to 0.9065 in
Cylinder head and valves	
Cylinder head warpage limit	0.002 in per 6 in (0.006 in overall)
Minimum valve margin	1/32-in
Valve stem diameter	0.311 to 0.312 in
Valve stem-to-guide clearance	0.001 to 0.003 in
Valve spring pressure	
Valve closed	80 to 90 lbs at 1.64 in
Valve open	200 lbs at 1.216 in
Valve spring free length	1.967 in
Valve spring installed height	Not available
Valve lifter	
Diameter	0.904 to 0.9045 in
Lifter-to-bore clearance	0.001 to 0.0025 in

Four-cylinder engine (continued)

Crankshaft and connecting rods

Connecting rod journal	
Diameter	2.0934 to 2.0955 in
Bearing oil clearance	
Desired	0.0015 to 0.0020 in
Allowable	0.001 to 0.0025 in
Connecting rod side clearance (end play)	0.0010 to 0.0019 in
Main bearing journal	
Diameter	2.4996 to 2.5001 in
Bearing oil clearance	
Desired	0.002 in
Allowable	0.001 to 0.0025 in
Crankshaft end play	0.0015 to 0.0065 in
Maximum taper and out-of-round (all journals)	0.0005 in

Pistons and rings

Piston-to-bore clearance	0.0013 to 0.0021 in
Piston ring end gap	
Compression rings	0.0010 to 0.020 in
Oil control ring (steel rail)	0.015 to 0.055 in
Piston ring side clearance	
Compression rings	0.001 to 0.0032 in
Oil control ring	0.001 to 0.0095 in
Piston diameter (1995 and later models)	
Letter code A	3.8755 to 3.8759 inches
Letter code B	3.8759 to 3.8763 inches
Letter code C	3.8763 to 3.8767 inches
Letter code D	3.8767 to 3.8771 inches
Letter code E	3.8771 to 3.8775 inches
Letter code F	3.8775 to 3.8779 inches

Torque specifications*

	Ft-lbs
Connecting rod cap nuts	33
Main bearing cap bolts	80

***Note:** Refer to Part A for additional torque specifications.*

V6 engine

General

Cylinder compression pressure	Not available
Maximum allowable variation between cylinders	30 psi
Oil pressure	Not available

Engine block

Cylinder bore diameter (standard)	3.503 to 3.506 in
Maximum allowable taper and out-of-round	0.001 in

Cylinder head and valves

Warpage limit	0.002 in per 6 in (0.006 in overall)
Minimum valve margin	1/32-in
Valve stem-to-guide clearance	0.001 to 0.002 in
Valve spring pressure	
Valve closed	88 lbs at 1.57 in
Valve open	195 lbs at 1.18 in
Valve spring free length	1.909 in
Valve spring installed height	1.57 in
Crankshaft and connecting rods	
Connecting rod journal	
Diameter	1.999 to 1.998 in
Bearing oil clearance	0.001 to 0.003 in
Connecting rod side clearance (end play)	0.006 to 0.017 in
Main bearing journal diameter	
1984	
Journals 1, 2 and 4	2.493 to 2.494 in
Journal 3	2.492 to 2.494 in
1985 and 1986 (all journals)	2.647 to 2.648 in
Main bearing oil clearance	0.0016 to 0.003 in
Crankshaft end play	0.002 to 0.006 in
Maximum taper and out-of-round (all journals)	0.001 in

Pistons and rings

Piston-to-bore clearance	0.0006 to 0.0016 in
Piston ring end gap	
Compression rings	0.0098 to 0.0196 in
Oil control ring (steel rail)	0.020 to 0.055 in
Piston ring side clearance	
Top compression ring	0.001 to 0.0027 in
Second compression ring	0.0015 to 0.0037 in
Oil control ring	0.0078 in maximum

Torque specifications*

	Ft-lbs
Connecting rod cap nuts	37
Main bearing cap bolts	70

*Note: Refer to Part B for additional torque specifications.

Inline six-cylinder engine

General

Cylinder compression pressure	120 to 150 psi
Maximum variation between cylinders	30 psi
Oil pressure	
At idle (600 rpm)	13 psi
Above 1600 rpm	37 to 75 psi

Cylinder head and valves

Warpage limit	0.002 in per 6 in
Minimum valve margin	1/32 in
Valve stem diameter	0.312 in
Valve stem-to-guide clearance	0.001 to 0.003 in
Valve spring pressure	
Valve open	205 to 220 lbs at 1.2 in
Valve closed	64 to 74 lbs at 1.625 in
Valve lifter	
Diameter	0.904 to 0.9045 in
Lifter-to-bore clearance	0.001 to 0.0025 in

Crankshaft and connecting rods

Connecting rod journal	
Diameter	2.0934 to 2.0955 in
Bearing oil clearance	
Desired	0.0015 to 0.002 in
Allowable	0.001 to 0.003 in
Connecting rod side clearance (end play)	0.010 to 0.019 in
Main bearing journals 1-6 diameter (all years)	2.4996 to 2.5001 in
Main bearing journal # 7 (1995 on)	2.4980 to 2.4995 in
Bearing oil clearance	
Desired	0.002 in
Allowable	0.001 to 0.0025 in
Crankshaft end play (at thrust bearing)	0.0015 to 0.0065 in
Maximum taper and out-of-round (all journals)	0.0005 in

Pistons and rings

Piston-to-bore clearance	
Desired	0.0012 to 0.0013 in
Allowable	0.0009 to 0.0017 in
Piston ring end gap	
Compression rings	0.010 to 0.020 in
Oil control ring (steel rails)	0.010 to 0.025 in
Piston ring side clearance	
Compression rings	
Desired	0.0017 in
Allowable	0.0017 to 0.0032 in
Oil control ring	
Desired	0.003 in
Allowable	0.001 to 0.008 in

Engine block

Maximum warpage	0.002 in per 6 in (0.008 in overall)
Cylinder bore diameter (standard)	3.8751 to 3.8775 in
Maximum taper and out-of-round	0.001 in

Inline six-cylinder engine

Torque specifications*	Ft-lbs
Main bearing cap bolts	80
Connecting rod cap nuts	33

***Note:** *Refer to Part C for additional torque specifications.*

1   General information

Included in this portion of Chapter 2 are the general overhaul procedures for the cylinder head(s) and internal engine components.

The information ranges from advice concerning preparation for an overhaul and the purchase of replacement parts to detailed, step-by-step procedures covering removal and installation of internal engine components and the inspection of parts.

The following Sections have been written based on the assumption that the engine has been removed from the vehicle. For information concerning in-vehicle engine repair, as well as removal and installation of the external components necessary for the overhaul, see Part A, B or C of this Chapter and Section 7 of this Part.

The Specifications included in this Part are only those necessary for the inspection and overhaul procedures which follow. Refer to Parts A, B and C for additional Specifications.

2   Engine overhaul - general information

Refer to illustrations 2.4a and 2.4b

It's not always easy to determine when, or if, an engine should be completely overhauled, as a number of factors must be considered.

High mileage is not necessarily an indication that an overhaul is needed, while low mileage doesn't preclude the need for an overhaul. Frequency of servicing is probably the most important consideration. An engine that's had regular and frequent oil and filter changes, as well as other required maintenance, will most likely give many thousands of miles of reliable service. Conversely, a neglected engine may require an overhaul very early in its life.

Excessive oil consumption is an indication that piston rings, valve seals and/or valve guides are in need of attention. Make sure that oil leaks aren't responsible before deciding that the rings and/or guides are bad. Perform a cylinder compression check to determine the extent of the work required (see Section 3).

Check the oil pressure with a gauge installed in place of the oil pressure sending unit **(see illustrations)** and compare it to the Specifications. If it's extremely low, the bearings and/or oil pump are probably worn out.

Loss of power, rough running, knocking or metallic engine noises, excessive valve train noise and high fuel consumption rates

2.4a Remove the oil pressure sending unit (arrow) . . .

2.4b . . . and connect a gauge to check oil pressure (inline six-cylinder engine shown) - the sending unit is located near the oil filter on all models

may also point to the need for an overhaul, especially if they're all present at the same time. If a complete tune-up doesn't remedy the situation, major mechanical work is the only solution.

An engine overhaul involves restoring the internal parts to the specifications of a new engine. During an overhaul, the piston rings are replaced and the cylinder walls are reconditioned (rebored and/or honed). If a rebore is done by an automotive machine shop, new oversize pistons will also be installed. The main bearings, connecting rod bearings and camshaft bearings are generally replaced with new ones and, if necessary, the crankshaft may be reground to restore the journals. Generally, the valves are serviced as well, since they're usually in less-than-perfect condition at this point. While the engine is being overhauled, other components, such as the distributor, starter and alternator, can be rebuilt as well. The end result should be a like new engine that will give many thousands of trouble free miles. **Note:** *Critical cooling system components such as the hoses, drivebelts, thermostat and water pump MUST be replaced with new parts when an engine is overhauled. The radiator should be checked carefully to ensure that it isn't clogged or leaking (see Chapter 3). Also, we don't recommend overhauling the oil pump - always install a new one when an engine is rebuilt.*

Before beginning the engine overhaul, read through the entire procedure to familiarize yourself with the scope and requirements of the job. Overhauling an engine isn't difficult if you have the right equipment and follow the instructions carefully, but it is time consuming. Plan on the vehicle being tied up for a minimum of two weeks, especially if parts must be taken to an automotive machine shop for repair or reconditioning. Check on

availability of parts and make sure that any necessary special tools and equipment are obtained in advance. Most work can be done with typical hand tools, although a number of precision measuring tools are required for inspecting parts to determine if they must be replaced. Often an automotive machine shop will handle the inspection of parts and offer advice concerning reconditioning and replacement. **Note:** *Always wait until the engine has been completely disassembled and all components, especially the engine block, have been inspected before deciding what service and repair operations must be performed by an automotive machine shop. Since the block's condition will be the major factor to consider when determining whether to overhaul the original engine or buy a rebuilt one, never purchase parts or have machine work done on other components until the block has been thoroughly inspected. As a general rule, time is the primary cost of an overhaul, so it doesn't pay to install worn or substandard parts.*

As a final note, to ensure maximum life and minimum trouble from a rebuilt engine, everything must be assembled with care in a spotlessly clean environment.

3   Cylinder compression check

Refer to illustration 3.6

1   A compression check will tell you what mechanical condition the upper end (pistons, rings, valves, head gaskets) of your engine is in. Specifically, it can tell you if the compression is down due to leakage caused by worn piston rings, defective valves and seats or a

3.6 A compression gauge with a threaded fitting for the spark plug hole is preferred over the type that requires hand pressure to maintain the seal

blown head gasket. **Note:** *The engine must be at normal operating temperature and the battery must be fully charged for this check. Also, if the engine is equipped with a carburetor, the choke valve must be all the way open to get an accurate compression reading (if the engine's warm, the choke should be open).*

2   Begin by cleaning the area around the spark plugs before you remove them (compressed air should be used, if available, otherwise a small brush or even a bicycle tire pump will work). The idea is to prevent dirt from getting into the cylinders as the compression check is being done.

3   Remove all of the spark plugs from the engine (see Chapter 1).

4   Block the throttle wide open.

5   Detach the coil wire from the center of the distributor cap and ground it on the engine block. Use a jumper wire with alligator clips on each end to ensure a good ground. On fuel-injected vehicles, the fuel pump circuit should also be disabled (see Chapter 4).

6   Install the compression gauge in the number one spark plug hole **(see illustration)**.

7   Crank the engine over at least seven compression strokes and watch the gauge. The compression should build up quickly in a healthy engine. Low compression on the first stroke, followed by gradually increasing pressure on successive strokes, indicates worn piston rings. A low compression reading on the first stroke, which doesn't build up during successive strokes, indicates leaking valves or a blown head gasket (a cracked head could also be the cause). Deposits on the undersides of the valve heads can also cause low compression. Record the highest gauge reading obtained.

8   Repeat the procedure for the remaining cylinders and compare the results to the Specifications.

9   Add some engine oil (about three squirts from a plunger-type oil can) to each cylinder, through the spark plug hole, and repeat the test.

10   If the compression increases after the oil is added, the piston rings are definitely worn. If the compression doesn't increase significantly, the leakage is occurring at the valves or head gasket. Leakage past the valves may be caused by burned valve seats and/or faces or warped, cracked or bent valves.

11   If two adjacent cylinders have equally low compression, there's a strong possibility that the head gasket between the is blown. The appearance of coolant in the combustion chambers or the crankcase would verify this condition.

12   If one cylinder is about 20 percent lower than the others, and the engine has a slightly rough idle, a worn exhaust lobe on the camshaft could be the cause.

13   If the compression is unusually high, the combustion chambers are probably coated with carbon deposits. If that's the case, the cylinder head(s) should be removed and decarbonized.

14   If compression is way down or varies greatly between cylinders, it would be a good idea to have a leak-down test performed by an automotive repair shop. This test will pinpoint exactly where the leakage is occurring and how severe it is.

4   Engine removal - methods and precautions

If you've decided that an engine must be removed for overhaul or major repair work, several preliminary steps should be taken.

Locating a suitable place to work is extremely important. Adequate work space, along with storage space for the vehicle, will be needed. If a shop or garage isn't available, at the very least a flat, level, clean work surface made of concrete or asphalt is required.

Cleaning the engine compartment and engine before beginning the removal procedure will help keep your tools and your hands clean.

An engine hoist or A-frame will also be necessary. Make sure the equipment is rated in excess of the combined weight of the engine and accessories. Safety is of primary importance, considering the potential hazards involved in lifting the engine out of the vehicle.

If the engine is being removed by a novice, a helper should be available. Advice and aid from someone more experienced would also be helpful. There are many instances when one person cannot simultaneously perform all of the operations required when lifting the engine out of the vehicle.

Plan the operation ahead of time. Arrange for or obtain all of the tools and equipment you'll need prior to beginning the job. Some of the equipment necessary to perform engine removal and installation safely and with relative ease are (in addition to an engine hoist) a heavy duty floor jack, complete sets of wrenches and sockets as described in the front of this manual, wooden blocks and

plenty of rags and cleaning solvent for mopping up spilled oil, coolant and gasoline. If the hoist must be rented, make sure that you arrange for it in advance and perform all of the operations possible without it beforehand. This will save you money and time.

Plan for the vehicle to be out of use for quite a while. A machine shop will be required to perform some of the work which the do-it yourselfer can't accomplish without special equipment. These shops often have a busy schedule, so it would be a good idea to consult them before removing the engine in order to accurately estimate the amount of time required to rebuild or repair components that may need work.

Always be extremely careful when removing and installing the engine. Serious injury can result from careless actions. Plan ahead, take your time and a job of this nature, although major, can be accomplished successfully.

5   Engine - removal and installation

Refer to illustrations 5.5a, 5.5b, 5.5c, 5.11, 5.12, 5.20, 5.24a and 5.24b

Warning: *The air conditioning system is under high pressure! Have a dealer service department or service station discharge the system before disconnecting any system hoses or fittings.*

Removal

1   Refer to Chapter 4 and relieve the fuel system pressure (fuel injected vehicles only), then disconnect the negative cable from the battery.

2   Cover the fenders and cowl and remove the hood (see Chapter 11). Special pads are available to protect the fenders, but an old bedspread or blanket will also work.

3   Remove the air cleaner assembly (see Chapter 4).

4   Drain the cooling system (see Chapter 1).

5   Label the vacuum lines, emissions system hoses, wiring connectors, ground strap

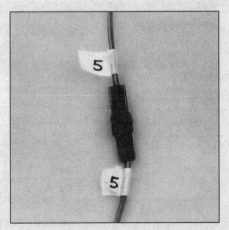

5.5a Label both ends of each wire before unplugging the connector

2D

5.5b Remove the attaching screw and unplug the firewall electrical connector

5.5c Remove the nut on the engine stud and disconnect the ground strap (arrow)

5.11 Set the power steering pump aside with the lines still connected - be sure it's upright so fluid won't spill

and fuel lines, to ensure correct reinstallation **(see illustration)**, then detach them **(see illustrations)**. If there's any possibility of confusion, make a sketch of the engine compartment and clearly label the lines, hoses and wires.

6   Label and detach all coolant hoses from the engine.

7   Remove the cooling fan, shroud and radiator (see Chapter 3).

8   Remove the drivebelt(s) (see Chapter 1).

9   **Warning:** *Gasoline is extremely flammable, so extra precautions must be taken when working on any part of the fuel system. DO NOT smoke or allow open flames or bare light bulbs near the vehicle. Also, don't work in a garage if a natural gas appliance with a pilot light is present. Disconnect the fuel lines running from the engine to the chassis (see Chapter 4). Plug or cap all open fittings/lines.*

10   Disconnect the throttle linkage (and TV linkage/cruise control cable, if equipped) from the engine (see Chapter 4).

11   On power steering equipped vehicles, unbolt the power steering pump (see Chapter 10). Leave the lines/hoses attached **(see illustration)** and make sure the pump is kept

in an upright position in the engine compartment (use wire or rope to restrain it out of the way).

12   On air conditioned vehicles, unbolt the compressor (see Chapter 3) and set it aside. Do not disconnect the hoses **(see illustration)**.

13   Drain the engine oil (see Chapter 1) and remove the oil filter.

14   Remove the starter motor (see Chapter 5).

15   Remove the alternator (see Chapter 5).

16   Unbolt the exhaust system from the engine (see Chapter 4).

17   If you're working on a vehicle with an automatic transmission, refer to Chapter 7 and remove the torque converter-to-driveplate fasteners.

18   Support the transmission with a jack. Position a block of wood between the jack and transmission to prevent damage to the transmission. Special transmission jacks with safety chains are available - use one if possible.

19   Attach an engine sling or a length of chain to the lifting brackets on the engine.

20   Roll the hoist into position and connect

the sling to it **(see illustration)**. Take up the slack in the sling or chain, but don't lift the engine. **Warning:** *DO NOT place any part of your body under the engine when it's supported only by a hoist or other lifting device.*

21   Remove the transmission-to-engine block bolts.

22   Remove the engine mount-to-frame bolts.

23   Recheck to be sure nothing is still connecting the engine to the transmission or vehicle. Disconnect anything still remaining.

24   Raise the engine slightly. Carefully work it forward to separate it from the transmission. If you're working on a vehicle with an automatic transmission, be sure the torque converter stays in the transmission (clamp a pair of vise-grips to the housing to keep the converter from sliding out). If you're working on a vehicle with a manual transmission, the input shaft must be completely disengaged from the clutch. Slowly raise the engine out of the engine compartment **(see illustrations)**. Check carefully to make sure nothing is hanging up.

25   Remove the flywheel/driveplate and mount the engine on an engine stand.

5.12 Unbolt the air conditioning compressor and set it out of the way

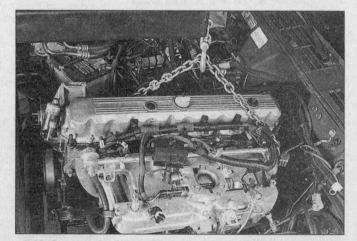

5.20 Connect the lifting sling to the hoist and take up the slack

5.24a Pull the engine forward as far as possible to clear the transmission . . .

5.24 b . . . then lift the engine high enough to clear the body

Installation

26 Check the engine and transmission mounts. If they're worn or damaged, replace them.

27 If you're working on a vehicle with a manual transmission, install the clutch and pressure plate (Chapter 7). Now is a good time to install a new clutch.

28 Carefully lower the engine into the engine compartment - make sure the engine mounts line up.

29 If you're working on a vehicle with an automatic transmission, guide the torque converter into the crankshaft following the procedure outlined in Chapter 7.

30 If you're working on a vehicle with a manual transmission, apply a dab of high-temperature grease to the input shaft and guide it into the crankshaft pilot bearing until the bellhousing is flush with the engine block.

31 Install the transmission-to-engine bolts and tighten them securely. **Caution:** *DO NOT use the bolts to force the transmission and engine together!*

32 Reinstall the remaining components in the reverse order of removal.

33 Add coolant, oil, power steering and transmission fluid as needed.

34 Run the engine and check for leaks and proper operation of all accessories, then install the hood and test drive the vehicle.

35 Have the air conditioning system recharged and leak tested.

6 Engine rebuilding alternatives

The do-it-yourselfer is faced with a number of options when performing an engine overhaul. The decision to replace the engine block, piston/connecting rod assemblies and crankshaft depends on a number of factors, with the number one consideration being the condition of the block. Other considerations are cost, access to machine shop facilities, parts availability, time required to complete

the project and the extent of prior mechanical experience on the part of the do-it-yourselfer.

Some of the rebuilding alternatives include:

Individual parts - If the inspection procedures reveal that the engine block and most engine components are in reusable condition, purchasing individual parts may be the most economical alternative. The block, crankshaft and piston/connecting rod assemblies should all be inspected carefully. Even if the block shows little wear, the cylinder bores should be surface honed.

Short block - A short block consists of an engine block with a crankshaft, camshaft and piston/connecting rod assemblies already installed. All new bearings are incorporated and all clearances will be correct. The existing valve train components, cylinder head(s) and external parts can be bolted to the short block with little or no machine shop work necessary.

Long block - A long block consists of a short block plus an oil pump, oil pan, cylinder head(s), rocker arm cover(s) and valve train components, timing sprockets and chain or gears and timing cover. All components are installed with new bearings, seals and gas-

kets incorporated throughout. The installation of manifolds and external parts is all that's necessary.

Give careful thought to which alternative is best for you and discuss the situation with local automotive machine shops, auto parts dealers and experienced rebuilders before ordering or purchasing replacement parts.

7 Engine overhaul - disassembly sequence

Refer to illustrations 7.3a, 7.3b, 7.3c and 7.5

1 It's much easier to disassemble and work on the engine if it's mounted on a portable engine stand. A stand can often be rented quite cheaply from an equipment rental yard. Before the engine is mounted on a stand, the flywheel/driveplate should be removed from the engine.

2 If a stand isn't available, it's possible to disassemble the engine with it blocked up on the floor. Be extra careful not to tip or drop the engine when working without a stand.

3 If you're going to obtain a rebuilt engine, all external components must come off first **(see illustrations)**, to be transferred to the

7.3a Inline six-cylinder engine - front view

7.3b Inline six cylinder engine - left side view (typical)

7.3c Inline six cylinder engine - right side view

replacement engine, just as they will if you're doing a complete engine overhaul yourself. These include:

Alternator and brackets
Emissions control components
Distributor, spark plug wires and spark
    plugs
Thermostat and housing cover
Water pump
Fuel injection components or carburetor
Intake/exhaust manifolds
Oil filter
Engine mounts
Clutch and flywheel/driveplate
Engine rear plate

Note: *When removing the external components from the engine, pay close attention to details that may be helpful or important during installation. Note the installed position of gaskets, seals, spacers, pins, brackets, washers, bolts and other small items.*

4   If you're obtaining a short block, which consists of the engine block, crankshaft, pistons and connecting rods all assembled, then the cylinder head(s), oil pan and oil pump will have to be removed as well. See Engine rebuilding alternatives for additional information regarding the different possibilities to be considered.

5   If you're planning a complete overhaul, the engine must be disassembled and the internal components **(see illustration)** removed in the following order:

Rocker arm cover(s)
Intake and exhaust manifolds
Rocker arms and pushrods
Cylinder head(s)
Valve lifters
Timing cover
Timing chain and sprockets
Camshaft
Oil pan
Oil pump
Piston/connecting rod assemblies
Crankshaft and main bearings

6   Before beginning the disassembly and overhaul procedures, make sure the following

7.5 Internal components - exploded view (inline six cylinder engine)

1   *Engine oil dipstick and tube*
2   *Oil filter by-pass plug*
3   *Build date code location*
4   *Ring Set*
5   *Piston*
6   *Pin set*
7   *Plug*
8   *Engine block*
9   *Oil channel plug*
10 *Camshaft*
11 *Connecting rod*
12 *Pin*
13 *Camshaft sprocket*
14 *Keys*
15 *Washer*
16 *Timing chain*
17 *Oil shedder (slinger)*
18 *Crankshaft sprocket*
19 *Crankshaft*
20 *Connecting rod bearing*
21 *Connecting rod and bearing cap*
22 *Main bearings*
23 *Vibration damper pulley*
24 *Washer*
25 *Vibration damper*
26 *Seal*
27 *Timing chain cover*
28 *Gasket*
29 *Main bearing cap (rear)*

30 *Main bearing cap seal kit (rear)*
31 *Pilot bushing (with manual transmission)*
32 *Bushing oil wick (with manual transmission*
33 *Flywheel and ring gear (with manual transmission)*
34 *Bearing set*
35 *Dowel*
36 *Plug*

8.2 A small plastic bag, with an appropriate label, can be used to store the valve train components so they can be kept together and reinstalled in the correct guide

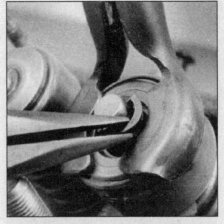

8.3 Use a valve spring compressor to compress the spring, them remove the keepers from the valve stem

8.4 If the valve won't pull through the guide, deburr the edge of the stem end and the area around the top of the keeper groove with a file

items are available. Also, refer to Engine overhaul - reassembly sequence for a list of tools and materials needed for engine reassembly.

> Common hand tools
> Small cardboard boxes or plastic bags for storing parts
> Gasket scraper
> Ridge reamer
> Vibration damper puller
> Micrometers
> Telescoping gauges
> Dial indicator set
> Valve spring compressor
> Cylinder surfacing hone
> Piston ring groove cleaning tool
> Electric drill motor
> Tap and die set
> Wire brushes
> Oil gallery brushes
> Cleaning solvent

8 Cylinder head - disassembly

Refer to illustrations 8.2, 8.3 and 8.4
Note: *New and rebuilt cylinder heads are commonly available for most engines at dealerships and auto parts stores. Due to the fact that some specialized tools are necessary for the disassembly and inspection procedures, and replacement parts may not be readily available, it may be more practical and economical for the home mechanic to purchase replacement head(s) rather than taking the time to disassemble, inspect and recondition the original(s).*
1   Cylinder head disassembly involves removal of the intake and exhaust valves and related components. If they're still in place, remove the rocker arm bolts or nuts, pivot and rocker arms from the cylinder head studs. Label the parts or store them separately so they can be reinstalled in their original locations.
2   Before the valves are removed, arrange

to label and store them, along with their related components, so they can be kept separate and reinstalled in the same valve guides they were removed from **(see illustration)**.
3   Compress the springs on the first valve with a spring compressor and remove the keepers **(see illustration)**. Carefully release the valve spring compressor and remove the retainer, the spring and the spring seat (if used).
4   Pull the valve out of the head, then remove the oil seal from the guide. If the valve binds in the guide (won't pull through), push it back into the head and deburr the area around the keeper groove with a fine file or whetstone **(see illustration)**.
5   Repeat the procedure for the remaining valves. Remember to keep all the parts for each valve together so they can be reinstalled in the same locations.
6   Once the valves and related components have been removed and stored in an organized manner, the head should be thoroughly cleaned and inspected. If a complete engine overhaul is being done, finish the engine disassembly procedures before beginning the cylinder head cleaning and inspection process.

9 Cylinder head - cleaning and inspection

Refer to illustrations 9.12, 9.14, 9.15a, 9.15b, 9.16, 9.17, 9.18 and 9.19
1   Thorough cleaning of the cylinder head(s) and related valve train components, followed by a detailed inspection, will enable you to decide how much valve service work must be done during the engine overhaul.
Note: *If the engine was severely overheated, the cylinder head is probably warped (see Step 12).*

Cleaning

2   Scrape all traces of old gasket material

and sealing compound off the head gasket, intake manifold and exhaust manifold sealing surfaces. Be very careful not to gouge the cylinder head. Special gasket removal solvents that soften gaskets and make removal much easier are available at auto parts stores.
3   Remove all built up scale from the coolant passages.
4   Run a stiff wire brush through the various holes to remove deposits that may have formed in them.
5   Run an appropriate size tap into each of the threaded holes to remove corrosion and thread sealant that may be present. If compressed air is available, use it to clear the holes of debris produced by this operation.
Warning: *Wear eye protection when using compressed air!*
6   Clean the rocker arm pivot bolt or stud threads with a wire brush.
7   Clean the cylinder head with solvent and dry it thoroughly. Compressed air will speed the drying process and ensure that all holes and recessed areas are clean. **Note:** *Decarbonizing chemicals are available and may prove very useful when cleaning cylinder heads and valve train components. They are very caustic and should be used with caution. Be sure to follow the instructions on the container.*
8   Clean the rocker arms, pivot balls or fulcrums, nuts or bolts and pushrods with solvent and dry them thoroughly (don't mix them up during the cleaning process). Compressed air will speed the drying process and can be used to clean out the oil passages.
9   Clean all the valve springs, spring seats, keepers and retainers (or rotators) with solvent and dry them thoroughly. Do the components from one valve at a time to avoid mixing up the parts.
10   Scrape off any heavy deposits that may have formed on the valves, then use a motorized wire brush to remove deposits from the valve heads and stems. Again, make sure the valves don't get mixed up.

2D

9.12 Check the cylinder head gasket surface for warpage by trying to slip a feeler gauge under the straightedge (see the Specifications for the maximum warpage allowed and use a feeler gauge of that thickness)

9.14 A dial indicator can be used to determine the valve stem-to-guide clearance (move the valve stem as indicated by the arrows)

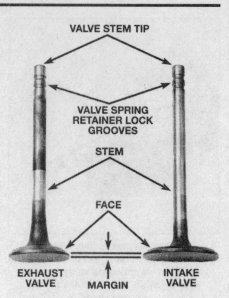

9.15a Check the valve for wear at the points shown here

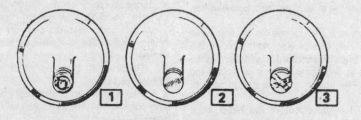

9.15b Valve stem tip wear patterns

1   *Proper tip pattern (rotator functioning properly)*
2   *No rotation pattern (replace rotator and check rotation)*
3   *Partial rotation pattern (replace rotator and check rotation)*

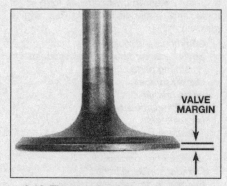

9.16 The margin width on each valve must be as specified (if no margin exists, the valve cannot be reused)

Inspection

Note: *Be sure to perform all of the following inspection procedures before concluding that machine shop work is required. Make a list of the items that need attention.*

Cylinder head

11   Inspect the head very carefully for cracks, evidence of coolant leakage and other damage. If cracks are found, check with an automotive machine shop concerning repair. If repair isn't possible, a new cylinder head should be obtained.
12   Using a straightedge and feeler gauge, check the head gasket mating surface for warpage **(see illustration)**. If the warpage exceeds the specified limit, it can be resurfaced at an automotive machine shop. **Note:** *If the V6 engine heads are resurfaced, the intake manifold flanges will also require machining.*
13   Examine the valve seats in each of the combustion chambers. If they're pitted, cracked or burned, the head will require valve service that's beyond the scope of the home mechanic.
14   Check the valve stem-to-guide clear-

ance by measuring the lateral movement of the valve stem with a dial indicator attached securely to the head **(see illustration)**. The valve must be in the guide and approximately 1/16-inch off the seat. The total valve stem movement indicated by the gauge needle must be divided by two to obtain the actual clearance. After this is done, if there's still some doubt regarding the condition of the valve guides they should be checked by an automotive machine shop (the cost should be minimal).

Valves

15   Carefully inspect each valve for uneven wear, deformation, cracks, pits and burned areas **(see illustrations)**. Check the valve stem for scuffing and galling and the neck for cracks. Rotate the valve and check for any obvious indication that it's bent. Look for pits and excessive wear on the end of the stem. The presence of any of these conditions indicates the need for valve service by an automotive machine shop.
16   Measure the margin width on each valve **(see illustration)**. Any valve with a margin narrower than specified will have to be replaced with a new one.

Valve components

17   Check each valve spring for wear (on the ends) and pits. Measure the free length and compare it to the Specifications **(see illustration)**. Any springs that are shorter than specified have sagged and should not be reused. The tension of all springs should be checked with a special fixture before deciding that they're suitable for use in a

9.17 Measure the free length of each valve spring with a dial or vernier caliper

9.18 Check each valve spring for squareness

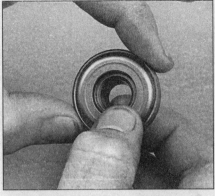

9.19 The exhaust valve rotators can be checked by turning the inner and outer sections in opposite directions to check for smooth movement and excessive play

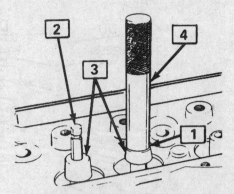

11.3 Make sure the new valve stem seals are seated against the tops of the valve guides

1   Valve seated in tool
2   End of valve stem - be sure to deburr this area before installing the seal
3   Seal
4   Valve seal installation tool (if you don't have this tool, a deep socket also will work)

2D

rebuilt engine (take the springs to an automotive machine shop for this check).

18   Stand each spring on a flat surface and check it for squareness **(see illustration)**. If any of the springs are distorted or sagged, replace all of them with new parts.

19   Check the spring retainers (or rotators) and keepers for obvious wear and cracks. Any questionable parts should be replaced with new ones, as extensive damage will occur if they fail during engine operation. Make sure the rotators operate smoothly with no binding or excessive play **(see illustration)**.

Rocker arm components

20   Check the rocker arm faces (the areas that contact the pushrod ends and valve stems) for pits, wear, galling, score marks and rough spots. Check the rocker arm pivot contact areas and pivot balls or fulcrums as well. Look for cracks in each rocker arm and nut or bolt.

21   Inspect the pushrod ends for scuffing and excessive wear. Roll each pushrod on a flat surface, like a piece of plate glass, to determine if it's bent.

22   Check the rocker arm bolt holes or studs in the cylinder heads for damaged threads and secure installation.

23   Any damaged or excessively worn parts must be replaced with new ones.

24   If the inspection process indicates that the valve components are in generally poor condition and worn beyond the limits specified, which is usually the case in an engine that's being overhauled, reassemble the valves in the cylinder head and refer to Section 10 for valve servicing recommendations.

10 Valves - servicing

1   Because of the complex nature of the job and the special tools and equipment needed, servicing of the valves, the valve seats and the valve guides, commonly known as a valve job, should be done by a professional.

2   The home mechanic can remove and

disassemble the head(s), do the initial cleaning and inspection, then reassemble and deliver it (or them) to a dealer service department or an automotive machine shop for the actual service work. Doing the inspection will enable you to see what condition the head and valve train components are in and will ensure that you know what work and new parts are required when dealing with an automotive machine shop.

3   The dealer service department, or automotive machine shop, will remove the valves and springs, recondition or replace the valves and valve seats, recondition the valve guides, check and replace the valve springs, spring retainers or rotators and keepers (as necessary), replace the valve seals with new ones, reassemble the valve components and make sure the installed spring height is correct. The cylinder head gasket surface will also be resurfaced if it's warped.

4   After the valve job has been performed by a professional, the head will be in like-new condition. When the head is returned, be sure to clean it again before installation on the engine to remove any metal particles and abrasive grit that may still be present from the valve service or head resurfacing operations. Use compressed air, if available, to blow out all the oil holes and passages.

11 Cylinder head - reassembly

Refer to illustrations 11.3, 11.5a, 11.5b, 11.5c, 11.6a, 11.6b, 11.8 and 11.9

1   Regardless of whether or not a head was sent to an automotive repair shop for valve servicing, make sure it's clean before beginning reassembly.

2   If a head was sent out for valve servicing, the valves and related components will already be in place. Begin the reassembly procedure with Step 8.

3   Install new seals on each of the intake valve guides. Using a hammer and a deep socket or seal installation tool, gently tap each seal into place until it's completely seated on the guide **(see illustration)**. Don't

twist or cock the seals during installation or they won't seal properly on the valve stems. The umbrella-type seals (if used) are installed over the valves after the valves are in place.

4   Beginning at one end of the head, lubricate and install the first valve. Apply moly-base grease or clean engine oil to the valve stem.

5   Drop the spring seat or shim(s) over the valve guide and set the valve springs, shield and retainer (or rotator) in place **(see illustrations)**.

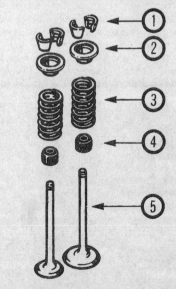

11.5a Valves and related components for inline four and six-cylinder engines - exploded view

1 Keepers	4 Umbrella-
2 Retainers	type oil seals
3 Springs	5 Valves

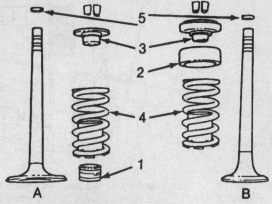
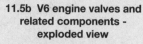

11.5b V6 engine valves and related components - exploded view

A   Intake valve
B   Exhaust valve
1   Umbrella seal (intake only)
2   Oil shedder (exhaust only)
3   Rotator/retainer
4   Spring
5   O-ring seal

11.5c Later model four-cylinder engines have progressively wound valve springs - install them with the closely wound coils toward the cylinder head

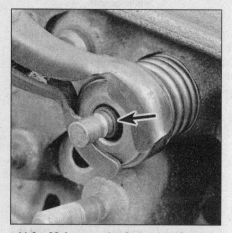

11.6a Make sure the O-ring seal (arrow) under the retainer is seated in the groove and not twisted before installing the keepers (V6 engine only)

11.6b Apply a small dab of grease to each keeper as shown here before installation - it will hold them in place on the valve stem as the spring is released

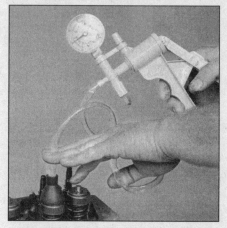

11.8 A special adapter and vacuum pump are required to check the O-ring valve stem seals for leaks

6   Compress the springs with a valve spring compressor and carefully install the O-ring oil seal (V6 only) in the lower groove of the valve stem. Make sure the seal isn't twisted - it must lie perfectly flat in the groove **(see illustration)**. Position the keepers in the upper groove, then slowly release the compressor and make sure the keepers seat properly. Apply a small dab of grease to each keeper to hold it in place if necessary **(see illustration)**.

7   Repeat the procedure for the remaining valves. Be sure to return the components to their original locations - don't mix them up!

8   On V6 engines, when all the valves are in place in both heads, the valve stem O-ring seals must be checked to make sure they don't leak. This procedure requires a vacuum pump and special adapter, so it may be a good idea to have it done by a dealer service department, repair shop or automotive machine shop. The adapter is positioned on each valve retainer or rotator and vacuum is applied with the hand pump **(see illustration)**. If the vacuum can't be maintained, the seal is leaking and must be checked/replaced before the head is installed on the engine.

9   Check the installed valve spring height with a ruler graduated in 1/32-inch increments or a dial caliper. If the head was sent out for service work, the installed height should be correct (but don't automatically assume that it is). The measurement is taken from the top of each spring seat or shim(s) to the top of the oil shield (or the bottom of the retainer/rotator, the two points are the same) **(see illustration)**. If the height is greater than specified, shims can be added under the springs to correct it. **Caution:** *Don't, under any circumstances, shim the springs to the point where the installed height is less than specified.*

10   Apply moly-base grease to the rocker arm faces and the pivots, then install the rocker arms and pivots on the cylinder head. Thread the bolts/nuts on three or four turns only (when the heads are installed on a V6 engine, the nuts will be tightened following a specific procedure).

12 Pistons/connecting rods - removal

Refer to illustrations 12.1, 12.3 and 12.6
Note: *Prior to removing the piston/connecting rod assemblies, remove the cylinder head(s), the oil pan and the oil pump by referring to the appropriate Sections in Chapter 2.*

1   Use your fingernail to feel if a ridge has formed at the upper limit of ring travel (about 1/4-inch down from the top of each cylinder).

11.9 Be sure to check the valve spring installed height (the distance from the top of the seat/shims to the top of the shield or the bottom of the retainer)

12.1 A ridge reamer is required to remove the ridge from the top of each cylinder - do this before removing the pistons!

12.3 Check the connecting rod side clearance with a feeler gauge as shown

12.6 To prevent damage to the crankshaft journals and cylinder walls, slip sections of hose over the rod bolts before removing the pistons

If carbon deposits or cylinder wear have produced ridges, they must be completely removed with a special tool **(see illustration)**. Follow the manufacturer's instructions provided with the tool. Failure to remove the ridges before attempting to remove the piston/connecting rod assemblies may result in piston breakage.

2   After the cylinder ridges have been removed, turn the engine upside-down so the crankshaft is facing up.

3   Before the connecting rods are removed, check the end play with feeler gauges. Slide them between the first connecting rod and the crankshaft throw until the play is removed **(see illustration)**. The end play is equal to the thickness of the feeler gauge(s). If the end play exceeds the service limit, new connecting rods will be required. If new rods (or a new crankshaft) are installed, the end play may fall under the specified minimum (if it does, the rods will have to be machined to restore it - consult an automotive machine shop for advice if necessary). Repeat the procedure for the remaining connecting rods.

4   Check the connecting rods and caps for identification marks. If they aren't plainly marked, use a small center punch to make the appropriate number of indentations on each rod and cap (1, 2, 3, etc., depending on the engine type and cylinder they're associated with).

5   Loosen each of the connecting rod cap nuts 1/2-turn at a time until they can be removed by hand. Remove the number one connecting rod cap and bearing insert. Don't drop the bearing insert out of the cap.

6   Slip a short length of plastic or rubber hose over each connecting rod cap bolt to protect the crankshaft journal and cylinder wall as the piston is removed **(see illustration)**.

7   Remove the bearing insert and push the connecting rod/piston assembly out through the top of the engine. Use a wooden hammer handle to push on the upper bearing surface in the connecting rod. If resistance is felt,

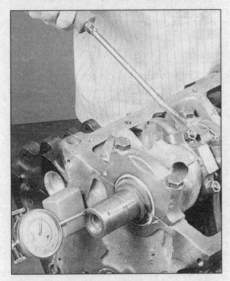

13.1 Checking crankshaft end play with a dial indicator

double-check to make sure that all of the ridge was removed from the cylinder.

8   Repeat the procedure for the remaining cylinders.

9   After removal, reassemble the connecting rod caps and bearing inserts in their respective connecting rods and install the cap nuts finger tight. Leaving the old bearing inserts in place until reassembly will help prevent the connecting rod bearing surfaces from being accidentally nicked or gouged.

10   Don't separate the pistons from the connecting rods (see Section 17 for additional information).

13   Crankshaft - removal

Refer to illustrations 13.1, 13.3, 13.4a, 13.4b and 13.4c
Note: *The crankshaft can be removed only after the engine has been removed from the vehicle. It's assumed that the flywheel or driveplate, vibration damper, timing chain, oil*

13.3 Checking crankshaft end play with a feeler gauge

pan, oil pump and piston/connecting rod assemblies have already been removed.

1   Before the crankshaft is removed, check the end play. Mount a dial indicator with the stem in line with the crankshaft and just touching one of the crank throws **(see illustration)**.

2   Push the crankshaft all the way to the rear and zero the dial indicator. Next, pry the crankshaft to the front as far as possible and check the reading on the dial indicator. The distance that it moves is the end play. If it's greater than specified, check the crankshaft thrust surfaces for wear. If no wear is evident, new main bearings should correct the end play.

3   If a dial indicator isn't available, feeler gauges can be used. Gently pry or push the crankshaft all the way to the front of the engine. Slip feeler gauges between the crankshaft and the front face of the thrust main bearing to determine the clearance **(see illustration)**.

4   Check the main bearing caps to see if they're marked to indicate their locations. They should be numbered consecutively from

13.4a Use a center punch or number stamping dies to mark the main bearing caps to ensure that they're reinstalled in their original locations on the block (make the punch marks near one of the bolt heads)

13.4b Mark the caps in order from the front of the engine to the rear (one mark for the front cap, two for the second one and so on) - the rear cap doesn't have to be marked since it cannot be installed in any other location

13.4c The arrow on the main bearing cap indicates the front of the engine

the front of the engine to the rear. If they aren't, mark them with number stamping dies or a center punch **(see illustrations)**. Main bearing caps generally have a cast-in arrow, which points to the front of the engine **(see illustration)**. Loosen the main bearing cap bolts 1/4-turn at a time each, until they can be removed by hand. Note if any stud bolts are used and make sure they're returned to their original locations when the crankshaft is reinstalled.

5   Gently tap the caps with a soft-face hammer, then separate them from the engine block. If necessary, use the bolts as levers to remove the caps. Try not to drop the bearing inserts if they come out with the caps.

6   Carefully lift the crankshaft out of the engine. It may be a good idea to have an assistant available, since the crankshaft is quite heavy. With the bearing inserts in place in the engine block and main bearing caps, return the caps to their respective locations on the engine block and tighten the bolts finger tight.

14   Engine block - cleaning

Refer to illustrations 14.1a, 14.1b, 14.8 and 14.10

1   Remove the core plugs by carefully driving one side of the core plug into the block with a blunt punch **(see illustration)**. Grasp the opposite edge of the core plug with locking pliers and lever the plug out of the block. On V6 engines, also remove the rear camshaft cover plate **(see illustration)**.

2   Using a gasket scraper, remove all traces of gasket material from the engine block. Be very careful not to nick or gouge the gasket sealing surfaces.

3   Remove the main bearing caps and separate the bearing inserts from the caps and the engine block. Tag the bearings, indicating which cylinder they were removed from and whether they were in the cap or the block, then set them aside.

4   Remove all of the threaded oil gallery

plugs from the block. The plugs are usually very tight - they may have to be drilled out and the holes retapped. Use new plugs when the engine is reassembled.

5   If the engine is extremely dirty it should be taken to an automotive machine shop to be steam cleaned or hot tanked.

6   After the block is returned, clean all oil holes and oil galleries one more time. Brushes specifically designed for this purpose are available at most auto parts stores. Flush the passages with warm water until the water runs clear, dry the block thoroughly and wipe all machined surfaces with a light, rust preventive oil. If you have access to compressed air, use it to speed the drying process and to blow out all the oil holes and galleries. **Warning:** *Wear eye protection when using compressed air!*

7   If the block isn't extremely dirty or sludged up, you can do an adequate cleaning job with hot soapy water and a stiff brush. Take plenty of time and do a thorough job. Regardless of the cleaning method used, be sure to clean all oil holes and galleries very thoroughly, dry the block completely and coat all machined surfaces with light oil.

14.1a The core plugs should be removed carefully by tapping one side in, then pull out the opposite side with locking pliers

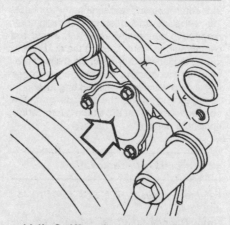

14.1b On V6 engines, remove the rear camshaft cover plate (arrow)

14.8 All bolt holes in the block - particularly the main bearing cap and head bolt holes - should be cleaned and restored with a tap (be sure to remove debris from the holes after this is done)

14.10 A large socket on an extension can be used to drive the new soft plugs into the bores

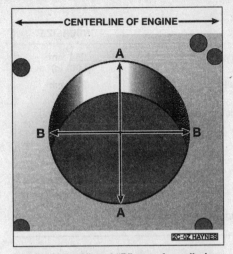

15.4a Use "A" and "B" axes for cylinder measurements as described in text

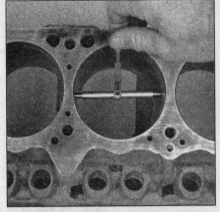

15.4b The ability to "feel" when the telescoping gauge is at the correct point will be developed over time, so work slowly and repeat the check until you are satisfied that the bore measurement is accurate

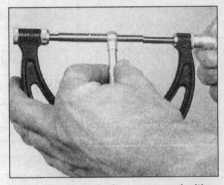

15.4c The gauge is then measured with a micrometer to determine the bore size

2D

8   The threaded holes in the block must be clean to ensure accurate torque readings during reassembly. Run the proper size tap into each of the holes to remove rust, corrosion, thread sealant or sludge and restore damaged threads **(see illustration)**. If possible, use compressed air to clear the holes of debris produced by this operation. Now is a good time to clean the threads o the head bolts and the main bearing cap bolts as well.

9   Reinstall the main bearing caps and tighten the bolts finger tight.

10   After coating the sealing surfaces of the new core plugs with Permatex no. 2 sealant, install them in the engine block **(see illustration)**. Make sure they're driven in straight and seated properly or leakage could result. Special tools are available for this purpose, but a large socket, with an outside diameter that will just slip into the core plug, a 1/2-inch drive extension and a hammer will work just as well.

11   Apply non-hardening sealant (such as Permatex no. 2 or Teflon pipe sealant) to the new oil gallery plugs and thread them into the holes in the block. Make sure they're tightened securely.

12   If the engine isn't going to be reassembled right away, cover it with a large plastic trash bag to keep it clean.

15   Engine block - inspection

Refer to illustrations 15.4a, 15.4b, 15.4c, 15.14a, 15.14b, 15.14c and 15.14d

1   Before the block is inspected, it should be cleaned as described in Section 14.

2   Visually check the block for cracks, rust and corrosion. Look for stripped threads in the threaded holes. It's also a good idea to have the block checked for hidden cracks by an automotive machine shop that has the special equipment to do this type of work. If defects are found, have the block repaired, if possible, or replaced.

3   Check the cylinder bores for scuffing and scoring.

4   Measure the diameter of each cylinder at the top (just under the ridge area), center

and bottom of the cylinder bore, parallel to the crankshaft axis **(see illustrations)**.

5   Next, measure each cylinder's diameter at the same three locations across the crankshaft axis. Compare the results to the specifications.

6   If the required precision measuring tools aren't available, the piston to-cylinder clearances can be obtained, though not quite as accurately, using feeler gauge stock. Feeler gauge stock comes in 12-inch lengths and various thicknesses and is generally available at auto parts stores.

7   To check the clearance, select a feeler gauge and slip it into the cylinder along with the matching piston. The piston must be positioned exactly as it normally would be. The feeler gauge must be between the piston and cylinder on one of the thrust faces (90~ to the piston pin bore).

8   The piston should slip through the cylinder (with the feeler gauge in place) with moderate pressure.

9   If it falls through or slides through easily, the clearance is excessive and a new piston will be required. If the piston binds at the lower end of the cylinder and is loose toward the top, the cylinder is tapered. If tight spots are encountered as the piston/feeler gauge is rotated in the cylinder, the cylinder is out-of-round.

10   Repeat the procedure for the remaining pistons and cylinders.

11   If the cylinder walls are badly scuffed or scored, or if they're out of-round or tapered beyond the limits given in the specifications, have the engine block rebored and honed at an automotive machine shop. If a rebore is done, oversize pistons and rings will be required.

12   If the cylinders are in reasonably good condition and not worn to the outside of the limits, and if the piston-to-cylinder clearances can be maintained properly, then they don't have to be rebored. Honing is all that's necessary (see Section 16).

13   Some four cylinder and inline six-cylin-

der engines are produced or rebuilt with oversized and undersized components, such as oversized cylinder bores, undersized main bearings and connecting rod journals, or oversized camshaft bearing bores.

14   Engines with such components are identified by a stamping and color coding system **(see illustrations)**.

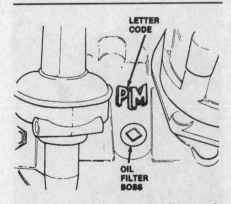

15.14a The engine component letter code for four-cylinder engines is located on the oil filter boss between the fuel pump and distributor

CODE	COMPONENT	UNDERSIZE
P	One or more connecting rod bearing journals	.254 mm (.010 in)
M	All crankshaft main bearing journals	.254 mm (.010 in)
PM	All crankshaft main bearing journals and one or more connecting rod journals	.254 mm (.010 in)
B	All cylinder bores	.254 mm (.010 in)
C	All camshaft bearing bores	.254 mm (.010 in)

15.14b Four-cylinder engine component codes

15.14c Inline six-cylinder component codes, (A) if any, are stamped on the boss (B) between the ignition coil (C) and distributor (D)

15.14d Inline six-cylinder engine component codes

Code Letter	Definition	
B	All cylinder bores	0.010-inch (0.254 mm) oversize
M	All crankshaft main bearing journals	0.010-inch (0.254 mm) undersize
P	All connecting rod bearing journals	0.010-inch (0.254 mm) undersize
C	All camshaft bearing bores	0.010-inch (0.254 mm) oversize

EXAMPLE: The code letters PM mean that the crankshaft main bearing journals and connecting rod journals are 0.010-inch undersized.

16 Cylinder honing

Refer to illustrations 16.3a and 16.3b

1   Prior to engine reassembly, the cylinder bores must be honed so the new piston rings will seat correctly and provide the best possible combustion chamber seal. **Note:** *If you don't have the tools or don't want to tackle the honing operation, most automotive machine shops will do it for a reasonable fee.*
2   Before honing the cylinders, install the main bearing caps and tighten the bolts to the specified torque.
3   Two types of cylinder hones are commonly available - the flex hone or "bottle brush" type and the more traditional surfacing hone with spring-loaded stones. Both will do the job, but for the less experienced mechanic the "bottle brush" hone will proba-

16.3a A "bottle brush" hone will produce better results if you have never done cylinder honing before

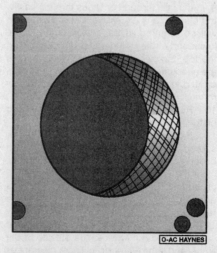

16.3b The cylinder hone should leave a smooth, crosshatch pattern with the lines intersecting at approximately a 60-degree angle

bly be easier to use. You'll also need some kerosene or honing oil, rags and an electric drill motor. Proceed as follows:

a)   *Mount the hone in the drill motor, compress the stones (if applicable) and slip it into the first cylinder* **(see illustration)**. *Be sure to wear safety goggles or a face shield!*

b)   *Lubricate the cylinder with plenty of honing oil or kerosene, turn on the drill and move the hone up-and-down in the cylinder at a pace that will produce a fine crosshatch pattern on the cylinder walls. Ideally, the crosshatch lines should intersect at approximately a 60° angle* **(see illustration)**. *Be sure to use plenty of lubricant and don't take off any more material than is absolutely necessary to produce the desired finish.* **Note:** *Piston ring manufacturers may specify a smaller crosshatch angle than the traditional 60° - read and follow any instructions included with the new rings.*

c)   *Don't withdraw the hone from the cylinder while it's running. Instead, shut off the drill and continue moving the hone up-and-down in the cylinder until it comes to a complete stop, then compress the stones and withdraw the hone. If you're using a "bottle brush" type hone, stop the drill motor, then turn the chuck in the normal direction of rotation while withdrawing the hone from the cylinder.*

d)   *Wipe the oil out of the cylinder and repeat the procedure for the remaining cylinders.*

4   After the honing job is complete, chamfer the top edges of the cylinder bores with a small file so the rings won't catch when the pistons are installed. Be very careful not to nick the cylinder walls with the end of the file.
5   The entire engine block must be washed again very thoroughly with warm, soapy water to remove all traces of the abrasive grit produced during the honing operation. **Note:** *The bores can be considered clean when a lint-free white cloth - dampened with clean engine oil- used to wipe them out doesn't pick up any more honing residue, which will*

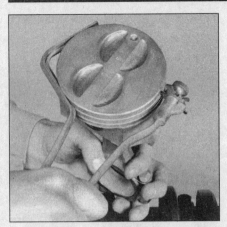

17.4a The piston ring grooves can be cleaned with a special tool, as shown here . . .

17.4b . . . or a section of broken ring

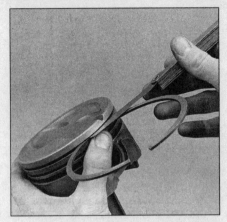

17.10 Check the ring side clearance with a feeler gauge at several points around the groove

17.11a Measure the piston diameter at a 90° angle to the piston pin and in line with it

2.8 LITER V-6 SERVICE PISTONS			
TYPE	**CODE**	**SIZE**	
STD.	S4	89.001-89.014 mm	3.5039-3.5044 in
	S5	89.014-89.027 mm	3.5044-3.5049 in
HI LIMIT	S6	89.027-89.040 mm	3.5049-3.5055 in
	S7	89.040-89.053 mm	3.5055-3.5060 in
.50 O.S. 0.020 In	1	89.501-89.514 mm	3.5236-3.5241 in
	2	89.514-89.527 mm	3.5241-3.5246 in
	3	89.527-89.540 mm	3.5246-3.5251 in
	4	89.540-89.553 mm	3.5251-3.5257 in
1.0 O.S. 0.040 In	1	90.001-90.014 mm	3.5435-3.5438 in
	2	90.014-90.027 mm	3.5438-3.5443 in
	3	90.027-90.040 mm	3.5443-3.5448 in
	4	90.040-90.053 mm	3.5448-3.5453 in

17.11b Hi-limit pistons for V6 engines are available to compensate for slight cylinder bore wear

2D

show up as gray areas on the cloth. Be sure to run a brush through all oil holes and galleries and flush them with running water.

6   After rinsing, dry the block and apply a coat of light rust preventive oil to all machined surfaces. Wrap the block in a plastic trash bag to keep it clean and set it aside until reassembly.

17 Pistons/connecting rods - inspection

Refer to illustrations 17.4a, 17.4b, 17.10, 17.11a and 17.11b

1   Before the inspection process can be carried out, the piston/connecting rod assemblies must be cleaned and the original piston rings removed from the pistons. **Note:** *Always use new piston rings when the engine is reassembled.*

2   Using a piston ring installation tool, carefully remove the rings from the pistons. Be careful not to nick or gouge the pistons in the process.

3   Scrape all traces of carbon from the top of the piston. A handheld wire brush or a piece of fine emery cloth can be used once

the majority of the deposits have been scraped away. Do not, under any circumstances, use a wire brush mounted in a drill motor to remove deposits from the pistons. The piston material is soft and may be eroded away by the wire brush.

4   Use a piston ring groove cleaning tool to remove carbon deposits from the ring grooves. If a tool isn't available, a piece broken off the old ring will do the job. Be very careful to remove only the carbon deposits - don't remove any metal and do not nick or scratch the sides of the ring grooves **(see illustrations)**.

5   Once the deposits have been removed, clean the piston/rod assemblies with solvent and dry them with compressed air (if available). Make sure the oil return holes in the back sides of the ring grooves are clear.

6   If the pistons and cylinder walls aren't damaged or worn excessively, and if the engine block is not rebored, new pistons won't be necessary. Normal piston wear appears as even vertical wear on the piston thrust surfaces and slight looseness of the top ring in its groove. New piston rings, however, should always be used when an engine is rebuilt.

7   Carefully inspect each piston for cracks around the skirt, at the pin bosses and at the ring lands.

8   Look for scoring and scuffing on the thrust faces of the skirt, holes in the piston

crown and burned areas at the edge of the crown. If the skirt is scored or scuffed, the engine may have been suffering from overheating and/or abnormal combustion, which caused excessively high operating temperatures. The cooling and lubrication systems should be checked thoroughly. A hole in the piston crown is an indication that abnormal combustion (preignition) was occurring. Burned areas at the edge of the piston crown are usually evidence of spark knock (detonation). If any of the above problems exist, the causes must be corrected or the damage will occur again. The causes may include intake air leaks, incorrect fuel/air mixture, incorrect ignition timing and EGR system malfunctions.

9   Corrosion of the piston, in the form of small pits, indicates that coolant is leaking into the combustion chamber and/or the crankcase. Again, the cause must be corrected or the problem may persist in the rebuilt engine.

10   Measure the piston ring side clearance by laying a new piston ring in each ring groove and slipping a feeler gauge in beside it **(see illustration)**. Check the clearance at three or four locations around each groove. Be sure to use the correct ring for each groove - they are different. If the side clearance is greater than specified, new pistons will have to be used.

11   Check the piston-to-bore clearance by measuring the bore (see Section 15) and the

18.6 Measure the diameter of each crankshaft journal at several points to detect taper and out-of-round conditions

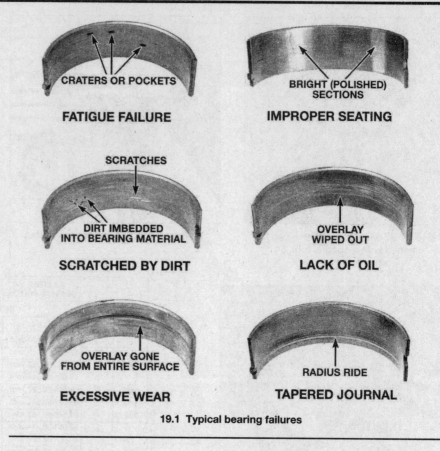

19.1 Typical bearing failures

piston diameter. Make sure the pistons and bores are correctly matched. Measure the piston across the skirt, at a 90° angle to and in line with the piston pin **(see illustration)**. Subtract the piston diameter from the bore diameter to obtain the clearance. If it's greater than specified, the block will have to be rebored and new pistons and rings installed **(see illustration)**. **Note:** *On 1995 and later four-cylinder engines, an accurate measurement of the outer diameter of the piston skirt cannot be achieved, since these pistons have a coating material applied to them after the final piston machining process. On these engines, measure the cylinder bore ID, then refer to this Chapter's Specifications to select the correct-size piston. Installing new pistons with this coating will require slightly more pressure than non-coated pistons. The bonding coating on the piston will give the appearance of a line-to-line fit with the cylinder bore.*

12   Check the piston-to-rod clearance by twisting the piston and rod in opposite directions. Any noticeable play indicates excessive wear, which must be corrected. The piston/connecting rod assemblies should be taken to an automotive machine shop to have the pistons and rods resized and new pins installed.

13   If the pistons must be removed from the connecting rods for any reason, they should be taken to an automotive machine shop. While they are there have the connecting rods checked for bend and twist, since automotive machine shops have special equipment for this purpose. **Note:** *Unless new pistons and/or connecting rods must be installed, do not disassemble the pistons and connecting rods.*

14   Check the connecting rods for cracks and other damage. Temporarily remove the rod caps, lift out the old bearing inserts, wipe the rod and cap bearing surfaces clean and inspect them for nicks, gouges and scratches. After checking the rods, replace the old bearings, slip the caps into place and tighten the nuts finger tight. **Note:** *If the engine is being rebuilt because of a connecting rod knock, be sure to install new rods.*

18 Crankshaft - inspection

Refer to illustration 18.6

1   Clean the crankshaft with solvent and dry it with compressed air (if available). Be sure to clean the oil holes with a stiff brush and flush them with solvent.

2   Check the main and connecting rod bearing journals for uneven wear, scoring, pits and cracks.

3   Rub a penny across each journal several times. If a journal picks up copper from the penny, it's too rough and must be reground.

4   Remove all burrs from the crankshaft oil holes with a stone, file or scraper.

5   Check the rest of the crankshaft for cracks and other damage. It should be magnafluxed to reveal hidden cracks - an automotive machine shop will handle the procedure.

6   Using a micrometer, measure the diameter of the main and connecting rod journals and compare the results to the specifications **(see illustration)**. By measuring the diameter at a number of points around each journal's circumference, you'll be able to determine whether or not the journal is out-of-round. Take the measurement at each end of the journal, near the crank throws, to determine if the journal is tapered.

7   If the crankshaft journals are damaged, tapered, out-of-round or worn beyond the limits given in the specifications, have the crankshaft reground by an automotive machine shop. Be sure to use the correct size bearing

inserts if the crankshaft is reconditioned.

8   Check the oil seal journals at each end of the crankshaft for wear and damage. If the seal has worn a groove in the journal, or if it's nicked or scratched, the new seal may leak when the engine is reassembled. In some cases, an automotive machine shop may be able to repair the journal by pressing on a thin sleeve. If repair isn't feasible, a new or different crankshaft should be installed.

9   Refer to Section 19 and examine the main and rod bearing inserts.

19 Main and connecting rod bearings - inspection

Inspection

Refer to illustration 19.1

1   Even though the main and connecting rod bearings should be replaced with new ones during the engine overhaul, the old bearings should be retained for close examination, as they may reveal valuable information about the condition of the engine **(see illustration)**.

2   Bearing failure occurs because of lack of lubrication, the presence of dirt or other foreign particles, overloading the engine and corrosion. Regardless of the cause of bearing failure, it must be corrected before the engine is reassembled to prevent it from happening again.

3   When examining the bearings, remove

Crankshaft Main Bearing Journal Color Code and Diameter	Corresponding Connecting Rod Bearing Insert Color Code	
	Upper Insert Size	Lower Insert Size
Yellow—53.2257–53.2079 mm 2.0955–2.0948 in. Standard	Yellow—Standard	Yellow—Standard
Orange—53.2079–53.1901 mm 2.0948–2.0941 in. Undersize (0.0178 mm or 0.0007 in.)	Yellow—Standard	Blue—Undersize 0.025 mm 0.001 in.
Black—53.1901–53.1723 mm 2.0941–2.0933 in. Undersize (0.0356 mm or 0.0014 in.)	Blue—Undersize 0.025 mm 0.001 in.	Blue—Undersize 0.025 mm 0.001 in.
Red—52.9717–52.9539 mm 2.0855–2.0848 in. Undersize (0.254 mm or 0.010 in.)	Red—Undersize 0.254 mm 0.010 in.	Red—Undersize 0.254 mm 0.010 in.

19.8a Connecting rod bearing selection chart (1984 through 1993 four-cylinder models and 1989 through 1993 inline-six models)

Crankshaft Journal		Corresponding Connecting Rod Bearing Insert	
Color Code	Diameter	Upper Insert Size	Lower Insert Size
Yellow	53.2257–53.2079 mm (2.0955–2.0948 in.)	Yellow - Standard	Yellow - Standard
Orange	53.2079–53.1901 mm (2.0948–2.0941 in.) 0.0178 mm (0.0007 in.) Undersize	Yellow - Standard	Blue - Undersize 0.025 mm (0.001 in.)
Blue	53.1901–53.1724 mm (2.0941–2.0934 in.) 0.0356 mm (0.0014 in.) Undersize	Blue - Undersize 0.025 mm (0.001 in.)	Blue - Undersize 0.025 mm (0.001 in.)
Red	52.9717–52.9539 mm (2.0855–2.0848 in.) 0.254 mm (0.010 in.) Undersize	Red - Undersize 0.254 mm (0.010 in.)	Red - Undersize 0.254 mm (0.010 in.)

19.8b Connecting rod bearing selection chart (1994 and later models)

2D

them from the engine block, the main bearing caps, the connecting rods and the rod caps and lay them out on a clean surface in the same general position as their location in the engine. This will enable you to match any bearing problems with the corresponding crankshaft journal.

4   Dirt and other foreign particles get into the engine in a variety of ways. It may be left in the engine during assembly, or it may pass through filters or the PCV system. It may get into the oil, and from there into the bearings. Metal chips from machining operations and normal engine wear are often present. Abrasives are sometimes left in engine components after reconditioning, especially when parts are not thoroughly cleaned using the proper cleaning methods. Whatever the source, these foreign objects often end up embedded in the soft bearing material and are easily recognized. Large particles will not embed in the bearing and will score or gouge the bearing and journal. The best prevention for this cause of bearing failure is to clean all parts thoroughly and keep everything spotlessly clean during engine assembly. Fre-

quent and regular engine oil and filter changes are also recommended.

5   Lack of lubrication (or lubrication breakdown) has a number of interrelated causes. Excessive heat (which thins the oil), overloading (which squeezes the oil from the bearing face) and oil leakage or throw off (from excessive bearing clearances, worn oil pump or high engine speeds) all contribute to lubrication breakdown. Blocked oil passages, which usually are the result of misaligned oil holes in a bearing shell, will also oil starve a bearing and destroy it. When lack of lubrication is the cause of bearing failure, the bearing material is wiped or extruded from the steel backing of the bearing. Temperatures may increase to the point where the steel backing turns blue from overheating.

6   Driving habits can have a definite effect on bearing life. Full throttle, low speed operation (lugging the engine) puts very high loads on bearings, which tends to squeeze out the oil film. These loads cause the bearings to flex, which produces fine cracks in the bearing face (fatigue failure). Eventually the bearing material will loosen in pieces and tear

away from the steel backing. Short trip driving leads to corrosion of bearings because insufficient engine heat is produced to drive off the condensed water and corrosive gases. These products collect in the engine oil, forming acid and sludge. As the oil is carried to the engine bearings, the acid attacks and corrodes the bearing material.

7   Incorrect bearing installation during engine assembly will lead to bearing failure as well. Tight fitting bearings leave insufficient bearing oil clearance and will result in oil starvation. Dirt or foreign particles trapped behind a bearing insert result in high spots on the bearing which lead to failure.

Bearing selection (four and inline six-cylinder engines)

Refer to illustrations 19.8a, 19.8b, 19.8c, 19.8d, 19.8e and 19.8f

8   If you have a four or inline six-cylinder engine and the oil clearances are incorrect (see Section 22) or you are replacing the original bearings, refer to the accompanying charts to select the correct new bearings (see illustrations).

Crankshaft No. 1 Main Bearing Journal Color Codes and Diameter in Inches (mm)	Cylinder Block No. 1 Main Bearing Bore Color Code and Size in Inches (mm)	Bearing Insert Color Code	
		Upper Insert Size	Lower Insert Size
Yellow — 2.5001 to 2.4996 (Standard) (63.5025 to 63.4898 mm)	Yellow — 2.6910 to 2.6915 (68.3514 to 68.3641 mm)	Yellow — Standard	Yellow — Standard
	Black — 2.6915 to 2.6920 (68.3641 to 68.3768 mm)	Yellow — Standard	Black — 0.001-inch Undersize (0.025 mm)
Orange — 2.4996 to 2.4991 (0.0005 Undersize) (63.4898 to 63.4771 mm)	Yellow — 2.6910 to 2.6915 (68.3514 to 68.3641 mm)	Yellow — Standard	Black — 0.001-inch Undersize — (0.001 mm)
	Black — 2.6915 to 2.6920 (68.3461 to 68.3768 mm)	Black — 0.001-inch Undersize (0.025 mm)	Black — 0.001-inch Undersize (0.025 mm)
Black — 2.4991 to 2.4986 (0.001 Undersize) (63.4771 to 63.4644 mm)	Yellow — 2.6910 to 2.6915 (68.3514 to 68.3641 mm)	Black — 0.001-inch Undersize — (0.025 mm)	Black — 0.001-inch Undersize — (0.025 mm)
	Black — 2.6915 to 2.6920 (68.3461 to 68.3768 mm)	Black — 0.001-inch Undersize (0.025 mm)	Green — 0.002-inch Undersize (0.051 mm)
Green — 2.4986 to 2.4981 (0.0015 Undersize) (63.4644 to63.4517 mm)	Yellow — 2.6910 to 2.6915 (68.3514 to 68.3641 mm)	Black — 0.001-inch Undersize — (0.025 mm)	Green — 0.002-inch Undersize — (0.051 mm)
Red — 2.4901 to 2.4896 (0.010 Undersize) (63.2485 to 63.2358 mm)	Yellow — 2.6910 to 2.6915 (68.3514 to 68.3641 mm)	Red — 0.010-inch Undersize (0.254 mm)	Red — 0.010-inch Undersize — (0.254 mm)

NOTE: With Green and Red Coded Crankshaft Journals, Use Yellow Coded Cylinder Block Bores Only.

Crankshaft Main Bearing Journal 2-3-4-5 Color Code and Diameter in Inches (Journal Size)	Bearing Insert Color Code	
	Upper Insert Size	Lower Insert Size
Yellow — 2.5001 to 2.4996 (Standard) (63.5025 to 63.4898 mm)	Yellow — Standard	Yellow — Standard
Orange — 2.4996 to 2.4991 (0.0005 Undersize) (63.4898 to 63.4771 mm)	Yellow — Standard	Black — 0.001-inch Undersize (0.025mm)
Black — 2.4991 to 2.4986 (0.001 Undersize) (63.4771 to 63.4644 mm)	Black — 0.001-inch Undersize (0.025 mm)	Black — 0.001-inch Undersize (0.025 mm)
Green — 2.4986 to 2.4981 (0.0015 Undersize) (63.4644 to 63.4517 mm)	Black — 0.001-inch Undersize (0.025 mm)	Green — 0.002-inch Undersize (0.051 mm)
Red — 2.4901 to 2.4896 (0.010 Undersize) (63.2485 to 63.2358 mm)	Red — 0.010-inch Undersize (0.054 mm)	Red — 0.010-inch Undersize (0.254 mm)

19.8c Main bearing selection chart - four-cylinder engines (1984 through 1993 models)

9   If the crankshaft has been reground, new undersize bearings must be installed. Disregard steps 11 through 13 and consult the automotive machine shop that reground the crankshaft. They will provide or help you select the correct size bearings.

10   Regardless of how the bearing sizes are determined, use the oil clearance measured with Plastigage (see Section 22) as a guide to ensure the proper size bearings are installed.

11   Four and inline six-cylinder engines are assembled at the factory with various sizes of color-coded bearing inserts as listed in **illustrations 19.8a through 19.8d.** The color code appears on the edge of the bearing insert.

12   The journal size codes are generally painted on the adjacent cheek toward the flanged end (rear) of the crankshaft, except for the rear main bearing journal, which is marked on the crankshaft rear flange.

13   To obtain a select fit, upper and lower bearing inserts of different sizes may be used as a pair. For example, a standard insert is sometimes used in combination with a 0.001-inch undersize insert to reduce clearance by 0.0005-inch. **Caution:** Never use a pair of bearing inserts with a greater size difference than 0.001-inch. When replacing inserts, the odd sized inserts must be either all on the top or all on the bottom.

20 Engine overhaul - reassembly sequence

Before beginning engine reassembly, make sure you have all the necessary new parts, gaskets and seals as well as the following items on hand:

Common hand tools
3/8-inch and 1/2-inch drive torque wrenches
Piston ring installation tool
Piston ring compressor
Vibration damper installation tool
Short lengths of rubber or plastic hose to fit over connecting rod bolts
Plastigage
Feeler gauges
A fine-tooth file
New engine oil
Engine assembly lube or moly-base grease
Gasket sealant
Thread locking compound

2   In order to save time and avoid problems, engine reassembly must be done in the following general order:

New camshaft bearings (must be done by automotive machine shop)
Piston rings
Crankshaft and main bearings
Piston/connecting rod assemblies
Oil pump
Camshaft and lifters
Oil pan
Timing chain and sprockets
Cylinder head(s), pushrods and rocker arms
Timing cover
Intake and exhaust manifolds
Rocker arm cover(s)
Engine rear plate
Flywheel/driveplate

Crankshaft No. 1 Main Bearing Journal Color Code and Diameter	Cylinder Block No. 1 Main Bearing Bore Color Code and Size	Bearing Insert Color Code	
		Upper Insert Size	Lower Insert Size
Yellow – 63.5025-63.4898 mm (2.5001-2.4996 in.) (Standard)	Yellow – 68.3514-68.3641 mm (2.6910-2.6915 in.)	Yellow – Standard	Yellow – Standard
	Black – 68.3641-68.3768 mm (2.6915-2.6920 in.)	Yellow – Standard	Black – 0.025 mm Undersize (0.001 in.)
Orange – 63.4898-63.4771 mm (2.4996-2.4991 in.) (0.0005 Undersize)	Yellow – 68.3514-68.3641 mm (2.6910-2.6915 in.)	Yellow – Standard	Black – 0.025 mm Undersize (0.001 in.)
	Black – 68.3641-68.3768 mm (2.6915-2.6920 in.)	Black – 0.025 mm Undersize (0.001 in.)	Black – 0.025 mm Undersize (0.001 in.)
Black – 63.4771-63.4644 mm (2.4991-2.4986 in.) (0.001 Undersize)	Yellow – 68.3514-68.3641 mm (2.6910-2.6915 in.)	Black – 0.025 mm Undersize (0.001 in.)	Black – 0.025 mm Undersize (0.001 in.)
	Black – 68.3641-68.3768 mm (2.6915-2.6920 in.)	Black – 0.025 mm Undersize (0.001 in.)	Green – 0.051 mm Undersize (0.002 in.)
Green – 63.4644-63.4517 mm (2.4986-2.4981 in.) (0.0015 Undersize)	Yellow – 68.3514-68.3641 mm (2.6910-2.6915 in.)	Black – 0.025 mm Undersize (0.001 in.)	Green – 0.051 mm Undersize (0.002 in.)
Red – 63.2485-63.2358 mm (2.4901-2.4986 in.) (0.010 Undersize)	Yellow – 68.3514-68.3641 mm (2.6910-2.6915 in.)	Red – 0.254 mm Undersize (0.010 in.)	Red – 0.254 mm Undersize (0.010 in.)

NOTE: With Green and Red Coded Crankshaft Journals, Use Yellow Coded Cylinder Block Bores Only.

Crankshaft Main Bearing Journals 2-6 Color Code and Diameter (Journal Size)	Bearing Insert Color Code	
	Upper Insert Size	Lower Insert Size
Yellow – 63.5025-63.4898 mm (2.5001-2.4996 in.) (Standard)	Yellow – Standard	Yellow – Standard
Orange – 63.4898-63.4771 mm (2.4996-2.4991 in.) (0.0005 Undersize)	Yellow – Standard	Black – 0.025 mm Undersize (0.001 in.)
Black – 63.4771-63.4644 mm (2.4991-2.4986 in.) (0.001 Undersize)	Black – 0.025 mm Undersize (0.001 in.)	Black – 0.025 mm Undersize (0.001 in.)
Green – 63.4644-63.4517 mm (2.4986-2.4981 in.) (0.0015 Undersize)	Black – 0.025 mm Undersize (0.001 in.)	Green – 0.051 mm Undersize (0.002 in.)
Red – 63.2485-63.2358 mm (2.4901-2.4966 in.) (0.010 Undersize)	Red – 0.054 mm Undersize (0.010 in.)	Red – 0.254 mm Undersize (0.010 in.)

Crankshaft Main Bearing Journal 7 Color Code and Diameter (Journal Size)	Bearing Insert Color Code	
	Upper Insert Size	Lower Insert Size
Yellow – 63.4873-63.4746 mm (2.4995-2.4990 in.) (Standard)	Yellow – Standard	Yellow – Standard
Orange – 63.4746-63.4619 mm (2.4990-2.4985 in.) (0.0005 Undersize)	Yellow – Standard	Black – 0.025 mm Undersize (0.001 in.)
Black – 63.4619-63.4492 mm (2.4985-2.4980 in.) (0.001 Undersize)	Black – 0.025 mm Undersize (0.001 in.)	Black – 0.025 mm Undersize (0.001 in.)
Green – 63.4492-63.4365 mm (2.4980-2.4975 in.) (0.0015 Undersize)	Black – 0.025 mm Undersize (0.001 in.)	Green – 0.051 mm Undersize (0.002 in.)
Red – 63.2333-63.2206 mm (2.4895-2.4890 in.) (0.010 Undersize)	Red – 0.254 mm Undersize (0.010 in.)	Red – 0.254 mm Undersize (0.010 in.)

19.8d Main bearing selection chart - inline six-cylinder engines (1989 through 1993 models)

2D

Crankshaft Journals #1 - #4		Corresponding Crankshaft Bearing Insert	
Color Code	Diameter	Upper Insert Size	Lower Insert Size
Yellow	63.5025-63.4898 mm (2.5001-2.4996 in.)	Yellow - Standard	Yellow - Standard
Orange	63.4898-63.4771 mm (2.4996-2.4991 in.) 0.0127 mm (0.0005 in.) Undersize	Blue - Undersize 0.025 mm (0.001 in.)	Yellow - Standard
Blue	63.4771-63.4644 mm (2.4991-2.4986 in.) 0.0254 mm (0.001 in.) Undersize	Blue - Undersize 0.025 mm (0.001 in.)	Blue - Undersize 0.025 mm (0.001 in.)
Green	63.4644-63.4517 mm (2.4986-2.4981 in.) 0.0381 mm (0.0015 in.) Undersize	Blue - Undersize 0.025 mm (0.001 in.)	Green - Undersize 0.051 mm (0.002 in.)
Red	63.2485-63.2358 mm (2.4901-2.4896 in.) 0.254 mm (0.010 in.) Undersize	Red - Undersize 0.254 mm (0.010 in.)	Red - Undersize 0.254 mm (0.010 in.)

Crankshaft Journals #5 Only		Corresponding Crankshaft Bearing Insert	
Color Code	Diameter	Upper Insert Size	Lower Insert Size
Yellow	63.4873-63.4746 mm (2.4995-2.4990 in.)	Yellow - Standard	Yellow - Standard
Orange	63.4746-63.4619 mm (2.4990-2.4985 in.) 0.0127 mm (0.0005 in.) Undersize	Blue - Undersize 0.025 mm (0.001 in.)	Yellow - Standard
Blue	63.4619-63.4492 mm (2.4985-2.4980 in.) 0.0254 mm (0.001 in.) Undersize	Blue - Undersize 0.025 mm (0.001 in.)	Blue - Undersize 0.025 mm (0.001 in.)
Green	63.4492-63.4365 mm (2.4980-2.4975 in.) 0.0381 mm (0.0015 in.) Undersize	Blue - Undersize 0.025 mm (0.001 in.)	Green - Undersize 0.051 mm (0.002 in.)
Red	63.2333-63.2206 mm (2.4895-2.4890 in.) 0.254 mm (0.010 in.) Undersize	Red - Undersize 0.254 mm (0.010 in.)	Red - Undersize 0.254 mm (0.010 in.)

19.8e Main bearing selection chart - four-cylinder engines (1994 and later models)

Crankshaft Journals #1 - #6		Corresponding Crankshaft Bearing Insert	
Color Code	Diameter	Upper Insert Size	Lower Insert Size
Yellow	63.5025-63.4898 mm (2.5001-2.4996 in.)	Yellow - Standard	Yellow - Standard
Orange	63.4898-63.4771 mm (2.4996-2.4991 in.) 0.0127 mm (0.0005 in.) Undersize	Blue - Undersize 0.025 mm (0.001 in.)	Yellow - Standard
Blue	63.4771-63.4644 mm (2.4991-2.4986 in.) 0.0254 mm (0.001 in.) Undersize	Blue - Undersize 0.025 mm (0.001 in.)	Blue - Undersize 0.025 mm (0.001 in.)
Green	63.4644-63.4517 mm (2.4986-2.4981 in.) 0.0381 mm (0.0015 in.) Undersize	Blue - Undersize 0.025 mm (0.001 in.)	Green - Undersize 0.051 mm (0.002 in.)
Red	63.2485-63.2358 mm (2.4901-2.4896 in.) 0.254 mm (0.010 in.) Undersize	Red - Undersize 0.254 mm (0.010 in.)	Red - Undersize 0.254 mm (0.010 in.)

Crankshaft Journals #7 Only		Corresponding Crankshaft Bearing Insert	
Color Code	Diameter	Upper Insert Size	Lower Insert Size
Yellow	63.4873-63.4746 mm (2.4995-2.4990 in.)	Yellow - Standard	Yellow - Standard
Orange	63.4746-63.4619 mm (2.4990-2.4985 in.) 0.0127 mm (0.0005 in.) Undersize	Blue - Undersize 0.025 mm (0.001 in.)	Yellow - Standard
Blue	63.4619-63.4492 mm (2.4985-2.4980 in.) 0.0254 mm (0.001 in.) Undersize	Blue - Undersize 0.025 mm (0.001 in.)	Blue - Undersize 0.025 mm (0.001 in.)
Green	63.4492-63.4365 mm (2.4980-2.4975 in.) 0.0381 mm (0.0015 in.) Undersize	Blue - Undersize 0.025 mm (0.001 in.)	Green - Undersize 0.051 mm (0.002 in.)
Red	63.2333-63.2206 mm (2.4895-2.4890 in.) 0.254 mm (0.010 in.) Undersize	Red - Undersize 0.254 mm (0.010 in.)	Red - Undersize 0.254 mm (0.010 in.)

2D

19.8f Main bearing selection chart - inline six-cylinder engines (1994 and later models)

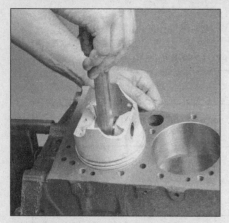

21.3 When checking piston ring end gap, the ring must be square in the cylinder bore (this is done by pushing the ring down with a piston as shown)

21 Piston rings - installation

Refer to illustrations 21.3, 21.4, 21.5, 21.9a, 21.9b and 21.12

1 Before installing the new piston rings, the ring end gaps must be checked. It's assumed that the piston ring side clearance has been checked and verified correct (Section 17).

2 Lay out the piston/connecting rod assemblies and the new ring sets so the ring sets will be matched with the same piston and cylinder during the end gap measurement and engine assembly.

3 Insert the top (number one) ring into the first cylinder and square it up with the cylinder walls by pushing it in with the top of the piston **(see illustration)**. The ring should be near the bottom of the cylinder, at the lower limit of ring travel.

4 To measure the end gap, slip feeler gauges between the ends of the ring until a gauge equal to the gap width is found **(see illustration)**. The feeler gauge should slide between the ring ends with a slight amount of drag. Compare the measurement to the

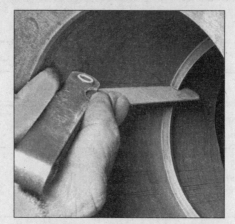

21.4 With the ring square in the cylinder, measure the end gap with a feeler gauge

Specifications. If the gap is larger or smaller than specified, double-check to make sure you have the correct rings before proceeding.

5 If the gap is too small, it must be enlarged or the ring ends may come in contact with each other during engine operation, which can cause serious damage to the engine. The end gap can be increased by filing the ring ends very carefully with a fine file. Mount the file in a vise equipped with soft jaws, slip the ring over the file with the ends contacting the file face and slowly move the ring to remove material from the ends. When performing this operation, file only from the outside in **(see illustration)**.

6 Excess end gap isn't critical unless it's greater than 0.040-inch. Again, double-check to make sure you have the correct rings for your engine.

7 Repeat the procedure for each ring that will be installed in the first cylinder and for each ring in the remaining cylinders. Remember to keep rings, pistons and cylinders matched up.

8 Once the ring end gaps have been checked/corrected, the rings can be installed on the pistons.

9 The oil control ring (lowest one on the piston) is usually installed first. It's normally

21.5 If the end gap is too small, clamp a file in a vise and file the ring ends (from the outside in only) to enlarge the gap slightly

composed of three separate components. Slip the spacer/expander into the groove **(see illustration)**. If an anti-rotation tang is used, make sure it's inserted into the drilled hole in the ring groove. Next, install the lower side rail. Don't use a piston ring installation tool on the oil ring side rails, as they may be damaged. Instead, place one end of the side rail into the groove between the spacer/expander and the ring land, hold it firmly in place and slide a finger around the piston while pushing the rail into the groove **(see illustration)**. Next, install the upper side rail in the same manner.

10 After the three oil ring components have been installed, check to make sure that both the upper and lower side rails can be turned smoothly in the ring groove.

11 The number two (middle) ring is installed next. It's usually stamped with a mark which must face up, toward the top of the piston. **Note:** *Always follow the instructions printed on the ring package or box - different manufacturers may require different approaches. Do not mix up the top and middle rings, as they have different cross sections.*

12 Use a piston ring installation tool and

21.9a Installing the spacer/expander in the oil control ring groove

21.9b DO NOT use a piston ring installation tool when installing the oil ring side rails

21.12 Installing the compression rings with a ring expander - the mark (arrow) must face up

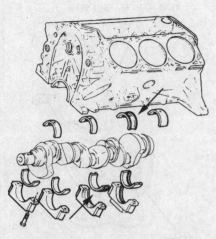

22.6 Be sure the thrust bearings (arrows) are installed in the proper location - number three (counting from the front of the engine) on all six cylinder engines and number two on four-cylinder engines (V6 shown)

22.11 Lay the Plastigage strips (arrow) on the main bearing journals, parallel to the crankshaft centerline

22.15 Compare the width of the crushed Plastigage to the scale on the envelope to determine the main bearing oil clearance (always take the measurement at the widest point of the Plastigage); be sure to use the correct scale - standard and metric scales are included

make sure the identification mark is facing the top of the piston, then slip the ring into the middle groove on the piston **(see illustration)**. Don't expand the ring any more than necessary to slide it over the piston.

13   Install the number one (top) ring in the same manner. Make sure the mark is facing up. Be careful not to confuse the number one and number two rings.

14   Repeat the procedure for the remaining pistons and rings.

22   Crankshaft - installation and main bearing oil clearance check

Refer to illustrations 22.6, 22.11, 22.15 and 22.22

1   Crankshaft installation is the first step in engine reassembly. It's assumed at this point that the engine block and crankshaft have been cleaned, inspected and repaired or reconditioned.

2   Position the engine with the bottom facing up.

3   Remove the main bearing cap bolts and lift out the caps. Lay them out in the proper order to ensure correct installation.

4   If they're still in place, remove the original bearing inserts from the block and the main bearing caps. Wipe the bearing surfaces of the block and caps with a clean, lint-free cloth. They must be kept spotlessly clean.

Main bearing oil clearance check

5   Clean the back sides of the new main bearing inserts and lay one in each main bearing saddle in the block. If one of the bearing inserts from each set has a large

groove in it, make sure the grooved insert is installed in the block. Lay the other bearing from each set in the corresponding main bearing cap. Make sure the tab on the bearing insert fits into the recess in the block or cap. **Caution:** *The oil holes in the block must line up with the oil holes in the bearing insert. Do not hammer the bearing into place and don't nick or gouge the bearing faces. No lubrication should be used at this time.*

6   The flanged thrust bearing must be installed in the proper cap and saddle. On four-cylinder engines it is number two (counting from the front of the engine); on all six-cylinder engines it's number three **(see illustration)**.

7   Clean the faces of the bearings in the block and the crankshaft main bearing journals with a clean, lint-free cloth.

8   Check or clean the oil holes in the crankshaft, as any dirt here can go only one way - straight through the new bearings.

9   Once you're certain the crankshaft is clean, carefully lay it in position in the main bearings.

10   Before the crankshaft can be permanently installed, the main bearing oil clearance must be checked.

11   Cut several pieces of the appropriate size Plastigage (they must be slightly shorter than the width of the main bearings) and place one piece on each crankshaft main bearing journal, parallel with the journal axis **(see illustration)**.

12   Clean the faces of the bearings in the caps and install the caps in their respective positions (don't mix them up) with the arrows pointing toward the front of the engine. Don't disturb the Plastigage.

13   Starting with the center main and working out toward the ends, tighten the main bearing cap bolts, in three steps, to the specified torque. Don't rotate the crankshaft at any time during this operation.

14   Remove the bolts and carefully lift off the main bearing caps. Keep them in order. Don't disturb the Plastigage or rotate the crankshaft. If any of the main bearing caps

are difficult to remove, tap them gently from side-to-side with a soft-face hammer to loosen them.

15   Compare the width of the crushed Plastigage on each journal to the scale printed on the Plastigage envelope to obtain the main bearing oil clearance **(see illustration)**. Check the Specifications to make sure it's correct.

16   If the clearance is not as specified, the bearing inserts may be the wrong size (which means different ones will be required). Before deciding that different inserts are needed, make sure that no dirt or oil was between the bearing inserts and the caps or block when the clearance was measured. If the Plastigage was wider at one end than the other, the journal may be tapered (refer to Section 18).

17   Carefully scrape all traces of the Plastigage material off the main bearing journals and/or the bearing faces. Use your fingernail or the edge of a credit card - don't nick or scratch the bearing faces.

Final crankshaft installation

18   Carefully lift the crankshaft out of the engine.

19   Clean the bearing faces in the block, then apply a thin, uniform layer of moly-base grease or engine assembly lube to each of the bearing surfaces. Be sure to coat the thrust faces as well as the journal face of the thrust bearing. On inline six-cylinder and 1984 V6 engines, install the rear main oil seal sections in the block and rear main bearing cap (see Section 23).

20   Make sure the crankshaft journals are clean, then lay the crankshaft back in place in the block.

21   Clean the faces of the bearings in the caps, then apply lubricant to them.

22   On four-cylinder and V6 engines, apply

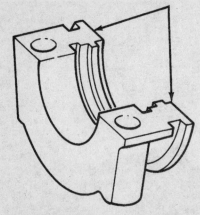

22.22 On four-cylinder and V6 engines, apply a thin coat of sealer to the chamfered area (arrows) of the rear main bearing cap

23.3 Tap around the outer edge of the new oil seal with a hammer and a blunt punch to seat it squarely in the bore

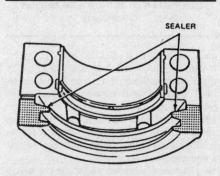

23.5 Rear main oil seal installation details - inline six-cylinder engine

23.7 Apply anaerobic sealant to the area shown on the rear main bearing cap prior to installation (don't get sealant in the grooves or on the seal)

a thin coat of sealer (Loctite 515 or equivalent) to the chamfered area of the rear main bearing cap **(see illustration)**. Install the caps in their respective positions with the arrows pointing toward the front of the engine.

23 Install the bolts.

24 Tighten all except the thrust bearing cap bolts to the specified torque (work from the center out and approach the final torque in three steps).

25 Tighten the thrust bearing cap bolts to 10-to-12 ft-lbs.

26 Tap the ends of the crankshaft forward and backward with a lead or brass hammer to line up the main bearing and crankshaft thrust surfaces.

27 Retighten all main bearing cap bolts to the specified torque, starting with the center main and working out toward the ends.

28 On manual transmission equipped models, install a new pilot bearing in the end of the crankshaft (see Chapter 8).

29 Rotate the crankshaft a number of times by hand to check for any obvious binding.

30 The final step is to check the crankshaft end play with a feeler gauge or a dial indicator as described in Section 13. The end play should be correct if the crankshaft thrust faces aren't worn or damaged and new bearings have been installed.

31 If you are working on an engine with a one-piece rear main oil seal, refer to Section 23 and install the new seal.

23 Rear main oil seal installation

Four-cylinder and 1985 and later V6 engines

Refer to illustration 23.3

Note: *1985 and later V6 engines with 11 mm thick one-piece rear main oil seals are externally identified by an oval cast into the left rear surface of the engine block next to the bellhousing.*

1 Clean the bore in the block/cap and the seal contact surface on the crankshaft. Check the crankshaft surface for scratches and nicks that could damage the new seal lip and cause oil leaks. If the crankshaft is damaged, the only alternative is a new or different crankshaft.

2 Apply a light coat of engine oil or multipurpose grease to the outer edge of the new seal. Lubricate the seal lip with moly-base grease or engine assembly lube.

3 On V6 engines, press the new seal into place with the special tool (if available) - see Part B, Section 20. The seal lip must face toward the front of the engine. If the special tool isn't available or if you have a four-cylinder engine, carefully work the seal lip over the end of the crankshaft and tap the seal in with a hammer and punch until it's seated in the bore **(see illustration)**.

Inline six-cylinder engines

Refer to illustrations 23.5 and 23.7

4 Inspect the rear main bearing cap and engine block mating surfaces, as well as the seal grooves, for nicks, burrs and scratches. Remove any defects with a fine file or deburring tool.

5 Install the semi-circular seal section in the block with the lip facing the front of the engine **(see illustration)**.

6 Repeat the procedure to install the other seal half in the rear main bearing cap.

7 During final installation of the crankshaft (after the main bearing oil clearances have been checked with Plastigage) as described in Section 22, apply a thin, even coat of anaerobic type gasket sealant to the chamfered areas of the cap or block **(see illustration)**. Don't get any sealant on the bearing face, crankshaft journal, seal ends or seal lips. Also, lubricate the seal lips with moly-base grease or engine assembly lube.

1984 V6 engine

Refer to illustration 23.9

8 1984 V6 engines were originally

equipped with two-piece seals. They have been superseded by a specially designed one-piece seal which slips over the end of the crankshaft and fits in the original seal grooves. This type of seal must be installed prior to final crankshaft installation. A two-piece neoprene seal is available from aftermarket manufacturers; however, this is recommended only for in-vehicle repair.

9 The replacement seal is lubricated and pre-mounted on an installation tool **(see illustration)**. Do not remove the seal from the tool prior to installation.

10 Apply a thin coat of Loctite 515 sealer (or equivalent) to the outside diameter of the replacement seal. The sealer bead should be no more than 1 mm (0.040-inch) thick.

11 Carefully push the replacement rear main seal and the tool onto the rear of the crankshaft as far as possible. Position the alignment arrow on the seal tool so it will be pointing down when installed.

12 Align the replacement seal with the seal groove in the block as you carefully install the crankshaft. Make sure the crankshaft is properly lubricated.

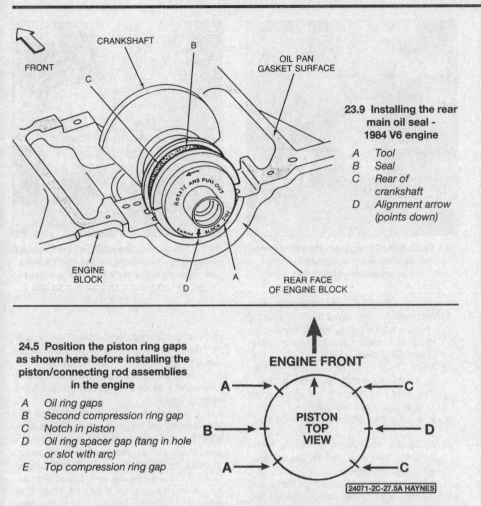

23.9 Installing the rear main oil seal - 1984 V6 engine

A Tool
B Seal
C Rear of crankshaft
D Alignment arrow (points down)

4 Clean the back side of the other bearing insert and install it in the rod cap. Again, make sure the tab on the bearing fits into the recess in the cap, and don't apply any lubricant. It's critically important that the mating surfaces of the bearing and connecting rod are perfectly clean and oil free when they're assembled.

5 Position the piston ring gaps at intervals around the piston (see illustration).

6 Slip a section of plastic or rubber hose over each connecting rod cap bolt.

7 Lubricate the piston and rings with clean engine oil and attach a piston ring compressor to the piston. Leave the skirt protruding about 1/4-inch to guide the piston into the cylinder. The rings must be compressed until they're flush with the piston.

8 Rotate the crankshaft until the number one connecting rod journal is at BDC (bottom dead center) and apply a coat of engine oil to the cylinder walls.

9 With the arrow or notch on top of the piston (see illustration) facing the front of the engine, gently insert the piston/connecting rod assembly into the number one cylinder bore and rest the bottom edge of the ring compressor on the engine block.

10 Tap the top edge of the ring compressor to make sure it's contacting the block around its entire circumference.

11 Gently tap on the top of the piston with the end of a wooden hammer handle (see illustration) while guiding the end of the connecting rod into place on the crankshaft journal. The piston rings may try to pop out of the ring compressor just before entering the cylinder bore, so keep some downward pressure on the ring compressor. Work slowly, and if any resistance is felt as the piston enters the cylinder, stop immediately. Find out what's hanging up and fix it before proceeding. Do not, for any reason, force the piston into the cylinder - you might break a ring and/or the piston.

12 Once the piston/connecting rod assembly is installed, the connecting rod bearing oil clearance must be checked before the rod cap is permanently bolted in place.

2D

24.5 Position the piston ring gaps as shown here before installing the piston/connecting rod assemblies in the engine

A Oil ring gaps
B Second compression ring gap
C Notch in piston
D Oil ring spacer gap (tang in hole or slot with arc)
E Top compression ring gap

ENGINE FRONT

A →   ← C

B → PISTON TOP VIEW ← D

A →   ← C

24071-2C-27.5A HAYNES

13 Remove the seal installation tool by rotating it to the left and pulling it off the crankshaft.

14 Apply a 1 mm (0.040-inch) bead of Loctite 515 sealer (or equivalent) to the end surfaces of the rear main bearing cap (see illustration 23.7). Do not get any sealer on the seal lips or on the crankshaft.

15 Install the main bearing caps (see Section 22).

24 Pistons/connecting rods - installation and rod bearing oil clearance check

Refer to illustrations 24.5, 24.9, 24.11, 24.13 and 24.17

1 Before installing the piston/connecting rod assemblies, the cylinder walls must be perfectly clean, the top edge of each cylinder must be chamfered, and the crankshaft must be in place.

2 Remove the cap from the end of the number one connecting rod (refer to the marks made during removal). Remove the original bearing inserts and wipe the bearing surfaces of the connecting rod and cap with a clean, lint-free cloth. They must be kept spotlessly clean.

Connecting rod bearing oil clearance check

3 Clean the back side of the new upper bearing insert, then lay it in place in the connecting rod. Make sure the tab on the bearing fits into the recess in the rod. Don't hammer the bearing insert into place and be very careful not to nick or gouge the bearing face. Don't lubricate the bearing at this time.

24.9 The arrow must point towards the front of the engine

24.11 The piston can be driven (gently) into the cylinder bore with the end of a wooden hammer handle

13   Cut a piece of the appropriate size Plastigage slightly shorter than the width of the connecting rod bearing and lay it in place on the number one connecting rod journal, parallel with the journal axis **(see illustration)**.

14   Clean the connecting rod cap bearing face, remove the protective hoses from the connecting rod bolts and install the rod cap. Make sure the mating mark on the cap is on the same side as the mark on the connecting rod.

15   Install the nuts and tighten them to the specified torque, working up to it in three steps. **Note:** *Use a thin-wall socket to avoid erroneous torque readings that can result if the socket is wedged between the rod cap and nut. If the socket tends to wedge itself between the nut and the cap, lift up on it slightly until it no longer contacts the cap. Do not rotate the crankshaft at any time during this operation.*

16   Remove the nuts and detach the rod cap, being very careful not to disturb the Plastigage.

17   Compare the width of the crushed Plastigage to the scale printed on the Plastigage envelope to obtain the oil clearance **(see illustration)**. Compare it to the Specifications to make sure the clearance is correct.

18   If the clearance is not as specified, the bearing inserts may be the wrong size (which means different ones will be required). Before deciding that different inserts are needed, make sure that no dirt or oil was between the bearing inserts and the connecting rod or cap when the clearance was measured. Also, recheck the journal diameter. If the Plastigage was wider at one end than the other, the journal may be tapered (refer to Section 18).

Final connecting rod installation

19   Carefully scrape all traces of the Plastigage material off the rod journal and/or bearing face. Be very careful not to scratch the bearing - use your fingernail or the edge of a credit card.

20   Make sure the bearing faces are perfectly clean, then apply a uniform layer of clean moly-base grease or engine assembly lube to both of them. You'll have to push the piston into the cylinder to expose the face of the bearing insert in the connecting rod - be sure to slip the protective hoses over the rod bolts first.

21   Slide the connecting rod back into place on the journal, remove the protective hoses from the rod cap bolts, install the rod cap and tighten the nuts to the specified torque. Again, work up to the torque in three steps.

22   Repeat the entire procedure for the remaining pistons/connecting rods.

24.13 Lay the Plastigage strips on each rod bearing journal, parallel to the crankshaft centerline

23   The important points to remember are

a) *Keep the back sides of the bearing inserts and the insides of the connecting rods and caps perfectly clean when assembling them.*

b) *Make sure you have the correct piston/rod assembly for each cylinder.*

c) *The notch or mark on the piston must face the front of the engine.*

d) *Lubricate the cylinder walls with clean oil.*

e) *Lubricate the bearing faces when installing the rod caps after the oil clearance has been checked.*

24   After all the piston/connecting rod assemblies have been properly installed, rotate the crankshaft a number of times by hand to check for any obvious binding.

25   As a final step, the connecting rod end play must be checked. Refer to Section 12 for this procedure.

26   Compare the measured end play to the Specifications to make sure it's correct. If it was correct before disassembly and the original crankshaft and rods were reinstalled, it should still be right. If new rods or a new crankshaft were installed, the end play may be inadequate. If so, the rods will have to be removed and taken to an automotive machine shop for resizing.

25   Initial start-up and break-in after overhaul

Warning: *Have a fire extinguisher handy when starting the engine for the first time.*

1   Once the engine has been installed in the vehicle, double-check the engine oil and coolant levels.

2   With the spark plugs out of the engine

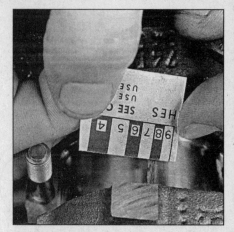

24.17 Measuring the width of the crushed Plastigage to determine the rod bearing oil clearance (be sure to use the correct scale - standard and metric scales are included)

and the ignition system disabled (see Section 3), crank the engine until oil pressure registers on the gauge or the light goes out.

3   Install the spark plugs, hook up the plug wires and restore the ignition system functions (Section 3).

4   Start the engine. It may take a few moments for the fuel system to build up pressure, but the engine should start without a great deal of effort. **Note:** *If backfiring occurs through the carburetor or throttle body, recheck the valve timing and ignition timing.*

5   After the engine starts, it should be allowed to warm up to normal operating temperature. While the engine is warming up, make a thorough check for fuel, oil and coolant leaks.

6   Shut the engine off and recheck the engine oil and coolant levels.

7   Drive the vehicle to an area with minimum traffic, accelerate at full throttle from 30 to 50 mph, then allow the vehicle to slow to 30 mph with the throttle closed. Repeat the procedure 10 or 12 times. This will load the piston rings and cause them to seat properly against the cylinder walls. Check again for oil and coolant leaks.

8   Drive the vehicle gently for the first 500 miles (no sustained high speeds) and keep a constant check on the oil level. It is not unusual for an engine to use oil during the break-in period.

9   At approximately 500 to 600 miles, change the oil and filter.

10   For the next few hundred miles, drive the vehicle normally. Do not pamper it or abuse it.

11   After 2000 miles, change the oil and filter again and consider the engine broken in.

Chapter 3
Cooling, heating and air conditioning systems

Contents

3

Specifications

General

Coolant capacity	See Chapter 1
Drivebelt tension	See Chapter 1
Radiator pressure cap rating	
Four-cylinder and V6 engines	12 to 15 psi
Inline six-cylinder engines	16 to 18 psi
Thermostat opening temperature	195-degrees F

Torque specifications

Ft-lbs

Four-cylinder engine

Fan and pulley-to-water pump hub bolts	15 to 22
Thermostat housing bolts	13
Water pump attaching bolts	13

V6 engine

Fan and pulley-to-water pump hub bolts	15 to 22
Thermostat housing attaching bolts	20 to 30
Water pump attaching bolts	
6mm	6 to 9
8mm	13 to 18
10mm	20 to 30

Inline six-cylinder engine

Fan and pulley-to-hub bolts	18
Thermostat housing bolts	13
Water pump attaching bolts	9 to 18

1   General information

Engine cooling system

All vehicles covered by this manual employ a pressurized engine cooling system with thermostatically controlled coolant circulation. An impeller type water pump mounted on the front of the block pumps coolant through the engine. The coolant flows around each cylinder and toward the rear of the engine. Cast-in coolant passages direct coolant around the intake and exhaust ports, near the spark plug areas and in close proximity to the exhaust valve guides.

A wax pellet type thermostat is located in a housing near the front of the engine. During warm up, the closed thermostat prevents coolant from circulating through the radiator. As the engine nears normal operating temperature, the thermostat opens and allows hot coolant to travel through the radiator, where it's cooled before returning to the engine.

The cooling system is sealed by a pressure type cap on the radiator or coolant reservoir, which raises the boiling point of the coolant and increases the cooling efficiency of the radiator. If the system pressure exceeds the cap pressure relief value, the excess pressure in the system forces the spring-loaded valve inside the cap off its seat. This allows either excess pressure to escape to the atmosphere (reservoir-mounted caps on inline six-cylinder vehicles) or coolant to escape through an overflow tube into a coolant reservoir (radiator-mounted caps on four-cylinder and V6 vehicles). When the system cools on four-cylinder and V6 vehicles, the excess coolant is automatically drawn from the reservoir back into the radiator.

On four-cylinder and V6 vehicles, the coolant reservoir does double duty as both the point at which fresh coolant is added to the cooling system to maintain the proper fluid level and as a holding tank for overheated coolant. On inline six-cylinder vehicles, the coolant reservoir is a pressurized part of the cooling system through which coolant flows in normal operation.

Heating system

The heating system consists of a blower fan and heater core located in the heater box, the hoses connecting the heater core to the engine cooling system and the heater/air conditioning control head on the dashboard. Hot engine coolant is circulated through the heater core. When the heater mode is activated, a flap door opens to expose the heater box to the passenger compartment. A fan switch on the control head activates the blower motor, which forces air through the core, heating the air.

Air conditioning system

The air conditioning system consists of a condenser mounted in front of the radiator, an evaporator mounted adjacent to the heater core, a compressor mounted on the

engine, a receiver-drier which contains a high pressure relief valve and the plumbing connecting all of the above components.

A blower fan forces the warmer air of the passenger compartment through the evaporator core (sort of a radiator-in-reverse), transferring the heat from the air to the refrigerant. The liquid refrigerant boils off into low pressure vapor, taking the heat with it when it leaves the evaporator.

2   Antifreeze - general information

Warning: *Do not allow antifreeze to come in contact with your skin or painted surfaces of the vehicle. Rinse off spills immediately with plenty of water. Antifreeze, if consumed, can be fatal to children and pets, so wipe up garage floor and drip pan coolant spills immediately. Keep antifreeze containers covered and repair leaks in your cooling system as soon as they are noticed.*

The cooling system should be filled with a water/ethylene glycol based antifreeze solution, which will prevent freezing down to at least -20-degrees F, or lower if local climate requires it. It also provides protection against corrosion and increases the coolant boiling point.

The cooling system should be drained, flushed and refilled at the specified intervals (see Chapter 1). Old or contaminated antifreeze solutions are likely to cause damage and encourage the formation of rust and scale in the system. Use distilled water with the antifreeze.

Before adding antifreeze, check all hose connections, because antifreeze tends to search out and leak through very minute openings. Engines don't normally consume coolant, so if the level goes down, find the cause and correct it.

The exact mixture of antifreeze-to-water which you should use depends on the relative weather conditions. The mixture should contain at least 50 percent antifreeze, but should never contain more than 70 percent antifreeze. Consult the mixture ratio chart on the antifreeze container before adding coolant. Hydrometers are available at most auto parts stores to test the coolant. Use antifreeze which meets the vehicle manufacturer's specifications.

3   Thermostat - check and replacement

Warning: *Do not remove the radiator cap, drain the coolant or replace the thermostat until the engine has cooled completely.*

Check

1   Before assuming the thermostat is to blame for a cooling system problem, check the coolant level, drivebelt tension (Chapter 1) and temperature gauge (or light) operation.

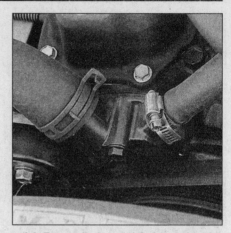

3.8   Four-cylinder models (shown) and inline six-cylinder engines have two hoses connected to the thermostat housing

2   If the engine seems to be taking a long time to warm up (based on heater output or temperature gauge operation), the thermostat is probably stuck open. Replace the thermostat with a new one.

3   If the engine runs hot, use your hand to check the temperature of the upper radiator hose. If the hose isn't hot, but the engine is, the thermostat is probably stuck closed, preventing the coolant inside the engine from escaping to the radiator. Replace the thermostat. **Caution:** *Don't drive the vehicle without a thermostat. The computer may stay in open loop and emissions and fuel economy will suffer.*

4   If the upper radiator hose is hot, it means that the coolant is flowing and the thermostat is open. Consult the Troubleshooting Section at the front of this manual for cooling system diagnosis.

Replacement

Refer to illustrations 3.8, 3.10 and 3.14

5   Disconnect the negative battery cable from the battery.

6   Drain the cooling system (see Chapter 1). If the coolant is relatively new or in good condition (see Chapter 1), save it and reuse it.

7   Follow the upper radiator hose to the engine to locate the thermostat housing.

8   Loosen the hose clamp(s), then detach the hose(s) from the fitting(s) **(see illustration)**. If a hose is stuck, grasp it near the end with a pair of Channelock pliers and twist it to break the seal, then pull it off. If a hose is old or deteriorated, cut it off and install a new one.

9   If the outer surface of the large fitting that mates with the hose is deteriorated (corroded, pitted, etc.) it may be damaged further by hose removal. If it is, the thermostat housing cover will have to be replaced.

10   Remove the bolts and detach the housing cover **(see illustration)**. If the cover is stuck, tap it with a soft-face hammer to jar it loose. Be prepared for some coolant to spill as the gasket seal is broken.

11   Note how it's installed (which end is fac-

3.10 After the hose(s) are removed, remove the bolts (arrows) and detach the housing from the engine - inline six-cylinder engine shown; others similar

3.14 Before installing the gasket, apply a thin, uniform layer of RTV sealant to both sides of it - inline six-cylinder engine shown; others similar

ing out), then remove the thermostat.

12   Stuff a rag into the engine opening, then remove all traces of old gasket material and sealant from the housing and cover with a gasket scraper. Remove the rag from the opening and clean the gasket mating surfaces with lacquer thinner or acetone.

13   Install the new thermostat in the housing. Make sure the correct end faces out - the spring end is normally directed into the engine.

14   Apply a thin, uniform layer of RTV sealant to both sides of the new gasket **(see illustration)** and position it on the housing.

15   Install the cover and bolts. Tighten the bolts to the specified torque.

16   Reattach the hose(s) to the fitting(s) and tighten the hose clamp(s) securely.

17   Refill the cooling system (see Chap-

ter 1).

18   Start the engine and allow it to reach normal operating temperature, then check for leaks and proper thermostat operation (as described in Steps 2 through 4).

4   Radiator - removal and installation

Refer to illustrations 4.5, 4.8a, 4.8b and 4.9
Warning: *Wait until the engine is completely cool before beginning this procedure.*

1   Disconnect the negative battery cable from the battery.

2   Drain the cooling system (see Chapter 1). If the coolant is relatively new or in good condition, save it and reuse it.

3   Loosen the hose clamps, then detach

the radiator hoses from the fittings on the radiator. If they're stuck, grasp each hose near the end with a pair of Channelock pliers and twist it to break the seal, then pull it off - be careful not to distort the radiator fittings! If the hoses are old or deteriorated, cut them off and install new ones.

4   On four-cylinder and V6 models, disconnect the reservoir hose from the radiator filler neck.

5   On four-cylinder and V6 models, remove the screws that attach the radiator fan shroud to the radiator **(see illustration)** and slide the shroud toward the engine. On inline six-cylinder models equipped with auxiliary electric cooling fans, remove the electric fan (see Section 5). On all inline six-cylinder models, remove the mounting bolts for the mechanical fan shroud, lift the shroud up until it clears the slots in the bottom bracket of the radiator, then push the shroud over the fan.

6   If the vehicle is equipped with an automatic transmission, disconnect the cooler lines from the rear of the radiator. Use a drip pan to catch spilled fluid.

7   Plug the lines and fittings.

8   Remove the radiator mounting bolts **(see illustrations)**.

3

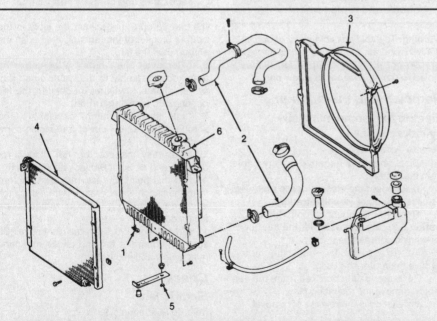

4.5 Radiator and related components - exploded view (four-cylinder and V6 engines)

1	Drain	3	Radiator fan shroud	5	E-clip ring
2	Radiator hoses	4	A/C condenser	6	Coolant reservoir

4.8a On inline six-cylinder models, use a Torx-bit to remove the hood safety catch, then unbolt the panel above the radiator

4.8b On some models, you must remove the bolt (arrow) to disconnect the air conditioning line bracket attached to the top of the radiator

4.9 When lifting the radiator out, be careful not to damage the cooling fins with the fan and fan shroud (arrows) - inline six-cylinder model shown

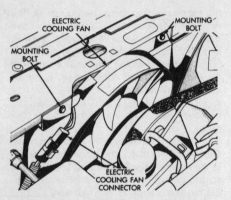

5.2 On inline six-cylinder engines equipped with electric cooling fans, disconnect the electrical connector, remove the mounting bolt, and lift the fan out

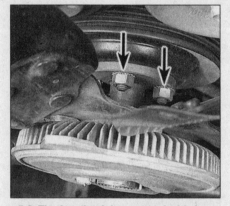

5.8 The fan clutch is mounted with four nuts; two are visible in this photo (arrows)

5.13 Disconnect the fan electrical connector and run jumper wires directly to the positive and negative terminals of the battery - be sure the clips don't touch each other and create a short circuit

9   Carefully lift out the radiator **(see illustration)**. Don't spill coolant on the vehicle or scratch the paint.

10   With the radiator removed, it can be inspected for leaks and damage. If it needs repair, have a radiator shop or dealer service department perform the work as special techniques are required.

11   Bugs and dirt can be removed from the front of the radiator with compressed air and a soft brush. Don't bend the cooling fins as this is done.

12   Check the radiator mounts for deterioration and make sure there's nothing in them when the radiator is installed.

13   Installation is the reverse of the removal procedure.

14   After installation, fill the cooling system with the proper mixture of antifreeze and water. Refer to Chapter 1 if necessary.

15   Start the engine and check for leaks. Allow the engine to reach normal operating temperature, indicated by the upper radiator hose becoming hot. Recheck the coolant level and add more if required.

16   If you're working on an automatic transmission equipped vehicle, check and add fluid as needed.

5   Engine cooling fan and clutch - check and replacement

Warning: *To avoid possible injury or damage, DO NOT operate the engine with a damaged fan. Do not attempt to repair fan blades - replace a damaged fan with a new one.*

Removal and installation

Electric fan (some inline six-cylinder models)

Refer to illustration 5.2

1   Disconnect the negative battery cable from the battery.

2   Disconnect the electric cooling fan wire harness connector **(see illustration)**.

3   Remove the mounting bolts **(see illustration 5.2)**, then carefully lift the fan out of the engine compartment.

4   To detach the fan from the motor, remove the motor shaft clip.

5   To remove the bracket from the fan motor, remove the mounting nuts.

6   Installation is the reverse of removal.

Mechanical fan with viscous clutch

Refer to illustration 5.8

7   Disconnect the negative battery cable. Remove the fan shroud mounting screws, lift

the shroud up until it clears the slots in the bottom bracket of the radiator, then push the shroud over the fan.

8   Remove the nuts attaching the fan/clutch assembly to the water pump hub or, on inline six-cylinder models, to the fan bearing hub **(see illustration)**.

9   Lift the fan/clutch assembly (and shroud, if necessary) out of the engine compartment.

10   Carefully inspect the fan blades for damage and defects. Replace it if necessary.

11   At this point, the fan may be unbolted from the clutch, if necessary. If the fan clutch is stored, position it with the radiator side facing down.

12   Installation is the reverse of removal. Be sure to tighten the fan and clutch mounting nuts evenly and securely.

Check

Electric fan

Refer to illustration 5.13

13   To test the motor, unplug the electrical connector at the motor and use jumper wires to connect the fan directly to the battery **(see illustration)**. If the fan still doesn't work,

6.1a The coolant temperature sending unit on four-cylinder engines is located near the left rear corner of the cylinder head (arrow)

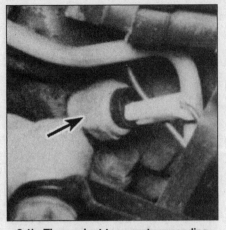

6.1b The coolant temperature sending unit on V6 engines (arrow) is located near the left front corner of the intake manifold

6.1c The coolant temperature sending unit on inline six-cylinder engines is located near the left rear corner of the cylinder head (arrow)

replace the fan motor.

14   If the motor tested OK, the fault lies in the coolant temperature switch or the wiring which connects the components. Carefully check all wiring and connections. If no obvious problems are found, further diagnosis should be done by a dealer service department or repair shop.

Mechanical fan with viscous clutch

15   Disconnect the negative battery cable and rock the fan back and forth by hand to check for excessive bearing play.
16   With the engine cold, turn the fan blades by hand. The fan should turn freely.
17   Visually inspect for substantial fluid leakage from the clutch assembly. If problems are noted, replace the clutch assembly.
18   With the engine completely warmed up, turn off the ignition switch and disconnect the negative battery cable from the battery. Turn the fan by hand. Some drag should be evident. If the fan turns easily, replace the fan clutch.

6   Coolant temperature sending unit - check and replacement

Refer to illustrations 6.1a, 6.1b and 6.1c
Warning: *Wait until the engine is completely cool before beginning this procedure.*
1   The coolant temperature indicator system is composed of a light or temperature gauge mounted in the instrument panel and a coolant temperature sending unit mounted on the engine **(see illustrations)**. Some vehicles have more than one sending unit, but only one is used for the indicator system.
Warning: *If the vehicle is equipped with an electric cooling fan, stay clear of the fan blades. The fan can come on at any time.*
2   If an overheating indication occurs, check the coolant level in the system and then make sure the wiring between the light or gauge and the sending unit is secure and all fuses are intact.

3   When the ignition switch is turned on and the starter motor is turning, the indicator light should be on (overheated engine indication).
4   If the light is not on, the bulb may be burned out, the ignition switch may be faulty or the circuit may be open. Test the circuit by grounding the wire to the sending unit while the ignition is on (engine not running for safety). If the gauge deflects full scale or the light comes on, replace the sending unit.
5   As soon as the engine starts, the light should go out and remain out unless the engine overheats. Failure of the light to go out may be due to a grounded wire between the light and the sending unit, a defective sending unit or a faulty ignition switch. Check the coolant to make sure it's the proper type. Plain water may have too low a boiling point to activate the sending unit.
6   If the sending unit must be replaced, simply unscrew it from the engine and install the replacement. Use sealant on the threads. Make sure the engine is cool before removing the defective sending unit. There will be some coolant loss as the unit is removed, so be prepared to catch it. Check the level after the replacement has been installed.

7   Coolant reservoir - removal and installation

Refer to illustrations 7.3a and 7.3b
Warning: *Wait until the engine is completely cool before beginning this procedure.*
1   Remove the reservoir filler cap and drain the cooling system (see Chapter 1) until the coolant level is below the bottom of the reservoir.
2   Loosen the hose clamps and detach the hose(s) from the reservoir.
3   Remove the screws or retaining strap and lift the reservoir from the vehicle **(see illustrations)**.
4   Installation is the reverse of removal.
5   Refill the cooling system and check for leaks.

8   Water pump - check

Refer to illustrations 8.4 and 8.5
1   A failure in the water pump can cause serious engine damage due to overheating.
2   There are three ways to check the operation of the water pump while it's installed on

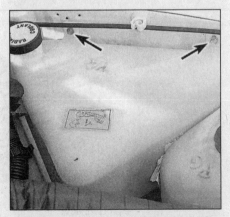

7.3a On four-cylinder and V6 models, the coolant reservoir is fastened by two screws (arrows)

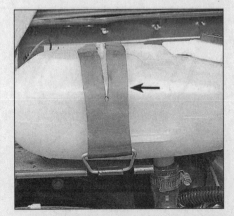

7.3b On inline six-cylinder models, the coolant reservoir is fastened by a rubber retaining strap (arrow)

3

8.4 The water pump weep hole (arrow) will drip coolant when the seal on the pump shaft fails (pump removed from engine for clarity)

8.5 Grasp the water pump flange and try to rock the shaft back and forth to check for play (fan and pulley shown removed for clarity)

9.4 On inline six-cylinder engines, loosen the pulley-to water pump bolts before removing the drivebelt

the engine. If the pump is defective, it should be replaced with a new or rebuilt unit.

3   With the engine running at normal operating temperature, squeeze the upper radiator hose. If the water pump is working properly, a pressure surge should be felt as the hose is released. **Warning:** *Keep your hands away from the fan blades!*

4   Water pumps are equipped with weep or vent holes. If a failure occurs in the pump seal, coolant will leak from the hole. In most cases you'll need a flashlight to find the hole on the water pump from underneath to check for leaks **(see illustration)**.

5   If the water pump shaft bearings fail there may be a howling sound at the front of the engine while it's running. Shaft wear can be felt if the water pump pulley is rocked up and down **(see illustration)**. Don't mistake drivebelt slippage, which causes a squealing sound, for water pump bearing failure.

9   Water pump - replacement

Refer to illustrations 9.4, 9.7a, 9.7b and 9.7c
Warning: *Wait until the engine is completely*

cool before beginning this procedure.

1   Disconnect the negative battery cable from the battery.

2   Drain the cooling system (see Chapter 1). If the coolant is relatively new or in good condition, save it and reuse it.

3   Remove the cooling fan and shroud (see Section 5).

4   On inline six-cylinder engines, loosen the pulley-to-water pump bolts **(see illustration)**. Remove the drivebelts (see Chapter 1) and the pulley at the end of the water pump shaft.

5   Loosen the clamps and detach the hoses from the water pump. If they're stuck, grasp each hose near the end with a pair of Channelock pliers and twist it to break the seal, then pull it off. If the hoses are deteriorated, cut them off and install new ones.

6   Remove all accessory brackets from the water pump. When removing the power steering pump and air conditioning compressor, don't disconnect the hoses. Tie the units aside with the hoses attached. Keep the power steering pump upright so fluid doesn't spill.

7   Remove the bolts and detach the water

pump from the engine. Note the locations of the various lengths and different types of bolts as they're removed to ensure correct installation **(see illustrations)**.

8   Clean the bolt threads and the threaded holes in the engine to remove corrosion and sealant.

9   Compare the new pump to the old one to make sure they're identical.

10   Remove all traces of old gasket material from the engine with a gasket scraper.

11   Clean the engine and new water pump mating surfaces with lacquer thinner or acetone.

12   Apply a thin coat of RTV sealant to the engine side of the new gasket.

13   Apply a thin layer of RTV sealant to the gasket mating surface of the new pump, then carefully mate the gasket and the pump. Slip a couple of bolts through the pump mounting holes to hold the gasket in place.

14   Carefully attach the pump and gasket to the engine and thread the bolts into the holes finger tight.

15   Install the remaining bolts (if they also hold an accessory bracket in place, be sure to reposition the bracket at this time). Tighten them to the specified torque in 1/4-turn incre-

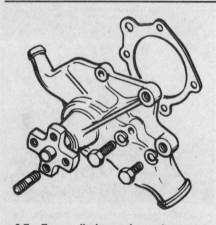

9.7a Four-cylinder engine water pump and gasket

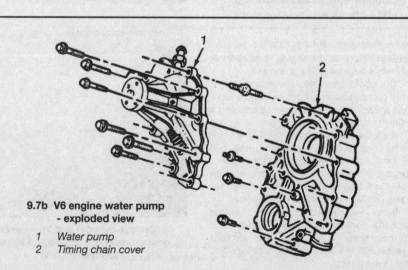

9.7b V6 engine water pump - exploded view

1   *Water pump*
2   *Timing chain cover*

9.7c Water pump bolt locations for the inline six-cylinder engine

ments. Note that the three sizes of bolts on V6 engines require different torques. Don't overtighten them or the pump may be distorted.
16   Reinstall all parts removed for access to the pump.
17   Refill the cooling system and check the drivebelt tension (Chapter 1). Run the engine and check for leaks.

10   Heater and air conditioner blower motor - removal and installation

Refer to illustrations 10.2 and 10.5

1984 through 1996 models
Removal
1   Disconnect the negative cable from the battery.
2   Disconnect the blower motor wires **(see illustration).**
3   Remove the blower motor mounting screws **(see illustration 10.2).**
4   Lift the blower motor from the vehicle. If you are not replacing the motor, you may disregard steps 5 through 7.
5   Detach the fan retainer clip from the fan hub **(see illustration).**

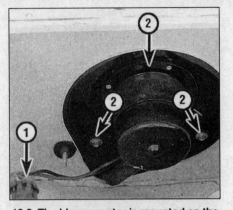

10.2 The blower motor is mounted on the engine-side of the firewall - disconnect the electrical connector (1), then remove the three blower motor mounting screws (2) - the top screw is hidden in this photo

6   Slip the fan off the old motor shaft, remove the motor-to-cover attaching nuts and pull the motor off the cover.
7   Mount the new motor on the cover and slip the fan onto the shaft. Install the retainer clip **(see illustration 10.5).**

Installation
8   Place the blower motor in position and install the mounting screws.
9   Reconnect the wires and battery cable. Test the motor in operation.

1997 and later models
10   If the vehicle is equipped with air conditioning, have the air conditioning system discharged.
11   Disconnect the cable from the battery negative terminal.
12   If the vehicle is equipped with air conditioning, loosen the accumulator retaining band screw, disconnect the accumulator inlet tube from the evaporator outlet tube and move the accumulator out of the way to access the blower motor.
13   Unplug the electrical connector from the blower motor.

14   Remove the three blower motor assembly retaining screws.
15   Remove the blower motor.
16   Installation is the reverse of removal. When you're done, take the vehicle back to the shop that discharged the air conditioning. Have the air conditioning system evacuated, charged and leak tested.

11   Heater core - replacement

Warning 1: *The air conditioning system is under high pressure. DO NOT disconnect any refrigerant fittings until after the system has been discharged by a dealer service department or service station.*
Warning 2: *Some models covered by this manual are equipped with Supplemental Restraint Systems, more commonly known as airbags. Always disable the airbag system before working in the vicinity of any airbag system components, to avoid the possibility of accidental deployment of the airbag(s), which could cause personal injury (see Chapter 12).*

1984 through 1996 models
Refer to illustrations 11.6, 11.7, 11.8, 11.11, 11.15, 11.17a and 11.17b
Removal
1   On air conditioned models, have the system discharged (see the Warning above).
2   Disconnect the negative cable from the battery.
3   Drain the cooling system (see Chapter 1) and disconnect the heater hoses at the heater core inlet and outlet.
4   Disconnect the blower motor electrical connector (see Section 10).
5   Remove the console (if equipped) and the lower instrument panel (see Chapter 11).
6   On air conditioned models, disconnect the A/C hoses from the expansion valve **(see illustration).**
7   Label and disconnect the electrical wires at the air conditioning relay, the air conditioning thermostat (if equipped) and the

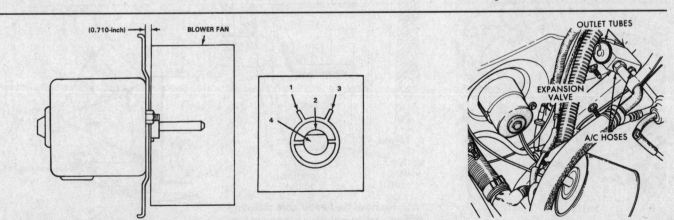

10.5 Pry the clip ears (1 and 3) off the flat surface (2) of the motor shaft (4) - bend the retaining clip like the original when installing it

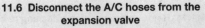

11.6 Disconnect the A/C hoses from the expansion valve

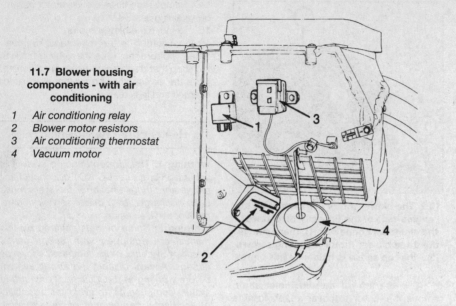

**11.7 Blower housing
components - with air
conditioning**

1   *Air conditioning relay*
2   *Blower motor resistors*
3   *Air conditioning thermostat*
4   *Vacuum motor*

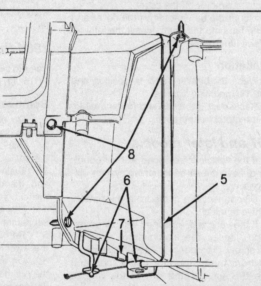

**11.8 Cut the retaining strap (5)
and remove the heater control
cable (6), the rear blower
housing flange clip (7) and the
retaining screws (8)**

blower motor resistors. Disconnect the vacuum hose at the vacuum motor **(see illustration)**.

8   Cut the retaining strap that holds the blower/evaporator housing to the heater core housing **(see illustration)**.

9   Disconnect the heater control cable **(see illustration 11.8)**.

10   Detach the clip at the rear of the blower housing flange and remove the retaining screws **(see illustration 11.8)**.

11   Remove the housing attaching nuts from the studs on the engine compartment side of the firewall **(see illustration)**.

12   Remove the evaporator drain tube.

13   Remove the right kick panel and then the instrument panel support bolt.

14   Gently pull out on the right side of the dash and rotate the housing down and toward the rear of the vehicle to disengage the housing studs from the dash panel. Then remove the blower/evaporator housing.

15   Remove the retaining screws **(see illustration)** and remove the heater core by pulling it straight down out of the housing.

Installation

16   Position the heater core in the housing and install the screws. Be sure to cement the seal in place to keep it from moving when the blower assembly is installed **(see illustration 11.15)**.

17   Reinstall the remaining parts in the reverse order of removal. Be sure to use the alignment tabs **(see illustrations)** to position the housings. **Note:** *When installing the blower/evaporator housing, avoid trapping wires between the housing fresh air inlet and the dash panel on the right side of the housing.*

18   Refill the cooling system.

19   Have the air conditioning system (if equipped) evacuated, recharged and leak tested.

20   Start the engine and check for proper operation.

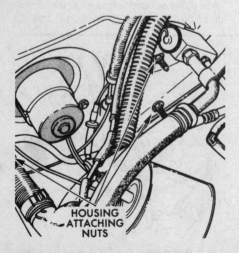

11.11 Remove the housing attaching nuts

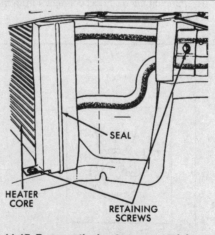

**11.15 Remove the heater core retaining
screws - when installing it, cement the
seal in place to keep it from moving when
the blower assembly is installed**

**11.17a Use the alignment tabs to position
the housing and . . .**

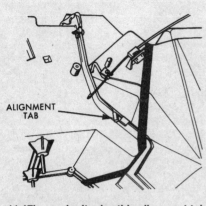

ALIGNMENT TAB

11.17b . . . don't miss this alignment tab

1997 and later models

Refer to illustrations 11.25, 11.29 and 11.36

21 If the vehicle is equipped with air conditioning, have the system discharged.
22 Disconnect the cable from the battery negative terminal.
23 Remove the instrument panel (see Section 23 in Chapter 11).
24 If the vehicle is not equipped with air conditioning, go to Step 26.
25 Disconnect the liquid refrigerant line fitting from the evaporator inlet tube **(see illustration)**. Disconnect the accumulator inlet tube refrigerant line fitting from the evaporator outlet tube **(see illustration 13.6)**. Plug the lines to prevent contamination
26 Drain the engine cooling system (see Chapter 1).
27 Disconnect the heater hoses from the heater core tubes.
28 Unplug the vacuum supply line and electrical connectors for the air conditioner and heater assembly.
29 Remove the five mounting nuts (located on the engine compartment side of the firewall) from the air conditioner and heater

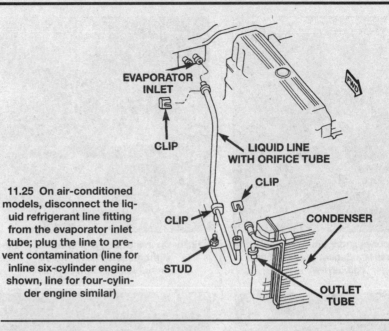

11.25 On air-conditioned models, disconnect the liquid refrigerant line fitting from the evaporator inlet tube; plug the line to prevent contamination (line for inline six-cylinder engine shown, line for four-cylinder engine similar)

housing mounting studs **(see illustration)**.
30 Pull the air conditioner and heater housing to the rear far enough to clear the mounting studs and the evaporator condensate drain tube to clear the dash panel holes. Remove the air conditioner and heater housing assembly.
31 Place the air conditioner and heater housing assembly on a work bench.
32 Unplug the vacuum harness connectors from the floor door actuator and, if applicable, the recirculation air door actuator.
33 Detach the vacuum harness from all retainer clips on the lower half of the air conditioner and heater housing.
34 Remove the blower motor assembly from the air conditioner and heater housing (see Section 10).
35 Carefully peel off the foam seal from the flange around the blower motor opening in the air conditioner and heater housing. If the

seal is deformed or damaged, replace it.
36 Pull the vacuum supply line and connector through the foam seal on the mounting flange for the heater core and evaporator coil tube **(see illustration)**.
37 If the vehicle is equipped with air conditioning, remove the screw that attaches the evaporator coil tube clamp and remove the clamp.
38 Carefully peel away the foam seal from the heater core and evaporator coil tube mounting flange. If the seal is deformed or damaged, replace it.
39 Pry off the two snap clips that attach the upper and lower halves of the air conditioner and heater housing.
40 Remove the 14 screws that attach the upper and lower halves of the air conditioner and heater housing.
41 Separate the two halves of the air conditioner and heater housing.

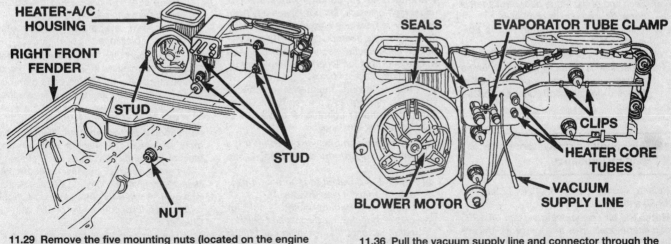

11.29 Remove the five mounting nuts (located on the engine compartment side of the firewall) from the air conditioner and heater housing mounting studs

11.36 Pull the vacuum supply line and connector through the foam seal on the mounting flange for the heater core and evaporator coil tube

12.7a On four-cylinder and V6 models, the sight glass is adjacent to the radiator cap (arrow)

12.7b On inline six-cylinder models, the sight glass is below the air conditioning compressor

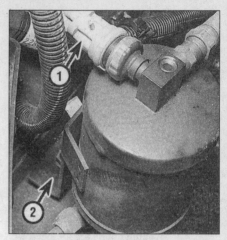

13.3 Unplug the electrical connector (1) and unbolt the mounting bracket (2) - early six-cylinder engine shown (others similar)

14.3 Disconnect the compressor clutch electrical connector (arrow) - inline six-cylinder engine shown

15.5a Disconnect the refrigerant lines from the condenser (inline six-cylinder model shown) - be sure to use a backup wrench to avoid bending the line

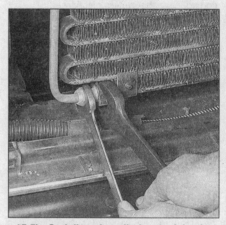

15.5b On inline six-cylinder models, the lower fitting can be reached through the grille opening

42   Remove the heater core assembly from the housing.
43   Installation is basically the reverse of removal, but during reassembly make sure that:

a) *Each end of the mode door pivot shaft is correctly seated in its pivot hole.*
b) *The blower motor venturi ring is correctly oriented as shown.*
c) *The evaporator coil tube rubber seal (if applicable) is correctly positioned in the grooves in both the upper and lower halves of the air conditioner and heater housing.*

12   Air conditioning system - check and maintenance

Refer to illustrations 12.7a, 12.7b, 12.9a, 12.9b and 12.11
Note: *The air conditioning system on 1995 and later models uses the non-ozone-depleting refrigerant, referred to as R-134a. The R-134a refrigerant and its lubricating oil are not compatible with the R-12 system, and under*

no circumstances should the two different types of refrigerant and lubricating oil be intermixed. If mixed, it could result in costly compressor failure due to improper lubrication.
Warning: *The air conditioning system is under high pressure. Do not loosen any hose fittings or remove any components until after the system has been discharged by a dealer service department or service station. Always wear eye protection when disconnecting air conditioning system fittings.*
1   The following maintenance checks should be performed on a regular basis to ensure that the air conditioner continues to operate at peak efficiency.

a) *Check the compressor drivebelt. If it's worn or deteriorated, replace it (see Chapter 1).*
b) *Check the drivebelt tension and, if necessary, adjust it (see Chapter 1).*
c) *Check the system hoses. Look for cracks, bubbles, hard spots and deterioration. Inspect the hoses and all fittings for oil bubbles and seepage. If there's any evidence of wear, damage or leaks, replace the hose(s).*

d) *Inspect the condenser fins for leaves, bugs and other debris. Use a "fin comb" or compressed air to clean the condenser.*
e) *Make sure the system has the correct refrigerant charge.*
f) *Check the evaporator housing drain tube for blockage.*

2   It's a good idea to operate the system for about 10 minutes at least once a month, particularly during the winter. Long term non-use can cause hardening, and subsequent failure, of the seals.
3   Because of the complexity of the air conditioning system and the special equipment necessary to service it, in-depth troubleshooting and repairs are not included in this manual (refer to the *Haynes Automotive Heating and Air Conditioning Repair Manual*). However, simple checks and component replacement procedures are provided in this Chapter.
4   The most common cause of poor cooling is simply a low system refrigerant charge. If a noticeable drop in cool air output occurs, the following quick check will help you deter-

15.5c On four-cylinder and V6 models, the refrigerant line connections are located next to the battery

15.6 On four-cylinder and V6 models, remove the screws (arrows) and pull off the cover to reach the mounting bolts

mine if the refrigerant level is low.

5   Warm the engine up to normal operating temperature.

6   Place the air conditioning temperature selector at the coldest setting and put the blower at the highest setting. Open the doors (to make sure the air conditioning system doesn't cycle off as soon as it cools the passenger compartment).

7   With the compressor engaged - the clutch will make an audible click and the center of the clutch will rotate - inspect the sight glass **(see illustrations)**. If the refrigerant looks foamy, it's low. Have the system charged by a dealer service department or automotive air conditioning shop.

8   If there's no sight glass, feel the inlet and outlet pipes at the compressor. One side should be cold and one hot. If there's no perceptible difference between the two pipes, there's something wrong with the compressor or the system. It might be a low charge - it might be something else. Take the vehicle to a dealer service department or an automotive air conditioning shop.

Adding refrigerant

Note: *Because of recent Federal regulations proposed by the Environmental Protection Agency, 14-ounce cans of refrigerant are not available to the do-it-yourselfer. Generally, it will be necessary to take your vehicle to a licensed air conditioning technician for charging. If you decide to ad refrigerant yourself from one of the 30-pound drums still available, you will need a manifold gauge set, all the necessary fittings, adapters and hoses and a copy of the Haynes Automotive Heating and Air Conditioning Manual before beginning.*

13   Air conditioning system receiver-drier or accumulator - removal and installation

Refer to illustration 13.3

Warning: *The air conditioning system is under high pressure. DO NOT disassemble any part of the system (hose, compressor, line fittings, etc.) until after the system has*

been depressurized by a dealer service department or service station.

1   Have the air conditioning system discharged (see **Warning** above).

2   Disconnect the negative battery cable from the battery.

3   Unplug the electrical connector from the pressure switch near the top of the receiver-drier **(see illustration)**. **Note:** *On four-cylinder and V6 models, the receiver-drier is mounted adjacent to the radiator. On inline six-cylinder models, it's below the air conditioning compressor.*

4   Disconnect the refrigerant line from the receiver-drier. Use a backup wrench to prevent twisting the tubing.

5   Plug the open fittings to prevent entry of dirt and moisture.

6   On 1984 through 1996 models, loosen the mounting bracket bolts and lift the receiver-drier out. On 1997 and later models, loosen the clamping band screw and remove the accumulator.

7   If a new receiver-drier is being installed, remove the Schrader valve and pour the oil out into a measuring cup, noting the amount. Add fresh refrigerant oil to the new receiver-drier equal to the amount removed from the old unit, plus one ounce.

8   Installation is the reverse of removal.

9   Take the vehicle back to the shop that discharged it. Have the A/C system evacuated, charged and leak tested.

14   Air conditioning system compressor - removal and installation

Refer to illustration 14.3

Warning: *The air conditioning system is under high pressure. DO NOT disassemble any part of the system (hoses, compressor, line fittings, etc.) until after the system has been depressurized by a dealer service department or service station.*

Note: *The receiver-drier (see Section 13) should be replaced whenever the compressor is replaced.*

1   Have the A/C system discharged (see

Warning above).

2   Disconnect the negative battery cable from the battery.

3   Disconnect the compressor clutch electrical connector **(see illustration)**.

4   Remove the drivebelt (see Chapter 1).

5   Disconnect the refrigerant lines from the top rear of the compressor. Plug the open fittings to prevent entry of dirt and moisture.

6   Unbolt the compressor from the mounting brackets and lift it out of the vehicle.

7   If a new compressor is being installed, follow the directions with the compressor regarding the draining of excess oil prior to installation.

8   The clutch may have to be transferred from the original to the new compressor.

9   Installation is the reverse of removal. Replace all O-rings with new ones specifically made for A/C system use and lubricate them with refrigerant oil.

10   Have the system evacuated, recharged and leak tested by the shop that discharged it.

15   Air conditioning system condenser - removal and installation

Refer to illustrations 15.5a, 15.5b, 15.5c and 15.6

Warning: *The air conditioning system is under high pressure. DO NOT disassemble any part of the system (hoses, compressor, line fittings, etc.) until after the system has been depressurized by a dealer service department or service station.*

Note: *The receiver-drier (Section 13) should be replaced whenever the condenser is replaced.*

1   Have the air conditioning system discharged (see **Warning** above).

2   Disconnect the negative battery cable from the battery.

3   Remove the radiator grille (Chapter 11).

4   Remove the radiator (see Section 4).

5   Disconnect the refrigerant lines from the condenser **(see illustrations)**.

6   Remove the mounting bolts from the condenser brackets **(see illustration)**.

3

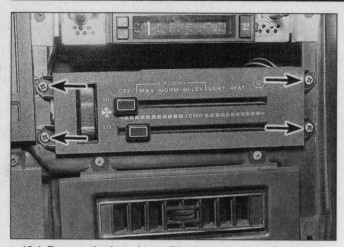

16.4 Remove the four air conditioning control panel mounting screws (arrows)

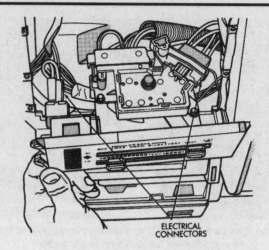

16.5 Unplug the electrical connectors

7   Lift the condenser out of the vehicle and plug the lines to keep dirt and moisture out.

8   If the original condenser will be reinstalled, store it with the line fittings on top to prevent oil from draining out.

9   If a new condenser is being installed, pour one ounce of refrigerant oil into it prior to installation.

10   Reinstall the components in the reverse order of removal. Be sure the rubber pads are in place under the condenser.

11   Have the system evacuated, recharged and leak tested by the shop that discharged it.

16   Air conditioner and heater control assembly - removal and installation

1984 through 1996 models

Refer to illustrations 16.4, 16.5, 16.6, 16.7 and 16.8

1   Disconnect the negative cable from the battery.

2   Detach the instrument panel bezel (see Chapter 11).

3   Remove the radio (see Chapter 12).

4   Remove the four mounting screws from the air conditioning/heater control panel **(see illustration)**.

5   Unplug the electrical connectors **(see illustration)**.

6   Detach the vacuum connector by bending back the locking tab **(see illustration)**.

7   Release the control cable locking tab with a screwdriver **(see illustration)**.

8   Detach the ring on the end of the control cable from the arm on the bottom of the control panel **(see illustration)**.

9   To install the control, reverse the removal procedure.

1997 and later models

10   Disconnect the cable from the negative battery cable.

11   Roll down the glove box from the instru-

ment panel (see Section 23 in Chapter 11).

12   Reach through the glove box opening in the instrument panel and unplug the two halves of the connector for the heater/air conditioning vacuum harness.

13   Remove the center bezel from the instrument panel (see Section 23 in Chapter 11).

14   Release the vacuum harness push-in retainer from the instrument panel (it's right under the air conditioner and heater control assembly).

15   Remove the four retaining screws from the air conditioner and heater control assembly.

16   Pull out the air conditioner and heater control assembly and unplug the electrical connectors from the backside of the control assembly.

17   Reach through the glove box opening and support the vacuum harness while removing the air conditioner and heater control assembly.

18   Installation is the reverse of removal.

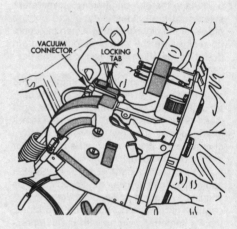

16.6 Bend back the locking tab to release the vacuum connector

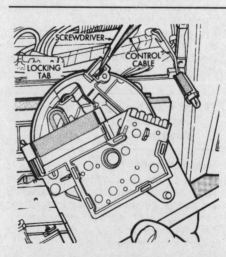

16.7 Release the control cable locking tab with a screwdriver

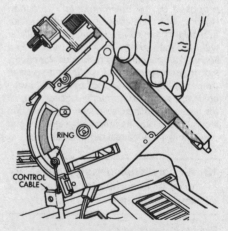

16.8 Detach the ring on the end of the control cable

Chapter 4
Fuel and exhaust systems

Contents

Specifications

Torque specifications
	Ft-lbs
Fuel pump (mechanical) mounting bolts	13 to 19
Carburetor-to-intake manifold mounting nuts	13 to 19
Throttle body mounting nuts (vehicles equipped with TBI)	16

1   General information

Fuel system

The fuel system consists of the fuel tank, the fuel pump, the fuel filter, a thermostatically controlled air cleaner assembly and a carburetor or fuel injection system.

1984 and 1985 models with a four-cylinder engine are equipped with a one-barrel Carter YFA feedback carburetor. Vehicles with this engine manufactured from 1986 through 1990 are equipped with a Throttle Body Injection (TBI) system. 1991 and later four-cylinder models are equipped with multi-point injection (MPI).

Vehicles with a V6 engine are equipped with either a 2SE (49-state and Canada) or E2SE (California) two-barrel feedback carburetor.

Vehicles with an inline six-cylinder engine are equipped with a Multi-Point Injection (MPI) system.

The fuel pumps of carbureted and fuel-injected vehicles are different: Carbureted vehicles use a mechanical pump driven by an eccentric lobe on the camshaft. Fuel injected engines use a gear/rotor type pump driven by an electric motor. The mechanical pump is mounted on the engine block; electric pumps are installed inside the fuel tank. Finally, it should be noted that the electric pumps used with TBI and MPI systems are different and cannot be interchanged.

Two different fuel feedback systems are used: One for four-cylinder engines and one for V6 engines sold in California. For more information on the fuel feedback systems, refer to Chapter 6.

Exhaust system

The basic exhaust system on all vehicles consists of a single or dual exhaust manifold, a front exhaust pipe, a catalytic converter, heat shield(s), and a muffler and tailpipe. The exhaust system is suspended from the underside of the vehicle and insulated from vibration by a series of rubber hangers.

2   Fuel pressure relief (fuel-injected vehicles)

Refer to illustrations 2.5 and 2.7
Warning: *Gasoline is extremely flammable, so extra precautions must be taken when working on any part of the fuel system. Do not smoke or allow open flames or bare light bulbs in or near the work area. Also, don't work in a garage where a natural gas appliance such as a water heater or clothes dryer is present.*

TBI-equipped vehicles

1   The fuel system of vehicles equipped with Throttle Body Injection (TBI) is only under pressure when the fuel pump is operating. As long as the fuel pump is not operating, TBI fuel system components can be removed without the need to release system pressure.

2.5 To relieve the fuel pressure on an MPI-equipped vehicle, remove the cap from the pressure test port, place some shop towels underneath to absorb sprayed fuel . . .

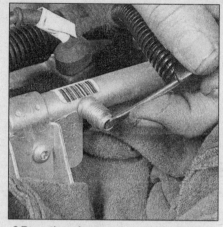

2.7 . . . then depress the valve in the test port with a small screwdriver or pin punch - wear safety goggles during pressure relief to protect your eyes from spraying fuel

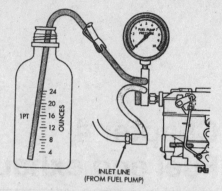

3.10 The proper setup for the fuel pump pressure and capacity tests (if you don't have a hose restrictor, a pair of vise grips with padded jaws will work just as well)

MPI-equipped vehicles

1995 and earlier models (equipped with a test port)

2   The MPI system is under constant fuel pressure ranging from 39 to 50 psi. Before attempting to service any fuel supply or return component on vehicles equipped with this system, release the fuel pressure.

3   Detach the cable from the negative battery terminal.

4   Remove the fuel tank filler cap to relieve fuel tank pressure.

5   Remove the cap from the pressure test port on the fuel rail **(see illustration)**.

6   Place shop towels under the pressure test port to absorb fuel when the pressure is released from the fuel rail.

7   Using a small screwdriver or pin punch, push the test port valve in to relieve fuel pressure **(see illustration)**. Absorb the spilled fuel with the shop towels. Install the cap on the pressure test port.

1996 and later models (no test port)

8   Later models may not have a test port. Wrap a shop towel around any fuel connector leading into the fuel rail. Carefully disconnect the fuel line (see Section 12).

9   Mop up any spilled fuel on engine surfaces and from below the vehicle before restarting the engine. It is best you let the vehicle stand until all the fuel evaporates.

10   Hang the shop towels outside away from all ignition sources until the fuel safely evaporates.

11   Be sure to attach the fuel line before starting the vehicle (see Section 12). Check to be sure there are no leaks.

3   Fuel pump/fuel pressure - check

Warning: *Gasoline is extremely flammable, so extra precautions must be taken when working on any part of the fuel system. Do not smoke or allow open flames or bare light*

bulbs in or near the work area. Also, don't work in a garage where a natural gas appliance such as a water heater or clothes dryer is present.

Mechanical pump (carburetor equipped vehicles)

Quick check

1   Detach the cable from the negative battery terminal.

2   Remove the air cleaner assembly (see Section 7).

3   Detach the fuel inlet fitting from the carburetor and place the end of the inlet line in a metal or plastic container.

4   Attach the cable to the negative battery terminal.

5   Hook up a remote starter switch, if available, in accordance with the manufacturer's instructions. If you don't have a remote starter switch, you will need an assistant for the following procedures.

6   Disable the ignition coil by detaching the primary lead wires (see Chapter 5).

7   With the fuel line directed into the container, have an assistant turn the ignition key to Start and crank the engine for about ten seconds.

8   Fuel should be emitted from the fuel line in well-defined spurts. If it isn't, there is a problem somewhere in the fuel delivery system. The following tests will determine where the problem lies.

Pressure test

Refer to illustration 3.10

Note: *You will need a fuel pressure gauge, a hose restrictor and a section of flexible hose to perform the following procedures.*

9   If the vehicle has a four-cylinder engine, detach the fuel return hose at the fuel filter and plug the fitting on the filter.

10   Attach a pressure gauge, restrictor and flexible hose between the fuel inlet fitting or the fuel filter and the carburetor **(see illustration)**.

11   Position the flexible hose and restrictor

so fuel can be discharged into a graduated container.

12   Reattach the coil primary lead wires (the engine must be operated for the following test).

13   Start the engine (or have your assistant start it), let it run at curb idle rpm, then discharge the fuel into the container by momentarily opening the hose restrictor.

14   Close the hose restrictor, allow the pressure to stabilize and note the pressure. The gauge should indicate 4-to-5 psi for four-cylinder engines and 6-to-7.5 psi for V6 engines.

   a) *If the pump pressure is not within specification, and the fuel lines are in satisfactory condition, the pump is defective and should be replaced.*

   b) *If the pump pressure is within specifications, perform the following tests for capacity and vacuum.*

Capacity test

15   Operate the engine at curb idle rpm.

16   Open the hose restrictor and allow fuel to discharge into a graduated container for 30 seconds, then close the restrictor. At least one pint of fuel should have been discharged.

   a) *If the pump volume is less than one pint, repeat the test with an auxiliary fuel supply and, for four-cylinder engines, a replacement fuel filter.*

   b) *If the pump volume conforms to the specified amount while using the auxiliary fuel supply, look for a restriction in the fuel supply line from the tank and check the tank vent to make sure it's working properly.*

Direct connection vacuum test (V6 engines)

17   You will need a vacuum gauge to perform the direct connection vacuum test. In this test, the vacuum test gauge is connected directly to the fuel pump inlet to test the pump's ability to create a vacuum.

18   Detach the fuel inlet line at the fuel pump.

19   Attach a vacuum gauge to the fuel pump inlet.

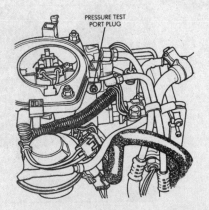

3.28 All system pressure tests on TBI-equipped vehicles require the removal of the pressure test port plug from the throttle body - in its place, install a special pressure test fitting (available at an auto parts department)

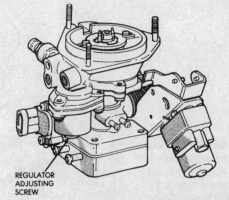

3.31 If the TBI system fuel pressure indicated on the test gauge is incorrect, adjust it by turning the regulator adjusting screw - turn the screw in (clockwise) to increase pressure or out (counterclockwise) to decrease pressure

20 Operate the engine at curb idle speed and note the vacuum gauge reading. It should indicate a vacuum of 10 in-Hg (the gauge will not indicate a vacuum until the fuel in the carburetor float bowl has been consumed and the pump begins to operate at full capacity).

21 If the pump vacuum is not within specification, the pump is defective. Replace it (see Section 4).

Indirect connection vacuum test (V6 engine)

22 You will need a vacuum gauge and a T-fitting to perform the indirect connection vacuum test. In this test, a vacuum gauge is connected by a T-fitting into the pump inlet to determine whether an obstruction exists in the fuel line or the in-tank fuel filter.

23 Detach the fuel inlet line at the fuel pump.

24 Install a T-fitting between the disconnected fitting and the fuel pump inlet. Connect a vacuum gauge to the T-fitting.

25 Operate the engine at a speed of 1500 rpm for 30 seconds and note the reading on the vacuum gauge (again, the gauge will not indicate any vacuum until the fuel in the carburetor float bowl has been consumed and the pump begins to operate at full capacity). The indicated vacuum should not exceed 3 in-Hg.

26 If the indicated vacuum exceeds the specified vacuum, check the fuel line for a restriction. A partially clogged in-tank fuel filter can also cause excess vacuum.

Electric pump (fuel injected vehicles)

TBI pressure test

Refer to illustrations 3.28 and 3.31

27 Detach the cable from the negative battery terminal.

28 Remove the pressure test port plug from the throttle body **(see illustration)**.

29 Install a pressure test fitting (available at a Jeep parts department) in place of the test port plug.

30 Attach a 0-to-30 psi fuel pressure gauge to the pressure test fitting (don't use a carburetor type 0-to-15 psi gauge).

31 Start the engine and let it idle. The pressure gauge should read 14-to-15 psi. If the pressure is incorrect, adjust it by turning the regulator adjusting screw **(see illustration)** to obtain the correct fuel pressure. Turn the screw at the bottom of the regulator in (clockwise) to increase pressure or out (counterclockwise) to decrease pressure.

32 If the fuel pressure is considerably higher than specified, and adjusting the regulator fails to lower it to the specified level, inspect the fuel return line for blockage.

33 If the fuel pressure is considerably below specification and adjusting the regulator fails to raise it to the specified level, momentarily pinch off the fuel return line and recheck the pressure.

  a) If the fuel pressure has risen, replace the pressure regulator (see Section 13).

  b) If the pressure has not risen, check the fuel filter (see Chapter 1) and fuel supply line for blockage.

MPI pressure test (1995 and earlier models)

Refer to illustrations 3.35, 3.36 and 3.41

34 The MPI fuel system employs a vacuum-assisted pressure regulator. Fuel pressure should be about eight to ten psi higher with the vacuum line attached to the regulator than with the vacuum line disconnected. Fuel system pressure should be 31 psi with the vacuum line attached to the regulator and 39 psi with the line detached.

35 On earlier models, attach a 0-to-60 psi fuel pressure gauge to the test port pressure fitting on the fuel rail **(see illustration)**. Later models without a fitting will require a special fixture available from Jeep. Follow the instructions that come with the adapter fixture.

36 Detach the vacuum tube from the fuel pressure regulator **(see illustration)**.

37 Start the engine.

38 Note the gauge reading. With the vacuum line detached, fuel pressure should be about 39 psi.

39 Attach the vacuum line to the pressure regulator. Note the gauge reading. Fuel pressure should be about 31 psi.

40 If the indicated fuel pressure is not about 8-to-10 psi higher with the vacuum line removed from the regulator, inspect the pressure regulator vacuum line for leaks, kinks and blockage.

41 If the fuel pressure is below the specified level, momentarily pinch off the hose section of the return line **(see illustration)**. **Warning:** *Don't allow the fuel pressure to exceed 60 psi.*

  a) *If the fuel pressure remains below specifications, inspect the fuel supply line, fuel filter (see Chapter 1) and fuel rail inlet (see Section 14) for blockage.*

  b) *If the fuel pressure rises, replace the regulator.*

  c) *If the fuel pressure is above specification, inspect the fuel return line for kinks and blockage.*

MPI fuel pressure leak down test (1995 and earlier models)

42 If an abnormally long cranking period is required to restart a hot engine after the vehicle has been shut down for a short period of time, the fuel pressure may be leaking past the fuel pressure regulator or the check valve in the outlet end of the fuel pump.

43 With the engine off, attach a gauge capable of reading 0-to-100 psi.

44 Start the vehicle and let the engine idle. Check the fuel pressure reading on the gauge. The fuel pressure should be within the specifications noted above.

45 Shut the engine off. Note the fuel pressure reading on the gauge. Leave the fuel pressure gauge connected. Allow the engine to sit for 30 minutes, then compare the fuel pressure reading on the gauge to the reading you took when you shut down the engine. A drop of 0-to-20 psi (to the 19-to-39 psi range) is acceptable. If the fuel pressure drop is within specification, the fuel pump outlet

4

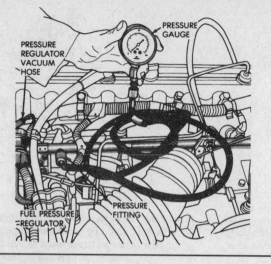

3.35 To test the MPI fuel system pressure, you will need to attach a 0-to-60 psi pressure gauge to the test port pressure fitting on the fuel rail

3.36 Detach the vacuum tube from the fuel pressure regulator, start the vehicle and note the gauge reading - with the vacuum line detached, fuel pressure should be about 39 psi

3.41 If the fuel pressure is below the specified level, momentarily pinch off the hose section of the fuel supply line - fuel pressure will rise to 95 psi when the fuel supply line is pinched, so turn the engine off immediately after pinching off the fuel supply line

check valve and the fuel pressure regulator are both operating normally. If the fuel pressure drop is greater than 20 psi, restart the vehicle, let the engine idle and momentarily pinch off the hose section of the fuel return line. The fuel pressure will rise to 95 psi when the fuel return line is pinched off, so shut the engine down immediately after pinching off the fuel return line. Note the pressure reading on the gauge. Allow the engine to sit for 30 minutes. Take another reading and compare it to the reading you took when you first shut down the engine.

a) If the fuel pressure has dropped about 20 psi or less, replace the fuel pressure regulator.
b) If the fuel pressure has dropped considerably more than 20 psi, fuel pressure is bleeding off past the outlet check valve in the fuel pump. Replace the pump.

MPI pressure test (1996 and later models)

Refer to illustration 3.47
Note: *In order to perform the fuel pressure*

test, you will need a fuel pressure gauge capable of measuring high fuel pressure. The fuel gauge must be equipped with the proper fittings or adapters required to attach it to the fuel rail.
46   Relieve the fuel pressure (see Section 2).
47   Remove the cap from the fuel pressure test port (if equipped) on the fuel rail and attach a fuel pressure gauge **(see Illustration).** **Note:** *A special adapter with a T-fitting will be required to perform the leakdown test described in Step 51.*
48   Start the engine and check the pressure on the gauge, comparing your reading with the pressure listed in this Chapter's Specifications.
49   If the fuel pressure is lower than specified, check the fuel lines and the fuel filter for restrictions. If no restriction is found, replace the fuel pump (see Section 4).
50   If fuel pressure is higher than specified, replace the fuel pressure regulator/fuel filter (see Section 15).
51   If the fuel pressure is within specifications, turn the engine off and monitor the fuel pressure for five minutes. The fuel pressure should not drop below 30 psi within five minutes. If it does, there is a leak in the fuel line, a fuel injector is leaking or the fuel pump module check valve is defective. To determine the problem area, perform the following:
Note: *A special fuel gauge adapter hose is required for the following tests, the fuel pressure gauge must be connected to the T-fitting* **(see Illustration 3.47).**
52   Install the adapter hose, then turn on the ignition key to power-up the fuel system. Clamp off the adapter hose between the T-fitting and the line to the fuel pump, then turn the ignition key off. If the pressure drops below 30 psi within five minutes, a fuel injector (or injectors) is leaking (or the fuel rail may be leaking, but such a leak would be very apparent).
53   With the adapter hose installed, clamp off the adapter hose between the T-fitting and the fuel rail. Turn the ignition key on to energize the fuel pump, then turn the key off. If the pressure drops below 30 psi within five

minutes, the fuel line is leaking, the check valve in the fuel pressure regulator is defective or the fuel pump is defective.

a) Check the entire length of the fuel line for leaks.
b) If no leaks are found in the fuel line and the pressure dropped quickly, replace the fuel pressure regulator/fuel filter.
c) If no leaks are found in the fuel line and the pressure dropped slowly, replace the fuel pump.

4   Fuel pump - removal and installation

Warning: *Gasoline is extremely flammable, so extra precautions must be taken when working on any part of the fuel system. Do not smoke or allow open flames or bare light bulbs in or near the work area. Also, don't work in a garage where a natural gas appliance such as a water heater or clothes dryer is present.*

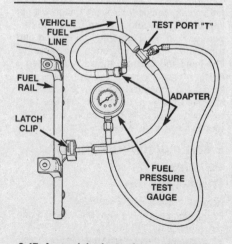

3.47 A special adapter hose with a T-fitting will be required for the fuel pressure checks

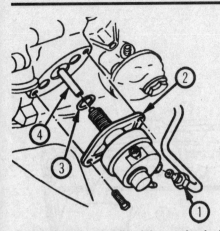

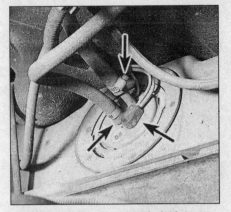

4.4a Installation details of the mechanical fuel pump used on V6 models

1　Inlet line
2　Fuel pump
3　Gasket
4　Actuating rod (this may fall out of engine block when fuel pump is removed)

4.4b Installation details of the mechanical fuel pump used on four-cylinder models

4.12 To remove the electric fuel pump from the fuel tank, detach the hoses and the electrical connector (arrows) - label the hoses so you can return them to their original locations

Mechanical pump

Refer to illustrations 4.4a and 4.4b

1　Detach the cable from the negative battery terminal.
2　Remove the fuel tank filler cap to relieve fuel pressure.
3　Wrap shop towels around the fuel pump inlet hose and outlet line fitting to absorb any fuel spilled during fuel pump removal.
4　Detach the fuel inlet hose and outlet line fitting from the fuel pump **(see illustrations)**.
5　Unscrew the fuel pump mounting bolts and remove the fuel pump and gasket.
6　Carefully scrape away any old gasket material from the fuel pump and engine block sealing surfaces.
7　Installation is the reverse of removal. Be sure to use a new gasket and tighten the fuel pump bolts to the specified torque.

Electric pump

Removal

Refer to illustrations 4.12, 4.15, 4.20a and 4.20b

Note: *The following procedure requires removal of the fuel tank for 1988 and earlier models.*

8　Remove the fuel tank filler cap to relieve fuel tank pressure.
9　Relieve the fuel system pressure (see Section 2).
10　Detach the cable from the negative battery terminal.
11　Raise the vehicle and place it securely on jackstands.
12　Disconnect the fuel vent, supply and return hoses from the fittings on the fuel pump/sending unit **(see illustration)**.
13　Detach the fuel pump/sending unit electrical harness connector from the main harness.
14　The fuel pump/sending unit assembly is located inside the fuel tank. It's held in place by a cam lock ring mechanism consisting of an inner ring with three locking cams and an outer ring with three retaining tangs.
15　To unlock the fuel pump/sending unit assembly, turn the inner ring counterclockwise until the locking cams are free of the retaining tangs. If the rings are locked together too tightly to release them by hand,

gently knock them loose with a wood dowel or a brass punch and hammer **(see illustration)**. **Warning:** *Do not use a steel punch to knock the lockrings loose. A spark could cause an explosion!*
16　Extract the fuel pump/sending unit assembly from the fuel tank. The fuel level float and sending unit are delicate. Do not bump them into the lockring during removal or the accuracy of the sending unit may be affected.
17　Inspect the condition of the gasket around the mouth of the lockring mechanism. If it's dried, cracked or deformed, replace it.
18　Inspect the inside of the tank. Have it cleaned by a radiator shop if sediment is present.
19　If you are replacing the fuel pump, make sure you get the right pump. The TBI and MPI pumps look alike, but they're NOT interchangeable.

Disassembly

20　Remove and discard the fuel pump inlet filter **(see illustrations)**.

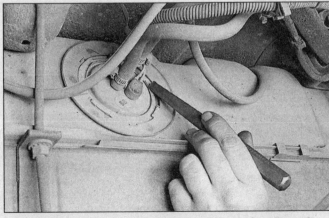

4.15 To loosen the lock ring for the fuel pump/sending unit assembly, turn it counterclockwise - if the ring is difficult to loosen, tap it loose with a wood dowel or brass punch and a small mallet (DO NOT use a steel punch, or you could cause an explosion!

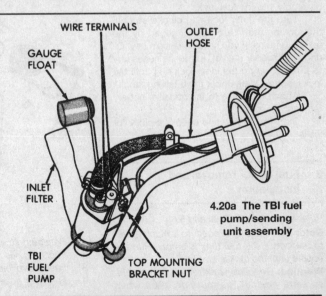

4.20a The TBI fuel pump/sending unit assembly

WIRE TERMINALS
GAUGE FLOAT
OUTLET HOSE
INLET FILTER
TBI FUEL PUMP
TOP MOUNTING BRACKET NUT

4

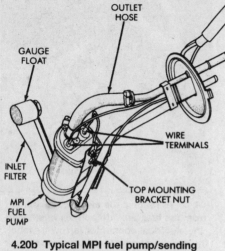

4.20b Typical MPI fuel pump/sending unit assembly (pre-1996 models)

21   Detach the fuel pump wires (the wire ends are different sizes and cannot be connected to the wrong terminal).
22   Detach the fuel pump outlet hose and clamp. Replace the hose if it shows signs of wear or deterioration.
23   Remove the fuel pump top mounting bracket nut. Remove the fuel pump.

Reassembly

24   Install a new inlet filter.
25   Place the fuel pump top mounting bracket over the top of the pump.
26   Position the fuel pump in the lower bracket. Slide the stud of the top bracket through the hole in the fuel pump side bracket. Tighten the fuel pump top mounting nut.
27   Install the fuel pump outlet hose. Secure it with new clamps.
28   Connect the wire terminals to the motor.

Installation

29   Insert the fuel pump/sending unit assembly into the fuel tank.
30   Turn the inner lock ring clockwise until the locking cams are fully engaged by the retaining tangs. If you have installed a new O-ring type rubber gasket, it may be necessary to push down on the inner lock ring until the locking cams slide under the retaining tangs.
31   Attach the cable to the negative battery terminal.
32   Start the engine and check carefully for leaks.

5   Fuel tank - removal and installation

Refer to illustrations 5.6a and 5.6b
Note: *The following procedure is much easier to perform if the fuel tank is empty. Run the engine until the tank is empty.*
Warning: *Gasoline is extremely flammable, so extra precautions must be taken when*

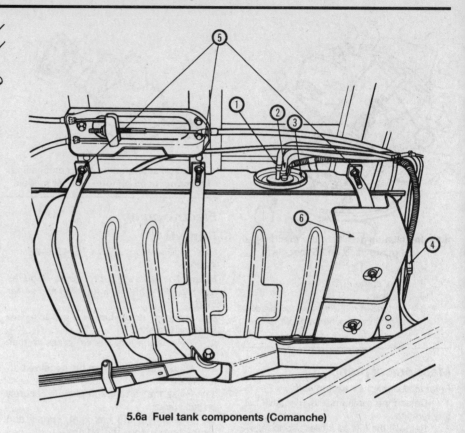

5.6a Fuel tank components (Comanche)

1	Fuel feed line	4	Sending unit/fuel pump electrical
2	Fuel return line		connector
3	Fuel gauge sending unit (and, on	5	Retaining strap nuts
	TBI/MPI models, fuel pump)	6	Protective shield

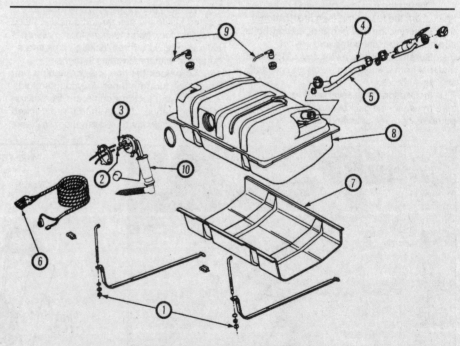

5.6b Fuel tank components (Cherokee)

1	Retaining strap nuts	5	Filler vent hose	8	Fuel tank
2	Fuel feed line	6	Fuel gauge sending unit	9	Vapor vent hose
3	Fuel return line		electrical connector	10	Fuel pump (TBI and
4	Fuel filler hose	7	Protective shield		MPI models)

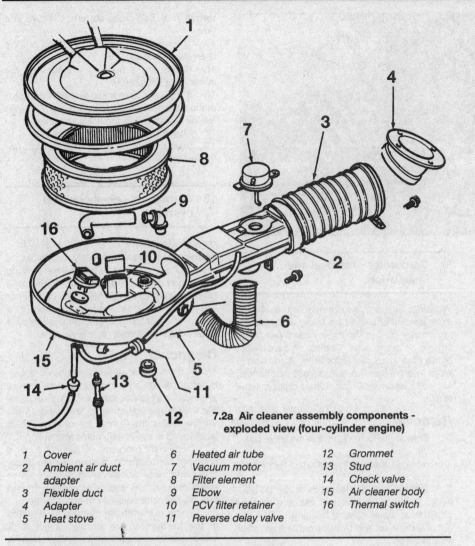

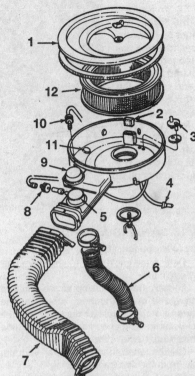

7.2a Air cleaner assembly components - exploded view (four-cylinder engine)

1	Cover	6	Heated air tube	12	Grommet
2	Ambient air duct	7	Vacuum motor	13	Stud
	adapter	8	Filter element	14	Check valve
3	Flexible duct	9	Elbow	15	Air cleaner body
4	Adapter	10	PCV filter retainer	16	Thermal switch
5	Heat stove	11	Reverse delay valve		

7.2b Air cleaner assembly components - exploded view (V6 engine)

1   Air cleaner cover
2   PCV valve filter
3   Thermal switch
4   Check valve
5   Vacuum motor
6   Heated air tube
7   Ambient air duct
8   Reverse delay valve
9   Trap door assembly
10   Reverse delay valve
11   Thermal vacuum switch
12   Filter element

working on any part of the fuel system. Do not smoke or allow open flames or bare light bulbs near the work area. Also, do not work in a garage if a natural gas-type appliance with a pilot light is present. While performing any work on the fuel tank, wear safety glasses and have a dry chemical (Class B) fire extinguisher on hand. If you spill any fuel on your skin, rinse it off immediately with soap and water.

1   Remove the fuel tank filler cap to relieve fuel tank pressure.
2   If the vehicle is fuel-injected, relieve the fuel system pressure (see Section 2).
3   Detach the cable from the negative terminal of the battery.
4   If the tank still has fuel in it, you can drain it at the fuel supply line after raising the vehicle.
5   Raise the vehicle and place it securely on jackstands.
6   Disconnect all hoses and the electrical connector for the fuel gauge sending unit and electric fuel pump (if equipped) **(see illustrations)**. Carefully label all hoses so you can reinstall them in their original locations.
7   Siphon the fuel from the tank at the fuel feed line, not the return line.
8   Support the fuel tank with a floor jack or

jackstands. Position a piece of wood between the jack head and the fuel tank to protect the tank.
9   Disconnect the fuel tank retaining straps and pivot them down until they are hanging out of the way.
10   Remove the tank from the vehicle.
11   Installation is the reverse of removal.

6   Fuel tank cleaning and repair - general information

1   All repairs to the fuel tank or filler neck should be carried out by a professional who has experience in this critical and potentially dangerous work. Even after cleaning and flushing of the fuel system, explosive fumes can remain and ignite during repair of the tank.
2   If the fuel tank is removed from the vehicle, it should not be placed in an area where sparks or open flames could ignite the fumes coming out of the tank. Be especially careful inside garages where a natural gas type appliance is located, because the pilot light could cause an explosion.

4

7   Air cleaner - removal and installation

Four-cylinder and V6 engines

Refer to illustrations 7.2a and 7.2b
1   Detach the cable from the negative terminal of the battery.
2   Detach the flexible ambient air duct from the air cleaner housing snorkel **(see illustrations)**.
3   Remove the wing nut from the cover and remove the cover and filter element. Inspect the element for contamination by dirt and moisture. Replace it if necessary (see Chapter 1).
4   Clearly label, then detach, all hoses from the air cleaner housing.
5   Remove the air cleaner housing assembly.
6   Installation is the reverse of removal.

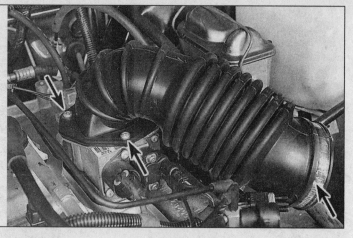

7.8 To remove the flexible air duct on inline six-cylinder engines, loosen the hose clamp at one end and remove the three attaching screws (arrows) from the other end - the third attaching screw is hidden by the duct in this photo

Inline six-cylinder engine

Refer to illustration 7.8

7   Detach the cable from the negative battery terminal.

8   Loosen the hose clamp at one end and remove the three attaching screws at the other end of the flexible air duct between the air cleaner housing and the throttle body **(see illustration)**. Remove the duct.

9   Clearly label, then detach, all vacuum hoses from the air cleaner housing.

10   Detach the heat riser tube from the air cleaner housing.

11   Remove the upper half of the air cleaner housing lid, then remove the air filter element.

12   Remove the two bolts and one nut from the floor of the air cleaner housing. Remove the housing.

13   Installation is the reverse of removal.

8   Carburetor - adjustments

1   The carburetors used on the vehicles covered by this manual are protected by a Federally-mandated extended warranty (at the time this manual was written, the warranty was for 5 years/50,000 miles, comes first - see your dealer for details).

2   We don't recommend carburetor adjustments while it's still under warranty. If you are having problems related to fuel delivery - and you have eliminated all other parts of the fuel delivery system as possible causes - take the vehicle to a dealer and have the carburetor professionally serviced.

3   Unless you really know what you're doing, we don't recommend you attempt to adjust an out-of-warranty carburetor either. The carburetors on these vehicles are expensive, complex and tricky to adjust. The adjustments are also totally interrelated, so it's impossible to do certain adjustments unless you have already done several others. Finally, by the time adjustments become necessary, the carburetor will probably be approaching the end of its service life and will be impossible to adjust to a like-new state of tune. At this point, you're better off buying a rebuilt or new carburetor.

9   Carburetor - removal and installation

Warning: *Gasoline is extremely flammable so extra precautions must be taken when working on any part of the fuel system. DO NOT smoke or allow open flames or bare light bulbs in or near the work area. Also, don't work in a garage if a natural gas appliance such as a water heater or clothes dryer is present.*

Removal

1   Detach the cable from the negative battery terminal.

2   Remove the fuel filler cap to relieve fuel tank pressure.

3   Remove the air cleaner from the carburetor. Be sure to label all vacuum hoses attached to the air cleaner housing.

4   Disconnect the throttle cable from the throttle lever.

5   If the vehicle is equipped with an automatic transmission, disconnect the TV cable from the throttle lever (see Chapter 7, Part B).

6   Clearly label all vacuum hoses and fittings, then disconnect the hoses.

7   Disconnect the fuel line from the carburetor.

8   Label the wires and terminals, then unplug all wire harness connectors.

9   Remove the mounting fasteners and detach the carburetor from the intake manifold. Remove the carburetor mounting gasket. Stuff a shop rag into the intake manifold openings.

Installation

10   Use a gasket scraper to remove all traces of gasket material and sealant from the intake manifold (and the carburetor, if it's being reinstalled), then remove the shop rag from the manifold openings. Clean the mating surfaces with lacquer thinner or acetone.

11   Place a new gasket on the intake manifold.

12   Position the carburetor on the gasket and install the mounting fasteners.

13   To prevent carburetor distortion or damage, tighten the fasteners to the specified

torque in a criss-cross pattern, 1/4-turn at a time.

14   The remaining installation steps are the reverse of removal.

15   Check and, if necessary, adjust the idle speed (see Chapter 1).

16   If the vehicle is equipped with an automatic transmission, refer to Chapter 7, Part B for the TV cable adjustment procedure.

17   Attach the negative battery cable.

18   Start the engine and check carefully for fuel leaks.

10   Carburetor - diagnosis and overhaul

Refer to illustrations 10.8a, 10.8b and 10.8c

Warning: *Gasoline is extremely flammable, so extra precautions must be taken when working on any part of the fuel system. DO NOT smoke or allow open flames or bare light bulbs in or near the work area. Also, don't work in a garage if a natural gas appliance such as a water heater or clothes dryer is present.*

Diagnosis

1   A thorough road test and check of carburetor adjustments should be done before any major carburetor service work. Specifications for some adjustments are listed on the Vehicle Emissions Control Information (VECI) label found in the engine compartment.

2   Carburetor problems usually show up as flooding, hard starting, stalling, severe backfiring and poor acceleration. A carburetor that's leaking fuel and/or covered with wet looking deposits definitely needs attention.

3   Some performance complaints directed at the carburetor are actually a result of loose, out-of-adjustment or malfunctioning engine or electrical components. Others develop when vacuum hoses leak, are disconnected or are incorrectly routed. The proper approach to analyzing carburetor problems should include the following items:

a) *Inspect all vacuum hoses and actuators for leaks and correct installation (see Chapters 1 and 6).*

b) *Tighten the intake manifold and carburetor mounting nuts/bolts evenly and securely.*

c) *Perform a cylinder compression test (see Chapter 2).*

d) *Clean or replace the spark plugs as necessary (see Chapter 1).*

e) *Check the spark plug wires (see Chapter 1).*

f) *Inspect the ignition primary wires.*

g) *Check the ignition timing (follow the instructions printed on the Emissions Control Information label).*

h) *Check the fuel pump and fuel pressure (see Section 3).*

i) *Check the heat control valve in the air cleaner for proper operation (see Chapter 1).*

j) *Check/replace the air filter element (see Chapter 1).*

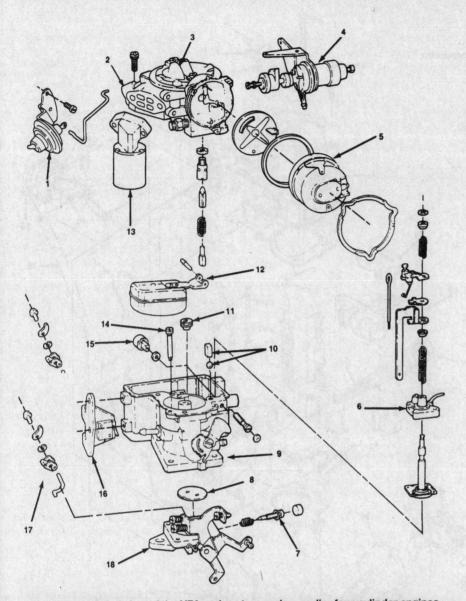

10.8a An exploded view of the YFA carburetor used on earlier four-cylinder engines

1	Vacuum break	10	Accelerator pump check ball and
2	Air horn		weight
3	Choke plate	11	Main metering jet
4	Sole-Vac throttle positioner	12	Float assembly
5	Choke assembly	13	Mixture control solenoid
6	Accelerator pump assembly	14	Low speed jet
7	Idle mixture screw with O-ring	15	Accelerator pump vent valve
8	Throttle plate	16	Wide open throttle (WOT) switch
9	Main body	17	Throttle shaft and lever
		18	Throttle body

k) Check the PCV system (see Chapter 6).
l) Check/replace the fuel filter (see Chapter 1). Also, the filter in the tank could be restricted.
m) Check for a plugged exhaust system.
n) Check EGR valve operation (see Chapter 6).
o) Check the choke-it should be completely open at normal engine operating temperature (see Chapter 1).

p) Check for fuel leaks and kinked or dented fuel lines (see Chapters 1 and 4).
q) Check accelerator pump operation with the engine off (remove the air cleaner cover and operate the throttle as you look into the carburetor throat - you should see a stream of gasoline enter the carburetor).
r) Check for incorrect fuel or bad gasoline.

s) Check the valve clearances (if applicable) and camshaft lobe lift (see Chapter 2).
t) Have a dealer service department or repair shop check the electronic engine and carburetor controls.

4   Diagnosing carburetor problems may require that the engine be started and run with the air cleaner off. While running the engine without the air cleaner, backfires are possible. This situation is likely to occur if the carburetor is malfunctioning, but just the removal of the air cleaner can lean the fuel/air mixture enough to produce an engine backfire. **Warning:** *Do not position any part of your body, especially your face, directly over the carburetor during inspection and servicing procedures. Wear eye protection!*

Overhaul

5   Once it's determined that the carburetor needs an overhaul, several options are available. If you're going to attempt to overhaul the carburetor yourself, first obtain a good quality carburetor rebuild kit (which will include all necessary gaskets, internal parts, instructions and a parts list). You'll also need some special solvent and a means of blowing out the internal passages of the carburetor with air.

6   An alternative is to obtain a new or rebuilt carburetor. They are readily available from dealers and auto parts stores. Make absolutely sure the exchange carburetor is identical to the original. A tag is usually attached to the top of the carburetor or a number is stamped on the float bowl. It will help determine the exact type of carburetor you have. When obtaining a rebuilt carburetor or a rebuild kit, make sure the kit or carburetor matches your application exactly. Seemingly insignificant differences can make a large difference in engine performance.

7   If you choose to overhaul your own carburetor, allow enough time to disassemble it carefully, soak the necessary parts in the cleaning solvent (usually for at least one-half day or according to the instructions listed on the carburetor cleaner) and reassemble it, which will usually take much longer than disassembly. When disassembling the carburetor, match each part with the illustration in the carburetor kit and lay the parts out in order on a clean work surface. Overhauls by inexperienced mechanics can result in an engine which runs poorly or not at all. To avoid this, use care and patience when disassembling the carburetor so you can reassemble it correctly.

8   Because carburetor designs are constantly modified by the manufacturer in order to meet increasingly more stringent emissions regulations, it isn't feasible to include a step-by-step overhaul of each type. You'll receive a detailed, well illustrated set of instructions with any carburetor overhaul kit; they will apply in a more specific manner to the carburetor on your vehicle. Exploded views of the three carburetors are included here **(see illustrations)**.

4

10.8b An exploded view of the Model 2SE carburetor used on 49-State V6 engines

Air horn components

1   Air horn long screw (2)
2   Air horn large screw
3   Air horn short screw (3)
4   Air horn medium screw
5   Vent stack assembly
6   Hot idle compensator screw (2)
7   Hot idle compensator
8   Hot idle compensator screw
9   Air horn assembly
10   Air horn gasket
11   Pump retainer
12   Pump stem seal
13   Stem seal retainer

Choke components

14   Primary vacuum break and bracket assembly
15   Vacuum break attaching screw
16   Air valve rod bushing
17   Air valve rod retainer
18   Vacuum break primary hose
19   Air valve rod
20   Fast idle cam rod
21   Intermediate choke shaft/lever/rod assembly
22   Intermediate choke shaft rod bushing
23   Intermediate choke shaft rod retainer
24   Secondary vacuum break and bracket assembly
25   Choke cover and coil assembly
26   Choke lever screw
27   Choke lever and contact assembly
28   Choke housing
29   Choke housing screw (2)
30   Stat cover retainer kit
31   Vacuum break attaching screw (2)

Float bowl components

32   Float bowl assembly
33   Fuel inlet nut
34   Fuel inlet nut gasket
35   Fuel inlet filter
36   Fuel filter spring
37   Float assembly
38   Float hinge pin
39   Float bowl insert
40   Needle and seat assembly
41   Pump return spring
42   Pump assembly
43   Main metering jet
44   Main metering assembly rod
45   Pump discharge ball
46   Pump discharge spring
47   Pump discharge spring retainer
48   Power piston assembly
49   Power piston spring

Throttle body components

50   Throttle body gasket
51   Throttle body assembly

52   Pump rod
53   Cam screw clip
54   Cam screw
55   Throttle stop screw spring
56   Throttle stop screw

57   Idle needle and spring
58   Throttle body attaching screw (4)
59   Idle speed kick actuator nut
60   Idle speed kick actuator retainer
61   Idle speed kick actuator

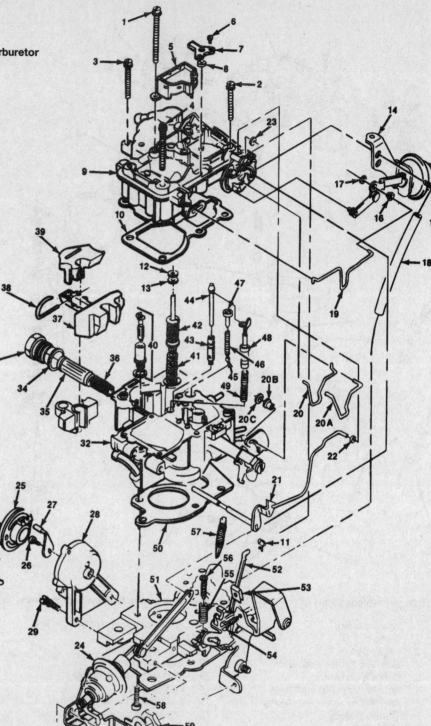

10.8c An exploded view of the E2SE carburetor used on California V6 engines

Air horn components

1   Mixture control (M/C) solenoid
2   Solenoid attaching screw assembly
3   M/C solenoid-to-air horn gasket
4   M/C solenoid spacer
5   M/C solenoid-to-float bowl seal
6   M/C solenoid seal retainer
7   Air horn assembly
8   Air horn-to-float bowl gasket
9   Air horn-to-float bowl short screw
10 Air horn-to-float bowl long screw
11 Air horn-to-float bowl large screw
12 Vent stack and screen assembly
13 Vent stack attaching screw
14 Pump stem seal
15 Pump stem seal retainer
16 TPS plunger seal
17 TPS plunger seal retainer
18 TPS actuator plunger

Choke components

19 Vacuum break and bracket assembly
20 Vacuum break primary hose
21 Vacuum break tee
22 Idle speed solenoid
23 Idle speed solenoid retainer
24 Idle speed solenoid attaching nut
25 Vacuum break bracket attaching screw
26 Air valve link
27 Air valve link bushing
28 Air valve link retainer
29 Fast idle cam link
29A Fast idle cam link
29B Link retainer
29C Link bushing
30 Vacuum break hose
31 Intermediate choke shaft/lever/link assembly
32 Intermediate choke link bushing
33 Intermediate choke link retainer
34 Secondary vacuum break and link assembly
35 Vacuum break attaching screw
36 Electric choke cover and coil assembly
37 Choke lever attaching screw
38 Choke coil lever assembly
39 Choke housing
40 Choke housing attaching screw
41 Choke cover retainer kit

Float bowl components

42 Fuel inlet nut
43 Fuel inlet nut gasket
44 Fuel inlet filter
45 Fuel filter spring
46 Float and lever assembly
47 Float hinge pin
48 Float bowl upper insert
48A Float bowl lower insert
49 Needle and seat assembly
50 Pump return spring
51 Pump plunger assembly
52 Primary metering jet assembly
53 Pump discharge ball retainer
54 Pump discharge spring
55 Pump discharge ball
56 TPS adjusting spring
57 Throttle Position Sensor (TPS)
58 Float bowl assembly
59 Float bowl gasket

Throttle body components

60 Pump link retainer
61 Pump link
62 Throttle body assembly
63 Cam screw clip
64 Fast idle cam screw
65 Idle needle and spring assembly
66 Throttle body-to-float bowl screw
67 Vacuum break bracket attaching screw
68 Idle stop screw
69 Idle stop screw spring
70 Insulator flange gasket

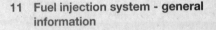

11   Fuel injection system - general information

Fuel-injected vehicles are equipped with either a Throttle Body Injection (TBI) system (four-cylinder engine) or a Multi-Point Injection (MPI) system (inline six-cylinder engine). Both systems use an Electronic Control Unit (ECU) to control pulse width.

The pulse width is the period of time during which the injector is energized (squirts fuel). The Electronic Control Unit (ECU), opens and closes the injector's ground path to control fuel injector pulse width and thus meter the amount of fuel available to the engine. By continually altering the pulse width, the ECU adjusts the air-fuel ratio for varying operating conditions. For more information about the ECU, see Chapter 6.

Throttle body injection (TBI) is a single-point system that injects fuel through one electrically operated fuel injector into the throttle body above the throttle plate.

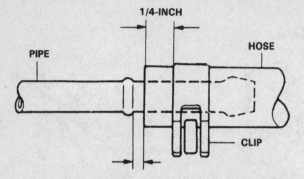

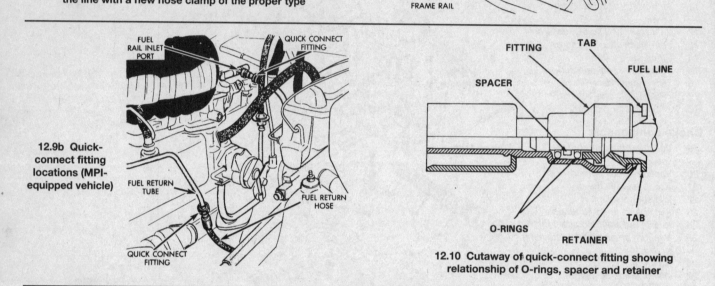

12.6 When attaching a section of rubber fuel hose to a metal fuel line, be sure to overlap the hose as shown and secure it to the line with a new hose clamp of the proper type

12.9a Typical under body quick-connect fittings (TBI-equipped vehicle)

12.9b Quick-connect fitting locations (MPI-equipped vehicle)

12.10 Cutaway of quick-connect fitting showing relationship of O-rings, spacer and retainer

Multi-Point Injection (MPI) is a multi-injector, sequential system: Fuel is injected into the intake manifold upstream of each intake valve in precisely metered amounts through electrically operated injectors. The injectors are energized in a specific sequence by the Electronic Control Unit (ECU). There is no injector in the throttle body itself, as in a TBI system; the six injectors are installed in the intake manifold. They receive pressurized fuel from a fuel rail attached to their upper ends.

System fuel pressure on both systems is provided by an in-tank electric fuel pump and is controlled by a spring and vacuum-assisted fuel pressure regulator.

12 Fuel lines and fittings - inspection and replacement

Warning: *Gasoline is extremely flammable, so extra precautions must be taken when working on any part of the fuel system. Don't smoke or allow open flames or bare light bulbs in or near the work area. Also, don't work in a garage where a natural gas appliance such as a water heater or clothes dryer is present.*

Inspection and replacement

Refer to illustration 12.6

1   Check the fuel lines and all fittings and connections for cracks, leakage and deformation.
2   Check the fuel tank vapor vent system hoses and connections for looseness, sharp bends and damage.
3   Check the fuel tank for deformation, cracks, fuel leakage and tank band looseness.
4   Check the filler neck for damage and fuel leakage.
5   Repair or replace any damaged or deteriorated hoses or lines. If your vehicle is equipped with fuel injection, see below for information on replacing fuel line fittings.
6   When attaching hoses to metal lines, overlap them as shown **(see illustration)**.

Fuel line fitting replacement (1993 and earlier fuel injected vehicles)

Refer to illustrations 12.9a, 12.9b, 12.10, 12.13, 12.14, 12.18a and 12.18b

7   Remove the fuel tank filler cap to relieve fuel tank pressure.
8   If the vehicle is equipped with Multi

Point Injection (MPI), relieve the system pressure before proceeding (see Section 2). If the vehicle is equipped with Throttle Body Injection (TBI), system pressure is bled off when the fuel pump is not operating - you can disconnect fuel hoses and lines as soon as you have turned off the engine.
9   All fuel injected engines use special quick-connect fuel line fittings. On TBI-equipped vehicles, the fittings are located at the throttle body ends of the nylon reinforced hoses which connect the throttle body to the fuel supply and return lines and under the vehicle along the left frame rail **(see illustration)**. On MPI-equipped engines, they're located at the fuel rail inlet port and at the connection between the fuel return line and the fuel return hose **(see illustration)**.
10   The fittings consist of two O-rings, a spacer between the two O-rings and a retainer **(see illustration)**.
11   Every time you disconnect a quick-connect fitting, you must replace the O-rings, spacer and retainer. These parts are available as a repair kit through any dealer parts department.
12   To disconnect a quick connect fitting, simply pinch the two retainer tabs together and pull the fitting apart. The retainer, O-rings and spacer will come out of the fitting when

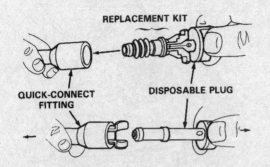

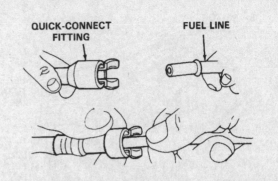

12.13 Insert the plug into the fitting until you hear a click, then withdraw and discard the plug

12.14 Once you've installed the new O-rings, spacer and retainer, push the fuel line into the fitting until you hear a click

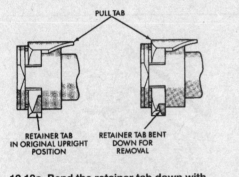

12.18a Bend the retainer tab down with your fingers

12.18b Push up on the bottom side of the pull tab, then release the retainer legs by pulling them away from each other

21   Inspect the connector body and retainer for damage. Inspect the inside of the fitting to ensure that it's free of dirt and/or obstructions. Replace as necessary.
22   Before installing the fuel line into the connector, wipe the fuel line end with a clean cloth, then apply clean engine oil to the outer surface of the fuel line.
23   To install the fuel line into the connector, align them and push the fuel line into place until it stops. Push the retainer down until it locks in place. Pull on the connector to ensure it's completely engaged.
24   After reassembly, run the engine and check for fuel leaks.

you pull the fuel lines apart. Discard these parts.
13   The replacement kit (O-rings, spacer and retainer) is installed on a disposable plastic plug. To replace these parts, push the disposable plug assembly into the quick connect fitting until you hear a "click" sound **(see illustration)**. Then grasp the end of the disposable plug and pull it out of the fitting.
14   Push the fuel line into the refurbished quick-connect fitting until you hear a "click" sound **(see illustration)**.
15   Verify that the connection is secure by pulling firmly back on the fuel line. It should be locked in place.

Fuel line fitting replacement (1994 and later fuel-injected vehicles)

Refer to illustrations 12.18a, 12.18b.
16   Relieve the system pressure before proceeding (see Section 2).
17   All fuel-injected engines use special quick-disconnect fuel line fittings. These fittings must be disconnected correctly; otherwise they will be damaged. **Caution:** If the retainer tab or removable tab is not bent down correctly prior to disconnecting the fitting, the connector could be damaged and must be replaced.
18   Bend the retainer tab down **(see illustration)**. While pushing up on the bottom side

of the pull tab, release the retainer legs by pulling them away from each other **(see illustration)**.
19   Raise the pull tab until the hooks on the retainer legs clear the stop on the fuel line nipple.
20   Grasp the connector and pull it straight off the fuel line. Some adhesion between the seal in the connector and the fuel line will occur over period of time. Twist the fitting on the fuel line, then push and pull the fitting until it moves freely.

13 Throttle body (TBI-equipped vehicles) - removal and installation

Refer to illustrations 13.2, 13.3, 13.6 and 13.8
1   Detach the cable from the negative battery terminal.
2   Detach the vacuum hoses from the throttle body upper bonnet **(see illustration)**. Release the mounting clips and remove the upper bonnet.
3   Remove the three mounting nuts and lift

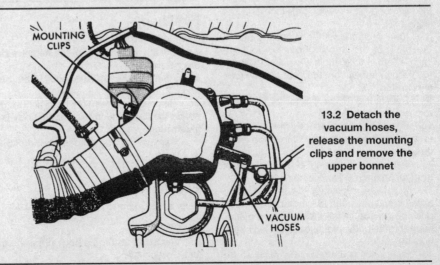

13.2 Detach the vacuum hoses, release the mounting clips and remove the upper bonnet

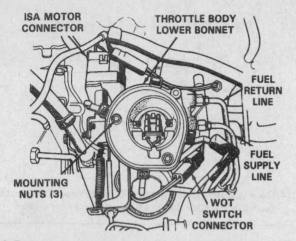

13.3 Remove the mounting nuts, then detach the lower bonnet

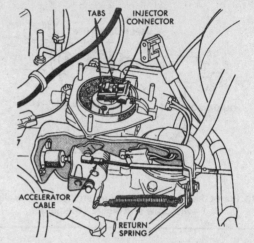

13.6 Remove the accelerator cable and return spring

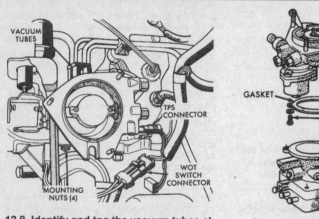

13.8 Identify and tag the vacuum tubes at the back of the throttle body, then detach them

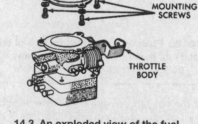

14.3 An exploded view of the fuel body assembly

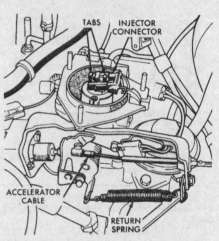

14.8 To remove the injector connector on TBI-equipped vehicles, compress the tabs and lift up

off the lower bonnet **(see illustration)**.
4   Disconnect the ISA motor harness connector **(see illustration 13.33)**.
5   Detach the fuel supply and return lines from the throttle body.
6   Remove the accelerator cable and return spring **(see illustration)**.
7   Detach the wire harness connector from the injector by compressing the lock tabs and lifting up.
8   Identify and label the vacuum tubes at the back of the throttle body so you can install them in their original locations. Detach the vacuum tubes from the throttle body **(see illustration)**.
9   Detach the TPS connector (two connectors on automatics).
10   Remove the throttle body mounting nuts **(see illustration 13.8)** and remove the throttle body.
11   Remove any old gasket material or dirt from the mating surfaces of the intake.
12   If you are replacing the throttle body assembly, remove the ISA motor and throttle position sensor. Install them on the replacement throttle body and adjust them (see Section 14).
13   Installation is the reverse of removal. Be

sure to use a new gasket and tighten the mounting nuts to the specified torque.

14   Throttle body injection (TBI) system - component replacement

Fuel body

Refer to illustration 14.3
1   Detach the cable from the negative battery terminal.
2   Remove the throttle body (see Section 13).
3   Remove the three Torx head screws that attach the fuel body to the throttle body **(see illustration)**.
4   Remove the original gasket and discard it.
5   Installation is the reverse of removal.

Fuel injector

Refer to illustrations 14.8, 14.9 and 14.10

Removal

6   Remove the throttle body upper bonnet (see Section 13).

7   Remove the throttle body lower bonnet (see Section 13).
8   Detach the injector connector **(see illustration)** by compressing the lock tabs and lifting up.
9   Remove the screws and lift off the retainer **(see illustration)**.
10   Using a small pair of pliers, gently grasp the center collar of the injector and carefully remove the injector by rocking it back and forth while lifting up **(see illustration)**. **Caution:** *Do not twist the injector during removal or its locating tab will be damaged.*
11   Remove and discard the centering ring and upper and lower O-rings **(see illustration 14.9)**. **Caution:** *Do not re-use these rings or fuel leakage and poor driveability may occur.*

Installation

12   Lubricate the replacement lower O-ring with light oil. Install the replacement lower O-ring in the bottom of the fuel injector housing bore.
13   Lubricate the replacement upper O-ring

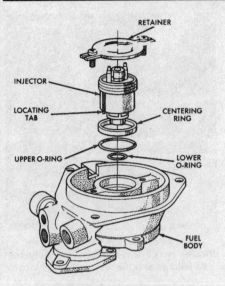

14.9 An exploded view of TBI fuel injector components

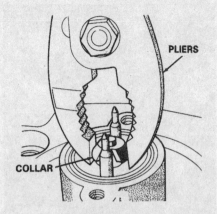

14.10 Gently grasp the center collar and carefully extract the injector by rocking it back and forth while lifting up - do not twist it during removal

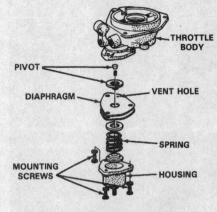

14.21 An exploded view of the fuel pressure regulator assembly

with light oil. Install it in the fuel injector housing bore.

14   Install the centering ring on top of the upper O-ring.

15   Align the locating tab on the bottom of the injector **(see illustration 14.9)** with the slot in the bottom of the housing and install the injector.

16   Install the retainer and tighten the screws securely.

17   Attach the injector connector.

18   Install the lower and upper bonnets.

19   Attach the negative battery cable.

Fuel pressure regulator

Refer to illustration 14.21

20   Remove the throttle body (see Section 13).

21   Remove the pressure regulator mount-

ing screws **(see illustration)**. **Warning:** *The regulator is under spring pressure. To prevent possible injury from the regulator flying off, keep the regulator forced against the throttle body while you're removing the screws.*

22   Remove the housing, spring, spring seats, diaphragm and pivot **(see illustration 14.21)**.

23   Remove any foreign material from the housing.

24   Installation is the reverse of removal. Make sure you install the diaphragm so its vent hole is aligned with the vent holes in the throttle body and housing.

25   After you have replaced the regulator and installed the throttle body on the engine, start the engine and check for leaks.

26   Adjust the fuel pressure (see Section 3).

Throttle position sensor (TPS)

Refer to illustration 14.28

27   Remove the upper and lower air inlet

bonnets (see Section 12).

28   Remove the throttle body (see Section 13). This is not absolutely necessary, but it makes it easier to get at the TPS. If you don't remove the throttle body, unplug the TPS connector **(see illustration)**. Note that the TPS used on vehicles with automatic transmissions has two connectors.

29   Remove the TPS mounting screws **(see illustration 14.28)**.

30   Remove the TPS from the throttle shaft lever.

31   Installation is the reverse of removal. Be sure the TPS arm is underneath the throttle shaft lever.

32   Have the TPS adjusted by a dealer service department.

Idle speed actuator (ISA) motor

Refer to illustration 14.33

33   Remove the ISA motor harness connector **(see illustration)**.

34   Detach the throttle return spring **(see**

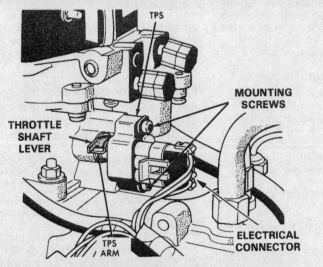

14.28 To remove the throttle position sensor (TPS), unplug the electrical connector and remove the two mounting screws

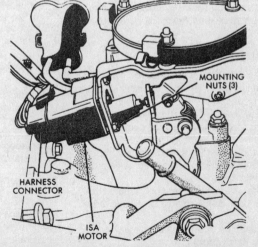

14.33 Disconnect the ISA motor harness connector; to remove the motor, unscrew the three mounting nuts - use a back-up wrench to avoid damaging the motor

4

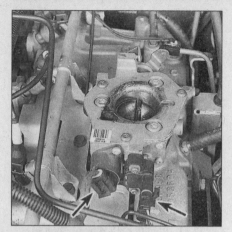

**15.4 Idle speed stepper motor connector
(left) - TPS connector (right)**

**15.5 Use a small screwdriver to pry the
MAP sensor vacuum tube from the back
of the throttle body**

**15.6 To detach the linkage from the
throttle arm, insert a screwdriver between
the linkage and the arm and pop it loose**

illustration 13.6).
35   Remove the three ISA motor-to-bracket
mounting nuts **(see illustration 14.33).** Use a
back-up wrench to prevent the studs which
hold the ISA motor together from turning.
Caution: *Don't attempt to remove the ISA
motor-to-bracket nuts without using a back-
up wrench on the stud nuts. ISA motor internal components may be damaged if the studs
disengage.*
36   Remove the ISA motor from the bracket.
37   Installation is the reverse of removal.
38   After you have replaced the ISA motor,
have it adjusted by a dealer service department.

---

15   Multi-Point Fuel Injection (MPI) -
component replacement

Warning: *Gasoline is extremely flammable,
so extra precautions must be taken when
working on any part of the fuel system. Don't
smoke or allow open flames or bare light
bulbs in or near the work area. Also, don't
work in a garage where a natural gas appliance such as a water heater or clothes dryer
is present.*

Throttle body

Refer to illustrations 15.4, 15.5, 15.6 and 15.7
1   Detach the cable from the negative battery terminal.
2   Relieve the fuel pressure (see Section 2).
3   Disconnect the flexible air duct from the
throttle body (see Section 7).
4   Detach the idle speed stepper motor
and throttle position sensor wire connectors
(see illustration).
5   Detach the MAP sensor vacuum tube
from the back of the throttle body **(see illustration).**
6   Detach the throttle linkage at the throttle
arm **(see illustration).**
7   If equipped an with automatic transmission, detach the line pressure cable at the
throttle arm **(see illustration).**

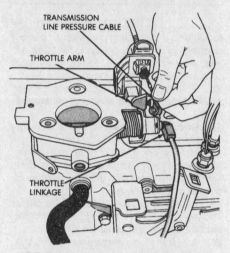

**15.7 If the vehicle is equipped with an
automatic transmission, detach the line
pressure cable at the throttle arm**

---

8   Remove the mounting bolts, throttle
body and gasket. Discard the old gasket and
clean the gasket mating surfaces on the
throttle body and intake manifold.
9   Installation is the reverse of removal.

Idle speed stepper motor

10   Detach the cable from the negative battery terminal.
11   Unplug the wire connector from the idle
speed stepper motor, remove the retaining
screws, then pull off the idle speed stepper
motor **(see illustration 15.4).**
12   Installation is the reverse of removal.

Throttle position sensor

Refer to illustration 15.15
13   Detach the cable from the negative battery terminal.
14   Unplug the TPS wire connector **(see
illustration 15.4).**
15   Bend back the TPS lock tabs, if
equipped **(see illustration).** Remove the
retaining screws, then detach the TPS from

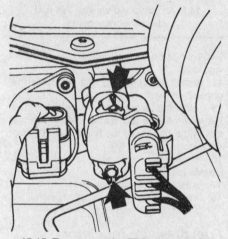

**15.15 To remove the TPS, remove the
retaining screws (arrows), then pull the
TPS from the throttle plate assembly - if
the screws are secured by lock tabs (as
shown here), bend the tabs back - use
new tabs when reinstalling the screws**

the throttle plate assembly.
16   Installation is the reverse of removal.
17   Have the TPS adjusted by a dealer service department.

Fuel injector rail assembly

*Refer to illustrations 15.21a, 15.21b, 15.23a,
15.23b, 15.23c, 15.24 and 15.27*
18   Remove the fuel filler cap to relieve fuel
tank pressure.
19   Relieve the fuel pressure (see Section 2).
20   Detach the cable from the negative battery terminal.
21   Numerically label, then unplug, the
injector harness connectors **(see illustrations).**
22   Detach the vacuum tube from the fuel
pressure regulator **(see illustration 3.36).**
23   Detach the fuel supply hose from the
fuel rail, the fuel return line from the intake
manifold, then the fuel return line from the
fuel pressure regulator **(see illustrations).**

15.21a Numerically label the injector harness connectors (arrows) so you don't mix them up during reassembly

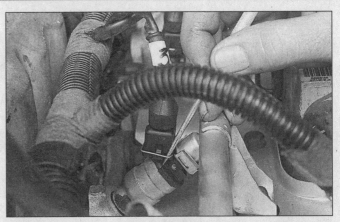

15.21b Use a scribe (shown) or a small screwdriver to pop loose the retaining clip that attaches the harness connector to each injector

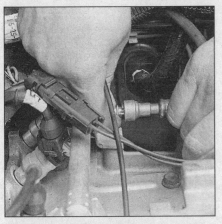

15.23a Detach the fuel supply hose from the fuel rail

15.23b Remove the nut from the fuel return line bracket and slide the bracket off its mounting stud . . .

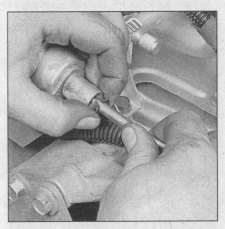

15.23c . . . then detach the fuel return line from the fuel pressure regulator (early model shown)

Refer to Section 12 for information on removing and replacing fuel line fittings.

24   Remove the fuel rail mounting bolts **(see illustration).**

25   If the vehicle is equipped with an automatic transmission, it may be necessary to remove the automatic transmission line pressure cable and bracket to remove the fuel rail

assembly **(see illustration 15.7).**

26   Remove the fuel rail by simultaneously rocking and pulling up on it until all injectors are out of the intake manifold. Work slowly and carefully so you don't damage the assembly.

27   Installation is the reverse of removal, but

be sure to do the following:

a) *Before attempting to reattach the fuel rail assembly mounting bolts, position the tip of each injector above its corresponding bore in the intake manifold* **(see illustration),** *then seat the injectors by pushing down on them. You must*

15.24 Remove the fuel rail mounting bolts (arrows)

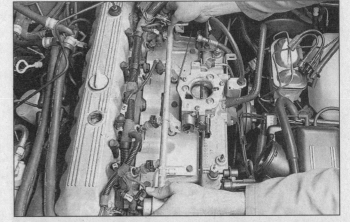

15.27 Before installing the fuel rail mounting bolts, position the tip of each injector over its respective bore in the intake manifold, then push down on the injectors to seat them completely

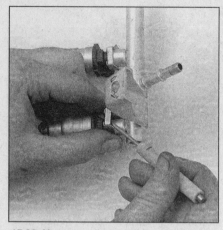

15.32 Use a small screwdriver (shown) or scribe to pop loose the clip that secures the injector to the fuel rail assembly

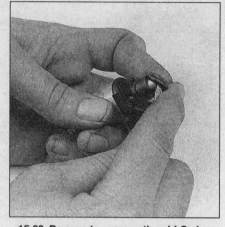

15.33 Be sure to remove the old O-ring and seal from the injector (a replacement kit, available at a dealer, includes six O-rings and seven seals - enough O-rings and seals for all six injectors and an extra seal for the fuel pressure regulator)

15.40 Remove the fuel pressure regulator retaining screws (arrows) (early models)

seat the injectors properly before you tighten the fuel rail mounting bolts.

b) When you reattach the fuel lines, be sure to use a new O-ring, spacer and retainer repair kit (see Section 12).

Fuel injector

Refer to illustrations 15.32 and 15.33

28   Remove the fuel filler cap to relieve fuel tank pressure.
29   Relieve the system fuel pressure (see Section 2).
30   Detach the cable from the negative battery terminal.
31   Remove the fuel injector rail assembly (see above).
32   Remove the retaining clip(s) that attach the injector(s) to the rail assembly **(see illustration)**.
33   If you are servicing a leaking injector, remove the old O-rings and seals **(see illustration)**.
34   An O-ring kit is available from a dealer parts department. This kit consists of 6 brown and 7 black seals. The brown seals fit on the injector tips; the black seals fit on the rail end

of the injectors. Do not switch the brown and black O-rings - they're different in design and composition. The extra black seal is for the fuel pressure regulator - see Step 42.
35   Because the O-ring kit includes enough O-rings for all six injectors (they're not available individually), it's a good idea to replace the upper and lower O-rings on all injectors, even if only one or two are actually leaking at the time you service them. Otherwise, you will probably find yourself repeating this entire procedure down the road - to fix other leaky injectors.
36   Installation is the reverse of removal. Be sure to coat the injector O-rings with a little fuel before installing the injectors.

Fuel pressure regulator

1995 and earlier models

Refer to illustrations 15.40, 15.41 and 15.42

37   Remove the fuel filler cap to relieve fuel tank pressure.

38   Relieve the system fuel pressure (see Section 2).
39   Detach the cable from the negative battery terminal.
40   Remove the fuel pressure regulator retaining (screws) or remove the clip **(see illustration)**.
41   Pull the regulator off the fuel rail **(see illustration)**.
42   Whenever you remove the fuel pressure regulator, always install a new O-ring **(see illustration)**. This O-ring is available as part of an O-ring kit for the injectors (see Step 34 above).
43   Installation is the reverse of removal.

1996 and later models

Refer to illustrations 15.44a and 15.44b

44   On 1996 and later models, the fuel pressure regulator is an integral part of the fuel pump/fuel sending unit assembly **(see illustration)**, which is mounted in the top of the fuel tank. On 1997 and later models, the fuel filter and pressure regulator are combined into a single component mounted on top of the fuel pump/fuel sending unit assembly **(see illustration)**.

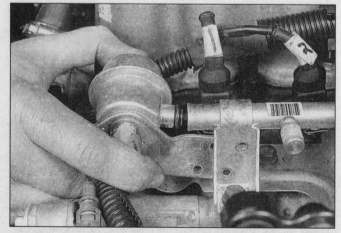

15.41 Pull the fuel pressure regulator off the end of the fuel rail (early models)

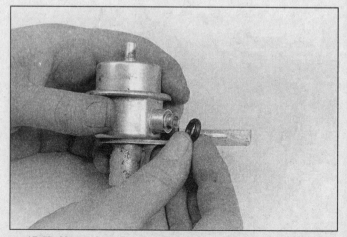

15.42 Always install a new O-ring on the pressure regulator (early models)

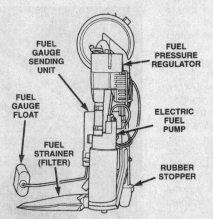

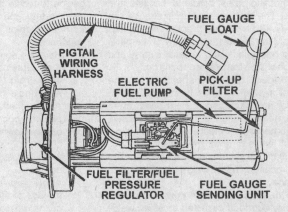

15.44a On 1996 models, the pressure regulator is mounted on top of the fuel pump; if either component is defective, the entire assembly must be replaced as a single unit

15.44b On 1997 and later models, the pressure regulator and the fuel filter are integrated into one unit that's mounted on top of the fuel pump; the filter/regulator unit can be replaced without removing the fuel pump assembly

1996 models

45   On these models, the fuel pressure regulator cannot be serviced separately from the fuel pump module. If the regulator is defective, the entire fuel pump module must be replaced (see Section 4).

1997 and later models

46   On these models, the fuel pressure regulator can be serviced without removing the fuel pump module.

47   Relieve the system fuel pressure (see Section 2).

48   Remove the fuel tank (see Section 5).

49   Wipe off the area around the fuel filter/pressure regulator.

50   Disconnect the fuel line from the pressure regulator (see Section 12).

51   Pry the retainer clamp loose from the tabs on top of the fuel pump module. Remove and discard the old clamp.

52   Carefully pry the filter/regulator out of the fuel pump module with a couple of screwdrivers.

53   Remove and discard the old gasket from the top of the fuel pump module.

54   Before discarding the old filter/regulator assembly, look for the two O-rings on the bottom of the old unit. If the smaller of the two O-rings (the lower one) is not on the filter/regulator, it may be in the fuel inlet passage inside the fuel pump module. Fish it out before proceeding.

55   Wipe out the recessed area in the top of the fuel pump where the filter/regulator is installed.

56   Look at the new filter/regulator unit. Make sure that both O-rings are in place. Apply a little clean engine oil to both O-rings. **Caution:** *Do NOT try to install the O-rings into the fuel pump separately from the filter/regulator; they'll be damaged when the filter/regulator is installed.*

57   Install a new gasket on top of the fuel pump module.

58   Place the new filter/regulator unit in position on top of the fuel pump module. Make sure that the fuel supply tube is aligned with the directional arrow on top of the fuel pump module **(see illustration 15.51)**. (The arrow should already be pointing forward; if it isn't, realign the pump so that it is). Push the filter/regulator unit into the fuel pump until it snaps into position (you should hear or feel a positive "click").

59   To install the new retainer clamp, place it in position, then push it down until locks into place. Make sure that both ends of the retainer are locked to the tabs on the fuel pump.

60   The remainder of installation is the reverse of removal.

16 Exhaust system servicing - general information

Refer to illustrations 16.1a and 16.1b
Warning: *Inspection and repair of exhaust system components should be done only after enough time has elapsed after driving the vehicle to allow the system components to cool completely. Also, when working under the vehicle, make sure it is securely supported on jackstands.*

1   The exhaust system **(see illustration)** consists of the exhaust manifold(s), the catalytic converter, the muffler, the tailpipe and

4

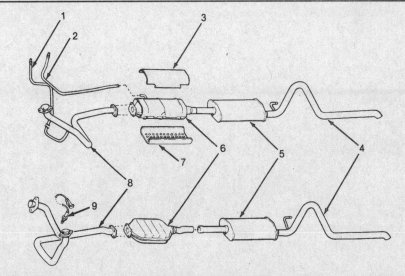

16.1a Exploded views of two typical exhaust systems (four-cylinder models above, V6 models below)

1	Upstream air injection tube (if equipped)	5	Muffler
2	Downstream air injection tube (if equipped)	6	Catalytic converter
3	Upper heat shield	7	Lower heat shield
4	Tail pipe	8	Exhaust pipe
		9	Oxygen sensor

all connecting pipes, brackets, hangers and clamps. The exhaust system is attached to the body with mounting brackets and rubber hangers **(see illustration)**. If any of the parts are improperly installed, excessive noise and vibration will be transmitted to the body.

2   Conduct regular inspections of the exhaust system to keep it safe and quiet. Look for any damaged or bent parts, open seams, holes, loose connections, excessive corrosion or other defects which could allow exhaust fumes to enter the vehicle. Deteriorated exhaust system components should not be repaired; they should be replaced with new parts.

3   If the exhaust system components are extremely corroded or rusted together, welding equipment will probably be required to remove them. The convenient way to accomplish this is to have a muffler repair shop remove the corroded sections with a cutting torch. If, however, you want to save money by doing it yourself (and you don't have a welding outfit with a cutting torch), simply cut off the old components with a hacksaw. If you have compressed air, special pneumatic cutting chisels can also be used. If you do

decide to tackle the job at home, be sure to wear safety goggles to protect your eyes from metal chips and work gloves to protect your hands.

4   Here are some simple guidelines to follow when repairing the exhaust system:

a) *Work from the back to the front when removing exhaust system components.*

b) *Apply penetrating oil to the exhaust system component fasteners to make them easier to remove.*

c) *Use new gaskets, hangers and clamps when installing exhaust systems components.*

d) *Apply anti-seize compound to the threads of all exhaust system fasteners during reassembly.*

e) *Be sure to allow sufficient clearance between newly installed parts and all points on the underbody to avoid overheating the floor pan and possibly damaging the interior carpet and insulation. Pay particularly close attention to the catalytic converter and heat shield.*

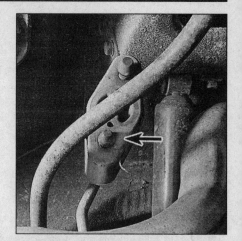

16.1b Whenever you raise the vehicle for any reason, be sure to inspect the rubber exhaust hangers (arrow) for deterioration - because they're subjected to heat and vibration, the hangers frequently wear out and break

Chapter 5
Engine electrical systems

Contents

5

1 General information

The engine electrical systems include all ignition, charging and starting components. Because of their engine-related functions, these components are discussed separately from chassis electrical devices such as the lights, the instruments, etc. (which are included in Chapter 12).

Always observe the following precautions when working on the electrical systems:

a) Be extremely careful when servicing engine electrical components. They are easily damaged if checked, connected or handled improperly.

b) Never leave the ignition switch on for long periods of time with the engine off.

c) Don't disconnect the battery cables while the engine is running.

d) Maintain correct polarity when connecting a battery cable from another vehicle during jump starting.

e) Always disconnect the negative cable first and hook it up last or the battery may be shorted by the tool being used to loosen the cable clamps.

It's also a good idea to review the

safety-related information regarding the engine electrical systems located in the Safety first section near the front of this manual before beginning any operation included in this Chapter.

3.2 Unscrew the nuts (1), remove the bolt (not visible in this photo) and lift off the locator bracket (2), then pull the hold-down clamp (3) off the battery

2 Battery - emergency jump starting

Refer to the Booster battery (jump) starting procedure at the front of this manual.

3 Battery - removal and installation

Refer to illustration 3.2

1   Caution: Always disconnect the negative cable first and hook it up last or the battery may be shorted by the tool being used to loosen the cable clamps. Disconnect both cables from the battery terminals.

2   Remove the battery hold down clamp (see illustration).

3   Lift out the battery. Be careful - it's heavy.

4   While the battery is out, inspect the carrier (tray) for corrosion (see Chapter 1).

5   If you are replacing the battery, make sure that you get one that's identical, with the same dimensions, amperage rating, cold cranking rating, etc.

6   Installation is the reverse of removal.

4 Battery cables - check and replacement

1   Periodically inspect the entire length of each battery cable for damage, cracked or burned insulation and corrosion. Poor battery cable connections can cause starting problems and decreased engine performance.

2   Check the cable-to-terminal connections at the ends of the cables for cracks, loose wire strands and corrosion. The presence of white, fluffy deposits under the insulation at the cable terminal connection is a sign that the cable is corroded and should be replaced. Check the terminals for distortion, missing mounting bolts and corrosion.

3   When removing the cables, **always disconnect the negative cable first and hook it up last** or the battery may be shorted by the tool used to loosen the cable clamps. Even if only the positive cable is being replaced, be sure to disconnect the negative cable from the battery first (see Chapter 1 for further information regarding battery cable removal).

4   Disconnect the old cables from the battery, then trace each of them to their opposite ends and detach them from the starter solenoid and ground terminals. Note the routing of each cable to ensure correct installation.

5   If you are replacing either or both of the old cables, take them with you when buying new cables. It is vitally important that you replace the cables with identical parts. Cables have characteristics that make them easy to identify: positive cables are usually red, larger in cross-section and have a larger diameter battery post clamp; ground cables are usually black, smaller in cross-section and have a slightly smaller diameter clamp for the negative post.

6   Clean the threads of the solenoid or ground connection with a wire brush to remove rust and corrosion. Apply a light coat of battery terminal corrosion inhibitor, or petroleum jelly, to the threads to prevent future corrosion.

7   Attach the cable to the solenoid or ground connection and tighten the mounting nut/bolt securely.

8   Before connecting a new cable to the battery, make sure that it reaches the battery post without having to be stretched.

9   Connect the positive cable first, followed by the negative cable.

5 Ignition system - general information

1984 and 1985 four-cylinder engine

The Computerized Emission Control Solid State Ignition (CEC/SSI) system used on 1984 and 1985 models with a four-cylinder engine utilizes a microcomputer unit (MCU) to monitor engine operating conditions. According to the operating conditions, the MCU advances or retards ignition timing as necessary.

1986 and later four-cylinder engines and 1986 through 1993 inline six-cylinder engines

1986, an ignition system designed for use with Throttle Body Injection (TBI) (which replaced the YFA carburetor that year) was installed on vehicles with a four-cylinder engine. This system, which is used on all 1986 and later four-cylinder vehicles, consists of a solid state ignition control module (ICM), a conventional distributor and rotor assembly, an Electronic Control Unit (ECU), a specially machined flywheel and a Top Dead Center (TDC) sensor located in the distributor.

The solid state ignition control module, which is located in the engine compartment on the right side of the shock tower area, consists of a solid state ignition circuit and an integrated ignition coil that can be removed and serviced separately if necessary. The ICM controls ignition advance/retard electronically. Signals from the ECU to the ICM determine the amount of ignition timing advance or retard needed to meet various engine load requirements.

The ECU gets its information regarding TDC, BDC and engine speed from a TDC sensor mounted on the flywheel/driveplate housing. The flywheel has two teeth machined off every 180-degrees to define a precise point 90-degrees before TDC and BDC. The sensor is non-adjustable; it is preset during engine assembly at the factory.

V6 engine

A High Energy Ignition (HEI) system with an externally mounted ignition coil is used on models with a V6 engine. There are two different methods for controlling spark timing.

a) The 49-state system uses conventional centrifugal and vacuum advance.
b) Spark timing is electronically controlled on California models by means of a special distributor actuated by the electronic control module (ECM). No vacuum or centrifugal advance is used.

1994 through 1999 inline six-cylinder engine

The ignition system used on inline six-cylinder models is similar in design and function to the system used on 1986 and later four-cylinder engines (see above).

2000 inline six-cylinder models

The 2000 models are equipped with a distributorless ignition system. The ignition system consists of the battery, ignition coils, spark plugs, camshaft position sensor, crankshaft position sensor and the Powertrain Control Module (PCM). The PCM controls the ignition timing and spark advance characteristics for the engine. The ignition timing is not adjustable.

The ignition system uses a "waste spark" method of spark distribution. The system utilizes three ignition coils in a coil rail assembly. Each coil is paired with the opposite cylinder in the firing order (1-6, 5-2, 3-4) so one cylinder under compression fires simultaneously with its opposing cylinder, where the piston is on the exhaust stroke. Since the cylinder on the exhaust stroke requires very little of the available voltage to fire its plug, most of the voltage is used to fire the plug of the cylinder on the compression stroke. Conventional ignition coils have one end of the secondary winding connected to the engine ground. On a waste spark system, neither end of the secondary winding is grounded - instead one end of the coil's secondary winding is directly attached to the spark plug and the other end is attached to the spark plug of the companion cylinder.

The PCM controls the ignition system by opening and closing the ignition coil ground circuit. The computerized ignition system provides complete control of the ignition timing by determining the optimum timing in response to engine speed, coolant temperature, throttle position and vacuum pressure in the intake manifold. These parameters are relayed to the PCM by the camshaft position sensor, crankshaft position sensor, throttle position sensor and the manifold absolute pressure sensor. The PCM and crankshaft position sensors are very important components of the ignition system. The ignition system will not operate and the engine will not start if the PCM or the crankshaft position sensor are defective. Refer to Chapter 6 for additional information on the various sensors.

6 Ignition system - check

All except 2000 inline six-cylinder models

Warning: *Because of the very high voltage generated by the ignition system, extreme care should be taken when this check is performed.*

1   If the engine turns over but won't start, disconnect the spark plug wire from any spark plug and attach it to a calibrated ignition tester (available at most auto parts stores). Connect the clip on the tester to a bolt or metal bracket on the engine. Make sure the tester is compatible with your ignition system if a universal tester isn't available. If you're unable to obtain a calibrated ignition tester, remove the wire from one of the spark plugs and, using an insulated tool, hold the end of the wire about 1/4-inch from a good ground.

2   Crank the engine and watch the end of the tester or spark plug wire to see if bright blue, well-defined sparks occur. If you're not using a calibrated tester, have an assistant crank the engine for you.

3   If sparks occur, sufficient voltage is reaching the plug to fire it (repeat the check at the remaining plug wires to verify that the distributor cap and rotor are OK). However, the plugs themselves may be fouled, so remove and check them as described in Chapter 1.

4   If no sparks or intermittent sparks occur,

remove the distributor cap and check the cap and rotor as described in Chapter 1. If moisture is present, dry out the cap and rotor, then reinstall the cap and repeat the spark test.

5   If there's still no spark, detach the coil secondary wire from the distributor cap and hook it up to the tester (reattach the plug wire to the spark plug), then repeat the spark check. Again, if you don't have a tester, hold the end of the wire about 1/4-inch from a good ground, using an insulated tool.

6   If sparks now occur, the distributor cap, rotor or plug wire(s) may be defective.

7   If no sparks occur, the coil-to-cap wire may be bad (check the resistance with an ohmmeter - resistance should be approximately 250 to 1000 ohms per inch of wire). If a known good wire doesn't make any difference in the test results, proceed to the next Step.

8   If there's still no spark, check the primary wire connections at the coil to make sure they're clean and tight. Check for voltage to the coil. Check the resistance of the coil also (see Section 7).

9   If there's still no spark, the pick-up coil or ignition module may be defective (see Sections 7 and 8).

2000 inline six-cylinder models

Refer to illustration 6.14

Caution: *Because of the ignition system design on these six-cylinder models, a special set of spark plug wire must be obtained or fabricated before the ignition system check can be performed. A test set can be fabricated from bulk spark plug wires and terminals available at most auto parts stores. As an alternative, individual jumper wires can be used to ground the spark plug terminals not being tested. Whichever method you choose, make sure all the spark plug terminals are either connected to a spark plug or grounded to the engine block before cranking the engine or damage to the ignition coils may result.*

10   Disable the fuel system by removing the fuel pump relay from the power distribution center (see Chapter 12). Remove the bolts retaining the ignition coil rail (see section 7). Carefully detach the coil rail from the spark plug, turn the coil rail over and place it in a secure location. Do not disconnect the electrical connector from the ignition coil rail.

11   Attach a calibrated ignition system tester (available at most auto parts stores) to one of the spark plug boots. Using a heavy gauge jumper wire, connect the clip on the tester to the bolt or bare metal bracket on the engine. Connect the remaining ignition coil spark plug terminals to the spark plugs or ground them to the engine. Crank the engine and watch the end of the tester to see if a bright blue, well-defined spark occurs (weak or intermittent spark is the same as no spark).

12   If a spark occurs, sufficient voltage is reaching the plug to fire it (repeat the check at the remaining spark plug terminals to verify that the ignition coils are good). If the ignition system is operating properly the problem lies elsewhere; i.e. a mechanical or fuel system problem. However, the spark plugs may be

6.14 Ignition coil rail electrical connector terminal identification (harness side) – 2000 six-cylinder engine

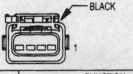

CAV	CIRCUIT	FUNCTION
1	K91 16TN/RD	COIL DRIVER NO. 1
2	A142 16DG/OR	AUTOMATIC SHUT DOWN RELAY OUTPUT
3	K92 16TN/PK	COIL DRIVER NO. 2
4	K93 16TN/OR	COIL DRIVER NO. 3

fouled, so remove and check them as described in Chapter 1.

13   If a spark occurs at one or more of the coils, remove the ignition coil rail and measure the primary resistance of the ignition coils (see Section 7). Check the spring inside each spark plug boot for corrosion or damage. Replace the ignition coil rail if defective.

14   If no spark occurs at all coils, disconnect the electrical connector from the coil rail and check for battery voltage to the ignition coil rail from the auto shut-down relay and power distribution center. Attach a 12 volt test light or voltmeter to a good engine ground point and check for power at the dark green/orange wire terminal of the ignition coil rail harness connector **(see illustration)**. Battery voltage should be available with the ignition key on. If there is no battery voltage present at the coil rail connector, check the auto shut-down relay (see Chapter 12) and circuits between the power distribution center and ignition coil rail (don't forget to check the fuses). Also check the auto shut-down relay control circuit from the relay to the PCM for continuity.

15   If the previous checks are correct, check for a trigger signal from the Powertrain Control Module. Attach a test light to the dark green/orange wire terminal and one of the other terminals of the coil rail harness connector. Crank the engine. The test light should blink with the engine cranking if a trigger signal is present. Check for a trigger signal between the dark green/orange wire

terminal and each of the other terminals in the connector. If a trigger signal is present at the coil rail connector, the Powertrain Control Module and the crankshaft position sensor are functioning properly. If a trigger signal is not present at the coil terminals, check the crankshaft position sensor (see Chapter 6). If the crankshaft position sensor is good, have the PCM checked by a dealer service department or other qualified repair shop.

7   Ignition coil - check and replacement

All except 2000 inline six-cylinder models

Refer to illustrations 7.2a, 7.2b, 7.4, 7.5 and 7.8

Check

1   Detach the cable from the negative battery terminal.

2   Locate the ignition coil **(see illustration)**. If you have difficulty finding the ignition coil, trace the lead from the center tower of the distributor cap back to the coil. You may find it easier to test the coil with it removed. If so, follow the removal procedure below. If you do not remove the coil, disconnect the wires from the high voltage and primary terminals **(see illustration)**.

7.2a The ignition coil on four and inline six-cylinder models is located on the right side of the engine compartment - when removing the coil, unscrew the mounting bracket bolts (arrows)

7.2b Locations of the primary (1) and high voltage (2) coil terminals - to check the coil, select the high resistance scale on the ohmmeter, touch the positive lead to the coil positive terminal and the negative lead to ground (the coil frame) - the indicated resistance should be infinite

5

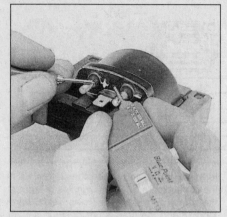

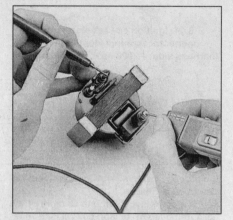

7.4 Using the low resistance scale, touch the positive lead to the positive primary coil terminal and the negative lead to the negative terminal - the indicated resistance should be zero or close to zero

7.5 Using the high resistance scale, touch the ohmmeter negative lead to the negative primary terminal and the positive lead to the high voltage terminal - the indicated resistance should be less than infinite

7.8 When removing the coil, detach the Ignition Control Module (ICM) wire connectors (if equipped)

3   Using an ohmmeter set on the high resistance scale, connect the positive lead to the primary terminal and the negative lead to ground **(see illustration 7.2b)**. The ohmmeter should indicate infinite resistance. If it doesn't, replace the coil.

4   Using the low resistance scale, connect the ohmmeter positive lead to the positive primary terminal and the negative lead to the negative terminal **(see illustration)**. The ohmmeter should indicate zero (or nearly zero) resistance. If it doesn't, replace the coil.

5   Using the high resistance scale, connect the ohmmeter negative lead to the negative primary terminal and the positive lead to the high voltage terminal **(see illustration)**. The ohmmeter should not indicate infinite resistance. If it does, replace the coil.

Replacement

6   Detach the cable from the negative battery terminal.

7   If you haven't already done so, detach the wires from the high voltage and primary terminals of the coil **(see illustration 7.2b)**.

8   Detach the Ignition Control Module (ICM) wire connectors, if equipped **(see illustration)**.

9   Remove the coil mounting bracket bolts **(see illustration 7.2a)** and detach the coil.

10   Installation is the reverse of removal.

2000 inline six-cylinder models
Check

11   Remove the ignition coil rail assembly (see *Replacement* procedure).

12   Clean the three outer cases and check for cracks or other damage. Clean the coil primary terminals and check the springs and spark plug boots for damage. If either the springs or spark plug boots are damaged, the coil rail assembly must be replaced.

13   Check the coil primary resistance. Connect an ohmmeter to terminals no. 2 and 1; 2 and 3; 2 and 4; of the coil connector. Each measurement is the primary resistance of one of the three coils in the coil rail assembly.

14   The coil resistance should be .97 to 1.18 ohms resistance for the primary side and 11,300 to 15,300 ohms for the secondary side. If either measured resistance value is not as specified, replace the ignition coil rail

assembly. **Note:** *The entire ignition coil rail assembly must be replaced if only one coil is defective.*

Replacement
Refer to illustrations 7.15 and 7.17
Note: *The ignition coil rail assembly consists of the three ignition coils, spark plug boots, springs, electrical connector and internal wiring. None of the components are serviced separately. The entire assembly is removed and replaced as a unit.*

15   Remove the ignition coil rail assembly mounting bolts **(see illustration)**.

16   Carefully work the spark plug boots off the spark plugs by alternately prying at one end of the coil rail, then the other. If difficulty is encountered, twist the spark plug boots with a spark plug boot removal tool to loosen the seal.

17   Position the coil rail to access the electrical connection. Disconnect the electrical connector from the coil rail by first pushing the slide tab up, then press the release lock and pull the connector off **(see illustration)**.

18   Remove the ignition coil rail assembly.

19   When installing the ignition coil rail, position all of the spark plug boots over the spark plugs and press the ignition coil rail down until the boots are completely seated.

20   Install the mounting bolts and tighten them in stages until secure.

21   Connect the electrical connector and slide the tab down until locked in place.

8   Ignition control module - check and replacement

Check

1   You need an approved ignition control module tester to test the ignition control modules used with any of the vehicles covered by this manual. Without this equipment, module testing is beyond the scope of the home mechanic. If you have access to the necessary tool, the instructions for testing the module are provided by the manufacturer.

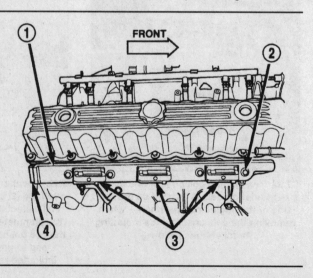

7.15 Ignition coil rail installation details - 2000 six-cylinder models

1   *Coil rail*
2   *Mounting bolts (1 of 4)*
3   *Ignition coils*
4   *Electrical connector*

FRONT

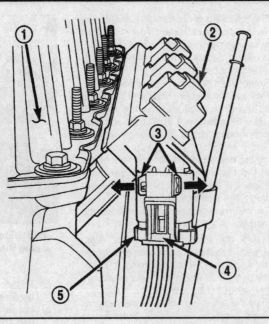

7.17 Ignition coil rail electrical connector details - 2000 six-cylinder models

1 Valve cover
2 Ignition coil rail
3 Slide tab
4 Release lock
5 Electrical connector

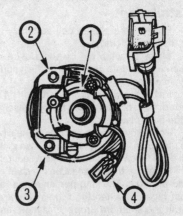

8.4 Ignition control module used on V6 models

1 Pickup coil assembly
2 Module attaching screws
3 Ignition control module
4 Pickup coil connector

Replacement

Refer to illustration 8.4

V6 models

2   Detach the cable from the negative battery terminal.
3   Remove the distributor cap and rotor (see Chapter 1).
4   Remove the two control module attaching screws **(see illustration)** and lift the module up.
5   Detach the pickup coil wire connector from the module **(see illustration 8.4)**. Note the wire colors - the connectors must not be interchanged.
6   Disconnect the wire harness connector.
7   If you will be reinstalling the old module, don't wipe the grease from the module or the distributor base. If you install a new module, a package of silicone dielectric compound will be included with it. Spread the compound on the metal face of the module and on the distributor base where the module seats. This compound is necessary for module insulation and cooling.
8   Installation is the reverse of removal.

Inline six-cylinder models

9   Remove the ignition coil (see Section 7).
10   Remove the two screws that attach the module to the coil frame. Pull off the module.
11   Installation is the reverse of removal.

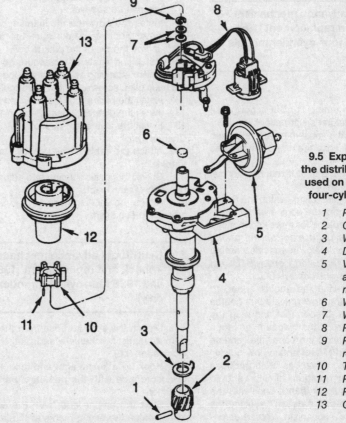

9.5 Exploded view of the distributor assembly used on 1984 and 1985 four-cylinder engines

1   Pin
2   Gear
3   Washer
4   Distributor body
5   Vacuum advance mechanism
6   Wick
7   Washers
8   Pickup coil
9   Pickup coil retainer
10 Trigger wheel
11 Pin
12 Rotor
13 Cap

9   Trigger wheel and/or pickup coil assembly - removal and installation (1984 and 1985 four-cylinder models only)

Refer to illustration 9.5

Removal

1   Detach the cable from the negative battery terminal.

2   The following procedure may be easier if you remove the distributor (see Section 13) but it's not absolutely necessary that you do so.
3   If you do decide to remove the distributor, place it in a bench vise.
4   Remove the distributor cap and rotor (see Chapter 1).
5   Using a small puller tool, remove the trigger wheel **(see illustration)**. Use a flat washer to prevent the puller from contacting the inner shaft. Remove the pin.
6   Remove the pickup coil assembly retainer and washers from the pivot pin on the base plate.
7   Remove the two pickup coil plate screws.
8   Lift the pickup coil assembly from the distributor housing.

5

Installation

9   Position the pickup coil assembly into the distributor housing.

10   Make sure the pin on the pickup coil assembly fits into the hole in the vacuum advance mechanism link.

11   Install the washers and retainer on the pivot pin to secure the pickup coil assembly to the base plate.

12   Position the wiring harness in the slot in the distributor housing. Install the two pickup coil plate screws and tighten them securely.

13   Install the trigger wheel on the shaft with hand pressure. The long portion of the teeth must face up. When the trigger wheel and slot in the shaft are properly aligned, use a pin punch and a small hammer to tap the pin into the locating groove in the trigger wheel and shaft. If the distributor is not installed in the engine, support the shaft while installing the trigger wheel pin.

14   Install the rotor and the distributor cap.

15   Attach the cable to the negative battery terminal.

---

10 Vacuum advance mechanism - check and replacement (1984 and 1985 four-cylinder models only)

Check

1   Hook up a timing light in accordance with the manufacturer's instructions.

2   Make sure the vacuum hose is attached to the vacuum advance mechanism **(see illustration 9.5).**

3   Run the engine until it reaches normal operating temperature.

4   With the engine idling, watch the timing marks at the front of the engine with the timing light (see Chapter 1 for more information). **Note:** *The white paint mark on the timing degree scale identifies the specified timing degrees Before Top Dead Center (BTDC) at 1600 rpm; it does not identify TDC.*

5   Slowly increase the engine speed to 2000 rpm, noticing how quickly the ignition timing advances. Disconnect and plug the vacuum hose from the vacuum advance mechanism. Again, increase the engine speed to 2000 RPM and notice how quickly the ignition timing advances. With the hose connected, the ignition timing should advance more quickly than it does with the hose disconnected. If it doesn't, replace the vacuum advance mechanism. **Note:** *A defective Micro Computer Unit (MCU) can also alter ignition timing.*

6   Unplug and reconnect the hose to the vacuum advance mechanism.

Replacement

7   Detach the cable from the negative battery terminal.

8   Remove the distributor cap and rotor (see Chapter 1).

9   The following procedure is easier if you remove the distributor (see Section 13), but it's not absolutely necessary that you do so.

10   If you do decide to remove the distributor, place it in a bench vise.

11   Remove the pickup coil (see Section 9).

12   Remove the vacuum hose and attaching screws from the vacuum advance mechanism **(see illustration 9.5).** Tilt the vacuum advance mechanism to disengage the link from the pickup coil pin, which protrudes through the distributor housing. If necessary, loosen the base plate screws to get more clearance. Lift the vacuum advance mechanism out of the distributor housing.

13   To calibrate a new vacuum advance mechanism:

a) *Insert an appropriately sized Allen wrench into the vacuum hose tube of the old vacuum advance mechanism; count the number of clockwise turns necessary to bottom the adjusting screw.*

b) *Turn the adjusting screw of the replacement vacuum advance mechanism clockwise to bottom it, then turn it counterclockwise the same number of turns you counted above.*

14   Install the advance mechanism on the distributor housing. Make sure the link is engaged on the pin of the pickup coil.

15   Install and tighten the vacuum advance mechanism attaching screws. If you loosened the base plate screws, tighten them too.

16   Install the pickup coil (see Section 9).

17   Install the distributor (see Section 13).

18   Install the distributor cap and rotor (see Chapter 1).

19   Attach the cable to the negative battery terminal.

20   Check the ignition timing and adjust it, if necessary (see Chapter 1).

21   Attach the vacuum hose to the vacuum advance mechanism.

---

11 Centrifugal advance mechanism - check and replacement (1984 and 1985 four-cylinder models only)

---

1   Detach the vacuum hose from the vacuum advance mechanism and plug it **(see illustration 9.5).**

2   Hook up a timing light and tachometer in accordance with the manufacturer's instructions.

3   Start the engine and, with the engine idling, watch the timing marks at the front of the engine with the timing light (see Chapter 1 for more information).

4   Slowly increase engine speed to 2000 rpm. Timing should advance smoothly as engine speed increases. If it advances unevenly, the centrifugal advance mechanism is faulty. Replace the distributor (see Section 12). You will need to switch the following parts to the replacement distributor:

a) *Cap and rotor (see Chapter 1).*

12.3 Before removing the pick-up coil, unplug the lead from the ignition module

b) *Trigger wheel and/or pickup coil assembly (see Section 9).*

c) *Vacuum advance mechanism (see Section 10).*

---

12 Ignition pick-up coil (V6 engines only) – check and replacement

---

Refer to illustrations 12.3, 12.11a, 12.11b, 12.11c, 12.14a and 12.14b

Check

1   Detach the cable from the negative terminal of the battery.

2   Remove the distributor cap and rotor (see Chapter 1).

3   Detach the pick-up coil leads from the module **(see illustration).**

4   Connect one lead of an ohmmeter to one terminal of the pick-up coil lead and the other to ground as shown. Flex the leads by hand to check for intermittent opens. The ohmmeter should indicate infinite resistance at all times. If it doesn't, the pick-up coil is defective and must be replaced.

5   Connect the ohmmeter leads to both terminals of the pick-up coil. Flex the leads by hand to check for intermittent opens. The ohmmeter should read one steady value between 500 and 1500 ohms as the leads are flexed by hand. If it doesn't, the pick-up coil is defective and must be replaced.

Replacement

6   Remove the distributor (see Section 13).

7   Remove the two rotor mounting screws and remove the rotor.

8   Disconnect the pick-up coil leads from the module.

9   On models with a Hall-effect switch, remove the switch retaining screws and the switch.

10   Mark the distributor gear and shaft so they can be reassembled in the same position.

11   Carefully mount the distributor in a soft-jawed vise and, using a hammer and punch,

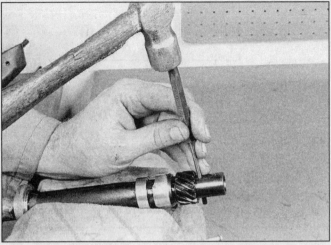

12.11a Mount the distributor shaft in a soft-jawed vise and, using a drift punch and hammer, knock out the roll pin

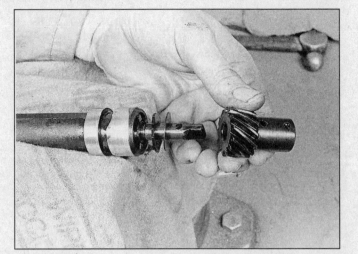

12.11b Remove the driven gear and spacer washers from the end of the shaft, making sure to note the order in which you remove any spacers

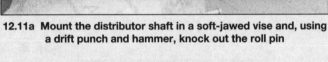

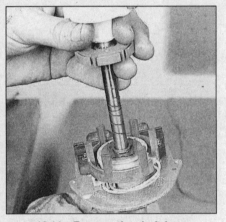

12.11c Remove the shaft from the distributor

12.14a To remove the pick-up coil from a V6 engine distributor, remove the retaining clip

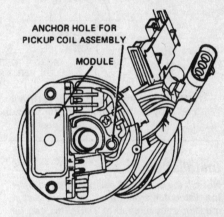

12.14b View of the distributor with the pick-up coil removed - when installing the pick-up coil make sure it engages with the anchor plate

5

remove the roll pin from the distributor shaft and gear **(see illustrations)**.
12   Remove the gear and washers from the shaft.
13   Carefully pull the shaft out through the top of the distributor.
14   Remove the "C" washer retaining ring at the center of the distributor **(see illustration)** and remove the pick-up coil **(see illustration)**.
15   Installation is the reverse of the removal procedure. Be sure to use a new roll pin when installing the drive gear on the shaft.

13 Distributor - removal and installation

Refer to illustrations 13.6, 13.10a, 13.10b, 13.10c, 13.11a, 13.11b, 13.11c, 13.11d, 13.13a, 13.13b, 13.13c, 13.14a and 13.14b

Removal

1   Detach the cable from the negative battery terminal.
2   Detach the primary lead from the coil.
3   Disconnect all wires from the distributor. If the wires don't unplug at the distributor,

follow the wires as they exit the distributor to find the connector.
4   Look for a raised "1" on the distributor cap. This marks the location for the number one cylinder spark plug wire terminal. If the cap does not have a mark for the number one terminal, locate the number one spark plug and trace the wire back to its corresponding terminal on the cap.
5   Remove the distributor cap (see Chapter 1) and turn the engine over until the rotor is pointing toward the number one spark plug terminal (see *locating TDC procedure* in Chapter 2).
6   Make a mark on the edge of the distributor base directly below the rotor tip and in line with it (if the rotor on your engine has more than one tip, use the center one for reference). Also, mark the distributor base and the engine block to ensure that the distributor is installed correctly **(see illustration)**.
7   Remove the distributor hold down bolt and clamp, then pull the distributor straight up to remove it. **Caution:** *DO NOT turn the crankshaft while the distributor is out of the engine, or the alignment marks will be useless.*

13.6 Before loosening the distributor hold-down bolt, make a mark on the edge of the distributor base directly below, and in line with, the rotor tip (1), then mark the distributor base and the engine block (2) to ensure that the distributor is installed correctly

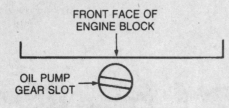

13.10a To seat the distributor on a 1984 through 1994 four-cylinder engine, use a screwdriver to rotate the oil pump gear so the slot is slightly past the three o'clock position . . .

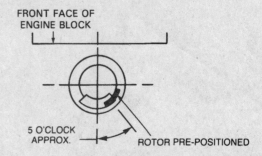

13.10b . . . then, with the distributor cap removed, install the distributor with the rotor pointing toward the five o'clock position

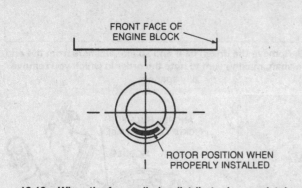

13.10c When the four-cylinder distributor is completely and correctly seated (engaged), the rotor should be at the six o'clock position

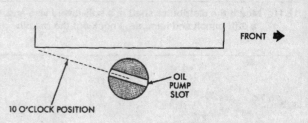

13.11a To seat the distributor on a 1995 and later four-cylinder engine, use a screwdriver to rotate the oil pump gear so the slot is at the ten o'clock position . . .

Installation

Note: *If the crankshaft has been moved while the distributor is out, the number one piston must be repositioned at TDC. This can be done by feeling for compression pressure at the number one plug hole as the crankshaft is turned. Once compression is felt, align the ignition timing zero mark with the pointer.*

8   Insert the distributor into the engine in exactly the same relationship to the block that it was in when removed.

9   To mesh the helical gears on the camshaft and the distributor, it may be necessary to turn the rotor slightly. Recheck the alignment marks between the distributor base and the block to verify that the distributor is in the same position it was in before removal. Also check the rotor to see if it's aligned with the mark you made on the edge of the distributor base.

Seating the distributor

If you have difficulty seating the distributor on the engine block, use the appropriate Step (10, 11 or 12 below) for the type of engine you have. If the distributor seats properly, with the flange flush with the engine block, proceed to Step 13.

10   If you have difficulty getting the distributor to seat properly on a 1984 through 1994 four-cylinder engine, use the following procedure:

a) *Make sure the number one piston is at TDC (see Chapter 2).*

b) *Look into the hole in the engine block where the distributor mounts. You should see the oil pump gear slot* (see illustration). *Put a screwdriver into the slot and rotate the oil pump gear so the gear slot is slightly past the three o'clock position.*

c) *With the distributor cap removed, install the distributor with the rotor pointing to the five o'clock position* (see illustration).

d) *With the distributor fully engaged in its correct location, the rotor should be pointing to the six o'clock position* (see illustration).

e) *If the distributor is still not installed correctly, repeat this procedure.*

11   If you are having difficulty getting the distributor to seat properly on a 1995 and later four-cylinder engine, use the following procedure:

a) *Make sure the number one cylinder is at TDC (see Chapter 2).*

b) *Look into the hole in the engine block where the distributor mounts. You should see the oil pump gear slot* (see illustration). *Put a screwdriver into the hole and rotate the oil pump gear so the gear slot is slightly before the ten o'clock position.*

c) *The pulse ring must be locked in place to hold the rotor in the number one cylinder firing position during installation. Lift the camshaft position sensor*

straight up and out of the distributor. Rotate the distributor shaft and install a 3/16 inch drift pin punch tool through the correct alignment hole in the plastic ring and into the mating hole in the distributor housing (see illustration). *This will prevent the distributor shaft and rotor from rotating during installation.*

d) *Install the rotor onto the shaft.*

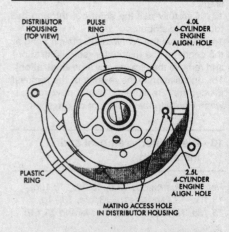

13.11b . . . then rotate the distributor shaft and install a 3/16 inch drift pin punch tool through the correct alignment hole in the plastic ring and into the mating hole in the distributor housing

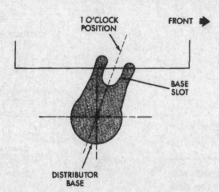

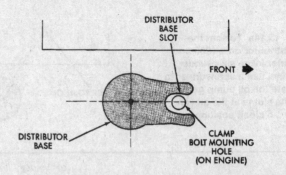

13.11c Visually line up the distributor with the centerline of the base slot in the one o'clock position

13.11d When the distributor is installed correctly, the centerline of the base slot will be aligned with the clamp bolt hole on the engine

e) *Visually line up the distributor with the centerline of the base slot in the one o'clock position* (see illustration).

f) *Install the distributor. It will rotate clockwise during installation and when the distributor is fully installed the centerline of the base slot will be aligned with the clamp bolt hole on the engine* (see illustration). *The rotor should also be pointing to slightly past (clockwise of) the three o'clock position.*

g) *If the distributor is still not installed correctly, repeat this procedure.*

12　If you have difficulty getting the distributor to seat properly on a V6 engine, use the following procedure:

a) *Remove the number one spark plug.*

b) *Place your finger over the number one spark plug hole and rotate the crankshaft slowly until you feel compression.*

c) *Align the timing index mark on the vibration damper with 0-degree (TDC) on the graduated timing degree scale on the timing gear cover.*

13.13a To seat the distributor on a 1994 and earlier inline six-cylinder engine, use a screwdriver to turn the oil pump gear shaft until the slot is slightly past the 11 o'clock position . . .

d) *Turn the rotor to point between the number one and number two spark plug terminals on the distributor cap.*

e) *Install the distributor.*

13　If you have difficulty getting the distributor to seat properly on a 1994 and earlier inline six-cylinder engine, use the following procedure:

a) *Look into the hole in the engine block where the distributor mounts. You should see the oil pump gear shaft* (see illustration). *Put a screwdriver into the slot in the oil pump gear shaft and rotate*

the shaft until the slot is slightly past the 11 o'clock position.

b) *If you removed it, Install the rotor.*

c) *Without engaging the distributor gear with the camshaft gear in the engine, position the distributor in the hole in the engine block. Be sure the distributor gasket is installed.*

d) *Visually line up the hold-down ear of the distributor housing with the hold-down clamp bolt hole* (see illustration).

e) *Turn the rotor to the four o'clock position* (see illustration).

5

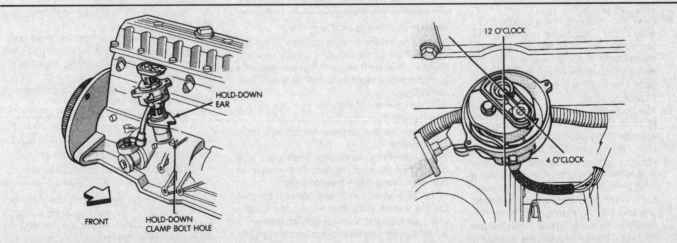

13.13b . . . visually line up the hold-down ear of the distributor housing with the hold-down clamp bolt hole . . .

13.13c . . . turn the rotor to the four o'clock position, then slide the distributor down into the block until it seats

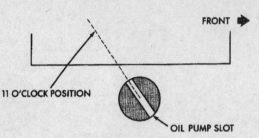

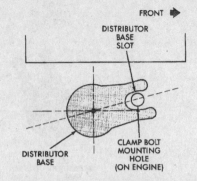

13.14a To seat the distributor on a 1995 and later inline six-cylinder engine, use a screwdriver to rotate the oil pump gear so the slot is at the eleven o'clock position

13.14b When the distributor is installed correctly, the centerline of the base slot will be aligned with the clamp bolt hole on the engine

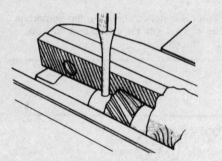

14.4 Use a small punch and hammer to drive out the retaining pin, then remove the distributor gear from the shaft

f) *Slide the distributor down into the engine block until it seats. Keep the hold-down ear aligned to the hole in the block.*

g) *The rotor should be in the five o'clock position with the trailing edge of the rotor blade lined up with the mark you scribed on the distributor housing prior to removal (the number one spark plug wire post location).*

h) *If the distributor still won't seat properly, repeat this procedure.*

14   If you are having difficulty getting the distributor to seat properly on a 1995 and later inline six-cylinder engine, use the following procedure:

a) *Make sure the number one cylinder is at TDC (see Chapter 2).*

b) *Look into the hole in the engine block where the distributor mounts. You should see the oil pump gear slot* **(see illustration)**. *Put a screwdriver into the hole and rotate the oil pump gear so the gear slot is slightly before the eleven o'clock position.*

c) *The pulse ring must be locked in place to hold the rotor in the number one cylinder firing position during installation. Lift the camshaft position sensor straight up and out of the distributor. Rotate the distributor shaft and install a 3/16 inch drift pin punch tool through the correct alignment hole in the plastic ring and into the mating hole in the distributor housing* **(see illustration 13.11b)**. *This will prevent the distributor shaft and rotor from rotating during installation.*

d) *Install the rotor onto the shaft.*

e) *Visually line up the distributor with the centerline of the base slot in the one o'clock position* **(see illustration 13.11c)**.

f) *Install the distributor. It will rotate clockwise during installation and when the distributor is fully installed, the centerline of the base slot will be aligned with the clamp bolt hole on the engine* **(see illustration)**. *The rotor should also be pointing to slightly past (clockwise of) the five o'clock position.*

g) *If the distributor is still not installed correctly, repeat this procedure.*

15   Place the hold-down clamp in position and loosely install the bolt.

16   On 1995 and later four-cylinder and inline 6-cylinder models, remove the pin punch tool from the distributor. Install the camshaft position sensor onto the distributor and align the wiring harness grommet with the notch in the distributor. Install the rotor.

17   Install the distributor cap.

18   Attach the electrical wires to the distributor.

19   Reattach the spark plug wires to the plugs (if removed).

20   Connect the cable to the negative terminal of the battery.

21   Check the ignition timing (see Chapter 1) and tighten the distributor hold down bolt securely.

14   Stator - replacement (1999 and earlier inline six-cylinder models)

Refer to illustrations 14.4 and 14.6

Note: *The stator is also referred to as the camshaft position sensor on early models. Refer to Chapter 6 for camshaft position sensor information relating to the 2000 six-cylinder models.*

1   Remove the distributor from the engine (see Section 13).

2   Remove the distributor cap and rotor (see Chapter 1).

3   Secure the distributor shaft gear in a bench vise. Wrap a rag around the shaft to prevent damage from the vise jaws.

4   Use a small punch and hammer to drive out the retaining pin, then remove the distributor gear from the shaft **(see illustration)**.

5   Remove the distributor shaft from the distributor housing.

6   Mark the location of the stator **(see illustration)** so you can return it to the same

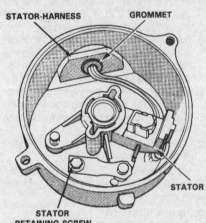

14.6 After marking the position of the stator assembly with paint or a scribe, remove the retaining screw, push the grommet through the distributor housing and remove the stator

location during reassembly.

7   Remove the stator retaining screw **(see illustration 14.6)**.

8   Remove the stator harness **(see illustration 14.6)** by pushing the grommet through the distributor housing. Remove the stator assembly.

9   Lift the stator (camshaft position sensor) from the distributor housing.

10   Installation is the reverse of removal.

15   Charging system - general information and precautions

The charging system includes the alternator, an internal voltage regulator, a charge indicator, the battery, a fusible link and the wiring between all the components. The charging system supplies electrical power for the ignition system, the lights, the radio, etc. The alternator is driven by a drivebelt at the front of the engine.

The purpose of the voltage regulator is to limit the alternator's voltage to a preset value. This prevents power surges, circuit overloads, etc., during peak voltage output.

The fusible link is a short length of insulated wire integral with the engine compartment wiring harness. The link is four wire gauges smaller in diameter than the circuit it protects. Production fusible links and their identification flags are identified by the flag color. See Chapter 12 for additional information regarding fusible links.

The charging system doesn't ordinarily require periodic maintenance. However, the drivebelt, battery, battery cables and connections should be inspected at the intervals outlined in Chapter 1.

The dashboard warning light should come on when the ignition key is turned to Start, then go off immediately. If it remains on, there is a malfunction in the charging system (see Section 16). Some vehicles are also equipped with a voltmeter. If the voltmeter indicates abnormally high or low voltage, check the charging system (see Section 16).

Be very careful when making electrical circuit connections to a vehicle equipped with an alternator and note the following:

a) *When reconnecting wires to the alternator from the battery, be sure to note the polarity.*

b) *Before using arc welding equipment to repair any part of the vehicle, disconnect the wires from the alternator and the battery terminals.*

c) *Never start the engine with a battery charger connected.*

d) *Always disconnect both battery leads before using a battery charger.*

e) *The alternator is turned by an engine drivebelt which could cause serious injury if your hands, hair or clothes become entangled in it with the engine running.*

f) *Because the alternator is connected directly to the battery, it could arc or cause a fire if overloaded or shorted out.*

g) *Wrap a plastic bag over the alternator and secure it with rubber bands before steam cleaning the engine.*

16 Charging system - check

Refer to illustration 16.5

1 If a malfunction occurs in the charging circuit, don't automatically assume that the alternator is causing the problem. First check the following items:

a) *Check the drivebelt tension and condition (Chapter 1). Replace it if it's worn or deteriorated.*

b) *Make sure the alternator mounting and adjustment bolts are tight.*

c) *Inspect the alternator wiring harness and the connectors at the alternator and voltage regulator. They must be in good condition and tight.*

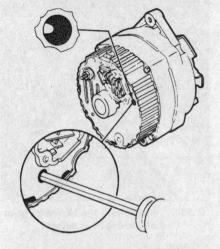

16.5 To test the alternator, locate the test hole in the back, ground the tab that's located inside the hole by inserting a screwdriver blade into the hole and touching the tab and the case at the same time

d) *Check the fusible link (if equipped) located between the starter solenoid and the alternator. If it's burned, determine the cause, repair the circuit and replace the link (the vehicle won't start and/or the accessories won't work if the fusible link blows). Sometimes a fusible link may look good, but still be bad. If in doubt, remove it and check for continuity with an ohmmeter or test light.*

e) *Start the engine and check the alternator for abnormal noises (a shrieking or squealing sound indicates a bad bearing).*

f) *Check the specific gravity of the battery electrolyte. If it's low charge the battery (doesn't apply to maintenance free batteries).*

g) *Make sure the battery is fully charged (one bad cell in a battery can cause overcharging by the alternator).*

h) *Disconnect the battery cables (negative first, then positive). Inspect the battery posts and the cable clamps for corrosion. Clean them thoroughly if necessary (see Chapter 1). Reconnect the cable to the positive terminal.*

i) *With the key off, connect a test light between the negative battery post and the disconnected negative cable clamp.*

  1) *If the test light does not come on, reattach the clamp and proceed to the next step.*

  2) *If the test light comes on, there is a short (drain) in the electrical system of the vehicle. The short must be repaired before the charging system can be checked.*

  3) *Disconnect the alternator wiring harness.*

   (a) *If the light goes out, the alternator is bad.*

  (b) *If the light stays on, pull each fuse until the light goes out (this will tell you which component is shorted).*

2 Using a voltmeter, check the battery voltage with the engine off. If should be approximately 12-volts.

3 Start the engine and check the battery voltage again. It should now be approximately 14-to-15 volts.

4 Locate the test hole in the back of the alternator. **Note:** *If there is no test hole, the vehicle is equipped with a newer CS type alternator. Further testing of this type of alternator must be done by a dealer service department or automotive electrical shop.*

5 Ground the tab that is located inside the hole by inserting a screwdriver blade into the hole and touching the tab and the case at the same time **(see illustration). Caution:** *Do not run the engine with the tab grounded any longer than necessary to obtain a voltmeter reading. If the alternator is charging, it is running unregulated during the test. This condition may overload the electrical system and cause damage to the components.*

6 The reading on the voltmeter should be 15 volts or higher with the tab in the test hole grounded.

7 If the voltmeter indicates low battery voltage, the alternator is faulty and should be replaced with a new one (see Section 17).

8 If the voltage reading is 15 volts or higher and a no charge condition is present, the regulator or field circuit is the problem. Remove the alternator (see Section 17) and have it checked further by an auto electric shop.

17 Alternator - removal and installation

Refer to illustrations 17.2, 17.3a and 17.3b

1 Detach the cable from the negative terminal of the battery.

2 Detach the electrical connectors from the alternator **(see illustration)**.

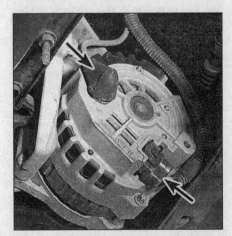

17.2 To remove the alternator, detach the electrical connectors (arrows) from below, . . .

5

17.3a ... loosen the alternator adjustment bolt (arrow) ...

17.3b ... and remove the alternator pivot bolt (arrow) from below

18.2 Mark the drive end frame and rectifier end frame assemblies with a scribe or paint before separating the two halves

18.3a With the through bolts removed, carefully separate the drive end frame and the rectifier end frame

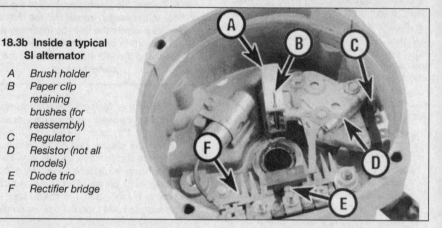

18.3b Inside a typical SI alternator

A   *Brush holder*
B   *Paper clip retaining brushes (for reassembly)*
C   *Regulator*
D   *Resistor (not all models)*
E   *Diode trio*
F   *Rectifier bridge*

3   Loosen the alternator adjustment and pivot bolts and detach the drivebelt **(see illustrations)**.

4   Remove the adjustment and pivot bolts and separate the alternator from the engine.

5   If you are replacing the alternator, take the old one with you when purchasing a replacement unit. Make sure the new/rebuilt unit looks identical to the old alternator. Look at the terminals - they should be the same in number, size and location as the terminals on the old alternator. Finally, look at the identification numbers - they will be stamped into the housing or printed on a tag attached to the housing. Make sure the numbers are the same on both alternators.

6   Many new/rebuilt alternators DO NOT have a pulley installed, so you may have to switch the pulley from the old unit to the new/rebuilt one. When buying an alternator, find out the shop's policy regarding pulleys - some shops will perform this service free of charge.

7   Installation is the reverse of removal.

8   After the alternator is installed, adjust the drivebelt tension (see Chapter 1).

9   Check the charging voltage to verify proper operation of the alternator (see Section 16).

18 Alternator brushes - replacement

Refer to illustrations 18.2, 18.3a, 18.3b, 18.4, 18.5, 18.6, 18.7 and 18.10

Note: *The following procedure applies only to SI type alternators. CS types have riveted housings and cannot be disassembled.*

1   Remove the alternator from the vehicle (Section 15).

2   Scribe or paint marks on the front and rear end frame housings of the alternator to facilitate reassembly **(see illustration)**.

3   Remove the four through-bolts holding the front and rear end frames together, then separate the drive end frame from the rectifier end frame **(see illustrations)**.

4   Remove the nuts holding the stator to the rear end frame and separate the stator from the end frame **(see illustration)**.

5   Remove the nuts attaching the diode trio to the rectifier bridge and remove the trio **(see illustration)**.

6   Remove the screws attaching the resis-

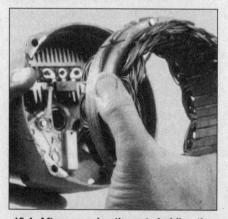

18.4 After removing the nuts holding the stator assembly to the end frame, remove the stator

18.5 Remove the nuts attaching the diode trio to the rectifier bridge and remove the trio

18.6 After removing the screws that attach the regulator and the resistor (if equipped) to the end frame, remove the brush holder

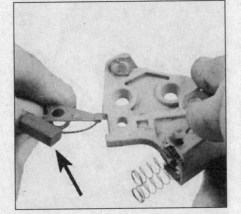

18.7 Slip the brush retainer off the brush holder and remove the brushes (arrow)

18.10 To hold the brushes in place during reassembly, insert a paper clip through the hole in the end frame nearest to the rotor shaft

tor (not used on all models), regulator and brush holder to the end frame and remove the brush holder **(see illustration)**.

7   Remove the brushes from the brush holder by slipping the brush retainer off the brush holder **(see illustration)**.

8   Remove the springs from the brush holder.

9   Installation is the reverse of the removal procedure, noting the following:

10   When installing the brushes in the brush holder, install the brush closest to the end frame first. Slip the paper clip through the rear of the end frame to hold the brush, then insert the second brush and push the paper clip in to hold both brushes while reassembly is completed **(see illustration)**. The paper clip should not be removed until the front and rear end frames have been bolted together.

19   Starting system - general information and precautions

The sole function of the starting system is to turn over the engine quickly enough to allow it to start.

The starting system consists of the battery, the starter motor, the starter solenoid, a remote starter relay (on some models) and the wires connecting them. A number of different starter motors - including Bosch, Delco, Mitsubishi and Motorcraft - are used on the vehicles covered by this manual. If you replace the starter motor, be sure to take the old starter with you to the parts department to make sure you get the right replacement. The solenoid is mounted directly on the starter motor or, on Motorcraft units, is a separate component located in the engine compartment.

The solenoid/starter motor assembly is installed on the lower part of the engine, next to the transmission bellhousing.

When the ignition key is turned to the Start position, the starter solenoid is actuated through the starter control circuit. The starter solenoid then connects the battery to the starter. The battery supplies the electrical energy to the starter motor, which does the actual work of cranking the engine.

The starter on a vehicle equipped with an automatic transmission can only be operated when the transmission selector lever is in Park or Neutral.

Always observe the following precautions when working on the starting system:

a) *Excessive cranking of the starter motor can overheat it and cause serious damage. Never operate the starter motor for more than 15 seconds at a time without pausing to allow it to cool for at least two minutes.*

b) *The starter is connected directly to the battery and could arc or cause a fire if mishandled, overloaded or shorted out.*

c) *Always detach the cable from the negative terminal of the battery before working on the starting system.*

20   Starter motor - testing in vehicle

Note: *Before diagnosing starter problems, make sure the battery is fully charged.*

1   If the starter motor does not operate at all when the switch is turned to Start, make sure the shift lever is in Neutral or Park (automatic transmission).

2   Make sure that the battery is charged and that all cables, both at the battery and starter solenoid terminals, are clean and secure.

3   If the starter motor spins but the engine is not cranking, the overrunning clutch in the starter motor is slipping and the starter motor must be replaced.

4   If, when the switch is actuated, the starter motor does not operate at all but the solenoid clicks, then the problem lies with either the battery, the starter relay (if equipped), the main solenoid contacts or the starter motor itself (or the engine is seized).

5   If the solenoid plunger cannot be heard when the switch is actuated, the battery is bad, the fusible link is burned (the circuit is open) or the solenoid itself is defective.

6   To check the solenoid, connect a jumper lead between the battery (+) and the ignition switch wire terminal (the small terminal) on the solenoid. If the starter motor now operates, the solenoid is OK and the problem is in the ignition switch, neutral start switch or the wiring.

7   If the starter motor still does not operate, remove the starter/solenoid assembly for disassembly, testing and repair.

8   If the starter motor cranks the engine at an abnormally slow speed, first make sure that the battery is charged and that all terminal connections are tight. If the engine is partially seized, or has the wrong viscosity oil in it, it will crank slowly.

9   Run the engine until normal operating temperature is reached, then disconnect the coil wire from the distributor cap and ground it on the engine.

10   Connect a voltmeter positive lead to the positive battery post and connect the negative lead to the negative post.

11   Crank the engine and take the voltmeter readings as soon as a steady figure is indicated. Do not allow the starter motor to turn for more than 15 seconds at a time. A reading of 9 volts or more, with the starter motor turning at normal cranking speed, is normal. If the reading is 9 volts or more but the cranking speed is slow, the motor is faulty. If the reading is less than 9 volts and the cranking speed is slow, the solenoid contacts are probably burned, the starter motor is bad, the battery is discharged or there is a bad connection.

21   Starter motor - removal and installation

Refer to illustrations 21.3, 21.4a and 21.4b
Note: *On some vehicles, it may be necessary to remove the exhaust pipe(s) or frame crossmember to gain access to the starter motor. In extreme cases it may even be necessary to unbolt the mounts and raise the engine slightly to get the starter out.*

5

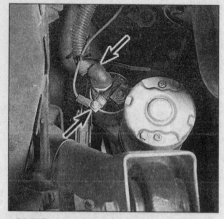

21.3 Before unbolting the starter motor assembly, detach the wires (arrows) from the solenoid (or from the starter motor itself on some models) - be sure the battery is disconnected!

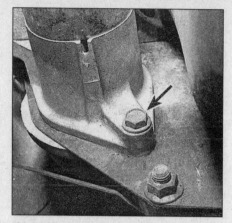

21.4a Lower starter mounting bolt (inline six-cylinder model shown) - view is from the rear, looking forward

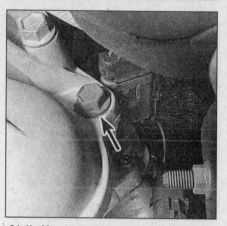

21.4b Upper starter mounting bolt (inline six-cylinder model shown) - view is from the front, looking to the rear

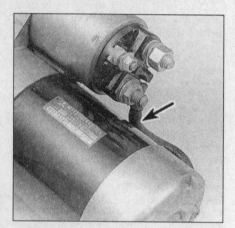

22.3 Before removing the solenoid from the starter motor, detach the starter motor terminal strap (arrow) - Mitsubishi starter shown

22.4 To detach the solenoid from the starter motor, remove the solenoid attaching screws (arrows) from the front of the starter's solenoid mount (Mitsubishi starter shown) - the screws on Delco starters are removed from the rear of the mount

22.5 Some solenoids, like this Delco unit, must be rotated clockwise before they can be disengaged from the starter motor

1   Detach the cable from the negative terminal of the battery.
2   Raise the vehicle and support it securely on jackstands.
3   Clearly label, then disconnect the wires from the terminals on the starter motor and solenoid (if mounted on the starter) **(see illustration)**.
4   Remove the mounting bolts **(see illustrations)** and detach the starter.
5   Installation is the reverse of removal.

22   Starter solenoid - removal and installation

1   Disconnect the cable from the negative terminal of the battery.

All solenoids except Motorcraft

Refer to illustrations 22.3, 22.4 and 22.5

2   Remove the starter motor (see Section 21).

3   Disconnect the strap from the solenoid to the starter motor terminal **(see illustration)**.
4   Remove the screws which secure the solenoid to the starter motor **(see illustration)**.
5   On some models, twist the solenoid in a clockwise direction to disengage the flange from the starter body **(see illustration)**.
6   Installation is the reverse of removal.

Motorcraft solenoid

Refer to illustration 22.7

7   Locate the solenoid - it's mounted in the engine compartment near the battery **(see illustration)**. Detach the electrical wires from the starter solenoid terminals. Label them to assure proper reassembly.
8   Remove the solenoid mounting bracket bolts and detach the solenoid.
9   Installation is the reverse of removal.

22.7 If you have a Motorcraft starter solenoid, it's located near the battery - to remove it, detach the wires and remove the mounting bracket bolts - be sure the battery is disconnected!

Chapter 6
Emissions and engine control systems

Contents

1 General information

Refer to illustrations 1.7, 1.8a, 1.8b, 1.8c, 1.8d, 1.8e, 1.8f, 1.8g, 1.8h and 1.8i

To prevent pollution of the atmosphere from incompletely burned and evaporating gases, and to maintain good driveability and fuel economy, a number of emission control systems are incorporated. They include the:

Computerized Emission Control (CEC) system
Exhaust Gas Recirculation (EGR) system
Air injection system
Fuel evaporative emission control system
Positive Crankcase Ventilation (PCV) system
Thermostatic Air Cleaner (TAC) system
Catalytic converter

All of these systems are linked, directly or indirectly, to the emission control system.

The Sections in this Chapter include general descriptions, checking procedures within the scope of the home mechanic and component replacement procedures (when possible) for each of the systems listed above.

Before assuming that an emissions control system is malfunctioning, check the fuel and ignition systems carefully. The diagnosis of some emission control devices requires specialized tools, equipment and training. If checking and servicing become too difficult or if a procedure is beyond your ability, con-

sult a dealer service department. Remember, the most frequent cause of emissions problems is simply a loose or broken vacuum hose or wire, so always check the hose and wiring connections first.

This doesn't mean, however, that emission control systems are particularly difficult to maintain and repair. You can quickly and easily perform many checks and do most of the regular maintenance at home with common tune-up and hand tools. **Note:** *Because of a Federally mandated extended warranty which covers the emission control system components (and any components which have a primary purpose other than emissions control but have significant effects on emis-*

sions), check with your dealer about warranty coverage before working on any emissions-related systems. Once the warranty has expired, you may wish to perform some of the component checks and/or replacement procedures in this Chapter to save money.

Pay close attention to any special precautions outlined in this Chapter. It should be noted that the illustrations of the various systems may not exactly match the system installed on your vehicle because of changes made by the manufacturer during production or from year-to-year.

A Vehicle Emissions Control Information label is located in the engine compartment **(see illustration).** This label contains impor-

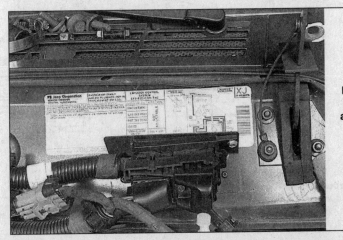

1.7 The Vehicle Emission Control Information (VECI) label contains tune-up specifications and vital information regarding the location of all emission control devices and the routing of all vacuum hoses

6

tant emissions specifications and adjustment information, as well as a vacuum hose schematic with emissions components identified. When servicing the engine or emissions systems, the VECI label in your particular vehicle should always be checked for up-to-date information.

Later 49-state models (1988 and 1989) are equipped with an emission maintenance timer and indicator light. The timer and light are there to tell you when to replace the oxygen sensor and PCV valve and perform other emission maintenance. See Chapter 1 for more information about the emission maintenance timer. The accompanying emission control system schematics **(see illustrations)** can help you troubleshoot the most common cause of emissions problems - loose, detached or misrouted vacuum hoses.

2   Positive Crankcase Ventilation (PCV) system

Refer to illustrations 2.5a and 2.5b

1   The Positive Crankcase Ventilation (PCV) system reduces hydrocarbon emissions by scavenging crankcase vapors. It does this by circulating fresh air from the air cleaner through the crankcase, where it mixes with blow-by gases and is then rerouted through a PCV valve to the intake manifold.

2   The main components of the PCV sys-

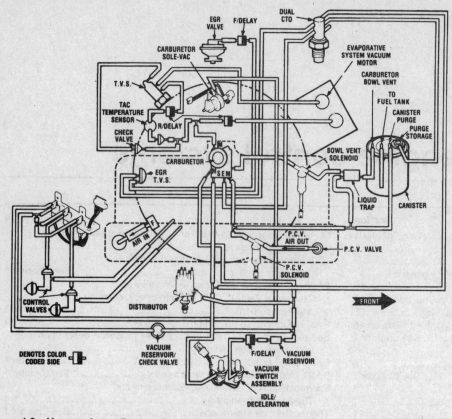

1.9a Vacuum hose diagram for carburetor-equipped, 49-state four-cylinder models (1984 and 1985)

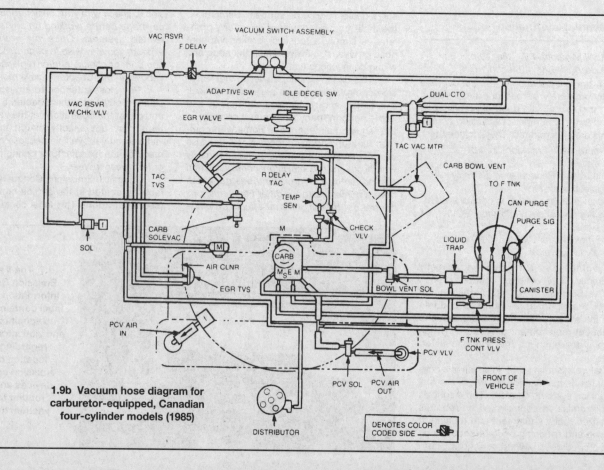

1.9b Vacuum hose diagram for carburetor-equipped, Canadian four-cylinder models (1985)

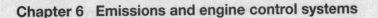

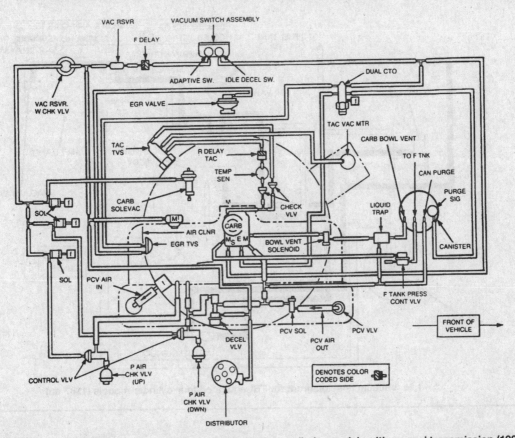

1.9c Vacuum hose diagram for carburetor-equipped, California four-cylinder models with manual transmission (1985)

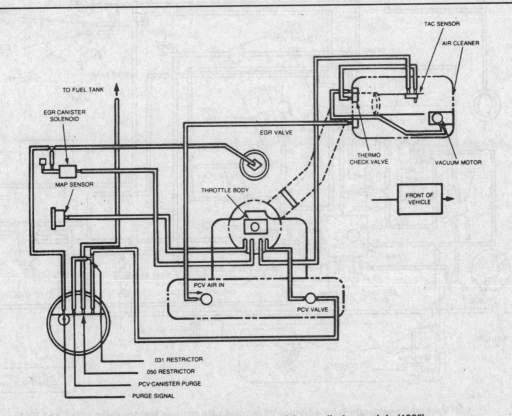

1.9d Vacuum hose diagram for TBI-equipped four-cylinder models (1986)

6

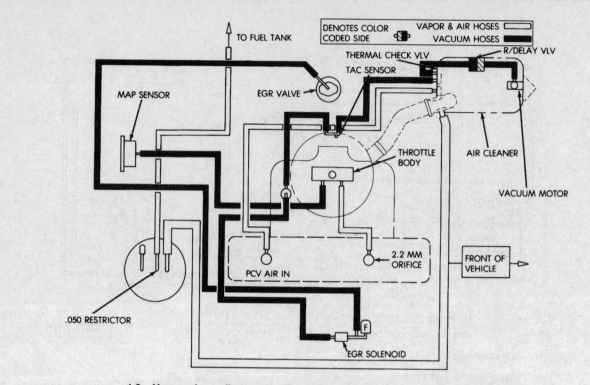

1.9e Vacuum hose diagram for TBI-equipped four-cylinder models (1987 on)

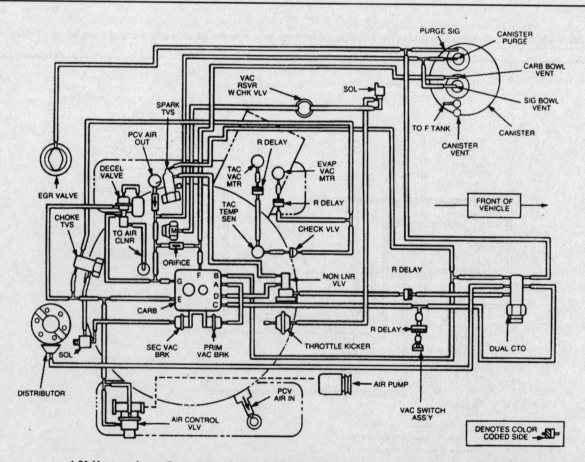

1.9f Vacuum hose diagram for 49-state V6 models with manual transmission (1985 and 1986)

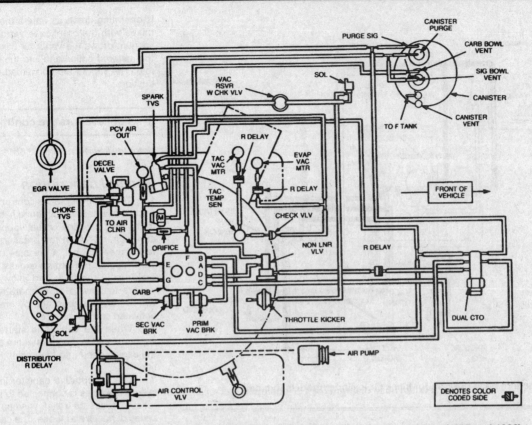

1.9g Vacuum hose diagram for 49-state V6 models with automatic transmission (1985 and 1986)

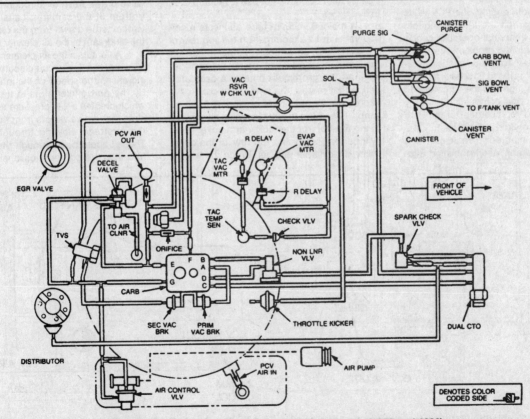

1.9h Vacuum hose diagram for Canadian V6 models (1985 and 1986)

6

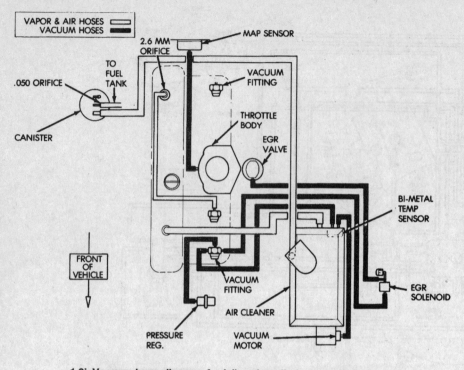

1.9i   Vacuum hose diagram for inline six-cylinder models (typical)

is operating, fresh air enters the engine and mixes with the crankcase vapors. Manifold vacuum draws the vapor/air mixture through the metered orifice and into the intake manifold. The vapors are consumed during combustion.

3   Fuel evaporative control system

Refer to illustrations 3.6a, 3.6b, 3.6c and 3.6d

General description

1   The fuel evaporative control system prevents the release of unburned hydrocarbons - from liquid gasoline or fuel vapors - into the atmosphere. When the pressure in the fuel tank exceeds 3 psi, a pressure relief/rollover valve opens, preventing excessive pressure build-up in the tank and allowing the fuel vapors to flow to an evaporative canister, where they are absorbed by granules of an activated carbon mixture.

2   When the vehicle is started, the fuel vapors stored in the canister are drawn out of the canister and into the engine for combustion.

3   The evaporative canister inlet on fuel-injected vehicles is connected to the fuel tank rollover/pressure relief valve(s) through a series of hoses and tubes. The canister outlet is connected to the air cleaner snorkel. When the engine is operating, the canister purge valve draws fresh air through the filter at the bottom of the canister, causing the stored vapors to be drawn from the canister and into the airstream in the air cleaner snorkel.

4   A venturi in the air cleaner assembly creates a purge line vacuum source. This venturi increases the speed of the intake air flowing by the purge inlet slots in the venturi wall, which creates a low pressure area at the inlet slots that draws vapors from the canister into the airstream flowing through the venturi. These vapors pass through the intake manifold into the combustion chambers, where

tem are the PCV valve, a fresh air filtered inlet and the vacuum hoses connecting these two components with the engine.

3   To maintain idle quality, the PCV valve restricts the flow when the intake manifold vacuum is high. If abnormal operating conditions (such as piston ring problems) arise, the system is designed to allow excessive amounts of blow-by gases to flow back through the crankcase vent tube into the air cleaner to be consumed by normal combustion.

4   Checking and replacement of the PCV valve and filter is covered in Chapter 1.

5   Later fuel-injected vehicles are equipped with a Crankcase Ventilation System **(see illustrations)**. The CCV system performs the same function as a conventional PCV system but does not use a vacuum controlled valve.

6   A molded vacuum tube connects manifold vacuum to a grommet on the top rear of the rocker arm cover (four-cylinder engine) or top front of the cover (inline six-cylinder engine). The grommet contains a calibrated orifice that meters the amount of crankcase vapors drawn out of the engine. A fresh air supply hose from the air cleaner is connected to the top front of the rocker arm cover (four-cylinder engine) or the top rear of the cover (inline six-cylinder engine). When the engine

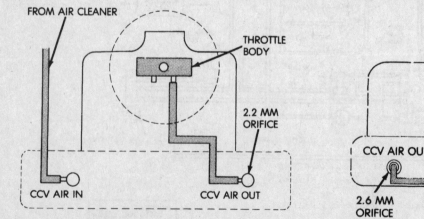

2.5a   Crankcase Ventilation (CCV) system used on fuel-injected four-cylinder engines

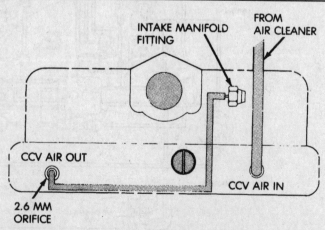

2.5b   Crankcase Ventilation (CCV) system used on fuel-injected inline six-cylinder engines

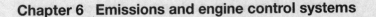

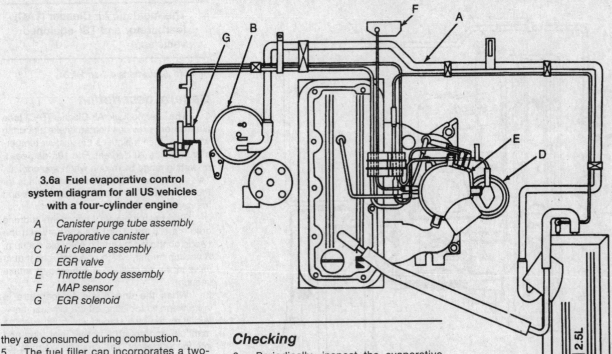

**3.6a Fuel evaporative control
system diagram for all US vehicles
with a four-cylinder engine**

A   Canister purge tube assembly
B   Evaporative canister
C   Air cleaner assembly
D   EGR valve
E   Throttle body assembly
F   MAP sensor
G   EGR solenoid

they are consumed during combustion.

5   The fuel filler cap incorporates a two-way relief valve that is closed during normal operating conditions. The relief valve is calibrated to open when the fuel tank has a pressure of 1.5 psi, or a vacuum of 1.8 in-Hg. Once the pressure or vacuum is relieved to the atmosphere, the valve returns to its normally closed position.

Checking

6   Periodically, inspect the evaporative system vacuum hoses to be sure they are attached and in good condition. Refer to the accompanying vacuum hose routing diagrams **(see illustrations)**.
7   See Chapter 1 for information regarding routine maintenance of the canister.

6

**3.6b Fuel evaporative control system
diagram for Canadian vehicles with a
four-cylinder engine**

A   Air cleaner assembly
B   EGR valve
C   Throttle body assembly
D   MAP sensor
E   EGR solenoid

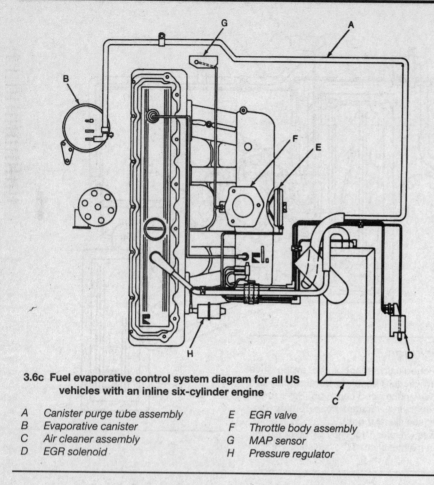

3.6c Fuel evaporative control system diagram for all US vehicles with an inline six-cylinder engine

A   Canister purge tube assembly
B   Evaporative canister
C   Air cleaner assembly
D   EGR solenoid

E   EGR valve
F   Throttle body assembly
G   MAP sensor
H   Pressure regulator

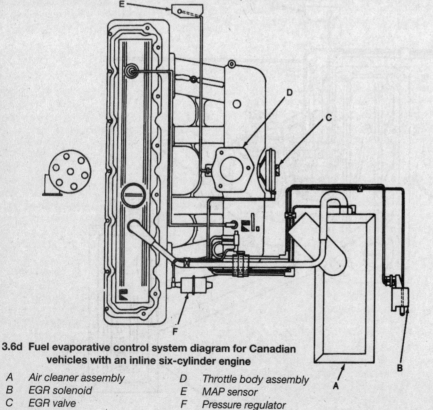

3.6d Fuel evaporative control system diagram for Canadian vehicles with an inline six-cylinder engine

A   Air cleaner assembly
B   EGR solenoid
C   EGR valve

D   Throttle body assembly
E   MAP sensor
F   Pressure regulator

4 Thermostatic Air Cleaner (TAC) (carburetor and TBI-equipped vehicles

Refer to illustrations 4.1 and 4.24

General description

1   The Thermostatic Air Cleaner (TAC) **(see illustration)** provides heated intake air during warm-up, then maintains the inlet air temperature within a 70-degrees F to 105-degrees F operating range by mixing warm and cool air. This allows leaner fuel/air mixture settings for the carburetor, which reduces emissions and improves driveability.

2   Two fresh air inlets - one warm and one cold - are used. The balance between the two is controlled by intake manifold vacuum. A vacuum motor, which operates a heat duct valve in the air cleaner, is actuated by intake vacuum.

3   When the underhood temperature is cold, warm air radiating off the exhaust manifold is routed by a shroud which fits over the manifold up through a hot air inlet tube and into the air cleaner. This provides warm air for the carburetor or throttle body, resulting in better driveability and faster warmup. As the temperature inside the air cleaner rises, the heat duct valve is gradually closed by the vacuum motor (which, in turn is controlled by a bi-metal temperature sensor inside the air cleaner) and the air cleaner draws air through a cold air duct instead. The result is a consistent intake air temperature. A time delay valve provides approximately 100 seconds delay before allowing the heat duct valve to completely close.

4   On some models, a trap door system opens in a similar manner to close off the air cleaner from the outside air when the engine is inoperative.

Checking

General operation

Note: *Make sure the engine is cold before beginning this test.*

5   Always check the vacuum source and the integrity of all vacuum hoses between the source and the vacuum motor before beginning the following test. Do not proceed until they're okay.

6   Apply the parking brake and block the wheels.

7   Detach the flexible duct (if equipped) from the air cleaner snorkel (see Chapter 4).

8   With the engine off, observe the heat duct valve inside the air cleaner snorkel. It should be fully open (heat off position). If it isn't, it might be binding or sticking. Make sure it's not rusted in an open or closed position by attempting to move it by hand. If it's rusted, it can usually be freed by cleaning and oiling the hinge. If it fails to work properly after servicing, replace it.

9   Start the engine. Note the position of the heat duct valve. Now it should be fully closed

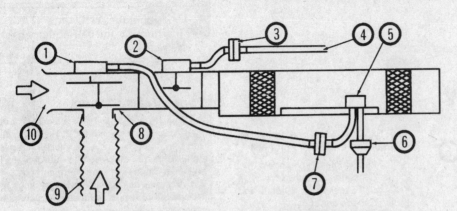

4.1 Cross-section diagram of the Thermostatic Air Cleaner (TAC) assembly used on carburetor and TBI-equipped models - shown in open (heat off) position

1 Heat duct valve vacuum motor	5 Bi-metal temperature sensor
2 Trap door vacuum motor (not on all models)	6 Check valve
3 Trap door time delay valve (not on all models)	7 Time delay valve
4 Vacuum source	8 Heat duct valve
	9 Heat duct
	10 Cool air inlet

4.24 To remove the heat duct valve vacuum motor, remove the air cleaner assembly, disconnect the hoses, then drill out the rivet (arrow)

to incoming air (heat on position).

10   Have an assistant rapidly depress and release the accelerator 1/2 to 3/4 of its travel. The heat duct valve should briefly remain stationary, move toward the heat off position, then back to the heat on position.

11   Loosely attach the flexible duct to the air cleaner and warm the engine to normal operating temperature. Remove the flexible duct and observe the heat duct valve. It should be either fully open or at a mixture position (half open) that provides the correct temperature for the inlet air to the carburetor or throttle body.

12   Stop the engine and connect the flexible duct to the air cleaner.

13   If the air valve does not function as described above, look for a mechanical bind in the linkage and disconnected vacuum hoses or air leaks at the vacuum motor, bi-metal sensor, time delay valve, check valve, intake manifold or vacuum hoses.

14   If the heat duct valve manually operates freely and you cannot find any hose disconnections or leaks, attach a hose from an intake manifold vacuum source directly to the vacuum motor and start the engine.

 a) If the heat duct valve closes, either the thermal switch, time delay valve or check valve is defective.

 b) If the heat duct valve does not close, replace the vacuum motor.

Bi-metal temperature sensor

15   Detach the two vacuum hoses from the sensor.

16   If necessary, remove the sensor (see the replacement procedure below) and cool it below 40-degrees F in a freezer.

17   Attach a vacuum pump to one of the sensor's vacuum fittings and a vacuum gauge to the other fitting.

18   Apply 14 in-Hg vacuum to the sensor.

19   With the switch at a temperature below 40-degrees F, the gauge should indicate a vacuum. Disconnect the vacuum pump momentarily to relieve the vacuum.

20   Warm the switch above 55-degrees F and again attempt to apply vacuum. There should not be steady vacuum reading on the gauge.

21   Replace the sensor if it's defective.

Component replacement

Heat duct valve vacuum motor

22   Remove the air cleaner (see Chapter 4).

23   Clearly label, then detach the vacuum hoses from the heat duct valve vacuum motor, bi-metal sensor and trap door vacuum motor (if equipped).

24   Drill out the rivet which secures the heat duct valve vacuum motor to the snorkel **(see illustration)**.

25   Lift the motor and tilt it to one side to detach the motor linkage from the heat duct valve assembly. Remove the motor.

26   Installation is the reverse of removal. If you don't have a rivet tool, you'll need a self-tapping sheet metal screw of the correct diameter to attach the vacuum motor to the snorkel. Make sure the rivet (or the screw) does not interfere with the movement of the heat duct valve.

Bi-metal temperature sensor

27   Remove the air cleaner (see Chapter 4).

28   Detach the two vacuum hoses from the sensor.

29   Pry up the tabs on the sensor retaining clip. Remove the clip, gasket and sensor from the air cleaner. Before removing the sensor, note its position in relation to the air cleaner to ensure proper reassembly.

30   Installation is the reverse of removal. Be sure to use a new gasket.

5   Air injection systems

General description

1   Four-cylinder and V6 models are equipped with air injection systems which help reduce hydrocarbons and carbon monoxide levels in the exhaust by injecting air into the exhaust stream. During cold engine operation, air is injected directly into the exhaust port; during normal operation, it's injected into the catalytic converter. Two types of air injection systems are used on these vehicles: An air pump system is used on V6 models and a Pulse Air system is used on four-cylinder models.

Air pump system

2   The air injection system on V6 models uses a belt-driven pump to pump air into the exhaust. The system includes the pump itself, an air management valve (or a trio of separate air control, air switching and air diverter valves on some models), an air injection manifold for each cylinder head (or, on California models, a single manifold on the right side of the engine), one or two check valves and the hoses connecting the components. The air management valve (or air control/switching/diverter valves), which is controlled by the vehicle's computer, directs the air to the correct location, depending on engine temperature and driving conditions. **Note:** *The computer may be referred to as the Electronic Control Module (ECM), Electronic Control Unit (ECU) or the MicroComputer Unit (MCU). During certain situations, such as deceleration, the air is diverted to the air cleaner to prevent backfiring from too much oxygen in the exhaust stream. The check valve(s) prevent exhaust gases from being forced back through the system.*

Pulse air system

3   The Pulse Air system uses the alternating pressure and vacuum pulsations created in the exhaust system - instead of an air

6

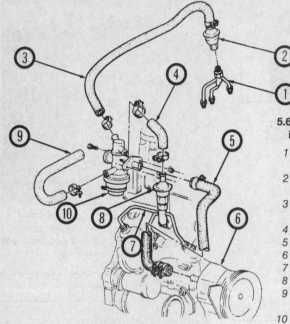

5.6 Exploded view of a typical air injection system (V6 models)

1   Left air injection manifold (49-state and Canada only)
2   Left check valve (49-state and Canada only)
3   Air hose to left check valve (49-state and Canada only)
4   Air hose to right check valve
5   Air pump outlet hose
6   Air pump
7   Right air injection manifold
8   Right check valve
9   Air hose from diverter valve to air cleaner
10 Diverter valve

5.11 To remove the air pump pulley, loosen the three bolts, remove the drivebelt, then remove the bolts

Component replacement

Refer to illustrations 5.11, 5.14, 5.19, 5.30 and 5.32

9   Always detach the cable from the negative terminal of the battery before replacing any of the following components.

Drivebelt (air pump system only)

10   Drivebelt replacement procedures are in Chapter 1.

Air pump pulley and filter

11   Compress the drivebelt to prevent the pulley from turning and loosen the pulley attaching bolts **(see illustration)**.
12   Remove the drivebelt (see Chapter 1).
13   Remove the pulley attaching bolts and detach the pulley.
14   To remove the air pump filter, grasp it firmly with needle nose pliers **(see illustration)** and pull it from the pump. **Caution:** *The filter will probably be damaged when you pull it off, so make sure no fragments fall into the air intake hose. Don't insert a screwdriver between the filter and the pump housing - you might damage the edge of the housing.*
15   To install the new filter, place it in position, place the pulley in position and tighten the pulley bolts evenly. As the pulley bolts are

pump- to draw air into the exhaust system. Air is supplied from the filtered side of the air cleaner through a hose to the air control valve, which is controlled by the MCU. When opened by the air switch solenoid, the air control valve allows air to flow to the air injection check valve, through which it enters the exhaust system. The following explains the components of the system:

a) *The air injection check valve is a reed valve that is opened and closed by the vacuum and pressure exhaust pulsations. During vacuum pulsations, atmospheric pressure opens the check valve and forces air into the exhaust system.*
b) *The air control valve controls the supply of filtered air routed to the air injection check valve. The valve is opened and closed by the air switch solenoid.*
c) *The air switch solenoid controls the air control valve by switching vacuum to the air control valve on and off during varying operating conditions. The solenoid is controlled by the MCU.*
d) *Vacuum is stored in the vacuum storage reservoir until it's released by the air switch solenoid.*
e) *Depending on operating conditions, the MCU switches the air injection point to and from the exhaust manifold and catalytic converter. The MCU does this by energizing and de-energizing the air switch solenoids.*

Checking

4   Because of the complexity of both the air pump and pulse air systems, it's difficult to diagnose either system at home. However, if you suspect the system is malfunctioning, the following simple tests will enable you to check each component.

5   Always begin the inspection of the system by carefully checking all hoses, vacuum lines and wires. Be sure they are in good condition and all connections are clean and tight. Make sure the pump drivebelt is in good condition and properly adjusted.

Air pump system

Refer to illustration 5.6
6   To check the pump:

a) *Start the engine and warm it up to normal operating temperature.*
b) *Locate the outlet hose attached to the air pump (see illustration) and, with the engine operating at about 1500 rpm, squeeze the hose. You should be able to feel the air pulsing through the hose.*
c) *Have an assistant increase the engine speed. There should be a proportional increase in air flow. If there is, the pump is working properly. If there isn't, the pump is malfunctioning. Replace it (see the component replacement procedure below).*

7   To test a check valve:

a) *Remove it (see the component replacement procedure below).*
b) *Clean it off thoroughly.*
c) *Try to blow through it from each end. Air should only pass through it in the direction of normal flow (towards the exhaust manifold). If the valve is open both ways or closed both ways, replace it.*

Pulse air system

8   Checking the pulse air system is beyond the scope of the home mechanic and can be hazardous under certain circumstances. Take the vehicle to a dealer service department or a certified emission control repair shop for testing.

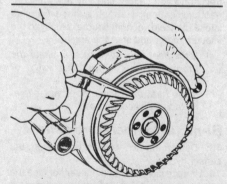

5.14 To remove the air pump filter, grasp it firmly with needle-nose pliers and pull it from the pump - be sure no filter fragments fall into the air intake hose!

5.19 To remove a check valve, loosen the hose clamp and detach the air hose from the valve, then unscrew the valve from the air injection manifold - be sure to use a backup wrench so you don't damage the manifold

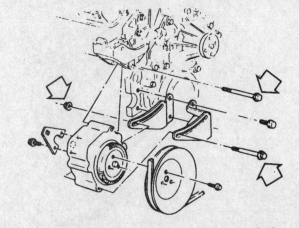

5.30 Air pump installation details - when removing only the pump, unscrew the two attaching bolts and nut (arrows)

tightened, the pulley will press the filter into the pump. **Caution:** *Do not attempt to install a filter by pressing or hammering it into place.*
16   The remainder of installation is the reverse of removal. **Note:** *It is normal for the new filter to have an interference fit with the pump housing. It may squeal in operation until it's worn in.*

Hoses and tubes

17   Before replacing any hose or tube, note how it is routed. Make a sketch or label the components to which it's attached.
18   Be sure to replace old hoses or tubes with new ones of the same size and material. Tighten all connections securely.

Check valve(s) (air pump system only)

19   Remove the hose clamp and detach the air hose from the check valve(s) **(see illustration)**.
20   Unscrew the check valve from the air injection manifold. Use a backup wrench so you don't bend or twist the manifold.
21   Installation is the reverse of removal. If you are replacing the valve, compare the new valve with the old one to be sure you have the correct replacement.

Diverter valve (air pump system only)

22   Detach the vacuum signal line and the air hoses from the diverter valve **(see illustration 5.6)**.
23   Remove the mounting bolts (the bolts may have tabbed lock washers which you will have to bend back).
24   Installation is the reverse of removal.

Air injection manifold(s) (air pump system only)

25   Detach the air hose(s) from the air injection manifold check valve(s) (see the check valve replacement procedure above).
26   Using a flare nut wrench (if available), unscrew the three fittings that attach each air injection manifold to its associated exhaust manifold **(see illustration 5.6)**.

27   Installation is the reverse of removal. Be sure to coat the threads of the air injection manifold fittings with Fel-Pro C 100 anti-seize compound or equivalent.

Air pump

28   Remove the drivebelt and pulley (see the air pump replacement procedure above).
29   Detach the air pump outlet hose **(see illustration 5.6)**.
30   Remove the attaching bolts and nut **(see illustration)** and remove the pump.
31   Installation is the reverse of removal.

Pulse air system

32   To replace pulse air system components, refer to the accompanying exploded view of the pulse air system **(see illustration)**.

5.32 Exploded view of a typical Pulse Air system

1	Air cleaner	7	Upstream check valve hose
2	Air cleaner-to-downstream air control valve vacuum hose	8	Downstream vacuum hose
3	Air cleaner-to-upstream air control valve vacuum hose	9	Downstream check valve hose
4	Upstream vacuum hose	10	Upstream check valve
5	Upstream air control valve	11	Downstream check valve
6	Downstream air control valve	12	Downstream tube-to-converter

A	Air switch solenoid (upstream)
B	Air switch solenoid (downstream)
C	Control wires from MCU (downstream)
D	Control wires from MCU (upstream)
E	Vacuum storage reservoir

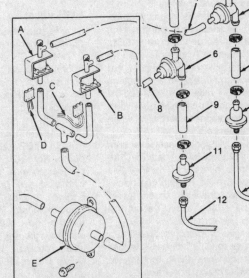

6

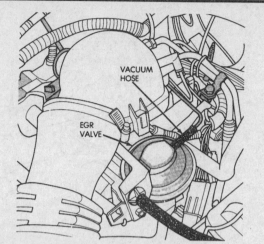

6.8a Location of the EGR valve on four-cylinder models - when checking the valve, remove the vacuum hose and connect a hand vacuum pump

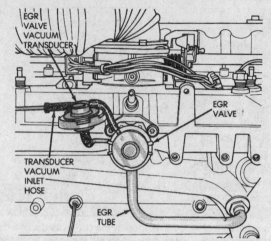

6.8b Location of the EGR valve on inline six-cylinder models - when checking the valve, detach the vacuum hose (arrow) - when removing the valve, detach the transducer vacuum inlet hose

6   Exhaust Gas Recirculation (EGR) system (1984 through 1990 models)

General description

1   High combustion temperatures raise the level of NOx (oxides of nitrogen) emissions in the exhaust. The Exhaust Gas Recirculation (EGR) system lowers NOx levels by allowing small amounts of exhaust gas into the combustion chamber to reduce combustion temperature.

2   The principal component of all EGR systems used on vehicles covered by this manual is the EGR valve itself. The EGR valve is a backpressure or vacuum controlled device which regulates the amount of exhaust gas bled into the intake. The EGR valve is usually open during warm engine operation and anytime the engine is running above idle speed. The amount of gas recirculated is controlled by variations in vacuum and exhaust backpressure.

3   The vacuum supply for the EGR valve is controlled by an EGR valve solenoid which is in turn controlled by the Electronic Control Module (ECM) or Electronic Control Unit (ECU). When energized by the ECM or ECU, the solenoid closes and prevents vacuum from reaching the EGR valve. When it's not energized, the solenoid is open and vacuum is applied to the EGR valve. The ECM or ECU monitors engine coolant temperature and other operating conditions to determine when EGR operation is desired.

4   On inline six-cylinder engines, a vacuum transducer is installed inline between the EGR valve solenoid and the EGR valve. The transducer is controlled by exhaust backpressure. When backpressure exceeds a preset value, the transducer allows vacuum to actuate the EGR valve.

5   Common engine problems associated with the EGR system are rough idling, stalling at idle, rough engine performance during light throttle application and stalling during deceleration.

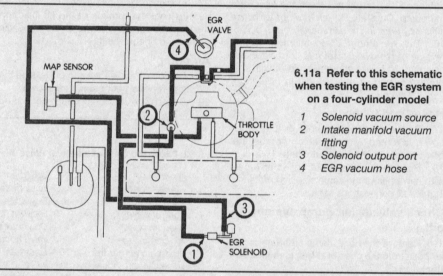

6.11a Refer to this schematic when testing the EGR system on a four-cylinder model

1   *Solenoid vacuum source*
2   *Intake manifold vacuum fitting*
3   *Solenoid output port*
4   *EGR vacuum hose*

Checking

Refer to illustrations 6.8a, 6.8b, 6.11a and 6.11b

EGR valve

6   Perform the EGR valve checking procedures in Chapter 1.

7   If the EGR valve appears to be in proper operating condition, carefully check all hoses connected to the valve for cracks, leaks and kinks. Replace or repair the valve and hoses as necessary.

8   With the engine idling at normal operating temperature, detach the vacuum hose from the EGR valve and connect a vacuum pump. **Note:** *On four-cylinder and inline six-cylinder engines, the EGR valves are located where shown* **(see illustrations)**. *On V6 engines, the EGR valve is located on the intake manifold, in front of the carburetor or throttle body. If the EGR valve is operating properly, the engine should stumble or stall when you apply vacuum. Replace the EGR valve with a new one if the idle remains unchanged.*

9   Because of the EGR system's interrela-

tionship with the ECM or ECU, further checks of the system are beyond the scope of the home mechanic. Have the system checked by a dealer service department or certified emission control repair shop.

EGR valve solenoid

10   Warm the engine up to normal operating temperature.

11   With the engine idling, detach the vacuum hose at the solenoid vacuum source **(see illustrations)** and attach a vacuum gauge to the hose. Verify the vacuum is at least 15 in-Hg.

a) *If vacuum is low, check the hose for kinks, twists or a loose connection at the manifold fitting.*

b) *If vacuum is okay, detach the gauge, reattach the hose and proceed to the next step.*

12   Detach the hose at the solenoid output port and attach a vacuum gauge to the port. The vacuum reading should be zero.

a) *If the vacuum reading is zero, leave the gauge connected and proceed to the next step.*

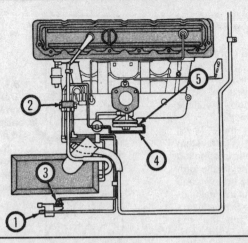

6.11b Refer to this schematic when testing the EGR system on an inline six-cylinder model

1 *Solenoid vacuum source*
2 *Vacuum connector*
3 *Solenoid output port*
4 *EGR vacuum hose*
5 *EGR valve nipple*

b) *If vacuum is present, have the solenoid and/or ECU checked by a dealer service department (special tools are required for further checking).*

13   Detach the electrical connector at the solenoid and determine whether there is vacuum at the solenoid output port. Vacuum should now be present at the output port.

a) *If vacuum is present, the EGR valve may be faulty. See the EGR checking procedure above.*
b) *If vacuum isn't present, replace the EGR valve solenoid.*

Component replacement

Refer to illustrations 6.19 and 6.24

EGR valve

14   Detach the vacuum hose from the EGR valve or EGR valve vacuum transducer **(see illustrations 6.8a and 6.8b)**.
15   Remove the two EGR valve mounting bolts.
16   Remove the EGR valve and gasket. Discard the gasket.
17   Clean the gasket mating surfaces of the EGR valve and intake manifold.
18   Installation is the reverse of removal. Be sure to use a new gasket.

EGR tube (four-cylinder engine)

19   Remove the EGR tube-to-exhaust manifold bolts **(see illustration)**.
20   Using a flare nut wrench (if available), unscrew the EGR tube nut at the intake manifold.
21   Remove the tube and discard the gasket.
22   Installation is the reverse of removal. Be sure to use a new gasket and tighten the bolts and nut securely.

EGR tube (inline six-cylinder engine)

23   Loosen the exhaust manifold mounting bolts (see Chapter 2, Part C).
24   Using a flare nut wrench (if available), loosen the EGR tube line nuts at the intake and exhaust manifolds **(see illustration)**.
25   Remove the EGR tube.
26   Install the EGR tube, but don't tighten the line nuts at this time.
27   Tighten the intake manifold line nut, then the exhaust manifold line nut. Tighten them to the torques specified in Chapter 2, Part C.
28   Tighten the exhaust manifold mounting bolts to the torque specified in Chapter 2, Part C.

7   Catalytic converter

Note: *Because of a Federally mandated extended warranty which covers emissions-related components such as the catalytic converter, check with a dealer service department before replacing the converter at your own expense.*

General description

1   The catalytic converter is an emission control device added to the exhaust system to reduce pollutants from the exhaust gas stream. Two types of converters are used. Your vehicle may be equipped with either of the two. The conventional oxidation catalyst reduces the levels of hydrocarbon (HC) and carbon monoxide (CO). The three-way catalyst lowers the levels of oxides of nitrogen (NOx) as well as hydrocarbons (HC) and carbon monoxide (CO).

Checking

2   The test equipment for a catalytic converter is expensive and highly sophisticated. If you suspect that the converter on your vehicle is malfunctioning, take it to a dealer or authorized emissions inspection facility for diagnosis and repair.
3   Whenever the vehicle is raised for servicing of underbody components, check the converter for leaks, corrosion, dents and other damage. Check the welds/flange bolts that attach the front and rear ends of the converter to the exhaust system. If damage is discovered, the converter should be replaced.
4   Although catalytic converters don't break too often, they do become plugged. The easiest way to check for a restricted converter is to use a vacuum gauge to diagnose the effect of a blocked exhaust on intake vacuum.

a) *Open the throttle until the engine speed is about 2000 RPM.*
b) *Release the throttle quickly.*
c) *If there is no restriction, the gauge will quickly drop to not more than 2 in Hg or more above its normal reading.*

6

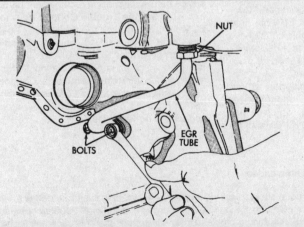

6.19 To remove the EGR tube on a four-cylinder engine, remove the bolts, then unscrew the nut with a flare nut wrench (if available)

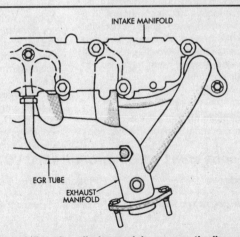

6.24 On an inline six-cylinder model, unscrew the line nuts with a flare nut wrench (if available)

d) *If the gauge does not show 5 in Hg or more above its normal reading, or seems to momentarily hover around its highest reading for a moment before it returns, the exhaust system, or the converter, is plugged (or an exhaust pipe is bent or dented, or the core inside the muffler has shifted).*

Component replacement

5   Refer to the information and diagrams in Chapter 4.

8   Computer Command Control System (CCCS) and trouble codes (V6 models) - description and trouble code retrieving

Refer to illustration 8.7

General description

This electronically controlled emissions and engine control system is linked with as many as nine other related emissions systems. It consists mainly of sensors and an Electronic Control Module (ECM). Completing the system are various engine components which respond to commands from the ECM.

In many ways, this system can be compared to the central nervous system in the human body. The sensors (nerves) constantly gather information and send this data to the ECM (brain), which processes the data and, if necessary, sends out a command for some type of vehicle (body) change.

Here's a specific example of how one portion of this system operates. An oxygen sensor, mounted in the exhaust manifold and protruding into the exhaust gas stream, constantly monitors the oxygen content of the exhaust gas as it travels through the exhaust pipe. If the percentage of oxygen in the exhaust gas is incorrect, an electrical signal is sent to the ECM. The ECM takes this information, processes it and then sends a command to the carburetor Mixture Control (M/C) solenoid, telling it to change the fuel/air mixture. To be effective, all this happens in a fraction of a second, and it goes on continuously while the engine is running. The end result is a fuel/air mixture which is constantly kept at a predetermined ratio, regardless of driving conditions.

Testing

One might think that a system which uses exotic electrical sensors and is controlled by an on-board computer would be difficult to diagnose. This is not necessarily the case.

The Computer Command Control System has a built-in diagnostic system which indicates a problem by flashing a CHECK ENGINE light on the instrument panel. When this light comes on during normal vehicle operation, a fault has been detected.

Perhaps more importantly, the ECM will recognize this fault in a particular system monitored by one of the various information sensors and store it in its memory in the form of a trouble code. Although the trouble code cannot reveal the exact cause of the malfunction, it greatly facilitates diagnosis as you or a professional mechanic can "tap into" the ECM's memory and be directed to the problem area.

Retrieving codes

To extract this information from the ECM memory, you must use a short jumper wire to ground terminals 6 and 7 on the diagnostic connector **(see illustration)**. The diagnostic connector is located in the engine compartment on the left (driver's side) fenderwell. **Caution:** *Do not start the engine with the terminals grounded.*

Turn the ignition to the On position - not the Start position. The CHECK ENGINE light should flash Trouble Code 12, indicating that the diagnostic system is working. Code 12 will consist of one flash, followed by a short pause, and then two flashes in quick succession. After a longer pause, the code will repeat itself two more times.

If no other codes have been stored, Code 12 will continue to repeat itself until the jumper wire is disconnected. If additional Trouble Codes have been stored, they will follow Code 12. Again, each Trouble Code will flash three times before moving on.

Once the code(s) have been noted, use the Trouble Code chart to locate the source of the fault.

It should be noted that the self-diagnosis feature built into this system does not detect all possible faults. If you suspect a problem with the Computer Command Control System, but the CHECK ENGINE light has not come on and no trouble codes have been stored, take the vehicle to a dealer service department or other repair shop for diagnosis.

Furthermore, when diagnosing an engine performance, fuel economy or exhaust emissions problem (which is not accompanied by a CHECK ENGINE light) do

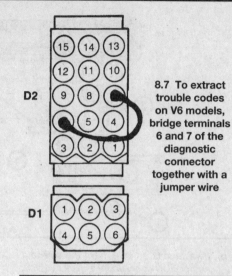

8.7 To extract trouble codes on V6 models, bridge terminals 6 and 7 of the diagnostic connector together with a jumper wire

not automatically assume the fault lies in this system. Perform all standard troubleshooting procedures, as indicated elsewhere in this manual, before turning to the Computer Command Control System .

Finally, since this is an electronic system, you should have a basic knowledge of automotive electronics before attempting any diagnosis. Damage to the ECM or related components can easily occur if care is not exercised.

Trouble Code Identification

Refer to the accompanying chart for a list of the typical trouble codes which may be encountered while diagnosing the Computer Command Control System. Also included are simplified troubleshooting procedures. If the problem persists after these checks have been made, the vehicle must be diagnosed by a professional mechanic who can use specialized diagnostic tools and advanced troubleshooting methods to check the system. Procedures marked with an asterisk (*) indicate component replacements which may not cure the problem in all cases. For this reason, you may want to seek professional advice before purchasing replacement parts.

To clear the Trouble Code(s) from the ECM memory, detach the negative (-) battery cable for at least 10 seconds. Disconnecting the power to the ECM to clear the memory can be an important diagnostic tool, especially on intermittent problems. **Caution:** *To prevent damage to the ECM, the ignition switch must be off when disconnecting or connecting power to the ECM.*

Trouble code chart - 1984 through 1986 V6 models

Trouble Code	Circuit or system	Probable cause
12 (one flash, pause, two flashes)	No reference pulses to ECM	This code should flash whenever the test terminal is grounded with the ignition On and the engine not running. If additional trouble codes are stored (indicating a problem), they will appear after this code has flashed three times. With the engine running, the appearance of this code indicates that no references from the distributor are reaching the ECM Carefully check the four-terminal EST connector are the distributor.

Trouble Code	Circuit or system	Probable cause
13 (one flash, pause, three flashes)	Oxygen sensor circuit	Check for a sticking or misadjusted throttle position sensor. Check the wiring and connectors from the oxygen sensor. Replace oxygen sensor* (see Chapter 1).
14 (one flash, pause, four flashes)	Coolant sensor circuit	If the engine is experiencing overheating problems the problem must be rectified before continuing (see Chapters 1 and 3). Check all wiring and connectors associated with the sensor. Replace the coolant sensor*.
15 (one flash, pause, five flashes)	Coolant sensor circuit (low temperature indicated)	See above.
21 (two flashes, pause, one flash)	TPS circuit (signal voltage high)	Check for sticking or misadjusted TPS. Check all wiring and connections at the TPS and at the ECM Adjust or replace TPS*.
23 (two flashes, pause, three flashes)	Mixture Control (M/C)	Check the electrical connections at the M/C solenoid (see solenoid circuit Chapter 4). If OK, clear the ECM memory and recheck for code(s) after driving the vehicle. Check wiring connections at the ECM. Check wiring from M/C solenoid.
34 (three flashes, pause, four flashes)	Manifold Absolute Pressure (MAP) sensor circuit	Check the hose to the MAP sensor for a leak. Check the wiring from the MAP sensor to the ECM. Check the connections at the ECM and the sensor. Replace the MAP sensor.*
41 (four flashes, pause, one flash)	No distributor signals	Check all wires and connections at the distributor. Check distributor pick-up coil connections (see Chapter 5).
42 (four flashes, pause, two flashes)	Bypass or EST problem	If the vehicle will start and run, check the wire leading to ECM terminal 12. An improper HEI module can also cause this trouble code.
44 (four flashes, pause, four flashes)	Lean exhaust	Check for a sticking M/C solenoid (Chapter 4). Check ECM wiring connections, particularly terminals 14 and 9. Check for vacuum leakage at carburetor base gasket, vacuum hoses or intake manifold gasket. Check for air leakage at air management system-to-exhaust ports and at decel valve Replace oxygen sensor.*
44 and 45 at the same time	Oxygen sensor or circuit	Check the oxygen sensor circuit. Replace the oxygen sensor.*
45 (four flashes, pause, five flashes)	Rich exhaust	Check for a sticking M/C solenoid (Chapter 4). Check wiring at M/C solenoid connector. Check the evaporative charcoal canister and its components for the presence of fuel (Chapters 1, 6). Replace oxygen sensor.*
51 (five flashes, pause, one flash)	PROM problem	Diagnosis should be performed by a dealer service department or other repair shop.
54 (five flashes, pause, four flashes)	Mixture control (M/C) solenoid	Check all M/C solenoid and ECM wires and connections Replace the M/C solenoid* (see Chapter 4).
55 (five flashes, pause, five flashes)	Reference voltage problem	Check for a short circuit to ground on the wire to ECM terminal 21. Possible faulty ECM or oxygen sensor.

6

9   Information sensors (V6 models)

Note: *After performing any checking procedure to any of the information sensors, be sure to clear the ECM of all trouble codes by disconnecting the cable from the negative terminal of the battery for at least ten seconds.*

Engine coolant temperature sensor

Refer to illustration 9.3

General description

1   The coolant sensor is a thermistor (a resistor which varies the value of its resistance in accordance with temperature changes). A failure in the coolant temperature sensor circuit should set either a Code 14 or a Code 15. These codes indicate a failure in the coolant temperature sensor circuit, so the appropriate solution to the problem will be either repair of a wire or replacement of the sensor.

Check

2   Unplug the electrical connector and use an ohmmeter to measure the resistance across the sensor terminals with the engine cold. Warm the engine up and take another measurement. If the difference in resistance readings is not approximately 500 ohms, the sensor is probably bad.

Replacement

Warning: *Wait until the engine is completely cool before beginning this procedure. Also, remove the radiator cap to release any residual pressure that may be present in the cooling system, then reinstall the cap. Read the antifreeze warning in Chapter 3.*

3   Prepare the new sensor for installation by wrapping the threads with Teflon sealing tape to prevent leakage and thread corrosion **(see illustration).**

4   To remove the sensor, release the lock-

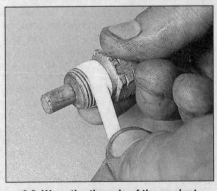

9.3 Wrap the threads of the coolant temperature sensor with Teflon tape to prevent leaks

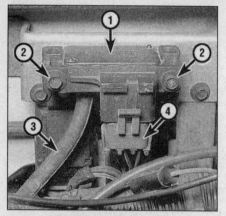

9.6 Typical MAP sensor installation details

1   MAP sensor
2   Mounting screw
3   Vacuum line
4   Electrical connector

ing tab **(see illustration)**, unplug the electrical connector, then carefully unscrew the sensor. Be prepared for coolant leakage and install the new sensor as quickly as possible. **Caution:** *Handle the coolant sensor with care. Damage to this sensor will affect the operation of the entire fuel injection system.*
5   Installation is the reverse of removal. Check the coolant level and add some, if necessary, to bring it to the proper level (see Chapter 1).

Manifold Absolute Pressure (MAP) sensor
Refer to illustration 9.6

General description
6   The Manifold Absolute Pressure (MAP) sensor **(see illustration)** monitors the intake manifold pressure changes resulting from changes in engine load and speed and converts the information into a voltage output. The ECM uses the MAP sensor to control fuel delivery and ignition timing.

Check
7   Unplug the electrical connector from the sensor and, using jumper wires, connect terminals A and C (the two outside terminals) to their corresponding terminals in the electrical connector. Connect the positive lead of a high-impedance digital voltmeter to terminal B (the center terminal) of the sensor and the negative lead to ground. With the ignition On (engine not running) the voltage reading should be about 4.5 to 5 volts. Start the engine and let it warm up. The reading should now be different from the original reading, and should fluctuate with changes in engine rpm. If it doesn't, check the vacuum hose for breaks, blockage and vacuum when the engine is running. If the hose and vacuum signal are OK, the sensor is probably bad. A malfunctioning MAP sensor will usually cause a Code 34 to be stored.

Replacement
8   To replace the sensor, detach the vacuum hose, unplug the electrical connector and remove the mounting screws **(see illustration 9.6)**. Installation is the reverse of removal.

Oxygen sensor

General description
9   The oxygen sensor is mounted in the exhaust system where it can monitor the oxygen content of the exhaust gas stream. By monitoring the voltage output of the oxygen sensor, the ECM will know what fuel mixture command to give the mixture control solenoid in the carburetor.
10   The oxygen sensor produces no voltage when it's below its normal operating temperature of about 600-degrees F. During this initial period before warm up, the ECM operates in open loop mode.
11   If the engine reaches normal operating temperature and/or has been running for two or more minutes, and if the oxygen sensor is producing a steady signal voltage between 0.35 and 0.55-volt, even though the TPS indicates the engine isn't at idle, the ECM will set a Code 13.
12   A delay of two minutes or more between engine start-up and normal operation or the sensor, followed by a low voltage signal or a short in the sensor circuit, will cause the ECM to set a Code 44. If a high voltage signal occurs, the ECM will set a Code 45.
13   When any of the above codes occur, the ECM operates in the open loop mode - that is, it controls fuel delivery in accordance with a programmed default value instead of feedback information from the oxygen sensor.

Check
14   The sensor can be checked with a high-impedance digital voltmeter. Warm up the engine to normal operating temperature, then turn the engine off. Unplug the oxygen sensor electrical connector and connect the positive probe of the voltmeter to the sensor side of the connector. **Caution:** *Don't let the sensor wire or the voltmeter lead touch the exhaust pipe or manifold. Ground the negative probe of the meter, turn the meter to the millivolt setting and start the engine.*
15   The reading on the voltmeter should fluctuate between 100 and 1,000 millivolts (0.1 and 1.0 volts). If the meter reading doesn't fluctuate, the sensor is probably bad (although a fuel system problem could be the cause).

Replacement
16   Refer to Chapter 1 for the oxygen sensor replacement procedure.

Throttle Position Sensor (TPS)
General description
17   The Throttle Position Sensor (TPS) is located on the carburetor float bowl.
18   By monitoring the output voltage from the TPS, the ECM can determine fuel delivery

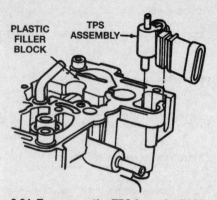

9.24 To remove the TPS from the E2SE carburetor, push up from the bottom of the electrical connector and remove the TPS from the float bowl (also remove the spring from the bottom of the float bowl)

based on throttle valve angle (driver demand). A broken or loose TPS can cause intermittent bursts of fuel from the injector and an unstable idle because the ECM thinks the throttle is moving.
19   A problem in any of the TPS circuits will set a Code 21. Once a trouble code is set, the ECM will use an artificial default value for TPS and some vehicle performance will return.

Check
20   Unplug the TPS electrical connector. Using a digital ohmmeter, connect the positive probe to the center terminal and the negative probe to each of the outside terminal (in turn).
21   With the ignition On slowly throttle lever until it is wide open, the resistance readings. If the resistance readings are jerky and inconsistent instead of gradual and progressive, replace the TPS with a new part.

Replacement
Refer to illustration 9.24
22   Detach the negative battery cable.
23   Unplug the TPS electrical connector.
24   Remove the top of the carburetor (see Chapter 4), then grasp the TPS assembly and lift it up out of the carburetor body **(see illustration)**.
25   Installation is the reverse of removal.

Adjustment
26   Carefully drill a small hole in the plug that covers the TPS adjustment screw, then pry the plug from its bore with a small pick or awl. **Caution:** *When drilling the hole, be careful not to damage the head of the adjusting screw.*
27   Disconnect the electrical connector from the TPS and connect the corresponding terminals of the connector and the TPS with jumper wires (you'll have to fabricate these using the proper terminals, which are generally available at auto parts stores or electronics supply stores).Strip a small portion from the center of the wires connected to terminals B and C.

28 Connect the probes of a high-impedance digital voltmeter to the stripped portion of the wires connecting terminals B and C.

29 Check to be sure that the air conditioner is turned off and the throttle lever is at its curb idle position. With the engine off and the ignition key turned to the On position, turn the TPS adjusting screw one way or the other until a reading of 0.026 volts is obtained.

30 Turn the ignition Off, remove the jumper wires and reconnect the electrical connector. Install a new plug in the hole over the adjustment screw. If a plug isn't available, plug the hole with RTV sealant.

10 Self-diagnosis codes (1991 through 1997 models) – description and trouble code retrieving

Note: *On 1990 and earlier models, with the exception of models equipped with a V6 engine, a special tool (Chrysler DRB-II or equivalent) is needed to access the self-diagnosis system.*

1 The computerized emissions control systems on 1991 through 1997 models are equipped with a self-diagnosis function that stores codes in the computer which you can retrieve. These codes indicate any areas of trouble in the system. Often, when a self-diagnosis code (sometimes referred to as a "trouble" code) is stored in the computer, a CHECK ENGINE or POWER LOSS light will illuminate on the dash panel.

2 Prior to obtaining the codes, set the parking brake and put the transmission in Park (automatic) or Neutral (manual). Raise the engine speed to approximately 2500 rpm and slowly let the speed down to idle. Also cycle the air conditioning system on briefly, then off (if equipped). Next, if the vehicle is equipped with an automatic transmission, with your foot on the brake, select each position on the transmission (Reverse, Drive, Low, etc.) and bring the shifter back to Park.

3 To display the codes, turn the ignition key ON, OFF, ON, OFF and finally ON. The codes will begin to flash on the CHECK ENGINE light. The light will blink the number of the first digit then pause and blink the number of the second digit. For example: Code 23, the MAT sensor circuit, would be indicated by two flashes, then a pause (about 2 seconds) followed by three flashes. If more than one code is being accessed, the pause between codes is about 3 or 4 seconds.

4 Certain criteria must be met for a fault code to be entered into the computer memory. The criteria may be a specific range of engine rpm, engine temperature or input voltage to the computer. It is possible that a fault code for a particular monitored circuit may not be entered into the memory despite a malfunction. This may happen because one of the fault code criteria has not been met.

5 The accompanying table is a list of the typical trouble codes which may be encountered while diagnosing the system. Also included in this Chapter are simplified testing procedures. If the problem persists after checks have been made, or if you are not certain which component is malfunctioning, more detailed service procedures will have to be performed by a dealer service department or other qualified repair shop. Procedures marked with an asterisk (*) indicate component replacements which may not cure the problem in all cases. For this reason, you may want to seek professional advice before purchasing replacement parts.

Trouble code chart - 1991 through 1997 models

Code	Circuit or system	Probable cause
11	No crank reference signal to computer	Check the circuit from the crankshaft position sensor to the PCM (computer). Replace the crankshaft position sensor.*
12	Battery disconnected	Battery voltage to PCM disconnected within last 50 key-on cycles.
13	Manifold Absolute Pressure (MAP) sensor	No change indicated between barometric pressure (engine off) and manifold vacuum (engine on). Check MAP sensor and circuit. Replace MAP sensor.*
14	Manifold Absolute Pressure (MAP) sensor	MAP sensor voltage too high or low. Check MAP sensor and circuit. Replace MAP sensor.
15	Vehicle speed sensor or circuit	No vehicle speed sensor signal detected by computer. Check circuit. Replace sensor.*
17	Engine running too cool	Check thermostat for proper operation. Check coolant sensor and circuit. Replace coolant sensor.*
21	Oxygen sensor or circuit	Check oxygen sensor and circuit. Replace sensor.*
22	Coolant temperature sensor or circuit	Check coolant temperature sensor circuit. Check sensor resistance. Replace sensor.*
23	Intake air temperature sensor or circuit	Check intake air temperature sensor circuit. Check sensor resistance. Replace sensor.*
24	Throttle Position Sensor (TPS) sensor or circuit	Check TPS sensor circuit. Check sensor voltage outputs. Replace sensor.*
25	Idle Air Control (IAC) circuit	Check IAC sensor circuit. Replace sensor.*
27	Fuel injector control circuit	Have the vehicle diagnosed at a dealer service department or other qualified repair shop.
31	Evaporative control circuit	Check the fuel evaporative control circuit and hoses (see Section 3).
33	Air conditioning clutch relay	Check the circuit to the A/C clutch for an open or short-circuit condition.
34	Cruise control circuit	Have the vehicle diagnosed at a dealer service department or other qualified repair shop.
35	Cooling fan relay circuit	Check the cooling fan circuit and relay. Replace relay.*

6

Trouble code chart - 1991 through 1997 models (continued)

Code	Circuit or system	Probable cause
37	Torque converter clutch	Have the vehicle diagnosed at a dealer service department or other qualified repair shop.
41	Alternator field not switching properly	Check the charging system (see Chapter 5). Further diagnosis should be performed by a dealer service department or other repair shop.
42	Automatic shutdown relay or control circuit	Have the vehicle diagnosed at a dealer service department or other qualified repair shop.
43	Misfire	Misfire in one or more cylinders.
44	Battery temperature sensor	Have the vehicle diagnosed at a dealer service department or other qualified repair shop.
46	Battery over voltage	Check the charging system (see Chapter 5).
47	Battery under voltage	Check the charging system (see Chapter 5).
51	Oxygen sensor - lean condition indicated	Check the oxygen sensor and circuit. Check for vacuum leaks. Replace oxygen sensor.*
52	Oxygen sensor - rich condition indicated	Check fuel injection system. Replace oxygen sensor.*
53	Powertrain Control Module (PCM) problem	Have the vehicle diagnosed at a dealer service department or other qualified repair shop.
54	Distributor sync pickup	Replace sync sensor in distributor.*
55	End of Code Output	Trouble code sequence complete.
62	Emissions Maintenance Reminder (EMR)	Have the vehicle diagnosed at a dealer service department or other qualified repair shop. mileage accumulator Or PCM
63	Controller failure EEPROM write denied	Have the vehicle diagnosed at a dealer service department or other qualified repair shop.
72	Catalytic converter circuit	Have the vehicle diagnosed at a dealer service department or other qualified repair shop
76	Fuel pump resistor bypass relay circuit	Have the vehicle diagnosed at a dealer service department or other qualified repair shop.
77	Cruise control system	Check power to cruise control solenoids.

11   Information sensors (1991 through 1997 models)

1   Information sensors inform the computer of the engine's current operating conditions so adjustments (air/fuel mixture, ignition timing, idle speed, etc.) can be made to achieve lower emissions and improved fuel economy. For more information on the computerized emissions control system, see Section 8. **Note:** *Because of a Federally mandated extended warranty which covers emissions-related components such as the information sensors, check with a dealer service department before replacing the sensors at your own expense.*

Oxygen sensor

Refer to illustration 11.2

General description

2   The Oxygen (O2) sensor is located in the exhaust manifold or exhaust down pipe **(see illustration)**. It provides a variable voltage signal (about 0 to 1 volt) to the computer, based on the oxygen content in the exhaust.

The computer uses this information to determine whether or not the fuel/air mixture needs to be altered by adjusting the pulse width of the fuel injectors. The oxygen sensor is normally replaced at the specified mileage (see Chapter 1).

11.2 An oxygen sensor located in the exhaust manifold (arrow)

Check and replacement

3   With the engine off and at normal operating temperature, disconnect the sensor electrical connector, insert the positive probe of a high-impedance voltmeter into the connector terminal (on the sensor side) and ground the negative lead. Start the engine and check to see if the reading fluctuates between 0.1 and 1.0 volts (100 to 1000 millivolts). If it doesn't, the sensor is faulty and should be replaced. On some models the oxygen sensor incorporates a heating element. With the engine cold, unplug the sensor connector and touch the ohmmeter leads across the white wire terminals. The reading should be between 5 and 7 ohms. The sensor is faulty if the ohmmeter displays an open circuit (infinity). Refer to Chapter 1 for the oxygen sensor replacement procedure.

Knock sensor

Refer to illustration 11.4

General description

4   A knock sensor **(see illustration)** is used on some models. The knock sensor is

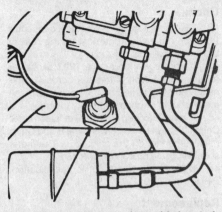

11.4 The knock sensor (arrow) is located on the engine block (on six-cylinder models - shown) or on the intake manifold (four-cylinder models)

on the left side of the engine block, above the oil pan on six-cylinder models and threaded into the intake manifold on four-cylinder models. The sensor is a piezoelectric crystal transducer that indicates detonation (knock) during engine operation by sending small electric impulses to the computer. The computer then retards the ignition to eliminate detonation (knock).

Check

5   A simple check of the knock sensor can be made by connecting a timing light in accordance with the tool manufacturers instructions, then starting the engine. Have an assistant strike the engine block near the sensor while you observe the timing marks. If the timing retards, the system is operating properly. If it doesn't, either the sensor or its circuit is faulty.

Replacement

6   If necessary for access to the sensor, raise the front of the vehicle and support it securely on jackstands.
7   Disconnect the knock sensor electrical connector.
8   Carefully unscrew the sensor.
9   Clean the threaded hole of any corrosion and old sealant. Use Teflon tape on the threads of the new sensor.
10   Install the new knock sensor into the engine block or intake manifold. Tighten it to 89 in-lbs. This torque is critical for the proper operation of the knock sensor.
11   Connect the electrical connector to the sensor.
12   Lower the vehicle.

Coolant Temperature Sensor (CTS)

Refer to illustration 11.13

General description

13   The CTS is a device that monitors the coolant temperature. It is normally mounted in the thermostat housing **(see illustration)**. As the coolant temperature changes, the

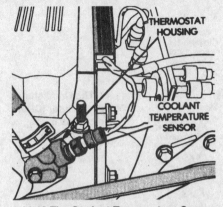

11.13 The Coolant Temperature Sensor (CTS) is normally mounted in the thermostat housing

resistance of the sensor changes, providing a varying input voltage to the computer. The computer uses information from the CTS to adjust fuel/air mixture, ignition timing and control EGR valve operation.

Check

14   Unplug the electrical connector from the sensor. Using an ohmmeter, check the resistance of the sensor with the engine cold - it should be approximately 7000 ohms or greater. Start the engine and allow it to reach normal operating temperature and recheck the resistance of the sensor. It should now be less than 1000 ohms.

Replacement

Warning: *Wait until the engine is completely cool (at least five hours after it has been run) before replacing the CTS.*

15   Prepare the new sensor by wrapping the threads with Teflon tape or applying a non-hardening thread sealant. Open the radiator cap to release any residual pressure. Squeeze the upper radiator hose and reinstall the cap (this will create a slight vacuum in the cooling system, which will minimize coolant loss).
16   Disconnect the electrical connector from the sensor. Place rags around the sensor to prevent spilling coolant. Unscrew the old sensor and immediately install the new one as quickly as possible, tightening it securely. Plug in the electrical connector. Check the coolant level and add some, if necessary (see Chapter 1).

Manifold Absolute Pressure (MAP) sensor

General description

17   The MAP sensor is mounted on the firewall, near the rear of the valve cover **(see illustration 9.8)** or on the throttle body. It is connected electrically to the computer, and on firewall mounted sensors, by a vacuum hose to the throttle body. The MAP sensor reads load (pressure) changes in the intake manifold. This causes electrical resistance changes in the MAP sensor, resulting in

changing input voltage to the computer. The computer uses this information (input voltage) to vary the air/fuel mixture.

Check (firewall-mounted models only)

Note: *On throttle-body mounted models, it is best to have the sensor checked by a dealer service department.*

18   Inspect the vacuum hose connections from the throttle body to the MAP sensor. Replace a cracked or broken hose with a new one.
19   Disconnect the vacuum hose from the MAP sensor and connect a hand vacuum pump. Apply vacuum. The sensor should hold vacuum. If it does not, replace the sensor.
20   Using a high-impedance digital voltmeter, probe the A and B terminals (marked on the MAP sensor body) and make sure that, with the ignition switch On and the engine Off, the voltage is between 4 and 5 volts. With the engine idling at normal operating temperature, the voltage should drop to around 2 volts. Check the voltage between the A and C terminals with the ignition On and the engine Off. Voltage should be approximately 5 volts. If the MAP sensor fails any of the tests, replace it with a new one.

Replacement

21   Disconnect the electrical connector from the MAP sensor.
22   Detach the vacuum hose from the MAP sensor.
23   Remove the two mounting screws.
24   Remove the MAP sensor from the firewall.
25   Installation is the reverse of removal.

Manifold Air Temperature (MAT) sensor

General description

Refer to illustration 11.26

26   The MAT sensor is installed in the intake manifold **(see illustration)**. Its sensor element extends into the air stream. The sensor resistance changes as the air stream temper-

11.26 The Manifold Air Temperature (MAT) sensor (arrow)

6

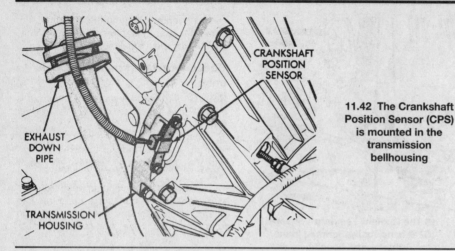

11.42 The Crankshaft Position Sensor (CPS) is mounted in the transmission bellhousing

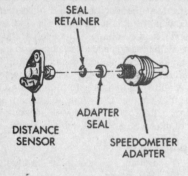

11.52 Vehicle speed sensor installation details

ature changes, sending a varied input voltage to the computer. The computer uses this information to alter the fuel/air mixture.

Check

27   Unplug the MAT sensor electrical connector and use a digital ohmmeter to check the sensor resistance. With the engine at normal operating temperature, the resistance of the sensor should be approximately 4000 ohms or less, decreasing as the air temperature in the intake manifold rises.

Replacement

28   Remove the air cleaner (if necessary).
29   Disconnect the electrical connector from the MAT sensor.
30   Unscrew the MAT sensor from the intake manifold.
31   Clean the threaded hole to remove any corrosion or old sealant.
32   Wrap the threads of the new MAT sensor with Teflon tape.
33   Install the sensor in the manifold and tighten it securely.

Throttle Position Sensor (TPS)

General description

34   The Throttle Position Sensor, which is mounted on the side of the throttle body and connected directly to the throttle shaft, senses throttle movement and position, then transmits an electrical signal to the computer. This signal enables the computer to determine when the throttle is closed, in its normal cruise condition or wide open.

Check

35   Using a high-impedance digital voltmeter, connect the positive probe to the center terminal and the negative probe to the ground terminal (black wire). (**Note:** *This is done by back probing the electrical connector (don't unplug the connector from the TPS).*
36   With the ignition On, slowly move the throttle lever until it is wide open and observe the reading on the meter. With the throttle fully closed it should read greater than 200 millivolts. With the throttle fully open, it should be a little less than 4.8 volts. If the voltage readings are incorrect, replace the TPS.

Replacement

37   Detach the negative battery cable.
38   Unplug the electrical connector from the TPS.
39   Remove the screws and detach the TPS from the throttle body.
40   Position the new TPS on the throttle body, making sure the slot in the TPS is aligned with the blade on the end of the throttle shaft.
41   Install the screws and tighten them securely. **Note:** *The TPS on 1991 and later models is non-adjustable.*

Crankshaft Position Sensor (CPS)

General description

Refer to illustration 11.42
42   The CPS, used on later fuel-injected models, is attached to the transmission bellhousing (**see illustration**). The CPS detects teeth on the flywheel as they pass the sensor during engine operation and provides the computer with information concerning engine speed and crankshaft angle. This information is used to advance or retard ignition timing and control fuel injector pulses.

Check

43   Disconnect the CPS electrical connector from the main wiring harness at the rear of the intake manifold.
44   If you're working on a 1992 or earlier model, connect an ohmmeter across terminals A and B (marked on the connector). The meter should read no continuity (open).
45   If you're working on a 1993 or later model, Check the resistance between the B and C terminals (marked on the connector. The meter should indicate no continuity (open).
46   If the readings are not to specification, replace the CPS.

Replacement

47   Disconnect the electrical connector from the CPS.
48   Detach the CPS by removing the two bolts.
49   Installation is the reverse of removal.

Vehicle speed sensor

Refer to illustration 11.52

General description

50   Later models are equipped with a vehicle speed sensor (instead of a speedometer cable) that sends a pulsating voltage signal to the computer that the computer converts to miles per hour.

Check

51   A problem with this sensor would most likely cause a code 15 to be stored in the computer's memory.

Replacement

52   To replace the vehicle speed sensor (located at the rear of the transfer case), raise the vehicle, support it securely on jackstands, disconnect the electrical connector from the sensor and unscrew it from the speedometer adapter (**see illustration**). Installation is the reverse of removal.

12   On-Board Diagnostics II (1998 and later models) - description and trouble code retrieval

Diagnostic tool information

Refer to illustrations 12.1, 12.3 and 12.4
1   A digital multimeter is necessary for checking fuel injection and emission related components (**see illustration**). A digital volt-ohmmeter is preferred over the older style analog multimeter for several reasons. The analog multimeter cannot display the volts-ohms or amps measurement in hundredths and thousandths increments. When working with electronic circuits, which are often very low voltage, this accurate reading is most important. Another good reason for the digital multimeter is the high impedance circuit. The digital multimeter is equipped with a high resistance internal circuitry (10 million ohms). Because a voltmeter is hooked up in parallel

12.1 Digital multimeters can be used for testing all types of circuits; because of their high impedance, they are much more accurate than analog meters for measuring low-voltage computer circuits

12.3 Scanners like the Actron Scantool and the AutoXray XP240 are powerful diagnostic aids; programmed with comprehensive diagnostic information, they can tell you just about anything you want to know about your engine management system

with the circuit when testing, it is vital that none of the voltage being measured should be allowed to travel the parallel path set up by the meter itself. This dilemma does not show itself when measuring larger amounts of voltage (9 to 12 volt circuits) but if you are measuring a low voltage circuit such as the oxygen sensor signal voltage, a fraction of a volt may be a significant amount when diagnosing a problem. However, there are several exceptions where using an analog voltmeter may be necessary to test certain sensors.

2   With the arrival of the On-Board Diagnostics II (OBD-II) system, a specially designed scanner is needed to extract trouble codes. Several tool manufacturers offer "generic" scan tools for the do-it-yourselfer. Ask the parts salesman at a local auto parts store for additional information concerning the availability of a generic scan tool for your vehicle.

3   Hand-held scanners **(see illustration)** are the most powerful and versatile tools for analyzing the engine management systems used on newer vehicles. Before purchasing a scan tool, carefully examine it to make sure that it's compatible with the year, make and model of the vehicle you are working on. Some scan tools are purpose-built to work with specific makes and models. Others use interchangeable cartridges to access the par-

ticular manufacturer (Ford, GM, Chrysler, etc.). Some scan tools and/or cartridges are designed to work with vehicles manufactured on a certain continent (Asia, Europe, North America).

4   Another type of code reader may be available for some models at parts stores **(see illustration)**. These tools simplify the procedure for extracting codes from the engine management computer by simply "plugging in" to the diagnostic connector on the vehicle wiring harness.

On-Board Diagnostic system general description

5   All models described in this manual are equipped with the second generation On-Board Diagnostics II (OBD-II) system. The systems consist of an onboard computer, known as the Powertrain Control Module (PCM), and information sensors, which monitor various functions of the engine and send data to the PCM.

6   Based on the data and the information programmed into the computer's memory, the PCM generates output signals to control various engine functions via control relays, solenoids and other output actuators. The

PCM is specifically calibrated to optimize the emissions, fuel economy and driveability of the vehicle.

7   Because of a Federally mandated warranty which covers the emissions system components and because any owner-induced damage to the PCM, the sensors and/or the control devices may void the warranty, it isn't a good idea to attempt diagnosis or replacement of the PCM at home while the vehicle is under warranty. Take the vehicle to a dealer service department if the PCM or a system component malfunctions.

Information sensors

8   **Battery temperature sensor** - The battery temperature sensor senses the temperature of the battery. The PCM uses this information in controlling the voltage output of the alternator.

9   **Camshaft position sensor** - The camshaft position sensor provides information on camshaft position. The PCM uses this information, along with the crankshaft position sensor information, to control fuel injection synchronization. The camshaft position sensor is located inside the distributor housing on 1998 and 1999 models and is also referred to as a stator. For more information on the camshaft position sensor on models through 1999, see Chapter 5. The camshaft position sensor on 2000 models is a separate unit installed where the distributor was located on earlier models. On 2000 models, see Section 13 for more information on the camshaft position sensor.

10   **Crankshaft position sensor** - The crankshaft position sensor senses crankshaft position (TDC) during each engine revolution. The PCM uses this information to control ignition timing and fuel injection synchronization.

11   **Engine coolant temperature sensor** - The engine coolant temperature sensor senses engine coolant temperature. The PCM uses this information to control the fuel mixture and ignition timing.

12   **Intake air temperature sensor** - The intake air temperature senses the temperature of the air entering the intake manifold. The PCM uses this information to control the fuel mixture.

13   **Manifold absolute pressure sensor** - The manifold absolute pressure monitors intake manifold pressure and ambient barometric pressure. The PCM uses this input signal to determine engine load and adjust the fuel mixture accordingly.

14   **Oxygen sensor** - The oxygen sensors generate a voltage signal that varies with the difference between the oxygen content of the exhaust and the oxygen in the surrounding air. The PCM uses this information to determine if the fuel system is running rich or lean.

15   **Power steering pressure switch** - The power steering pressure switch is used on four-cylinder models to detect high line pressure in the power steering system. The PCM uses this input signal to adjust the idle speed under increased engine loads during low-

12.4 Trouble code tools simplify the task of extracting the trouble codes

6

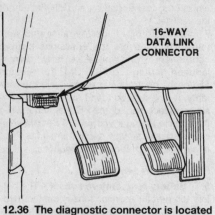

16-WAY DATA LINK CONNECTOR

12.36 The diagnostic connector is located under the left end of the dash

speed vehicle maneuvers.

16   **Throttle position sensor** - The throttle position sensor senses throttle movement and position. This signal enables the PCM to determine when the throttle is closed, in a cruise position, or wide open. The PCM uses this information to control fuel delivery and ignition timing.

17   **Transmission pressure and temperature sensors** - The transmission pressure and temperature sensor signals are used by the PCM to control shift operation (four-speed electronic automatic transmission only).

18   **Vehicle speed sensor** - The vehicle speed sensor provides information to the PCM to indicate vehicle speed. Only 1997 models are equipped with a vehicle speed sensor. On 1998 and later models, the PCM receives vehicle speed information from the antilock brake system rear wheel speed sensor.

19   **Miscellaneous PCM inputs** - In addition to the various sensors, the PCM monitors various switches and circuits to determine vehicle operating conditions. The switches and circuits include:
a) *Air conditioning system*
b) *Auto shutdown relay*
c) *Battery voltage*
d) *Brake On/Off switch*
e) *Cruise control system*
f) *Engine oil pressure*
g) *EVAP leak detection pump operation*
h) *Fuel level*
i) *Ignition switch*
j) *Overdrive switch*
k) *Park/neutral position switch*
l) *Sensor signal and ground circuits*

Output actuators

20   **Air conditioning clutch relay** - The PCM controls the operation of the air conditioning compressor clutch with the air conditioning clutch relay.

21   **Automatic shutdown relay** - The automatic shut down relay supplies battery power to the fuel injectors, ignition coil and oxygen sensor heaters. The PCM controls the opera-

tion of the automatic shutdown relay.

22   **Check Engine light** - The PCM will illuminate the Check Engine light if a malfunction in the electronic engine control system occurs.

23   **Cruise control vacuum and vent solenoids** - The cruise control system operation is controlled by the PCM.

24   **Engine cooling fan relay (four-cylinder only)** - The engine cooling fan, on four-cylinder models, is controlled by the PCM according to information received from the engine coolant temperature sensor.

25   **EVAP canister purge solenoid** - The evaporative emission canister purge solenoid is a solenoid valve, operated by the PCM to purge the fuel vapor canister and route fuel vapor to the intake manifold for combustion.

26   **EVAP leak detection pump** - Some models are equipped with a self-diagnostic EVAP leak detection system. The system checks the integrity of the EVAP system when the engine is started cold. The PCM controls the operation of the EVAP leak detection pump.

27   **Fuel injectors** - The PCM opens the fuel injectors individually in firing order sequence. The PCM also controls the time the injector is held open (pulse width). The pulse width of the injector (measured in milliseconds) determines the amount of fuel delivered. For more information on the fuel delivery system and the fuel injectors, including injector replacement, refer to Chapter 4.

28   **Fuel pump relay** - The fuel pump relay is activated by the PCM with the ignition switch in the Start or Run position. When the ignition switch is turned on, the relay is activated to supply initial line pressure to the system. Refer to Chapter 12 or your owner's manual for more information on relay location. For more information on fuel pump check and replacement, refer to Chapter 4.

29   **Idle air control valve** - The idle air control valve controls the amount of air allowed to bypass the throttle plate when the throttle valve is closed or at idle position. The more air allowed to bypass the throttle plate, the higher the idle speed. The idle air control valve opening and the resulting idle speed is controlled by the PCM.

30   **Ignition coil** - The PCM controls the ignition coil by grounding the primary circuit to the ignition coil which generates high voltage in the secondary circuit, thus sending spark from the ignition coil to the distributor. The PCM controls ignition timing depending on engine operation conditions. Refer to Chapter 5 for more information on the ignition coil.

31   **Tachometer** - The PCM operates the tachometer from the information received from the crankshaft position sensor.

32   **Transmission shift control solenoids** - The PCM receives input signals from various sensors and switches such as the vehicle speed sensor, transmission temperature sensor, throttle position sensor and manifold absolute pressure sensor to determine shifting points, required line pressure and torque converter lock-up operations of the transmis-

sion (four-speed electronic automatic transmission only).

33   **Voltage regulator** - The PCM controls the charging system voltage by grounding the alternator field driver circuit depending on electrical load requirements. The voltage regulator circuitry is fully contained with in the PCM. Any failure in the voltage regulator circuitry requires replacement of the PCM.

Obtaining diagnostic trouble codes

Refer to illustration 12.36

34   The PCM will illuminate the CHECK ENGINE light (also known as the Malfunction Indicator Lamp) on the dash if it recognizes a fault in the system. The light will remain illuminated until the problem is repaired and the code is cleared or the PCM does not detect any malfunction for several consecutive drive cycles.

35   The diagnostic codes for the On Board Diagnostic (OBD) system can only be extracted from the PCM using a SCAN tool. If the SCAN tool isn't available, have the vehicle diagnosed by a dealer service department or other qualified repair facility.

36   The SCAN tool is programmed to interface with the OBD system by plugging into the diagnostic connector **(see illustration)**. When used, the SCAN tool has the ability to diagnose specific driveability problems and it allows freeze frame data to be retrieved from the PCM stored memory. Freeze frame data is an OBD II PCM feature that records all related sensor and actuator activity on the PCM data stream whenever an engine fault is detected and a trouble code is set. This ability to look at the circuit conditions and values when the malfunction occurs provides a valuable tool when trying to diagnose intermittent driveability problems. If the tool is not available and intermittent driveability problems exist, have the vehicle checked at a dealer service department or other qualified repair shop. **Note:** *The SCAN tool trouble code designation is referred to as a P0 or P1 code. Refer to the "Generic Scan Tool Code" column of the trouble code chart for the code identification.*

Clearing diagnostic trouble codes

37   After the system has been repaired, the codes must be cleared from the PCM memory with the SCAN tool. **Caution:** *Do not disconnect the battery from the vehicle in an attempt to clear the codes. If necessary, have the codes cleared by a dealer service department or other qualified repair facility.*

38   Always clear the codes from the PCM before starting the engine after a new electronic emission control component is installed onto the engine. The PCM stores the operating parameters of each sensor. The PCM may set a trouble code if a new sensor is allowed to operate before the parameters from the old sensor have been erased.

Check Engine Light Flash Code	Generic Scan Tool Code	DRB Scan Tool Display	Description of Diagnostic Trouble Code
12			Battery power to PCM was disconnected
54	P0340	No Cam Signal at PCM	No camshaft signal detected during engine cranking.
53	P0601	Internal Controller Failure	PCM Internal fault condition detected.
47	P0162	Charging System Voltage Too Low	Battery voltage sense input below target charging during engine operation. Also, no significant change detected in battery voltage during active test of generator output circuit.
46	P1594	Charging System Voltage Too High	Battery voltage sense input above target charging voltage during engine operation.
42	P1388	Auto Shutdown Relay Control Circuit	An open or shorted condition detected in the auto shutdown relay circuit.
41	P0622	Generator Field Not Switching Properly	An open or shorted condition detected in the generator field control circuit.
37	P0743	Torque Converter Clutch Soleniod/Trans Relay Circuits	An open or shorted condition detected in the torque converter part throttle unlock solenoid control circuit (3 speed auto RH trans. only).
35	P1491	Rad Fan Control Relay Circuit	An open or shorted condition detected in the low speed radiator fan relay control circuit (2.5L only).
34	P1595	Speed Control Solenoid Circuits	An open or shorted condition detected in the Speed Control vacuum or vent solenoid circuits.
33	P0645	A/C Clutch Relay Circuit	An open or shorted condition detected in the A/C clutch relay circuit.
31	P0443	EVAP Purge Solenoid Circuit	An open or shorted condition detected in the duty cycle purge solenoid circuit.
27	P0203 or P0202 or P0201	Injector #3 Control Circuit	Injector #3 output driver does not respond properly to the control signal.
		Injector #2 Control Circuit	Injector #2 output driver does not respond properly to the control signal.
		Injector #1 Control Circuit	Injector #1 output driver does not respond properly to the control signal.
25	P0505	Idle Air Control Motor Circuits	A shorted or open condition detected in one or more of the idle air control motor circuits.
24	P0122 or P0123	Throttle Position Sensor Voltage Low	Throttle position sensor input below the minimum acceptable voltage
		Throttle Position Sensor Voltage High	Throttle position sensor input above the maximum acceptable voltage.
22	P0117	ECT Sensor Voltage Too Low	Engine coolant temperature sensor input below minimum acceptable voltage.

12.39 Diagnostic trouble codes (1998 and later models) (continued on the following pages)

Diagnostic trouble code identification

Refer to illustration 12.39

39 The accompanying list of diagnostic trouble codes is a compilation of all the codes that may be encountered (see Illustration). Not all codes pertain to all models and not all codes will illuminate the Check Engine light when set. The codes listed under the "Check Engine Light Flash Code" column are codes that may be displayed by the Check Engine light on 1997 models only (see Step 35). All other models require a SCAN tool to access the diagnostic trouble codes.

6

Check Engine Light Flash Code	Generic Scan Tool Code	DRB Scan Tool Display	Description of Diagnostic Trouble Code
22	or P0118	ECT Sensor Voltage Too High	Engine coolant temperature sensor input above maximum acceptable voltage.
17	1281	Engine Is Cold Too Long	Engine did not reach operating temperature within acceptable limits.
14	P0107 or P0108	MAP Sensor Voltage Too Low MAP Sensor Voltage Too High	MAP sensor input below minimum acceptable voltage. MAP sensor input above maximum acceptable voltage.
13	P1297	No Change in MAP From Start to Run	No difference recognized between the engine MAP reading and the barometric (atmospheric) pressure reading from start-up.
11	P0320	No Crank Reference Signal at PCM	No crank reference signal detected during engine cranking.
	P0351	Ignition Coil #1 Primary Circuit	Peak primary circuit current not achieved with maximum dwell time.
42	P1389	No ASD Relay Output Voltage at PCM	An Open condition Detected In The ASD Relay Output Circuit.
63	P1696	PCM Failure EEPROM Write Denied	Unsuccessful attempt to write to an EEPROM location by the PCM.
37	P0753	Trans 3-4 Shift Sol/Trans Relay Circuits	Current state of output port for the solenoid is different from expected state.
23	P0112 or P0113	Intake Air Temp Sensor Voltage Low Intake Air Temp Sensor Voltage High	Intake air temperature sensor input below the maximum acceptable voltage. Intake air temperature sensor input above the minimum acceptable voltage.
27	P0204	Injector #4 Control Circuit	Injector #4 output driver does not respond properly to the control signal.
21	P0132	Left Upstream O2S Shorted to Voltage	Oxygen sensor input voltage maintained above the normal operating range.
	P0154	O2 2/1 Signal Inactive	No signal at O2 2/1 sensor
	P0152	O2 2/1 Shorted High	Oxygen sensor input voltage sustained above the normal operating range.
53	PO600	PCM Failure SPI Communications	PCM internal fault condition detected
27	P0205 or P0206	Injector #5 Control Circuit Injector #6 Control Circuit	Injector #5 output driver does not respond properly to the control signal. Injector #6 output driver does not respond properly to the control signal.
45	P0712 or	Trans Temp Sensor Voltage Too Low	Voltage less than 1.55 volts.

Diagnostic trouble codes (1998 and later models) (continued)

Check Engine Light Flash Code	Generic Scan Tool Code	DRB Scan Tool Display	Description of Diagnostic Trouble Code
45	P0713	Trans Temp Sensor Voltage Too High	Voltage greater than 3.76 volts.
27	P0207	Injector #7 Control Circuit	Injector #7 output driver does not respond properly to the control signal.
	or		
	P0208	Injector #8 Control Circuit	Injector #8 output driver does not respond properly to the control signal.
77	P1683	SPD CTRL PWR RLY; or S/C 12V Driver CKT	Malfuntion detected with power feed to speed control servo solenoids
34	P1596	MUX S/C Switch High	Speed control switch input above the maximum acceptable voltage.
	or		
	P1597	MUX S/C Switch Low	Speed control switch input below the minimum acceptable voltage.
42	P1282	Fuel Pump Relay Control Circuit	An open or shorted condition detected in the fuel pump relay control circuit.
21	P0133 or P0152	O2 1/1 Slow Response	Oxygen sensor response slower than minimum required switching frequency.
	or		
	P0135	O2 1/1 Heater Circuit	Upstream oxygen sensor heating element circuit malfunction
	P0139	O2 1/1 Slow Response	Oxygen sensor response slower than minimum required switching frequency.
	P0141	O2 1/2 Heater Circuit	Oxygen sensor heating element circuit malfunction.
43	P0300	Multiple Cylinder Mis-fire	Misfire detected in multiple cylinders.
	or		
	P0301	Cylinder #1 Mis-fire	Misfire detected in cylinder #1.
	or		
	P0302	Cylinder #2 Mis-fire	Misfire detected in cylinder #2.
	or		
	P0303	Cylinder #3 Mis-fire	Misfire detected in cylinder #3.
	or		
	P0304	Cylinder #4 Mis-fire	Misfire detected in cylinder #4.
72	P0420	Catalyst 1/1 Effic	Catalyst efficiency below required level.
31	P0441	Incorrect Purge Flow	Insufficient or excessive vapor flow detected during evaporative emission system operation.
37	P1899	P/N Switch Stuck in Park or in Gear	Incorrect input state detected for the Park/Neutral switch, auto. trans. only.
	P0551	Pwr Steering Sw Perf	Power steering high pressure seen at high speed (2.5L only).
52	P0172	Left Bank or Fuel System Rich	A rich air/fuel mixture has been indicated by an abnormally lean correction factor.
51	P0171	Right Rear (or just) Fuel System Lean	A lean air/fuel mixture has been indicated by an abnormally rich correction factor.

Diagnostic trouble codes (1998 and later models) (continued)

6

Check Engine Light Flash Code	Generic Scan Tool Code	DRB Scan Tool Display	Description of Diagnostic Trouble Code
	P0175	Fuel System 2/1 Rich	A rich air/fuel mixture has been indicated by an abnormally lean correction factor.
	P0174	Fuel System 2/1 Lean	A lean air/fuel mixture has been indicated by an abnormally lean correction factor.
	P0153	O2 2/1 Slow Response	Oxygen sensor response slower than minimum required switching frequency.
	P0159	O2 2/1 Slow Response	Oxygen sensor response slower than minimum required switching frequency.
	P0155	O2 2/1 Heater circuit	Oxygen sensor heater element malfunction.
	P0161	O2 2/1 Heater circuit	Oxygen sensor heater element malfunction.
21	P0138	Left Bank Downstream or Downstream and Pre-Catalyst O2S Shorted to Voltage	Oxygen sensor input voltage maintained above the normal operating range.
	P0158	O2 2/2 Shorted High	Oxygen sensor input voltage maintained above the normal operating range.
17	P0125	Closed Loop Temp Not Reached	Engine does not reach 20°F within 5 minutes with a vehicle speed signal.
24	P0121	TPS Voltage Does Not Agree With MAP	TPS signal does not correlate to MAP sensor
14	P1296	No 5 Volts To MAP Sensor	5 Volt output to MAP sensor open.
25	P1294	Target Idle Not Reached	Actual idle speed does not equal target idle speed.
37	P1756 or P1757	Governor Pressure Not Equal to Target @ 15-20 PSI Governor Pressure Above 3 PSI In Gear With 0 MPH	Governor sensor input not between 10 and 25 psi when requested. Governor pressure greater than 3 psi when requested to be 0 psi.
37	P0740	Torq Conv Clu, No RPM Drop At Lockup	Relationship between engine speed and vehicle speed indicates no torque converter clutch engagement (auto. trans. only).
42	P0462 or P0463 or P0460	Fuel Level Sending Unit Volts Too Low Fuel Level Sending Unit Volts Too High Fuel Level Unit No Change Over Miles	Open circuit between PCM and fuel gauge sending unit. Circuit shorted to voltage between PCM and fuel gauge sending unit. No movement of fuel level sender detected.
44	P1493 or P1492	Ambient/Batt Temp Sen VoltsToo Low Ambient/Batt Temp Sensor VoltsToo High	Battery temperature sensor input voltage below an acceptable range. Battery temperature sensor input voltage above an acceptable range.

Diagnostic trouble codes (1998 and later models) (continued)

Check Engine Light Flash Code	Generic Scan Tool Code	DRB Scan Tool Display	Description of Diagnostic Trouble Code
21	P0131	Left Bank and Upstream O2S Shorted to Ground	O2 sensor voltage too low, tested after cold start.
	or		
	P0137	Downstream, Left Bank Downstream and Pre-Catalyst O2S Shorted to Ground	O2 sensor voltage too low, tested after cold start.
11	P1391	Intermittent Loss of CMP or CKP	Intermittent loss of either camshaft or crankshaft position sensor
31	P0442	Evap Leak Monitor Small Leak Detected	A small leak has been detected by the leak detection monitor
	or		
	P0455	Evap Leak Monitor Large Leak Detected	The leak detection monitor is unable to pressurize Evap system, indicating a large leak.
45	P0711	Trans Temp Sensor, No Rise After Start	Sump temp did not rise more than 16°F within 10 minutes when starting temp is below 40°F or sump temp is above 260°F with coolant below 100°F.
37	P0783	3-4 Shift Sol, No RPM Drop @ 3-4 Shift	The ratio of engine rpm/output shaft speed did not change beyond on the minimum required.
15	P0720	Low Ouput Spd Sensor RPM Above 15 mph	Output shaft speed is less than 60 rpm with vehicle speed above 15 mph.
45	P1764	Governor Pessure Sensor Volts Too Low	Voltage less than .10 volts.
	or		
	P1763	Governor Pressure Sensor Volts Too HI	Voltage greater than 4.89 volts.
	or		
	P1762	Governor Press Sen Offset Volts Too Lo or High	Sensor input greater or less than calibration for 3 consecutive Neutral/Park occurances.
37	P0748	Governor Pressure Sol Control/Trans Relay Circuits	Current state of solenoid output port is different than expected.
37	P1765	Trans 12 Volt Supply Relay Ctrl Circuit	Current state of solenoid output port is different than expeted.
43	P0305	Cylinder #5 Mis-fire	Misfire detected in cylinder #5.
	or		
	P0306	Cylinder #6 Mis-fire	Misfire detected in cylinder #6.
	or		
	P0307	Cylinder #7 Mis-fire	Misfire detected in cylinder #7.
	or		
	P0308	Cylinder #8 Mis-fire	Misfire detected in cylinder #8.
	P0432	Catalyst 2/1 EFFIC	Catalyst 2/1 efficiency below required level
	P0151	O2 2/1 Voltage Low	Oxygen sensor input voltage maintained below normal operating range.

Diagnostic trouble codes (1998 and later models) (continued)

6

Check Engine Light Flash Code	Generic Scan Tool Code	DRB Scan Tool Display	Description of Diagnostic Trouble Code
	P0157	O2 2/1 Voltage Low	Oxygen sensor input voltage maintained below normal operating range.
31	P1495 or P1494	Leak Detection Pump Solenoid Circuit Leak detection pump SW or mechanical fault	Leak detection pump solenoid circuit fault (open or short) Leak detection pump switch does not respond to input.
11	P1398	Mis-fire Adaptive Numerator at Limit	CKP sensor target windows have too much variation
31	P1486	Evap leak monitor pinched hose found	Plug or pinch detected between purge solenoid and fuel tank
45	P0751	O/D Switch Pressed (LO) More Than 5 Min	Overdrive Off switch input too low for more than 5 minutes.
	P0147	O2 1/3 Heater Circuit	Oxygen sensor heater element malfunction.
21	P0133 or P1195	Cat Mon slow O2 1/1	A slow switching oxygen sensor has been detected in bank 1/1 during catalyst monitor test.
	P0153 or P1196	Cat Mon slow O2 2/1	A slow switching oxygen sensor has been detected in bank 2/1 during catalyst monitor test.
	P0129 or P1197	Cat Mon slow O2 1/2	A slow switching oxygen sensor has been detected in bank 1/2 during catalyst monitor test.
15		No vehicle speed sensor signal	No Vehicle speed sensor signal detected during driving conditions.

Diagnostic trouble codes (1998 and later models) (continued)

13 Camshaft position sensor (2000 six-cylinder models) - general information and replacement

General information

1   The camshaft position sensor, in conjunction with the crankshaft position sensor, determines the timing for the fuel injection on each cylinder. The camshaft position sensor is a Hall-Effect device. The sensor is located on top of the oil pump drive and is triggered by a pulse ring.

Replacement

Refer to illustrations 13.3 and 13.5
Note: *The camshaft position sensor is located on top of the oil pump drive assembly. It is possible to replace the camshaft position sensor without removing the oil pump drive assembly. If the oil pump drive assembly must be removed, the mounting flange must be precisely marked in relation to the engine*

block or it will be necessary to reset the camshaft position sensor with a scan tool.
2   Disconnect the electrical connector from the camshaft position sensor.
3   If removing the sensor only, remove the two mounting bolts and remove the sensor from the oil pump drive **(see illustration)**. Install the new sensor, tighten the bolts securely and connect the electrical connector.
4   If removing the oil pump drive assembly, position the number one cylinder at TDC on the compression stroke (see Chapter 2B).
5   Remove the sensor. Insert an appropriate size alignment pin (such as a drill bit) through the holes in the sensor base and pulse ring **(see illustration)**. If the holes do not align, rotate the crankshaft until they do.
6   Mark the position of the oil pump drive mounting flange base to the engine block to ensure the oil pump drive can be re-installed in exactly the same position as originally installed.
7   Remove the oil pump drive hold-down bolt and withdraw the oil pump drive assembly. Remove and discard the O-ring.

8   If the crankshaft has been moved while the oil pump drive is out, the number one piston must be repositioned at TDC. This can be done by feeling for compression pressure at the number one spark plug hole as the crankshaft is rotated. Once compression is felt, continue rotating the crankshaft until the mark on the crankshaft damper is aligned with the zero or TDC mark on the timing indicator (see Chapter 2B).
9   Turn the pulse ring until the hole in the pulse ring aligns with the hole in the housing and install the alignment pin the holes **(see illustration 13.5)**. Install a new O-ring on the oil pump drive assembly.
10   Insert the oil pump drive into the engine block and align the marks made in Step 6.
11   Because of the helical-cut drive gear, the assembly will rotate clockwise as the gears engage. Be sure to compensate by starting the installation with the drive gear assembly positioned counterclockwise from the desired finished position. Installing the oil pump drive assembly is very similar to installing a distributor, see Chapter 5 for

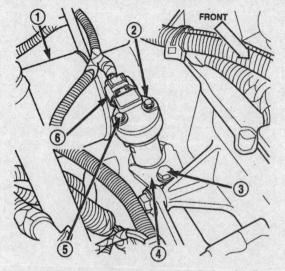

13.3 Camshaft position sensor installation details – 2000 six-cylinder models

1   Oil filter
2   Camshaft position sensor
3   Clamp bolt
4   Oil pump drive clamp
5   Camshaft position sensor mounting bolts
6   Electrical connector

13.5 Before removing the camshaft position sensor on a 2000 six-cylinder model, position the engine with the number one cylinder at TDC and install an appropriate size alignment pin through the hole in the base of the pulse ring

1   Pulse ring
2   Alignment pin
3   Oil pump driveshaft assembly

additional information, if necessary. The oil pump drive must be installed in the same position as originally installed before removal. When properly installed, a line drawn through the center of the camshaft position sensor and electrical connector should be parallel with the centerline of the engine. **Note:** *If the alignment marks are lost or poor driveability symptoms appear after installing the oil pump drive, take the vehicle to a dealer service department or other properly equipped repair facility and have the fuel system synchroniza-tion reset with a scan tool.*

12   Install the oil pump drive hold-down clamp and tighten the bolt to 17 ft-lbs.

13   The remainder of installation is the reverse of removal.

6

Notes

Chapter 7 Part A
Manual transmission

Contents

Specifications

Torque specifications

	Ft-lbs
Clutch housing-to-engine bolts (all models)	28
Shift lever cover retaining bolts	
T4/5 transmission	13
AX 4/5 transmission	13
BA 10/15 transmission	17

1 General information

All vehicles covered in this manual come equipped with either a four or five-speed manual transmission or an automatic transmission. All information on the manual transmission is included in this Part of Chapter 7. Information on the automatic transmission can be found in Part B of this Chapter.

Both four and five-speed manual transmissions are used in these models. On five-speed transmissions, the 5th gear is an over-drive.

Due to the complexity, unavailability of replacement parts and the special tools necessary, internal repair by the home mechanic is not recommended. The information in this Chapter is limited to general information and removal and installation of the transmission.

Depending on the expense involved in having a faulty transmission overhauled, it may be a good idea to replace the unit with either a new or rebuilt one. Your local dealer or transmission shop should be able to supply you with information concerning cost, availability and exchange policy. Regardless of how you decide to remedy a transmission problem, you can still save a lot of money by removing and installing the unit yourself.

2 Oil seal replacement

Refer to illustrations 2.6, 2.11, 21.13a and 2.13b

1   Oil leaks frequently occur due to wear of the extension housing oil seal and/or the speedometer drive gear oil seal and O-ring. Replacement of these seals is relatively easy, since the repairs can usually be performed without removing the transmission or transfer case (4WD models) from the vehicle.

2   The extension housing oil seal is located at the extreme rear of the transmission or transfer case, where the driveshaft is attached. If leakage at the seal is suspected, raise the vehicle and support it securely on jackstands. If the seal is leaking, transmission lubricant will be built up on the front of the driveshaft and may be dripping from the rear of the transmission or transfer case.

3   Refer to Chapter 8 and remove the driveshaft.

4   Using a soft-faced hammer, carefully tap the dust shield (if equipped) to the rear and remove it from the transmission. Be careful not to distort it.

5   Using a screwdriver or pry bar, carefully pry the oil seal out of the rear of the transmission or transfer case. Do not damage the splines on the transmission output shaft.

6   If the oil seal cannot be removed with a screwdriver or pry bar, a special oil seal removal tool (available at auto parts stores)

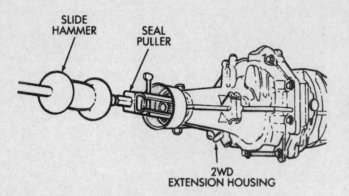

2.6 On some models, a slide hammer with a special seal puller may be required to remove the oil seal

2.11 Before loosening the retaining bolt, mark the relationship between the speedometer adapter housing and the transmission or transfer case with white paint - the adapter is offset and must be installed in the same relationship (retaining bolt shown removed for clarity)

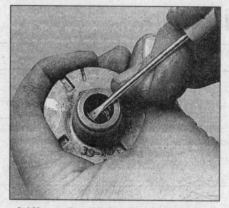

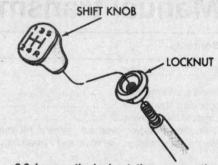

2.13a Use a small screwdriver to remove the O-ring from the groove . . .

2.13b . . . then lift the adapter seal from the inside of the housing

3.2 Loosen the locknut, then remove the shift knob

will be required **(see illustration)**.

7   Using a large section of pipe or a very large deep socket as a drift, install the new oil seal. Drive it into the bore squarely and make sure it's completely seated.

8   Reinstall the dust shield by carefully tapping it into place. Lubricate the splines of the transmission output shaft and the outside of the driveshaft sleeve yoke with lightweight grease, then install the driveshaft. Be careful not to damage the lip of the new seal.

9   The speedometer cable and driven gear housing is located on the side of the extension housing. Look for transmission oil around the cable housing to determine if the seal and O-ring are leaking.

10   Disconnect the speedometer cable.

11   Mark the relationship of the speedometer adapter to the extension housing **(see illustration)**.

12   Remove the bolt and withdraw the adapter.

13   Install a new O-ring and adapter seal on the adapter and reinstall the adapter and cable assembly on the extension housing **(see illustrations)**.

3   Manual transmission shift lever - removal and installation

All models except NV3550 transmission

Refer to illustrations 3.2, 3.3, 3.5a and 3.5b

1   Place the shift lever in Neutral.

2   Loosen the locknut, then remove the

3.3 Use a screwdriver to detach the bezel from the console, then lift the boot off

shift knob **(see illustration)**.

3   Use a screwdriver to disengage the boot retainer bezel from the console, then lift off the boot **(see illustration)**.

4   Remove the screws and lift out the center portion of the console.

5   Remove the five cover retaining bolts and lift the shift lever straight up and out of the transmission **(see illustrations)**. On some later models it may also be necessary to remove a snap-ring and spring washer before detaching the lever.

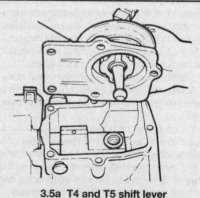

3.5a T4 and T5 shift lever installation details

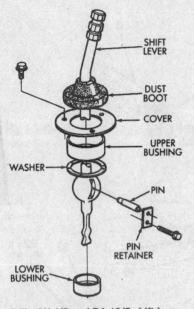

3.5b   AX 4/5 and BA 10/5 shift lever installation details

4.2   Pry on the transmission mount with a prybar or large screwdriver to check for excessive looseness

6   Installation is the reverse of removal. Tighten the shift lever cover retaining bolts to the specified torque.

NV3550 transmission

7   Shift the transmission into first or third gear.
8   Remove the center console and shift boot.
9   Install nuts on two M6 x 1.0 bolts and thread the bolts into the threaded holes at the base of the shift lever.
10   Tighten the nuts equally until the shift lever loosens on the shift tower stub shaft.
11   Remove the shift lever from the shift tower.
12   Installation is the revere of removal.

4   transmission mount - check and replacement

Refer to illustration 4.2

1   Insert a large screwdriver or prybar into the space between the transmission extension housing and the crossmember (2WD models) or the transfer case and the crossmember (4WD models). Try to pry the transmission or transfer case up slightly.
2   The transmission should not move away from the insulator much at all **(see illustration)**.
3   To replace the mount, remove the nuts attaching the insulator to the crossmember and the bolts attaching the insulator to the transmission.
4   Raise the transmission or transfer case slightly with a jack and remove the insulator, noting which holes are used in the crossmember for proper alignment during installation.
5   Installation is the reverse of the removal procedure. Be sure to tighten the nuts/bolts securely.

5   Manual transmission - removal and installation

Note: *On 4WD models, the transmission and transfer case are removed as a unit.*

Removal

1   Disconnect the negative cable from the battery.
2   Working inside the vehicle, remove the shift lever (see Section 3). On 4WD models, disconnect the transfer case shift linkage.
3   Raise the vehicle and support it securely on jackstands.
4   Disconnect the speedometer cable and wire harness connectors from the transmission and transfer case. Unbolt the clutch release cylinder from the clutch housing (external cylinder) or disconnect the hydraulic line at the transmission (see Chapter 8).
5   Remove the driveshaft(s) (see Chapter 8). Use a plastic bag to cover the end of the transmission to prevent fluid loss and contamination.
6   Remove the exhaust system components as necessary for clearance (see Chapter 4).
7   Support the engine. This can be done from above with an engine hoist, or by placing a jack (with a block of wood as an insulator) under the engine oil pan. The engine should remain supported at all times while the transmission is out of the vehicle.
8   Support the transmission with a jack - preferably a special jack made for this purpose. Safety chains will help steady the transmission on the jack. **Warning:** *On 4WD models, do not attempt to remove the transmission/transfer case assembly without a transmission jack and safety chains. The assembly is very heavy and awkward to remove.*
9   Remove the rear transmission (or transfer case) support-to-crossmember nuts and bolts.
10   Remove the nuts from the crossmember bolts. Raise the transmission slightly and remove the crossmember.
11   Remove the bolts securing the clutch housing to the engine.
12   Make a final check that all wires and hoses have been disconnected from the transmission and then move the transmission and jack toward the rear of the vehicle until the transmission input shaft is clear of the clutch housing. On 2WD models, keep the transmission level as this is done. On some 4WD models, the transmission/transfer case assembly is too long to be removed with the engine level. You may need to lower the rear of the engine to provide sufficient clearance for removal. The fan could contact the shroud during this operation, so it is a good idea to place a block of wood between the rear of the cylinder head(s) and the firewall to limit the distance the rear of the engine drops.
13   Once the input shaft is clear, lower the transmission and remove it from under the vehicle. **Caution:** *Do not depress the clutch pedal while the transmission is out of the vehicle.*
14   The clutch components can be inspected by removing the clutch housing from the engine (see Chapter 8). In most cases, new clutch components should be routinely installed if the transmission is removed.

Installation

15   If removed, install the clutch components (see Chapter 8).
16   With the transmission secured to the jack as on removal, raise the transmission into position behind the engine and then carefully slide it forward, engaging the input shaft with the clutch plate hub. Do not use excessive force to install the transmission - if the input shaft does not slide into place, readjust the angle of the transmission so it is level and/or turn the input shaft so the splines engage properly with the clutch.
17   Install the clutch housing-to-engine bolts. Tighten the bolts to the specified torque.
18   Install the crossmember and transmission support. Tighten all nuts and bolts securely.
19   Remove the jacks supporting the transmission and the engine.
20   Install the various items removed previously, referring to Chapter 8 for the installation of the driveshaft and Chapter 4 for information regarding the exhaust system components. If the hydraulic line was disconnected, bleed the clutch hydraulic system (see Chapter 8).
21   Make a final check that all wires, hoses

7A

6.4a T4 manual transmission - exploded view

1 Transmission cover
2 Shift rail
3 1st-2nd gear shift fork
4 Selector plate
5 Selector arm interlock plate and pin
6 3rd-4th gear shift fork
7 Plug
8 Thrust washer, rear bearing and cup
9 1st gear
10 1st gear blocking ring
11 Output shaft
12 Reverse sliding gear and insert springs
13 1st gear pin
14 Synchronizer insert
15 2nd gear blocking ring
16 2nd gear
17 Thrust washer
18 Snap-ring
19 3rd gear
20 3rd gear blocking ring
21 3rd-4th gear synchronizer spring, hub, insert and sleeve
22 4th gear blocking ring
23 Needle thrust bearing and race
24 Clutch shaft needle roller bearing
25 Countershaft needle thrust bearing
26 Countershaft rear bearing
27 Countershaft rear spacer
28 Countershaft gear
29 Countershaft front thrust washer
30 Countershaft front bearing
31 Offset lever
32 Damper sleeve
33 Roll pin
34 Detent spring
35 Detent ball
36 Adapter housing
37 Identification tag
38 Adapter housing seal
39 Vent
40 Reverse idler gear shaft
41 Roll pin
42 Reverse idler gear
43 Bushing
44 Reverse gear lever fork
45 Reverse gear lever pivot pin bolt
46 Back-up light switch
47 Drain plug
48 Transmission case

49 Clutch shaft
50 Front bearing
51 Front bearing race

52 Shim
53 Oil seal
54 Front bearing cap

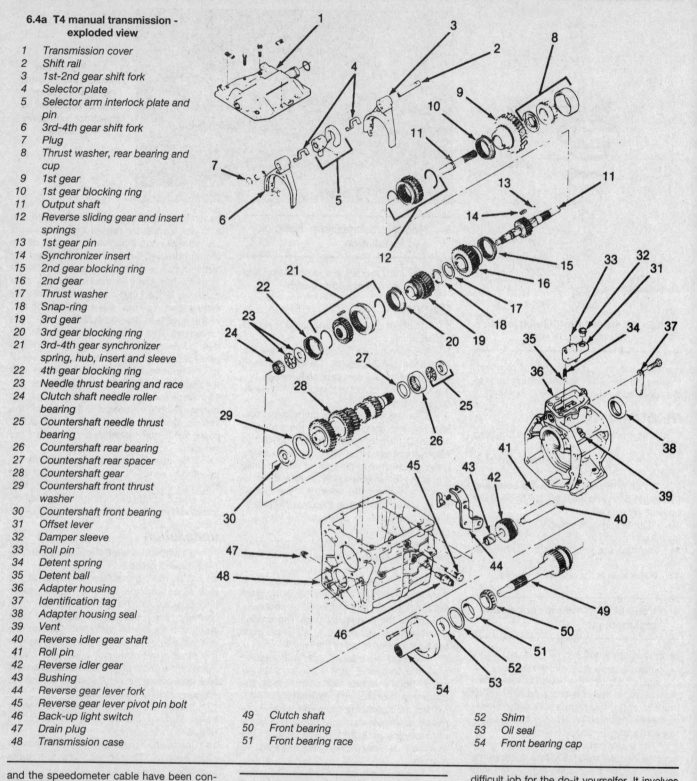

and the speedometer cable have been con-nected and that the transmission has been filled with lubricant to the proper level (see Chapter 1). Lower the vehicle.
22 Working inside the vehicle, install the shift lever (see Section 3). On 4WD models, connect the transfer case shift linkage.
23 Connect the negative battery cable. Road test the vehicle for proper operation and check for leakage.

6   Manual transmission overhaul - general information

Refer to illustrations 6.4a, 6.4b, 6.4c, 6.4d, 6.4e, 6.4f and 6.4g
Note: *The manufacturer doesn't provide exploded view drawings of the NV3550 trans-mission.*

Overhauling a manual transmission is a difficult job for the do-it yourselfer. It involves the disassembly and reassembly of many small parts. Numerous clearances must be precisely measured and, if necessary, changed with select fit spacers and snap-rings. As a result, if transmission problems arise, it can be removed and installed by a competent do-it-yourselfer, but overhaul should be left to a transmission repair shop. Rebuilt transmissions may be available - check with your dealer parts depart-

6.4b T5 Manual transmission - exploded view

1   O-ring
2   Transmission cover
3   Shift rail
4   1st-2nd gear shift fork
5   Selector plate
6   Selector arm, interlock plate and pin
7   3rd-4th gear shift fork
8   Plug
9   Snap-ring
10 5th driven gear
11 Thrust washer, rear bearing and cup
12 1st gear
13 1st gear blocking ring
14 Output shaft
15 Reverse sliding gear and insert springs
16 Roll pin
17 Synchronizer insert
18 2nd gear blocking ring
19 2nd gear and thrust washer
20 Snap-ring
21 3rd gear
22 3rd gear blocking ring
23 3rd-4th gear synchronizer springs, hub, insert and sleeve
24 4th gear blocking ring
25 Needle thrust bearing and race
26 Clutch shaft needle roller bearing
27 Funnel
28 Snap-ring
29 Thrust bearing race
30 Countershaft needle thrust bearing and race
31 Insert retainer
32 5th gear synchronizer sleeve and insert springs
33 5th gear synchronizer insert, hub and blocking ring
34 5th gear
35 Snap-ring and spacer
36 Countershaft rear bearing and spacer
37 Countershaft gear
38 Countershaft front bearing and thrust washer
39 Offset lever
40 Damper sleeve
41 Detent spring and ball
42 Identification tag
43 Adapter housing seal
44 Adapter housing
45 Vent
46 Roll pin
47 5th gear shift fork
48 5th gear and reverse gear rail
49 5th gear and reverse gear shift lever
50 Roll pin

51 Reverse idler gear, bushing and shaft
52 5th gear lever pivot pin bolt and back-up light switch
53 Transmission case
54 Drain plug

55 Clutch shaft
56 Front bearing
57 Front bearing cap oil seal, shim and race
58 Front bearing cap

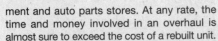

7A

ment and auto parts stores. At any rate, the time and money involved in an overhaul is almost sure to exceed the cost of a rebuilt unit.

Nevertheless, it's not impossible for an inexperienced mechanic to rebuild a transmission if the special tools are available and the job is done in a deliberate step-by-step manner so nothing is overlooked.

The tools necessary for an overhaul include internal and external snap-ring pliers, a bearing puller, a slide hammer, a set of pin punches, a dial indicator and possibly a hydraulic press. In addition, a large, sturdy workbench and a vise or transmission stand will be required.

During disassembly of the transmission, make careful notes of how each piece comes off, where it fits in relation to other pieces and what holds it in place. Exploded views are included (see illustrations) to show where the parts go - but actually noting how they are installed when you remove the parts will make it much easier to get the transmission back together.

Before taking the transmission apart for repair, it will help if you have some idea what area of the transmission is malfunctioning. Certain problems can be closely tied to specific areas in the transmission, which can make component examination and replacement easier. Refer to the Troubleshooting section at the front of this manual for information regarding possible sources of trouble.

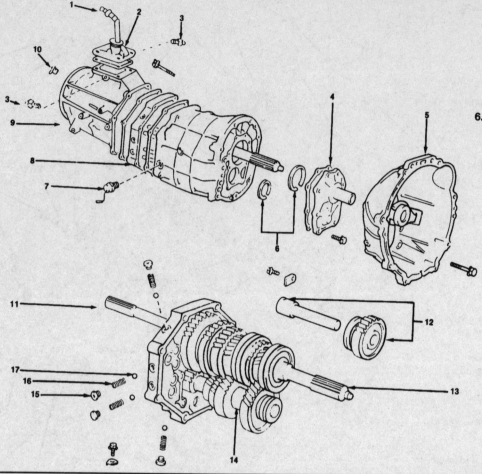

6.4c AX 4/5 Manual transmission - exploded view

1   *Shift lever*
2   *Shift lever retainer*
3   *Restrict pins*
4   *Front bearing retainer*
5   *Clutch housing*
6   *Snap-ring*
7   *Back-up light switch*
8   *Intermediate plate*
9   *Adapter housing*
10 *Adapter screw plug*
11 *Output shaft*
12 *Reverse idler gear*
13 *Input shaft*
14 *Counter gear*
15 *Straight screw plug*
16 *Spring*
17 *Locking ball*

6.4d AX 4/5 manual transmission output shaft, countershaft and synchronizer - exploded view

1   *Output shaft*
2   *Snap-ring*
3   *Snap-ring*
4   *5th gear*
5   *Rear bearing retainer*
6   *Synchronizer ring 4th gear*
7   *Input shaft*
8   *Counter gear*
9   *Counter rear bearing*
10 *Snap-ring*
11 *Spacer*
12 *Needle roller bearing 5th gear*
13 *Counter 5th gear*
14 *Hub sleeve no. 3*
15 *Synchronizer ring 5th gear*
16 *Gear spline piece no. 5*

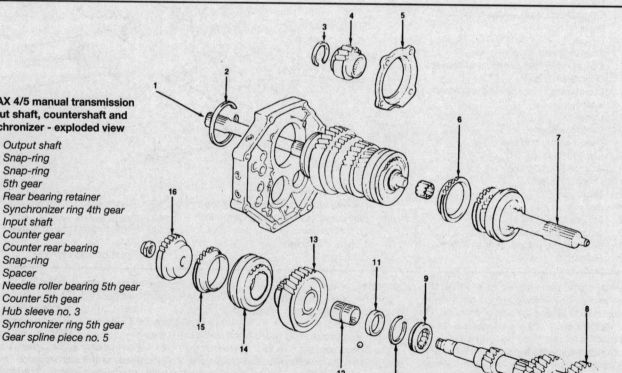

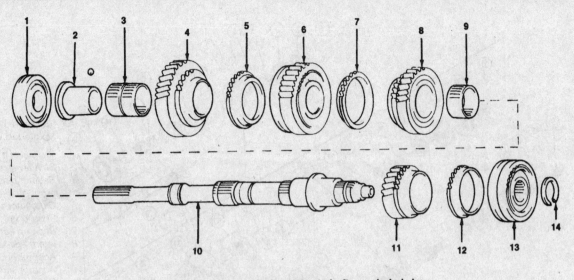

6.4e AX 4/5 manual transmission output shaft - exploded view

1	Rear bearing	6	Hub sleeve no. 1	11	3rd gear
2	Inner race	7	2nd gear synchronizer ring	12	3rd gear synchronizer ring
3	1st gear needle roller bearing	8	2nd gear	13	Hub sleeve no. 2
4	1st gear	9	2nd gear needle roller bearing	14	Snap-ring
5	1st gear synchronizer ring	10	Output shaft		

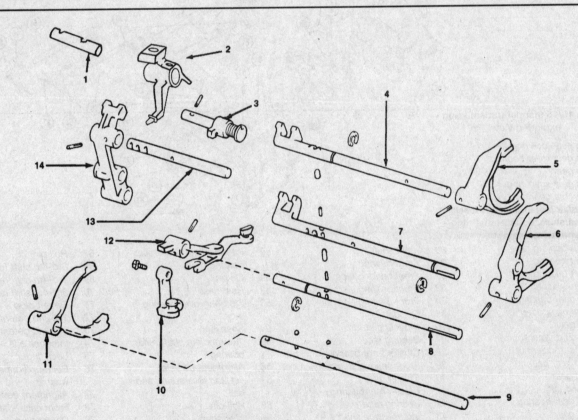

6.4f AX 5 manual transmission shift components - exploded view

1	Shift lever shaft	6	Shift fork no. 1	11	Shift fork no. 3
2	Shift lever housing	7	Shift fork shaft no. 1	12	Reverse shift arm and fork
3	Reverse restrict pin	8	Shift fork shaft no. 3	13	Shift fork shaft no. 5
4	Shift fork shaft no. 2	9	Shift fork shaft no. 4	14	Reverse shift head
5	Shift fork no. 2	10	Reverse shift arm bracket		

7A

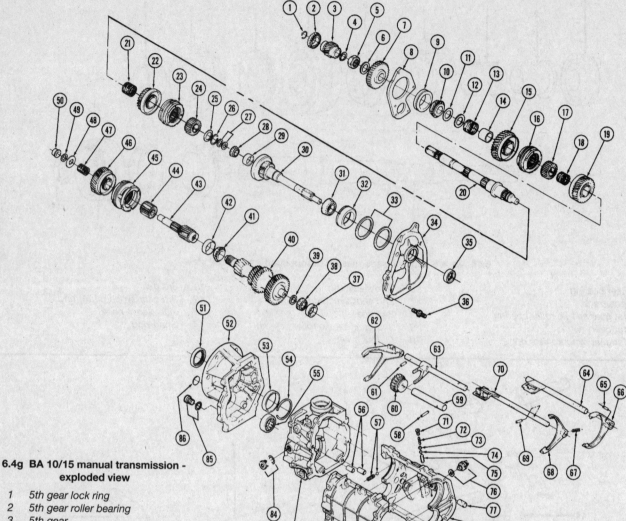

**6.4g BA 10/15 manual transmission -
    exploded view**

1	5th gear lock ring					
2	5th gear roller bearing					
3	5th gear					
4	5th gear snap-ring					
5	Reverse gear nut					
6	Washer (not used in all models)					
7	Reverse gear					
8	Rear bearing retainer					
9	Rear bearing race	30	Input shaft	50	Intermediate shaft roller	67	Roll pin
10	Rear bearing	31	Front bearing		bearing race	68	3rd-4th shift fork
11	Rear bearing shim	32	Front bearing race	51	Oil seal	69	3rd-4th shift rail lock pin
12	Shim washer	33	Front bearing shims	52	Rear case	70	3rd-4th shift rail
13	1st gear bearing	34	Front bearing retainer	53	5th gear roller bearing	71	Detent plug
14	Bearing spacer	35	Oil seal		race	72	1st-2nd detent spring
15	1st gear	36	Mounting studs	54	Snap-ring	73	1st-2nd detent ball
16	1st-2nd synchronizer	37	Bearing race	55	Intermediate shaft roller	74	Interlock ball
17	Synchronizer hub	38	Cluster front bearing		bearing	75	Lock finger
18	2nd gear bearing	39	Front bearing shim	56	Alignment dowels	76	Back-up light switch and
19	2nd gear	40	Cluster gear	57	1st-2nd lock spring and		washer
20	Mainshaft	41	Cluster rear bearing		ball	77	Alignment dowels
21	3rd gear bearing	42	Bearing race	58	Roll pin	78	Front case - left half
22	3rd gear	43	Intermediate shaft	59	Idler shaft	79	Detent plug
23	3rd-4th synchronizer	44	Synchronizer hub	60	Reverse idler gear	80	5th-Reverse detent spring
24	Synchronizer hub	45	5th gear synchronizer	61	Roll pin	81	5th-Reverse detent ball
25	Spring washer	46	5th intermediate gear	62	5th-Reverse shift fork	82	Front case - right half
26	Lock ring	47	5th intermediate gear	63	5th-reverse shift rail	83	Intermediate case
27	Pilot bearing shims		bearing	64	1st-2nd shift rail	84	Fill plug and washer
28	Pilot bearing	48	Thrust washer	65	Roll pin	85	Drain plug and washer
29	Pilot bearing race	49	End play shim	66	1st-2nd shift fork	86	Access plug

Chapter 7 Part B
Automatic transmission

Contents

Specifications

Torque specifications
Ft-lbs (unless otherwise indicated)

Transmission housing-to-engine bolts	30
Torque converter-to-driveplate bolts	40
Three-speed transmission front band	
Adjusting screw (with adapter)	36 in-lbs
Adjusting screw (without adapter)	72 in-lbs
Locknut	30
Three-speed transmission rear band	
Adjusting screw	41 in-lbs
Locknut	35
Neutral start switch	
Three-speed	24
Four-speed	
Adjusting bolt	9
Attaching nut	5
Transmission pan bolts	10

1 General information

All vehicles covered in this manual come equipped with a four or five-speed manual transmission or a three or four-speed automatic transmission. All information on the automatic transmission is included in this Part of Chapter 7. Information on the manual transmission can be found in Part A of this Chapter.

Due to the complexity of the automatic transmissions covered in this manual and the need for specialized equipment to perform most service operations, this Chapter contains only general diagnosis, routine maintenance, adjustment and removal and installation procedures.

If the transmission requires major repair work, it should be left to a dealer service department or an automotive or transmission repair shop. You can, however, remove and install the transmission yourself and save the expense, even if the repair work is done by a transmission shop.

2 Diagnosis - general

Note: *Automatic transmission malfunctions may be caused by five general conditions: poor engine performance, improper adjustments, hydraulic malfunctions, mechanical malfunctions or malfunctions in the computer or its signal network. Diagnosis of these problems should always begin with a check of the easily repaired items: fluid level and condition (Chapter 1), shift linkage adjustment and throttle linkage adjustment. Next, perform a road test to determine if the problem has been corrected or if more diagnosis is necessary. If the problem persists after the preliminary tests and corrections are completed, additional diagnosis should be done by a dealer service department or transmission repair shop. Refer to the Troubleshooting Section at the front of this manual for information on symptoms of transmission problems.*

Preliminary checks

1 Drive the vehicle to warm the transmission to normal operating temperature.

2 Check the fluid level as described in Chapter 1:

a) If the fluid level is unusually low, add enough fluid to bring the level within the designated area of the dipstick, then check for external leaks (see below).

b) If the fluid level is abnormally high, drain off the excess, then check the drained fluid for contamination by coolant. The presence of engine coolant in the automatic transmission fluid indicates that a failure has occurred in the internal radiator walls that separate the coolant from the transmission fluid (see Chapter 3).

c) If the fluid is foaming, drain it and refill the transmission, then check for coolant in the fluid or a high fluid level.

3 Check the engine idle speed. **Note:** *If the engine is malfunctioning, do not proceed with the preliminary checks until it has been repaired and runs normally.*

4 Check the throttle valve linkage for freedom of movement. Adjust it if necessary (Section 4). **Note:** *The throttle valve linkage*

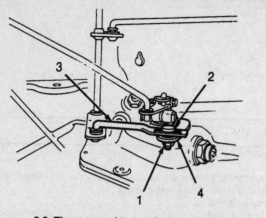

3.3 Three-speed transmission shift linkage

1	Jam nut	3	Shift rod
2	Trunnion	4	Bellcrank

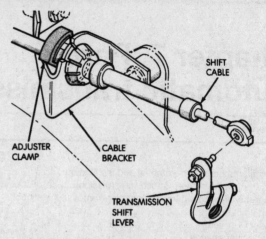

3.4 Four-speed transmission shift linkage

may function properly when the engine is shut off and cold, but it may malfunction once the engine is hot. Check it cold and at normal engine operating temperature.

5   Inspect the shift control linkage (see Section 3). Make sure it's properly adjusted and the linkage operates smoothly.

Fluid leak diagnosis

6   Most fluid leaks are easy to locate visually. Repair usually consists of replacing a seal or gasket. If a leak is difficult to find, the following procedure may help.

7   Identify the fluid. Make sure it's transmission fluid and not engine oil or brake fluid (automatic transmission fluid is a deep red color).

8   Try to pinpoint the source of the leak. Drive the vehicle several miles, then park it over a large sheet of cardboard. After a minute or two, you should be able to locate the leak by determining the source of the fluid dripping onto the cardboard.

9   Make a careful visual inspection of the suspected component and the area immediately around it. Pay particular attention to gasket mating surfaces. A mirror is often helpful for finding leaks in areas that are hard to see.

10   If the leak still cannot be found, clean the suspected area thoroughly with a degreaser or solvent, then dry it.

11   Drive the vehicle several miles at normal operating temperature and varying speeds. After driving the vehicle, visually inspect the suspected component again.

12   Once the leak has been located, the cause must be determined before it can be properly repaired. If a gasket is replaced but the sealing flange is bent, the new gasket will not stop the leak. The bent flange must be straightened.

13   Before attempting to repair a leak, check to make sure the following conditions are corrected or they may cause another leak. **Note:** *Some of the following conditions cannot be fixed without highly specialized tools and expertise. Such problems must be*

referred to a transmission repair shop or a dealer service department.

Gasket leaks

14   Check the pan periodically. Make sure the bolts are tight, no bolts are missing, the gasket is in good condition and the pan is flat (dents in the pan may indicate damage to the valve body inside).

15   If the pan gasket is leaking, the fluid level or the fluid pressure may be too high, the vent may be plugged, the pan bolts may be too tight, the pan sealing flange may be warped, the sealing surface of the transmission housing may be damaged, the gasket may be damaged or the transmission casting may be cracked or porous. If sealant instead of gasket material has been used to form a seal between the pan and the transmission housing, it may be the wrong sealant.

Seal leaks

16   If a transmission seal is leaking, the fluid level or pressure may be too high, the vent may be plugged, the seal bore may be damaged, the seal itself may be damaged or improperly installed, the surface of the shaft protruding through the seal may be damaged or a loose bearing may be causing excessive shaft movement.

17   Make sure the dipstick tube seal is in good condition and the tube is properly seated. Periodically check the area around the speedometer gear or sensor for leakage. If transmission fluid is evident, check the O-ring for damage.

Case leaks

18   If the case itself appears to be leaking, the casting is porous and will have to be repaired or replaced.

19   Make sure the oil cooler hose fittings are tight and in good condition.

Fluid comes out the vent pipe or fill tube

20   If this condition occurs, the transmission is overfilled, there is coolant in the fluid, the case is porous, the dipstick is incorrect, the

vent is plugged or the drain back holes are plugged.

---

3   Shift linkage or cable - check and adjustment

Refer to illustrations 3.3, 3.4 and 3.13

Check

1   Firmly apply the parking brake and try to momentarily operate the starter in each shift lever position. The starter should operate in Park and Neutral only. If the starter operates in any position other than Park or Neutral, adjust the shift linkage (see below). If, after adjustment, the starter still operates in positions other than Park or Neutral, the neutral start switch is defective or in need of adjustment (see Section 6).

Shift Linkage Adjustment (1984 through 1993 models)

2   Place the shift lever in Park. Raise the vehicle and support it securely on jackstands.

3   On three-speed transmissions, loosen the shift rod jam nut, remove the lockpin and disengage the trunnion and shift rod at the bellcrank **(see illustration)**.

4   On four-speed transmissions, release the adjuster clamp to unlock the shift cable, then unsnap the shift cable from the cable bracket **(see illustration)**.

5   Be sure the shift lever on the transmission is all the way to the rear, in the last detent. This is the Park position.

6   Make sure the park lock is engaged by trying to rotate the driveshaft. The driveshaft will not rotate if the park lock is functioning properly.

7   On three-speed transmissions, slide the shift rod trunnion forward or backward in the shift rod slot, as necessary, so the pin fits freely in the bellcrank arm. Tighten the jam nut securely while holding the shift rod so it doesn't turn as the jam nut is tightened. When properly adjusted, there should be no lash in the shift linkage. Pull down on the shift

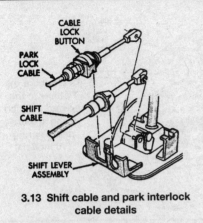

3.13 Shift cable and park interlock cable details

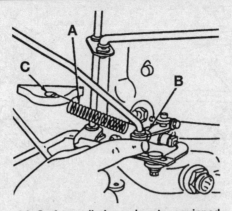

4.3 On four-cylinder carburetor equipped models, hook one end of the TV linkage return spring (A) over the end of the throttle control lever (B) and the other end through the hole in the bellhousing boss (C)

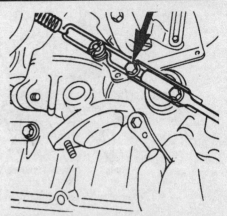

4.5 Use a box end wrench to loosen the adjusting link bolt (arrow)

rod while pushing up on the outer part of the bellcrank to eliminate lash.

8   On four-speed transmissions, snap the shift cable into the bracket, then press down on the adjuster clamp until it snaps in place to lock the shift cable.

9   Lower the vehicle.

10   With the parking brake firmly applied, make sure the engine starts only in Park and Neutral. If the linkage appears to be adjusted properly, but the starter operates in positions other than Park or Neutral, check the neutral start switch (see Section 6).

11   Check the steering wheel lock to make sure it operates smoothly.

Shift Cable adjustment (1994 and later models)

12   Place the shift lever in Park. Raise the vehicle and support it securely on jackstands.

13   At transmission end of the cable, release the cable adjuster clamp to unlock the cable from the shift lever (**see illustration**).

14   Unsnap the shift cable from the transmission cable bracket (**see illustrations 3.13**).

15   Be sure the shift lever on the transmission is all the way to the rear, in the detent. This is the Park position.

16   Make sure the park lock is engaged by trying to rotate the driveshaft. The driveshaft will not rotate if the park lock is functioning properly.

17   Snap the cable back into the transmission shift lever. Make sure it is locked in place.

18   Lock the shift cable into the transmission bracket by pressing the cable adjuster clamp down until it snaps into place.

19   Lower the vehicle.

20   With the parking brake firmly applied, make sure the engine starts only in Park or Neutral. If the cable is adjusted appears to be adjusted properly, but the starter operates in positions other than Park or Neutral, check the neutral start switch (see Section 6).

Park interlock cable adjustment (1994 and later models)

21   On floor shift models, remove the center

console (see Chapter 11).

22   Place the shift lever in Park. Raise the vehicle and support it securely on jackstands.

23   Turn the ignition switch to the Lock position.

24   Pull the cable lock button up to release the cable (**see illustration 3.13**).

25   Pull the cable forward, release the cable and press the cable lock button down until it snaps in place.

26   Check the adjustment as follows:

a) Check the movement of the release shift handle button (floor shift) or release lever (column shift). You should be able to press the button inward or move the column lever.

b) Turn the ignition switch to the On position.

c) Press the floor shift lever release button or move the column lever, then shift to Neutral.

d) Try to turn the ignition switch to the Lock position. If the cable adjustment is correct, the ignition switch cannot be turned to the Lock position.

e) Try to turn the ignition switch to the D position. If the cable adjustment is correct, the ignition switch cannot be turned to the D position.

27   Move the shift lever back to the Park position and check the ignition switch operation. You should be able to turn the switch to the Lock position and the shift lever release button or lever should not move.

28   Lower the vehicle.

29   On floor shift models, install the center console.

4   Throttle valve (TV) linkage - adjustment

1   The throttle linkage on these models can be adjusted to correct harsh, delayed or erratic shifting and lack of kickdown.

3-speed transmission (1984 through 1993 models)

Four-cylinder engine (carburetor equipped models only)

Refer to illustrations 4.3 and 4.5

Note: *On four-cylinder fuel injected models, TV linkage adjustment requires special equipment. The job should be left to a dealer service department repair shop.*

2   Remove the air cleaner assembly, then remove the spark plug wire separator from the throttle bracket and secure it out of the way.

3   From under the vehicle, hold the throttle control lever all the way forward, against the stop, and secure it with a spring (**see illustration**).

4   In the engine compartment, block the throttle open and set the carburetor linkage completely off the fast idle cam. On air conditioning equipped models, turn the ignition switch on to energize the throttle stop solenoid.

5   Loosen the adjusting link bolt (**see illustration**).

6   Remove the load on the throttle valve linkage by pulling on the link and tightening the link bolt.

7   Turn the ignition off, install the spark plug wires and separator, connect the throttle stop solenoid on air conditioned models, install the air cleaner and remove the spring from the throttle control lever. Test drive the vehicle and check for proper shifting operation, readjusting as necessary.

V6 engine

Refer to illustrations 4.8 and 4.12

8   Remove the throttle control rod spring. Use the spring to hold the adjusting link against the nylon stop, in the forward position (**see illustration on following page**).

9   Block the choke open and set the throttle off the fast idle cam. On models equipped with a throttle-operated solenoid valve, turn the ignition switch on, to energize the solenoid, and open the throttle halfway to lock it. Return the throttle to the idle position.

7B

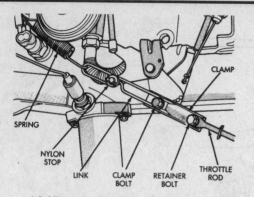

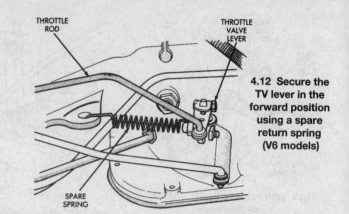

4.12 Secure the TV lever in the forward position using a spare return spring (V6 models)

4.8 V6 engine TV rod adjustment link details

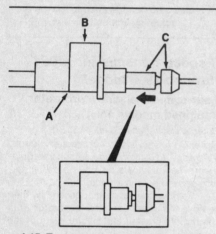

4.15 To retract the cable adjuster (A), press the button (B), then push the cable plunger (C) in

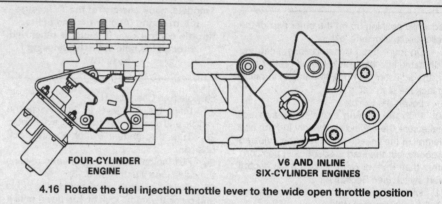

4.16 Rotate the fuel injection throttle lever to the wide open throttle position

10   Raise the vehicle and support it securely with jackstands.

11   Loosen the retainer and clamp bolts on the throttle control adjusting link **(see illustration 4.8)**.

12   Use a spare spring to hold the transmission throttle valve lever all the way forward, against the stop **(see illustration)**. Hook the spring to the torque converter boss.

13   Push on the end of the link to eliminate any lash, pull the clamp to the rear so the bolt bottoms in the rear of the rod slot. Tighten the clamp bolt. Pull the throttle control rod to the rear until the rod bolt bottoms in the front rod slot and tighten the retainer bolt.

14   Install the throttle control rod spring, remove the spring from the transmission lever and lower the vehicle.

Four-speed transmission

Refer to illustrations 4.15 and 4.16

15   With the ignition switch in the Off position, retract the cable adjuster by depressing the button, then pushing the plunger in **(see illustration)**.

16   Rotate the throttle lever on the fuel injection unit to the wide open throttle position **(see illustration)**.

17   Hold the throttle lever open and allow the cable plunger **(see illustration 4.15)** to extend all the way, which automatically adjusts the cable. Once the plunger is fully extended, release the throttle lever.

Three-speed transmission (1994 and later models)

Refer to illustrations 4.19 and 4.20

18   Place the transmission in the Park position.

19   Working within the engine compartment, press the cable release button **(see illustration)**.

20   Push the cable conduit back into the cable adjuster body as far as possible **(see illustration)**.

21   Rotate the throttle body lever to the wide open position. At this point the cable will automatically ratchet to the correct adjustment position.

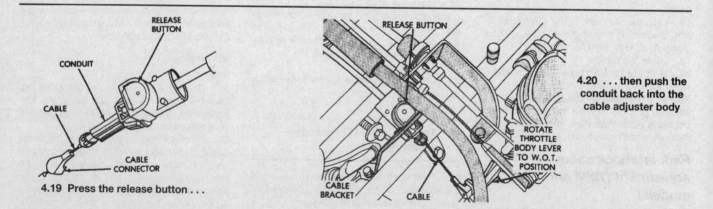

4.19 Press the release button . . .

4.20 . . . then push the conduit back into the cable adjuster body

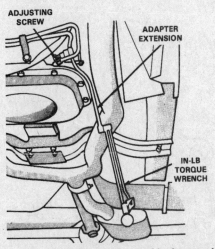

5.5 An adapter extension and inch-pound torque wrench are required to adjust the front transmission band

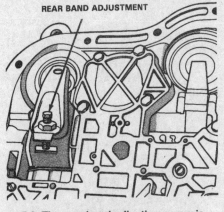

5.8 The rear band adjusting screw is accessible after removing the transmission pan

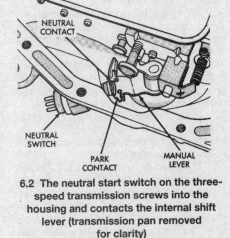

6.2 The neutral start switch on the three-speed transmission screws into the housing and contacts the internal shift lever (transmission pan removed for clarity)

5 Bands - adjustment (three-speed transmission)

Refer to illustrations 5.5 and 5.8

1   The transmission bands should be adjusted at the specified intervals or when the automatic upshifts or downshifts become consistently harsh and/or erratic.

2   Raise the vehicle and support it on jackstands.

Front band

3   The front band adjusting screw is located on the left side of the transmission, just above the throttle valve control levers.

4   Loosen the adjusting screw locknut and back it off five turns. Check the adjusting screw to make sure it turns freely in the case, lubricating it if necessary.

5   Using an in-lb torque wrench, adapter extension and a 5/16-inch socket, tighten the adjusting screw to the specified torque **(see illustration)**. If the adapter tool is not used, the alternate torque must be used.

6   On all 1984 through 1993 models, back the screw out two full turns. On 1994 and later four-cylinder models, back the screw out two and one-half turns. On 1994 and later inline six-cylinder models, back the screw out two and one-quarter turns.

7   Tighten the adjusting screw locknut to the specified torque while holding the screw with a socket or wrench so it does not rotate.

Rear band

8   The rear band adjusting screw is accessible after removing the oil pan **(see illustration)**. Consequently, it's very convenient to make this adjustment at the time of the transmission fluid and filter change (see Chapter 1).

9   Remove the oil pan.

10   Loosen the adjusting screw locknut.

11   Tighten the adjusting screw with an in-lb torque wrench and a 1/4-inch socket to the specified torque.

12   On all 1984 through 1993 models, back the screw out seven full turns. On 1994 and later four-cylinder models, back the screw out seven full turns. On 1994 and later inline six-cylinder models, back the screw out four full turns.

13   Install the locknut, hold the screw with a socket or wrench so it cannot turn and tighten the locknut to the specified torque.

14   Install the transmission pan and lower the vehicle.

6 Neutral start switch - check, adjustment and replacement

Three-speed transmission

Refer to illustration 6.2

1   The neutral start and back-up light switches are combined into one unit. The switch has three terminals with the neutral switch being the center one. A ground for the starter solenoid circuit is provided through the gearshift lever in only the Neutral and Park positions.

2   Raise the vehicle and place it securely on jackstands. Remove the wiring connector from the neutral switch **(see illustration)** and test the switch for continuity. Continuity should exist between the center terminal and transmission case only when the gearshift is in Neutral and Park. If the switch appears to be faulty, check the shift linkage (see Section 3) before replacing the switch.

3   Prior to replacing the switch, place a container under it to catch the transmission fluid.

4   Remove the switch and allow the fluid to drain into the container.

5   Place the gearshift lever in Park and Neutral and check the lever finger position and lever and shaft alignment with the switch opening.

6   Install the switch and seal, tightening the bolts to the specified torque.

7   Test the switch for continuity and plug in the connector.

8   Lower the vehicle and check the transmission fluid level (see Chapter 1), adding the specified fluid as necessary.

Four-speed transmission

Refer to illustrations 6.9, 6.10 and 6.13

9   Raise the vehicle and place it securely on jackstands. Unplug the connector and check for continuity, referring to **the accompanying illustration**. Replace the switch if

7B

SWITCH TERMINAL IDENTIFICATION		B	C	A	E	G	H
	P	O—	—O				
	R			O—	—O		
	N	O—	—O				
	D						
	3			O—	——	—O	
	1-2			O—	——	——	—O

6.9 The terminals for the four-speed transmission neutral start switch connector are shown above - the chart shows when continuity should exist between the terminals: when in Park or Neutral, continuity should exist between B and C; when in Reverse, between A and E; when in 3rd, between A and G; when in 1st or 2nd, between A and H

continuity is not as specified.

10   To replace the switch, unplug the connector, pry up the washer tabs and remove the attaching nut and adjusting bolt **(see illustration)**. Lift the switch off the shaft.

11   Disconnect the shift linkage on the left side of the transmission and rotate the shift lever all the way to the rear (Park), then two detent positions forward to Neutral.

12   Place the switch in position and install the nut and bolt finger tight. Tighten the attaching nut to the specified torque but do not bend the tabs down.

13   Make sure the transmission is in Neutral and rotate the switch until the neutral standard line on the housing lines up with the vertical groove in the manual valve shaft **(see illustration)**. Tighten the adjusting bolt to the specified torque and bend the washer tabs over the attaching nut.

14   Plug in the electrical connector, connect the shift linkage, lower the vehicle and check the switch operation to make sure the starter operates in Park or Neutral only. Be sure to firmly apply the parking brake when making this check.

7   Automatic transmission - removal and installation

Note: *On 4WD models, the transmission and transfer case are removed together, as an assembly.*

Removal

1   Disconnect the negative cable from the battery.

2   Raise the vehicle and support it securely on jackstands.

3   Drain the transmission fluid (see Chapter 1), then reinstall the pan.

4   Remove the torque converter cover.

5   Mark the relationship of the torque converter to the driveplate with white paint so they can be installed in the same position.

6   Remove the torque converter-to-driveplate bolts. Turn the crankshaft for access to each bolt. Turn the crankshaft in a clockwise direction only (as viewed from the front).

7   Remove the starter motor (see Chapter 5).

8   Remove the driveshaft(s) (see Chapter 8).

9   Disconnect the speedometer cable.

10   Detach the wire harness connectors from the transmission.

11   On models so equipped, disconnect the vacuum hoses.

12   Remove any exhaust components which will interfere with transmission removal (see Chapter 4).

13   Disconnect the TV linkage rod or cable.

14   Disconnect the shift linkage. On 4WD models, also disconnect the transfer case shift linkage.

15   Support the engine with a jack. Use a block of wood under the oil pan to spread the load.

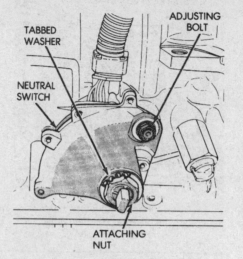

6.10 Four-speed neutral start switch mounting details

16   Support the transmission and transfer case (4WD models) with a jack - preferably a jack made for this purpose. Safety chains will help steady the transmission on the jack. **Warning**: *On 4WD models, you must use safety chains because the transmission and transfer case assembly is very heavy and awkward to remove.*

17   Remove the rear mount-to-crossmember bolts.

18   Raise the transmission and transfer case (4WD models) slightly, then remove the four crossmember-to-frame bolts and detach the crossmember.

19   Remove the bolts securing the transmission to the engine.

20   Lower the transmission slightly and disconnect and plug the transmission fluid cooler lines.

21   Remove the transmission dipstick tube.

22   Move the transmission to the rear to disengage it from the engine block dowel pins and make sure the torque converter is detached from the driveplate. Secure the torque converter to the transmission so it won't fall out during removal.

Installation

23   Prior to installation, make sure the torque converter hub is securely engaged in the pump.

24   With the transmission and transfer case (4WD models) secured to the jack, raise it into position. Be sure to keep it level so the torque converter does not slide forward. Connect the transmission fluid cooler lines.

25   Turn the torque converter to line up its bolt holes with the holes in the driveplate. The white paint mark on the torque converter and the driveplate made in Step 5 must line up.

26   Move the transmission and transfer case (4WD models) forward carefully until the dowel pins and the torque converter are engaged.

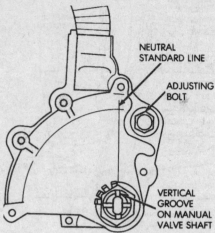

6.13 With the transmission in Neutral, the neutral standard line on the switch must be aligned with the vertical groove in the manual valve shaft

27   Install the transmission housing-to-engine bolts. Tighten them to the specified torque.

28   Install the torque converter-to-driveplate bolts. Tighten the bolts to the specified torque.

29   Install the rear mount-to-crossmember and crossmember-to-frame bolts. Tighten the bolts securely.

30   Remove the jacks supporting the transmission and transfer case (4WD models) and the engine.

31   Install the dipstick tube.

32   Install the starter motor (see Chapter 5).

33   Connect the vacuum hose(s) (if equipped).

34   Connect the shift and TV linkage. On 4WD models, connect the transfer case shift linkage.

35   Plug in the transmission wire harness connectors.

36   Install the torque converter cover.

37   Install the driveshaft(s).

38   Connect the speedometer cable.

39   Adjust the shift linkage.

40   Install any exhaust system components that were removed or disconnected.

41   Lower the vehicle.

42   Fill the transmission with the specified fluid (see Chapter 1), run the engine and check for fluid leaks.

8   Brake/transmission shift interlock cable - removal, installation and adjustment

Removal and installation

1   Later models are equipped with a brake/transmission shift interlock cable, which prevents the transmission from being shifted out of Park or Neutral unless the

brake pedal is depressed. Here's how it works: When you depress the brake pedal, a switch down at the pedal closes the circuit to a solenoid that releases the park/brake interlock cable. If you don't depress the brake pedal, the solenoid locks the park/brake interlock cable, which prevents the shift lever from being moved. This safety system is virtually trouble-free. However, the following procedure is included in the event that the cable should ever break, or should you have to disconnect it in order to service something else.

2 Remove the steering column lower shroud (see illustration 9.18 in Chapter 12), remove the headlight switch knob and shaft (see Section 13 in Chapter 12) and the steering column opening cover (see illustration 16.14 in Chapter 11).

3 Remove the cable tie that secures the brake/transmission shift interlock cable to the steering column. The cable tie is located in front of the solenoid.

4 Unplug the electrical connector from the solenoid.

5 With the ignition key in the unlocked position, disengage the lock tab which connects the end of the cable to the steering column and detach the cable from the steering column.

6 Remove the center console assembly (see Chapter 11).

7 Disconnect the cable eyelet from the bellcrank on the shift lever.

8 Disconnect the cable from the shift lever bracket (see illustration 3.13).

9 Remove the cable.

10 Installation is the reverse of removal. Be sure to adjust the cable before installing the center console.

Adjustment

11 Shift the transmission into Park. If you have just installed or replaced the cable, the center console and gear position indicator trim panel should still be off. If they're not, remove them now (see Chapter 11).

12 Pull up the cable lock button to release the park/brake interlock cable (see illustration 3.13).

13 Turn the ignition switch to the LOCK position.

14 Using a suitable spacer, create a one-millimeter gap between the shifter pawl and the top of the shift gate.

15 Pull the cable forward, then release the cable and press down the cable lock button until it snaps into place.

16 Check your adjustment as follows:

a) Verify that you cannot press in the release button in the shift handle (floor shift) or move the release lever (column shift).

b) Turn the ignition switch to the RUN position.

c) Verify that you cannot shift out of PARK unless the brake pedal is applied.

d) With the transmission shift lever out of PARK, release the brake pedal and the release button or release lever, then shift through all the gears. Verify that the ignition key doesn't go to the LOCK position.

e) Verify that the shift lever can be returned to the PARK position without having to apply the brake.

f) Move the shift lever back to the PARK position and verify that when the ignition key is turned to the LOCK position, the shift lever cannot be moved out of PARK.

g) If the system operates as described, the adjustment is correct. If it doesn't, readjust the cable.

17 Install the center console.

7B

Notes

Chapter 7 Part C
Transfer case

Contents

Specifications

Torque specifications — Ft-lbs

Transfer case-to-transmission nuts	30
Universal joint flange nut	35
Universal joint strap-to-transfer case yoke bolts	14
Rear crossmember-to-chassis sill bolts	30
Transmission support-to-crossmember nuts	33

1 General information

The transfer case is a device which passes the power from the engine and transmission to the front and rear driveshafts.

Several transfer cases were used on these models: New process 207/231 (Command Trac) and 228/229/242 (Selec-Trac). Command-Trac is a part-time transfer case with three operating ranges: 2WD high, 4WD high (full-time 4WD) 4WD high lock (part-time 4WD) and 4WD low lock (part-time 4WD).

2 Shift linkage adjustment

Refer to illustrations 2.2 and 2.4

1   Remove the shift lever boot or move the carpeting aside for access to the lever.
2   Position the shift lever as far to the rear as possible and insert a 1/8-inch thick shim between the shift lever and the shift gate **(see illustration)**.
3   Raise the vehicle and support it securely on jackstands.
4   Under the vehicle, loosen the shift linkage trunnion lock bolt and make sure the shift rod fits freely in the trunnion and shift lever **(see illustration)**. Tighten the bolt securely.
5   Lower the vehicle, remove the shim and reinstall any components which were removed.

3 Transfer case - removal and installation

Removal

1   Disconnect the negative cable from the battery.

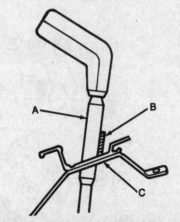

2.2 Pull the shift lever (A) back and insert the shim (B) between the lever and the shift gate (C)

2   Raise the vehicle and support it securely on jackstands.
3   Drain the transfer case lubricant (see Chapter 1).
4   Disconnect the speedometer cable, shift lever, vacuum/vent lines and wire harness connectors from the transfer case.
5   Remove the driveshafts (see Chapter 8).
6   Remove the exhaust system components as necessary for clearance (see Chapter 4).
7   Support the transmission with a jack or jackstand. The transmission should remain supported at all times while the transfer case is out of the vehicle.
8   Support the transfer case with a jack - preferably a special jack made for this purpose. Safety chains will help steady the transfer case on the jack.
9   Remove the rear transmission support-

2.4 Loosen the lock bolt (D) on the trunnion (E) and adjust the rod (F) so it fits freely in the trunnion and the shift lever (G)

to-crossmember nuts and bolts.
10   Remove the crossmember bolts. Raise the transfer case slightly and remove the crossmember.
11   Remove the nuts securing the transmission to the transfer case. Lower the transmission sufficiently to allow access to the upper nuts.
12   Make a final check that all wires and hoses have been disconnected from the transfer case and then move the transfer case and jack toward the rear of the vehicle until the transfer case is clear of the transmission. Keep the transfer case level as this is done.
13   Once the input shaft is clear, lower the transfer case and remove it from under the vehicle.

Installation

14   With the transfer case secured to the jack as on removal, raise it into position behind the transmission and then carefully slide it forward, engaging the input shaft with the transmission output shaft. Do not use excessive force to install the transfer case - if the input shaft does not slide into place, readjust the angle so it is level and/or turn the input shaft so the splines engage properly

with the transmission.

15   Install the transmission-to-transfer case nuts. Tighten the nuts to the specified torque.
16   Install the crossmember and transmission support. Tighten all nuts and bolts securely.
17   Remove the jacks supporting the transmission and the transfer case.
18   Install the various items removed previously, referring to Chapter 8 for the installation of the driveshafts and Chapter 4 for infor-

mation regarding the exhaust system components.
19   Make a final check that all wires, hoses and the speedometer cable have been connected and that the transmission has been filled with lubricant to the proper level (see Chapter 1). Lower the vehicle.
20   Connect the negative battery cable. Road test the vehicle for proper operation and check for leakage.

4.4a Typical Command-Trac transfer case (New Process 207) - exploded view

1   Main driveshaft
2   Case housing
3   Oil pump housing seal
4   Oil pump housing
5   Oil pump
6   Speedometer drive gear
7   Mainshaft rear bearing retainer
8   Case vent connector
9   Bolt
10   Mainshaft rear bearing
11   Mainshaft rear bearing retaining ring
12   Mainshaft extension
13   Hex bolt
14   Case mainshaft extension bushing
15   Mainshaft extension seal
16   Case oil plug
17   Hex bolt
18   Housing alignment dowel washer
19   Housing alignment dowel
20   Front output shaft pilot bearing
21   Front output shaft
22   Planet gear assembly carrier
23   Planet gear carrier retaining ring thrust washer
24   Planet gear carrier retaining ring
25   Planet gear carrier annulus gear
26   Main driveshaft synchronizer retainer ring
27   Main driveshaft assembly synchronizer
28   Synchronizer strut
29   Synchronizer strut spring
30   Synchronizer stop ring
31   Drive chain sprocket bearing
32   Drive chain sprocket
33   Drive chain sprocket thrust washer
34   Input main drive gear thrust washer
35   Input drive gear pilot bearing
36   Cup plug
37   Input main drive assembly gear
38   Input drive gear thrust
39   Input drive gear thrust bearing washer
40   Low range lock plate
41   Vacuum four wheeldrive switch
42   Four wheel drive indicator light switch seal
43   Oil access hole plug
44   Case housing (front half)
45   Input drive bearing
46   Input drive gear seal
47   Hex bolt
48   Front output driveshaft flange yoke
49   Front output driveshaft yoke nut
50   Front output driveshaft yoke (rubber)
51   Front output driveshaft yoke deflector
52   Front output shaft seal

53   Front output shaft bearing retainer ring
54   Front output shaft bearing
55   Shift sector spring screw
56   Screw
57   Shift sector and shaft oil seal
58   Shift sector and shaft retainer
59   Shifter shaft lever
60   Shift shaft lever nut
61   Shift sector assembly spring
62   Range fork bushing
63   Fork end pad
64   Range shift fork pin

65   Range shift fork center
66   Range shift assembly fork
67   Mode shift fork bracket pin
68   Mode shift fork center pad
69   Mode shift assembly fork
70   Mode shift fork spring cup
71   Mode shift fork spring
72   Mode shift fork assembly bracket
73   Shift fork shaft
74   Shift sector
75   Shift selector shaft spacer
76   Drive chain

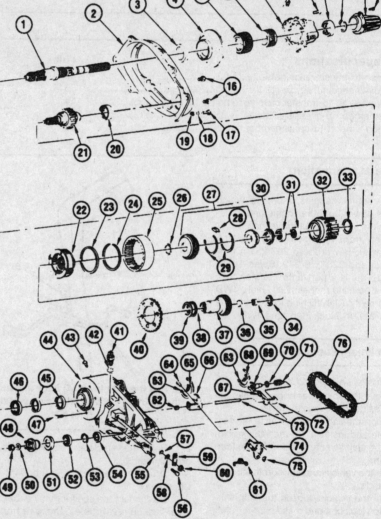
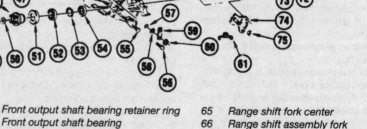

4 Transfer case overhaul - general information

Refer to illustrations 4.4a, 4.4b, 4.4c, 4.4d and 4.4e

Overhauling a transfer case is a difficult job for the do-it-yourselfer. It involves the disassembly and reassembly of many small parts. Numerous clearances must be precisely measured and, if necessary, changed with select fit spacers and snap-rings. As a result, if transfer case problems arise, it can be removed and installed by a competent do-it-yourselfer, but overhaul should be left to a transmission repair shop. Rebuilt transfer cases may be available - check with your dealer parts department and auto parts stores. At any rate, the time and money involved in an overhaul is almost sure to exceed the cost of a rebuilt unit.

Nevertheless, it's not impossible for an inexperienced mechanic to rebuild a transfer case if the special tools are available and the job is done in a deliberate step-by-step manner so nothing is overlooked.

The tools necessary for an overhaul include internal and external snap-ring pliers, a bearing puller, a slide hammer, a set of pin punches, a dial indicator and possibly a hydraulic press. In addition, a large, sturdy workbench and a vise or transmission stand will be required.

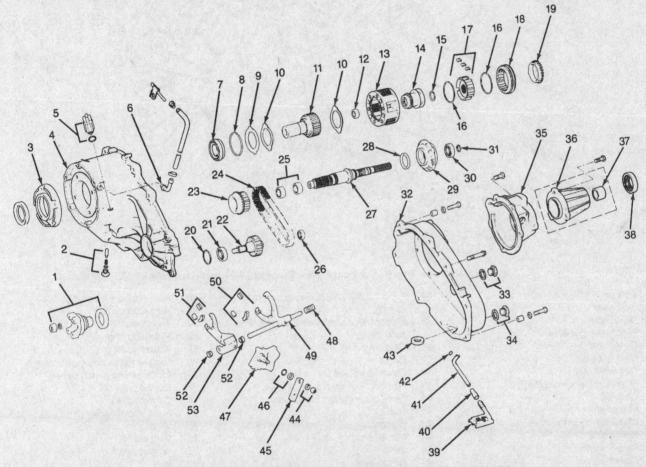

4.4b Typical Command-Trac transfer case (New Process 231) - exploded view

1	Front yoke nut, seal washer, yoke and oil seal	
2	Shift detent plug, spring and pin	
3	Front retainer and seal	
4	Front case	
5	Vacuum switch and seal	
6	Vent assembly	
7	Input gear bearing and snap-ring	
8	Low range gear snap-ring	
9	Input gear retainer	
10	Low range gear thrust washers	
11	Input gear	
12	Input gear pilot bearing	
13	Low range gear	
14	Range fork shift hub	
15	Synchronizer hub snap-ring	
16	Synchronizer hub springs	
17	Synchronizer hub and inserts	
18	Synchronizer sleeve	
19	Synchronizer stop ring	
20	Sap-ring	
21	Output shaft front bearing	
22	Output shaft (front)	
23	Drive sprocket	
24	Drive chain	
25	Drive sprocket bearings	
26	Output shaft rear bearing	
27	Mainshaft	
28	Oil seal	
29	Oil pump assembly	
30	Rear bearing	
31	Snap-ring	
32	Rear case	
33	Fill plug and gasket	
34	Drain plug and gasket	
35	Rear retainer	
36	Extension housing	
37	Bushing	
38	Oil seal	
39	Oil pickup screen	
40	Tube connector	
41	Oil pickup tube	
42	Pickup tube O-ring	
43	Magnet	
44	Range lever nut and washer	
45	Range lever	
46	O-ring and seal	
47	Sector	
48	Mode spring	
49	Mode fork	
50	Mode fork inserts	
51	Range fork inserts	
52	Range fork bushings	
53	Range fork	

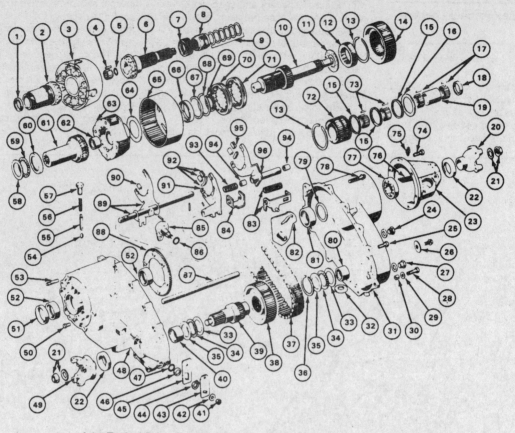

4.4c Typical Selec-Trac transfer case (New Process 228) - exploded view

1	Spacer	34	Front output shaft bearing assembly thrust	65	Annulus gear assembly
2	Side gear			66	Annulus bushing
3	Differential	35	Front output shaft bearing assembly race (thin)	67	Thrust washer
4	Pilot bearing rollers			68	Retaining ring
5	O-ring seal	36	Retaining ring	69	Thrust bearing
6	Rear output shaft	37	Chain	70	High range sliding clutch sleeve
7	Oil pump	38	Driven sprocket	71	Mode sliding clutch sleeve
8	Speedometer drive gear	39	Front output shaft	72	Carrier
9	Shim kit	40	Front output shaft front bearing	73	Carrier rollers
10	Mainshaft	41	Nut	74	Rear retainer bolt
11	Mainshaft thrust washer	42	Washer	75	Vent
12	Spline gear	43	Mode lever	76	Vent seal
13	Retaining ring	44	Snap-ring	77	Output bearing
14	Sprocket	45	Range lever	78	Bolt
15	Spacer	46	O-ring retainer	79	Seal
16	Sprocket thrust washer	47	O-ring seal	80	Front output shaft rear bearing
17	Side gear roller	48	Front case half	81	Output shaft inner bearing
18	Spacer (short)	49	Front output yoke	82	Range sector
19	Spacer (long)	50	Low range plate bolt	83	Range bracket (outer) and spring
20	Rear yoke	51	Input shaft oil seal	84	Range bracket (inner)
21	Nut and seal washer	52	Input shaft bearing	85	Mode selector
22	Seal	53	Stud	86	O-ring seal
23	Rear retainer	54	Ball	87	Range rail
24	Plug assembly	55	Plunger half dowel	88	Low range lockout plate
25	Bolt	56	Plunger spring	89	Mode fork, rail and pin
26	Identification tag	57	Screw	90	Mode fork pad
27	Plug assembly	58	Input race	91	Range fork
28	Dowel bolt	59	Input thrust bearing	92	Range fork pads
29	Dowel bolt washer	60	Input race (thick)	93	Range bracket spring (inner)
30	Case half dowel	61	Input shaft	94	Locking fork bushing
31	Rear case half	62	Input bearing	95	Locking fork pads
32	Magnet	63	Planetary gear assembly	96	Locking fork
33	Front output shaft bearing assembly race (thick)	64	Input gear thrust washer		

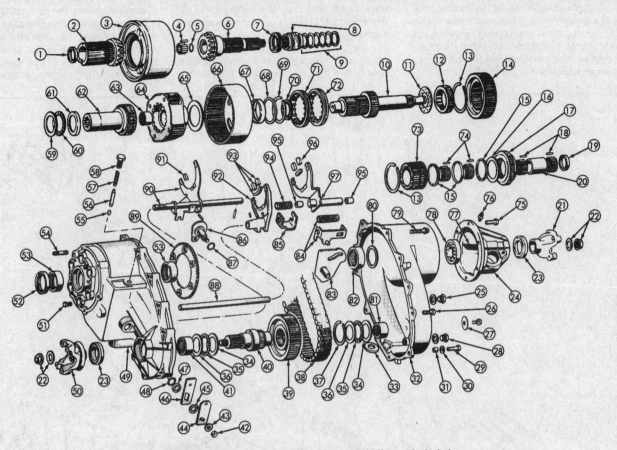

4.4d Typical Selec-Trac transfer case (New Process 229) - exploded view

1	Spacer	34	Bearing race (thick)	66	Annulus gear
2	Side gear	35	Bearing	67	Annulus bushing
3	Viscous coupling	36	Bearing race (thin)	68	Thrust washer
4	Pilot bearing rollers (15)	37	Retaining ring	69	Retaining ring
5	O-ring	38	Drive chain	70	Thrust bearing
6	Rear output shaft	39	Driven sprocket	71	Range sleeve
7	Oil pump	40	Front output shaft	72	Mode sleeve
8	Speedometer drive gear	41	Front output shaft front bearing	73	Carrier
9	Shims	42	Nut	74	Carrier bearings
10	Mainshaft	43	Washer	75	Rear retainer bolt
11	Mainshaft thrust washer	44	Mode lever	76	Vent
12	Spline gear	45	Snap-ring	77	Vent seal
13	Retaining ring	46	Range lever	78	Output bearing
14	Sprocket	47	O-ring retainer	79	Bolt
15	Spacer	48	O-ring	80	Seal
16	Sprocket thrust washer	49	Front case half	81	Front output shaft rear bearing
17	Clutch gear	50	Front yoke	82	Output shaft bearing
18	Mainshaft bearings	51	Bolt	83	Range sector
19	Bearing spacers (two short)	52	Input gear oil seal	84	Range bracket (outer) and spring
20	Bearing spacer (one long)	53	Input gear bearing	85	Range bracket (inner)
21	Rear yoke	54	Stud	86	Mode sector
22	Nut and seal washer	55	Detent ball	87	O-ring
23	Seal	56	Pin	88	Range rail
24	Rear retainer	57	Spring	89	Low range lockout plate
25	Plug assembly	58	Screw	90	Mode fork, rail and pin
26	Bolt	59	Bearing race (thin)	91	Mode fork pad
27	Identification tag	60	Thrust bearing	92	Range fork
28	Plug	61	Bearing race (thick)	93	Range fork pads
29	Dowel bolt	62	Input gear	94	Range bracket spring (inner)
30	Dowel bolt washer	63	Pilot bearing	95	Locking fork bushing
31	Case half dowel	64	Planetary gear	96	Locking fork pads
32	Rear case half	65	Input gear thrust washer	97	Locking fork
33	Magnet				

7C

During disassembly of the transfer case, make careful notes of how each piece comes off, where it fits in relation to other pieces and what holds it in place. Exploded views are included **(see illustrations)** to show where the parts go - but actually noting how they are installed when you remove the parts will make it much easier to get the transfer case back together.

Before taking the transfer case apart for repair, it will help if you have some idea what area of the transfer case is malfunctioning.

Certain problems can be closely tied to specific areas in the transfer case, which can make component examination and replacement easier. Refer to the *Troubleshooting* section at the front of this manual for information regarding possible sources of trouble.

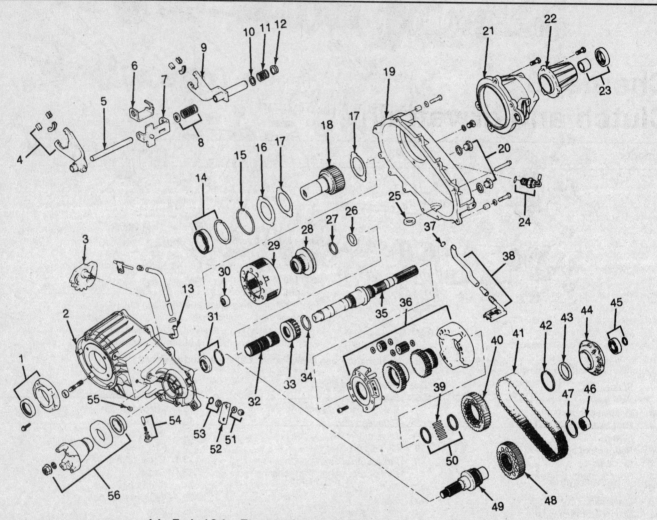

4.4e Typical Selec-Trac transfer case (New Process 242) - exploded view

1 Front bearing retainer and seal	21 Rear bearing retainer	39 Mainshaft bearing rollers
2 Front case	22 Extension housing	40 Drive sprocket
3 Shift sector	23 Bushing and oil seal	41 Drive chain
4 Low range fork and inserts	24 Vacuum switch	42 Snap-ring
5 Shift rail	25 Magnet	43 Oil pump seal
6 Shift bracket	26 Thrust ring	44 Oil pump
7 Slider bracket	27 Snap-ring	45 Rear bearing and snap-ring
8 Bushing and spring	28 Shift sleeve	46 Front output shaft rear bearing
9 Mode fork and inserts	29 Low range gear	47 Snap-ring
10 Bushing	30 Pilot bushing (input gear/mainshaft)	48 Driven sprocket
11 Fork spring	31 Front output shaft front bearing and	49 Front output shaft
12 Bushing	snap-ring	50 Mainshaft bearing spacers
13 Vent tube assembly	32 Intermediate clutch shaft	51 Shift lever washer and nut
14 Input gear bearing and snap-ring	33 Shift sleeve	52 Shift lever
15 Low range gear snap-ring	34 Snap-ring	53 Sector O-ring and seal
16 Retainer, low range gear	35 Mainshaft	54 Detent pin, spring and plug
17 Thrust washer, low range gear	36 Differential assembly	55 Seal plug
18 Input gear	37 Oil pump tube O-ring	56 Front yoke nut, seal washer, yoke,
19 Rear case	38 Oil pump pickup tube and screen	slinger and oil seal
20 Drain/fill plugs		

Chapter 8
Clutch and drivetrain

Contents

Specifications

Clutch

Fluid type	See Chapter 1

Torque specifications
Ft-lbs

Pressure plate-to-flywheel bolts	
Four-cylinder and V6 engines	23
Inline six-cylinder engine..	40
Clutch/brake pedal pivot pin nut	20

Drivetrain

Torque specifications
Ft-lbs (unless otherwise indicated)

Driveshaft U-joint strap bolts...	14
Driveshaft-to-transfer case flange bolts	22
Front axle shift motor housing bolts	108 in-lbs
Front axleshaft hub nut...	175
Front hub assembly-to-steering knuckle bolts...........................	75
Front differential pinion shaft nut	210
Brake backing plate nuts...	32
Rear axle U-bolt nuts..	44 to 59

---

1 General information

The information in this Chapter deals with the components from the rear of the engine to the drive wheels, except for the transmission and transfer case, which are dealt with in the previous Chapter. For the purposes of this Chapter, these components are grouped into three categories; clutch, driveshaft and axles. Separate Sections within this Chapter offer general descriptions and checking procedures for components in each of the three groups.

Since nearly all the procedures covered in this Chapter involve working under the vehicle, make sure it's securely supported on sturdy jackstands or on a hoist where the vehicle can be easily raised and lowered.

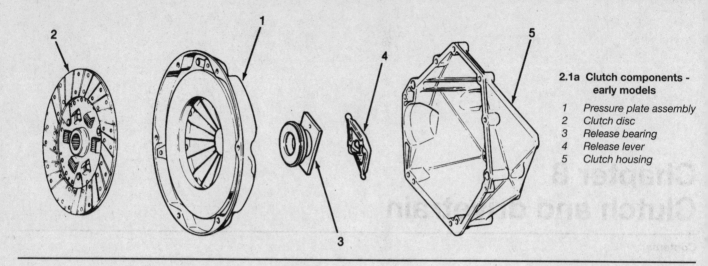

**2.1a Clutch components -
early models**

1   *Pressure plate assembly*
2   *Clutch disc*
3   *Release bearing*
4   *Release lever*
5   *Clutch housing*

2   Clutch - description and check

Refer to illustrations 2.1a and 2.1b

1   All vehicles with a manual transmission use a single dry plate, diaphragm spring type clutch **(see illustrations)**. The clutch disc has a splined hub which allows it to slide along the splines of the transmission input shaft. The clutch and pressure plate are held in contact by spring pressure exerted by the diaphragm in the pressure plate.

2   The clutch release system is operated by hydraulic pressure. The hydraulic release system consists of the clutch pedal, a master cylinder and fluid reservoir, the hydraulic line, a release (or slave) cylinder which actuates the clutch release lever (models with externally mounted release cylinder) and the clutch release (or throwout) bearing. Later models use a hydraulic release bearing, which is a combination release bearing/slave cylinder assembly mounted inside the clutch housing.

3   When pressure is applied to the clutch pedal to release the clutch, hydraulic pressure is exerted against the release bearing, either through the release lever or the hydraulic release bearing. The bearing pushes against the fingers of the diaphragm spring of the pressure plate assembly, which in turn releases the clutch plate.

4   Terminology can be a problem when discussing the clutch components because common names are in some cases different from those used by the manufacturer. For example, the driven plate is also called the clutch plate or disc, the clutch release bearing is sometimes called a throwout bearing, the release cylinder is sometimes called the operating or slave cylinder.

5   Other than to replace components with obvious damage, some preliminary checks should be performed to diagnose clutch problems.

a)   *The first check should be of the fluid level in the clutch master cylinder. If the fluid level is low, add fluid as necessary and inspect the hydraulic system for leaks. If the master cylinder reservoir has run dry, bleed the system as described in Section 8 and retest the clutch operation.*

b)   *To check "clutch spin down time," run the engine at normal idle speed with the transmission in Neutral (clutch pedal up - engaged). Disengage the clutch (pedal down), wait several seconds and shift the transmission into Reverse. No grinding noise should be heard. A grinding noise would most likely indicate a problem in the pressure plate or the clutch disc (assuming the transmission is in good condition).*

c)   *To check for complete clutch release, run the engine (with the parking brake applied to prevent movement) and hold the clutch pedal approximately 1/2-inch from the floor. Shift the transmission between 1st gear and Reverse several*

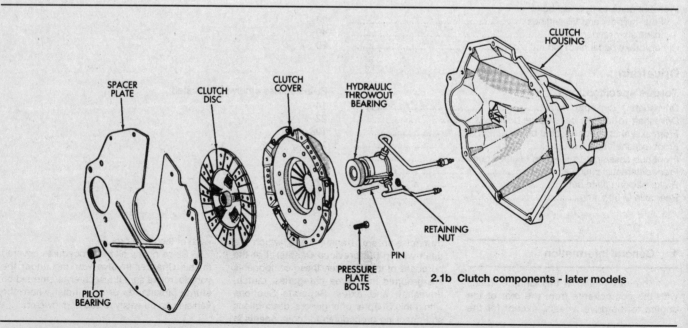

2.1b Clutch components - later models

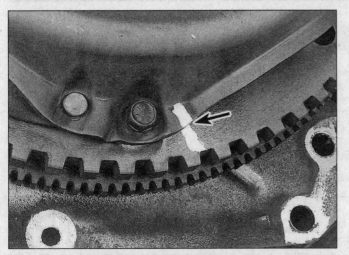

3.5 Mark the relationship of the pressure plate to the flywheel (arrow) (in case you are going to reuse the same pressure plate)

3.8 Check the surface of the flywheel for cracks, hot spots (dark colored areas) and other obvious defects; resurfacing by a machine shop will correct minor defects - the surface on this flywheel is in fairly good condition; however, resurfacing is always a good idea

times. If the shift is rough, component failure is indicated. Check the release cylinder pushrod travel (externally mounted release cylinder models only). With the clutch pedal depressed completely, the release cylinder pushrod should extend substantially. If it doesn't, check the fluid level in the clutch master cylinder.
d) Visually inspect the pivot bushing at the top of the clutch pedal to make sure there is no binding or excessive play.

3 Clutch components - removal, inspection and installation

Warning: Dust produced by clutch wear and deposited on clutch components may contain asbestos, which is hazardous to your health. DO NOT blow it out with compressed air and DO NOT inhale it. DO NOT use gasoline or petroleum-based solvents to remove the dust. Brake system cleaner should be used to flush the dust into a drain pan. After the clutch components are wiped clean with a rag, dispose of the contaminated rags and cleaner in a covered, marked container.

Removal
Refer to illustration 3.5

1 Access to the clutch components is normally accomplished by removing the transmission, leaving the engine in the vehicle. If, of course, the engine is being removed for major overhaul, then check the clutch for wear and replace worn components as necessary. However, the relatively low cost of the clutch components compared to the time and trouble spent gaining access to them warrants their replacement anytime the engine or transmission is removed, unless they are new or in near perfect condition. The following procedures are based on the assumption the engine will stay in place.
2 Referring to Chapter 7 Part A, remove the transmission from the vehicle. Support

the engine while the transmission is out. Preferably, an engine hoist should be used to support it from above. However, if a jack is used underneath the engine, make sure a piece of wood is positioned between the jack and oil pan to spread the load. **Caution:** The pickup for the oil pump is very close to the bottom of the oil pan. If the pan is bent or distorted in any way, engine oil starvation could occur.
3 Remove the release cylinder (models with externally mounted cylinders) (see Section 7). On models with a hydraulic release bearing assembly, disconnect the hydraulic line at the clutch housing.
4 To support the clutch disc during removal, install a clutch alignment tool through the clutch disc hub.
5 Carefully inspect the flywheel and pressure plate for indexing marks. The marks are usually an X, an O or a white letter. If they cannot be found, scribe marks yourself so the pressure plate and the flywheel will be in the same alignment during installation **(see illustration)**.
6 Turning each bolt a little at a time, loosen the pressure plate-to flywheel bolts.

Work in a criss-cross pattern until all spring pressure is relieved. Then hold the pressure plate securely and completely remove the bolts, followed by the pressure plate and clutch disc.

Inspection
Refer to illustrations 3.8, 3.10 and 3.12
7 Ordinarily, when a problem occurs in the clutch, it can be attributed to wear of the clutch driven plate assembly (clutch disc). However, all components should be inspected at this time.
8 Inspect the flywheel for cracks, heat checking, grooves and other obvious defects **(see illustration)**. If the imperfections are slight, a machine shop can machine the surface flat and smooth, which is highly recommended regardless of the surface appearance. Refer to Chapter 2 for the flywheel removal and installation procedure.
9 Inspect the pilot bearing (see Section 5).
10 Inspect the lining on the clutch disc. There should be at least 1/16-inch of lining above the rivet heads. Check for loose rivets, distortion, cracks, broken springs and other obvious damage **(see illustration)**. As men-

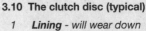

3.10 The clutch disc (typical)
1 **Lining** - will wear down in use
2 **Rivets** - secure the lining and will damage the pressure plate or flywheel surface if allowed to contact it
3 **Marks** - "Flywheel side" or something similar

8

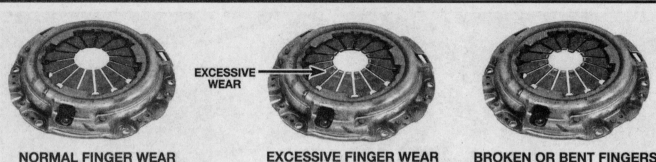

NORMAL FINGER WEAR   EXCESSIVE FINGER WEAR   BROKEN OR BENT FINGERS

EXCESSIVE WEAR

3.12   Replace the pressure plate if excessive wear or damage is noted

tioned above, ordinarily the clutch disc is routinely replaced, so if in doubt about the condition, replace it with a new one.

11   Check the condition of the release bearing following the procedure in Section 4.

12   Check the machined surfaces and the diaphragm spring fingers of the pressure plate **(see illustration)**. If the surface is grooved or otherwise damaged, replace the pressure plate. Also check for obvious damage, distortion, cracking, etc. Light glazing can be removed with medium grit emery cloth. If a new pressure plate is required, new and factory-rebuilt units are available.

Installation

Refer to illustration 3.14

13   Before installation, clean the flywheel and pressure plate machined surfaces with lacquer thinner or acetone. It's important that no oil or grease is on these surfaces or the lining of the clutch disc. Handle the parts only with clean hands.

14   Position the clutch disc and pressure plate against the flywheel with the clutch held in place with an alignment tool **(see illustration)**. Make sure it's installed properly (most replacement clutch plates will be marked "flywheel side" or something similar - if not marked, install the clutch disc with the damper springs toward the transmission).

15   Tighten the pressure plate-to-flywheel bolts only finger tight, working around the

pressure plate.

16   Center the clutch disc by ensuring the alignment tool extends through the splined hub and into the pilot bearing in the crankshaft. Wiggle the tool up, down or side-to-side as needed to bottom the tool in the pilot bearing. Tighten the pressure plate-to-flywheel bolts a little at a time, working in a criss-cross pattern to prevent distorting the cover. After all of the bolts are snug, tighten them to the specified torque. Remove the alignment tool.

17   Using high-temperature grease, lubricate the inner groove of the release bearing (see Section 4). Also place grease on the release lever contact areas (on models with release levers) and the transmission input shaft bearing retainer.

18   Install the clutch release bearing as described in Section 4.

19   Install the transmission, release cylinder (externally mounted type only) and all components removed previously. Tighten all fasteners to the proper torque specifications.

4   Clutch release bearing - removal, inspection and installation

Warning: *Dust produced by clutch wear and deposited on clutch components may contain asbestos, which is hazardous to your health. DO NOT blow it out with compressed*

air and DO NOT inhale it. DO NOT use gasoline or petroleum-based solvents to remove the dust. Brake system cleaner should be used to flush the dust into a drain pan. After the clutch components are wiped clean with a rag, dispose of the contaminated rags and cleaner in a covered, marked container.

Removal

1   Following the appropriate procedure outlined in Chapter 7, remove the transmission.

Externally mounted release cylinder

2   Unbolt the release cylinder from the clutch housing and pull it out of its recess. Hang it out of the way with a piece of wire (it's not necessary to disconnect the hose).

3   Remove the release lever from the ball stud, slide the bearing off the transmission input shaft bearing retainer and separate the bearing from the lever.

Release bearing assembly

Refer to illustrations 4.4 and 4.5

4   Unbolt the insulator plate from the clutch housing and slide the plate and rubber insulator off the lines **(see illustration)**.

5   Remove the pressed metal retaining nut from the base of the hydraulic release bearing **(see illustration)** and slide the release bearing assembly off the transmission input shaft bearing retainer.

3.14   Center the clutch disc in the pressure plate with an alignment tool before the bolts are tightened

4.4   Remove the bolts (arrows), then slide the insulator plate and rubber insulator back on the lines

4.5   Remove the hydraulic release bearing retaining nut - be sure to use a new nut upon installation

5.9 Pack the recess behind the bearing with heavy grease and force it out hydraulically with a steel rod slightly smaller than the bore in the bearing - when the hammer strikes the rod, the bearing will pop out of the crankshaft

Inspection

6   Hold the center portion of the bearing stationary and rotate the outer portion while applying pressure. If the bearing doesn't turn smoothly or if it's noisy, replace it with a new one. Wipe the bearing with a clean rag and inspect it for damage and wear. Don't immerse the bearing in solvent - it is sealed for life and immersion would ruin it.

7   Hydraulic release bearings must also be checked for fluid leakage where the rubber seal meets the cylinder housing. If any leakage is evident, replace the entire assembly.

Installation

8   Installation for either style bearing is the reverse of the removal procedure, with a couple of points which must be noted.

9   Lightly lubricate the transmission input shaft bearing retainer with high-temperature grease. On hydraulic release bearing models, install a new pressed metal retaining nut. **Note:** *When new, the hydraulic release bearing is retained in the compressed position by nylon straps - they are designed to break the first time the clutch pedal is depressed, so there is no need to remove them.*

10   With the non-hydraulic release bearing (externally mounted release cylinder), lubricate the ends of the release lever and the ball socket with high-temperature grease. Also pack the inner groove of the bearing with the same grease. When installing the bearing and lever, make sure the bearing retaining spring clips are engaged with the lever ends.

11   Install the transmission (see Chapter 7). On hydraulic release bearing models, connect the hydraulic line and bleed the system as described in Section 8.

5   Pilot bearing - inspection and replacement

Refer to illustrations 5.9, 5.10a and 5.10b

1   The clutch pilot bearing is a needle roller

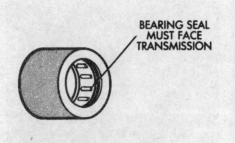

5.10a The pilot bearing incorporates an O-ring seal which cannot be replaced separately. If there is any indication that the seal is leaking, or if the bearing is dry, replace it. The bearing must be installed with the seal towards the transmission

type bearing which is pressed into the rear of the crankshaft. Its primary purpose is to support the front of the transmission input shaft. The pilot bearing should be inspected whenever the clutch components are removed from the engine. Due to its inaccessibility, if you are in doubt as to its condition, replace it with a new one. **Note:** *If the engine has been removed from the vehicle, disregard the following steps which do not apply.*

2   Remove the transmission (refer to Chapter 7 Part A).

3   Remove the clutch components (Section 3).

4   Inspect for any excessive wear, scoring, lack of grease, dryness or obvious damage. If any of these conditions are noted, the bearing should be replaced. A flashlight will be helpful to direct light into the recess.

5   Removal can be accomplished with a special puller and slide hammer, but an alternative method also works very well.

6   Find a solid steel bar which is slightly smaller in diameter than the bearing. Alternatives to a solid bar would be a wood dowel or a socket with a bolt fixed in place to make it solid.

7   Check the bar for fit - it should just slip into the bearing with very little clearance.

8   Pack the bearing and the area behind it (in the crankshaft recess) with heavy grease. Pack it tightly to eliminate as much air as possible.

9   Insert the bar into the bearing bore and strike the bar sharply with a hammer which will force the grease to the back side of the bearing and push it out **(see illustration)**. Remove the bearing and clean all grease from the crankshaft recess.

10   To install the new bearing, pack the inside of the bearing and lightly lubricate the outside surface with wheel bearing grease. Drive the bearing into the recess with an alignment tool or bushing driver. The seal must face out and the bearing must go in perfectly straight **(see illustrations)**.

11   Install the clutch components, transmission and all other components removed previously, tightening all fasteners properly.

5.10b Tap the bearing in using a bushing driver or a socket

6   Clutch master cylinder - removal, overhaul and installation

Note: *Before beginning this procedure, contact local parts stores and dealer service departments concerning the purchase of a rebuild kit or a new master cylinder. Availability and cost of the necessary parts may dictate whether the cylinder is rebuilt or replaced with a new one. If it's decided to rebuild the cylinder, inspect the bore as described in Step 12 before purchasing parts.*

Removal

Refer to illustrations 6.3 and 6.5

1   Disconnect the cable from the negative battery terminal.

2   Remove the clutch master cylinder reservoir cap, and, using a suction gun or large syringe, suck out as much fluid as possible.

3   Disconnect the hydraulic line from the cylinder, using a flare nut wrench, if available. Some models have a separable reservoir, which, when removed, makes access to the fitting much easier **(see illustration)**.

4   Remove the lower master cylinder mounting nut.

6.3 To avoid rounding off the corners of the fitting nut (arrow), use a flare nut wrench when loosening it

8

6.5 Remove the cotter pin (1) (barely visible in this photo), plastic washer and spring washer, then disconnect the pushrod from the clutch pedal pin; next, hold the upper mounting nut (2) with a wrench while an assistant unscrews the mounting bolt from the engine side of the firewall

5　Working under the dash, disconnect the pushrod from the clutch pedal. It is retained by a cotter pin, plastic washer and spring washer **(see illustration)**.

6　Also from under the dash, remove the cylinder upper mounting nut **(see illustration 6.5)**. Rather than attempt to loosen the nut from this position, it is much easier to simply hold the nut while an assistant unscrews the bolt from the engine side of the firewall.

7　Pull the cylinder out of the hole in the firewall. Be careful not to let any of the fluid drip onto the vehicle's paint, as it will damage it.

Overhaul

Refer to illustrations 6.10, 6.11, 6.14a and 6.14b

8　On models with a removable reservoir,

6.10 Push in on the pushrod and remove the snap-ring with a pair of snap-ring pliers

detach the reservoir from the cylinder body by loosening the clamp.

9　Mount the cylinder in a vise, with the vise jaws clamping on the mounting flange. It's a good idea to line the jaws of the vise with wood or rags to prevent damage to the flange surface.

10　Pry the dust boot from the cylinder. Push in on the pushrod to depress the plunger, then remove the snap-ring with a pair of snap-ring pliers **(see illustration)**. Remove the pushrod and boot. Discard the snap-ring.

11　Pull the plunger assembly from the bore **(see illustration)**. If it's hard to pull out, remove the cylinder from the vise, turn it so the flanged end is down and tap it sharply against a block of wood.

6.11 Pull the plunger assembly from the bore

12　Inspect the bore of the cylinder for scratches, score marks, pitting and ridges. The surface must be smooth to the touch. If the bore isn't perfectly smooth, the master cylinder must be replaced with a new or factory rebuilt unit.

13　If the cylinder will be rebuilt, use the parts contained in the rebuild kit and follow any specific instructions which accompany the kit. Wash all parts to be re-used with brake cleaner or clean brake fluid. DO NOT use petroleum-based solvents.

14　Compress the spring and, using a small screwdriver, pry up on the valve stem retainer tab to release the retainer, spring and valve stem assembly **(see illustrations)**.

15　Remove the plunger, then remove and discard the seals from the plunger.

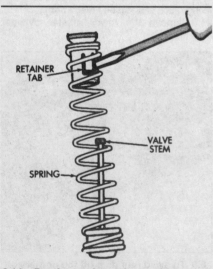

6.14a Pry the retainer tab out with a small screwdriver to release the valve stem

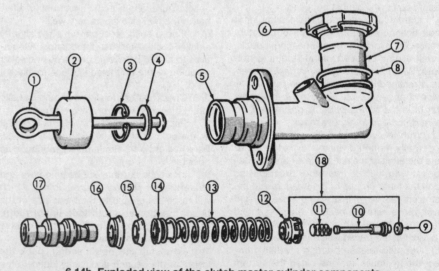

6.14b Exploded view of the clutch master cylinder components

1	Pushrod	7	Reservoir	13	Plunger spring
2	Dust boot	8	Retaining clamp	14	Valve stem retainer
3	Snap-ring	9	Stem tip seal	15	Plunger rear seal
4	Washer	10	Valve stem	16	Plunger front seal
5	Master cylinder body	11	Retainer spring	17	Plunger
6	Reservoir cap	12	Spring retainer	18	Valve stem assembly

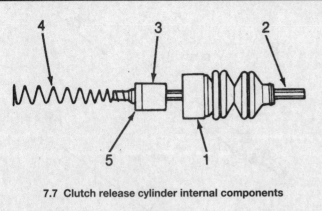

7.7 Clutch release cylinder internal components

1	Boot	3	Plunger	5	Seal
2	Pushrod	4	Spring		

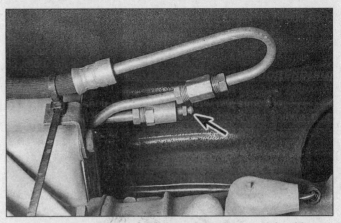

8.5 Location of the bleeder screw (arrow) on models with a hydraulic clutch release bearing

16   Separate the spring retainer and the valve stem assembly from the plunger spring.

17   Remove the valve stem from the spring retainer. Remove the stem tip seal and spring washer from the stem. Discard the seal.

18   Install the new plunger seals on the plunger, making sure the lips of the seals face the small end of the plunger.

19   Install the new valve stem tip seal and spring washer onto the valve stem. The shoulder of the stem tip seal must fit into the undercut at the end of the stem.

20   Place the spring retainer on the valve stem and over the spring washer, making sure the large end of the retainer is facing the seal end of the valve stem.

21   Slide the stem and spring retainer assembly into place on the plunger spring.

22   Holding the plunger in one hand, compress the spring against the plunger and guide the valve stem into the hole in the retainer. When the end of the stem passes through the hole, bend the tab on the retainer in to lock the stem and retainer to the plunger.

23   Lubricate the cylinder bore and plunger assembly with clean brake fluid. Install the assembly in the cylinder bore.

24   Place a new seal, snap-ring and dust boot on the pushrod. Lubricate the end of the pushrod, the seal and the lips of the dust boot with the grease provided in the rebuild kit. Install the pushrod, depress the plunger and seat the snap-ring in its groove.

25   Slide the pushrod seal (not shown in **illustration 6.14b**) and dust boot up against the end of the cylinder and stretch the boot over the cylinder body.

Installation

26   Position the cylinder against the firewall and install the lower mounting nut, but don't tighten it fully yet.

27   Connect the hydraulic line to the cylinder, tightening the fitting by hand only. Since the cylinder is still loose, it can be wiggled around slightly to make it easier to align the fitting threads.

28   Install the upper mounting bolt, again using an assistant to tighten the bolt while the nut is held stationary.

29   Connect the pushrod to the clutch pedal, install the washers and a new cotter pin.

30   Tighten the lower mounting nut.

31   Tighten the hydraulic line fitting securely.

32   Install the fluid reservoir if it was previously removed.

33   Fill the reservoir with brake fluid conforming to DOT 3 specifications and bleed the system as described in Section 8.

7   Clutch release cylinder - removal, overhaul and installation

Note: *This procedure applies to externally mounted release cylinders only (the hydraulic release bearing mounted inside the clutch housing is not serviceable). Before beginning this procedure, contact local parts stores and dealer service departments concerning the purchase of a rebuild kit or a new release cylinder. Availability and cost of the necessary parts may dictate whether the cylinder is rebuilt or replaced with a new one. If it's decided to rebuild the cylinder, inspect the bore as described in Step 8 before purchasing parts.*

Removal

1   Disconnect the negative cable from the battery.

2   Raise the vehicle and support it securely on jackstands.

3   Disconnect the hydraulic line at the release cylinder. If available, use a flare nut wrench on the fitting, which will prevent the fitting from being rounded off. Have a small can and rags handy, as some fluid will be spilled as the line is removed.

4   Remove the two release cylinder mounting bolts.

5   Remove the release cylinder.

Overhaul

Refer to illustration 7.7

6   Separate the rubber boot from the cylinder.

7   Pull the pushrod, boot, plunger and spring from the cylinder **(see illustration)**.

8   Carefully inspect the bore of the cylinder. Check for deep scratches, score marks and ridges. The bore must be smooth to the touch. If any imperfections are found, the release cylinder must be replaced with a new one.

9   Using the new parts in the rebuild-kit, assemble the components using plenty of fresh brake fluid for lubrication. Note the installed direction of the spring and the seal (the seal lips must point toward the spring).

Installation

10   Install the release cylinder on the clutch housing. Make sure the pushrod is seated in the release fork pocket.

11   Connect the hydraulic line to the release cylinder. Tighten the fitting.

12   Fill the clutch master cylinder with brake fluid conforming to DOT 3 specifications.

13   Bleed the system as described in Section 8.

14   Lower the vehicle and connect the negative battery cable.

8   Clutch hydraulic system - bleeding

Refer to illustration 8.5

1   The hydraulic system should be bled to remove all air whenever any part of the system has been removed or if the fluid level has fallen so low that air has been drawn into the master cylinder. The procedure is very similar to bleeding a brake system.

2   Fill the master cylinder with new brake fluid conforming to DOT 3 specifications. **Caution:** *Don't re-use any of the fluid coming from the system during the bleeding operation. Also, don't use fluid which has been inside an open container for an extended period of time.*

3   Raise the vehicle and place it securely on jackstands to gain access to the release cylinder (or bleeder screw), which is located on the left side of the clutch housing.

8

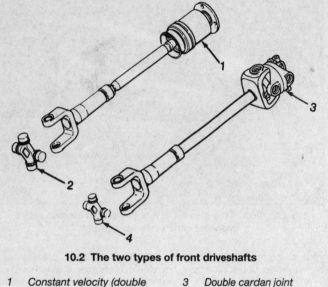

10.2 The two types of front driveshafts

1   *Constant velocity (double*   3   *Double cardan joint*
    offset) type joint   4   *U-joint*
2   *U-joint*

**10.3 The two types of rear driveshafts - the type 1 shaft (top) has
a welded yoke at each end, while the type 2 shaft (bottom) has a
welded yoke at one end and a splined slip yoke at the other**

4   On models with an externally mounted release cylinder, remove the two release cylinder mounting bolts and pull the cylinder from the clutch housing. Remove the pushrod from the cylinder, then compress the plunger into the cylinder with a universal gear puller or a C-clamp and socket.

5   Remove the dust cap which fits over the bleeder screw **(see illustration)** and push a length of plastic hose over the screw. Place the other end of the hose in a clear container with about two inches of brake fluid. The hose end must be in the fluid at the bottom of the container.

6   Have an assistant depress the clutch pedal and hold it. Open the bleeder screw on the release cylinder, allowing fluid to flow through the hose. Close the bleeder screw when the flow of bubbles or old fluid ceases. Once closed, have your assistant release the pedal.

7   Continue this process until all air is evacuated from the system, indicated by a solid stream of fluid being ejected from the bleeder screw each time with no air bubbles in the hose or container. Keep a close watch on the fluid level inside the master cylinder - if the level drops too low, air will be sucked back into the system and the process will have to be started all over again.

8   Install the pushrod, mount the release cylinder (externally mounted cylinder only) and lower the vehicle. Top off the fluid level and check carefully for proper operation before placing the vehicle in normal service.

9   Clutch pedal - removal and installation

Removal

1   Disconnect the cable from the negative

battery terminal.

2   Remove the cotter pin and washer and detach the clutch master cylinder pushrod from the clutch pedal **(see illustration 6.5)**. Remove the pedal return spring.

3   Remove the pivot bolt nut from the right end of the pivot bolt, then slide the bolt out. **Note:** *While doing this, slide another bolt in from the opposite side to support the brake pedal.*

4   Remove the clutch pedal and inspect it for wear and distortion. Check the bushings for excessive wear, replacing parts as necessary.

Installation

5   Lubricate the pedal bushings with multipurpose grease and position the pedal in its bracket. Install the bolt from the left side, forcing out the bolt that was installed to support the brake pedal. Install the nut and tighten it securely.

6   Connect the clutch master cylinder pushrod to the pedal and install the washer and a new cotter pin. Install the pedal return spring.

10   Driveshafts, differentials and axles - general information

Refer to illustrations 10.2 and 10.3

There are two types of front (4WD only) and rear driveshafts available in the vehicles covered in this manual. On 4WD models, their designated application is determined by the model of transfer case installed.

Both style front driveshafts are equipped with a slip yoke and single cardan universal joint at the lower (front axle) end. The type 1 front driveshaft employs a double offset constant velocity joint at the upper

(transfer case) end, where the type 2 shaft uses a double cardan universal joint **(see illustration)**. The slip yokes and universal joints are equipped with grease fittings and require periodic lubrication (see Chapter 1).

The rear driveshaft on Command-Trac vehicles is a one piece shaft with welded yokes at each end. Vehicles with the Selec-Trac system employ a shaft with a welded yoke at the rear and a slip yoke at the front **(see illustration)**.

The front and rear driveshafts are finely balanced during production and whenever they are removed or disassembled, they must be reassembled and installed in the exact manner and positions they were originally in, to avoid excessive vibration.

Two types of differentials are used with these vehicles - standard differentials and as an option, Trac-Loc differentials. The latter style has the ability to transfer torque from one wheel to the other if traction is lost and the wheel begins to spin.

The front axle shafts (4WD models) are equipped with a universal joint at the end of each axle, which allows the front wheels to turn right and left while transmitting torque. Vehicles equipped with Command-Trac four-wheel drive are fitted with single cardan U-joints. Selec-Trac vehicles utilize a constant velocity style joint. Each system (except for later Selec-Trac equipped vehicles) has a front axle disconnect feature which disengages the right front axle via a splined shift collar which joins the right-side intermediate axle with the outer axle. The system is controlled by a vacuum motor and shift fork, actuated when the transfer case lever is shifted into one of the 4WD ranges. When disengaged, the differential is allowed to freewheel instead of turning the ring gear, pinion and front driveshaft, thereby saving unnecessary wear on these components during two-wheel drive operation.

The front axle housing on 4WD models is held in alignment with the body by four control arms and a track rod.

The rear axles are the semi-floating type, supported on the outer ends by the rear wheel bearings which are pressed onto the axleshafts. The axle housing is held in alignment to the body by the suspension leaf springs.

Because of the complexity and critical nature of the differential adjustments, as well as the special equipment needed to perform the operations, disassembly of the differential should be done by a dealer service department or repair shop.

11   Driveline inspection

1   Raise the rear of the vehicle and support it securely on jackstands.
2   Crawl under the vehicle and visually inspect the driveshaft. Look for any dents or cracks in the tubing. If any are found, the driveshaft must be replaced.
3   Check for any oil leakage at the front and rear of the driveshaft. Leakage where the driveshaft enters the transmission or transfer case indicates a defective rear transmission or transfer case seal. Leakage where the driveshaft enters the differential indicates a defective pinion seal. For these repair operations refer to Chapter 7 and Section 14, respectively.
4   While under the vehicle, have an assistant turn the rear wheel so the driveshaft will rotate. As it does, make sure the universal joints are operating properly without binding, noise or looseness.
5   The universal joint can also be checked with the driveshaft motionless, by gripping your hands on either side of the joint and attempting to twist the joint. Any movement at all in the joint is a sign of considerable wear. Lifting up on the shaft will also indicate movement in the universal joints.
6   Finally, check the driveshaft mounting bolts at the ends to make sure they are tight.
7   On 4WD models, the above driveshaft checks should be repeated on the front driveshaft. In addition, check for grease leakage around the sleeve yoke, indicating failure of the yoke seal.
8   Check for leakage where the driveshafts connect to the transfer case and front differential. Leakage indicates worn oil seals.
9   Also check for leakage at the ends of the axle housings (at the drum brake backing plates), which would indicate a defective axle seal.

12   Driveshafts - removal and installation

Note: *Whenever a driveshaft is removed, new U-joint straps must be used during installation.*

Rear driveshaft

Refer to illustration 12.3

Removal

1   Disconnect the negative cable from the battery.
2   Raise the vehicle and support it securely on jackstands. Place the transmission in Neutral with the parking brake off.
3   Using a scribe, white paint or a hammer and punch, place marks on the driveshaft and the differential yoke in line with each other **(see illustration)**. This is to make sure the driveshaft is reinstalled in the same position to preserve the balance. On vehicles with Selec-Trac 4WD (welded yokes at both ends of the driveshaft), also mark the relationship of the front driveshaft yoke to the transfer case yoke.
4   Remove the rear universal joint bolts and straps **(see illustration 12.3)**. Turn the driveshaft (or tires) as necessary to bring the bolts into the most accessible position. Remove the front U-joint bolts and straps on Selec-Trac models.
5   Tape the bearing caps to the spider to prevent the caps from coming off during removal.
6   Lower the rear of the driveshaft and then slide the front out of the transmission or transfer case.
7   To prevent loss of fluid and protect against contamination while the driveshaft is out, wrap a plastic bag over the transmission or transfer case housing and hold it in place with a rubber band (2WD and Command-Trac 4WD vehicles only).

Installation

8   Remove the plastic bag from the transmission or transfer case and wipe the area clean. Inspect the oil seal carefully. Procedures for replacement of this seal can be found in Chapter 7.
9   Slide the front of the driveshaft into the

transmission (2WD) or transfer case (Command-Trac models only). On Selec-Trac equipped vehicles, connect the front of the driveshaft to the transfer case yoke (making sure the match marks line up) and install the U-joint straps and bolts. Tighten the bolts to the specified torque.
10   Raise the rear of the driveshaft into position, checking to be sure the marks are in alignment. If not, turn the rear wheels to match the pinion flange and the driveshaft.
11   Remove the tape securing the bearing caps and install new straps and bolts. Tighten the bolts to the specified torque.

Front driveshaft (4WD)

Refer to illustration 12.12

12   Using white paint, chalk or a scribe, mark the relationship of the upper universal joint flange to the transfer case companion flange **(see illustration)**. Also mark the relationship of the lower U-joint to the front differential pinion shaft yoke.
13   Remove the lower U-joint strap bolts. Unbolt the upper U-joint flange from the transfer case companion flange **(see illustration 12.12)** and remove the shaft from the vehicle. Tape the bearing caps to the spider to prevent the caps from falling off.
14   Installation is the reverse of the removal procedure. Be sure to tighten the fasteners to the specified torque.

13   Universal joints - replacement

Refer to illustrations 13.2, 13.4 and 13.9
Note: *A press or large vise will be required for this procedure. It may be a good idea to take the driveshaft to a repair or machine shop where the universal joints can be replaced for you, normally at a reasonable charge.*
1   Remove the driveshaft as outlined in the previous Section.

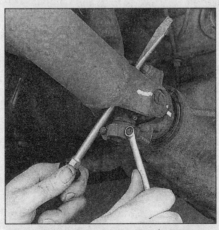

12.3 Before removing the bolts, mark the relationship of the driveshaft yoke to the differential pinion shaft yoke - to prevent the driveshaft from turning when loosening the flange bolts, insert a screwdriver through the yoke

12.12 On the upper U-joint, mark the relationship of the U-joint joint flange to the transfer case companion flange - when removing the bolts, insert a screwdriver through the U-joint to prevent the driveshaft from turning

8

13.2 A pair of needle-nose pliers can be used to remove the universal joint snap-rings

13.4 To press the universal joint out of the driveshaft yoke, set it up in a vise with the small socket pushing the joint and bearing cap into the large socket

13.9 If the snap-ring will not seat in the groove, strike the yoke with a hammer. This will relieve tension that has set up in the yoke, and slightly spring the yoke ears. This should also be done if the joint feels tight when assembled.

Single cardan U-joint (front or rear driveshaft)

2   Using a small pair of pliers, remove the snap-rings from the spider **(see illustration)**.
3   Supporting the driveshaft, place it in position on either an arbor press or on a workbench equipped with a vise.
4   Place a piece of pipe or a large socket with the same inside diameter over one of the bearing caps. Position a socket which is of slightly smaller diameter than the cap on the opposite bearing cap **(see illustration)** and use the vise or press to force the cap out (inside the pipe or large socket), stopping just before it comes completely out of the yoke. Use the vise or large pliers to work the cap the rest of the way out.
5   Transfer the sockets to the other side and press the opposite bearing cap out in the same manner.
6   Pack the new universal joint bearings with grease. Ordinarily, specific instructions for lubrication will be included with the universal joint servicing kit and should be followed carefully.
7   Position the spider in the yoke and partially install one bearing cap in the yoke. If the replacement spider is equipped with a grease fitting, be sure it's offset in the proper direction (toward the driveshaft).
8   Start the spider into the bearing cap and then partially install the other cap. Align the spider and press the bearing caps into position, being careful not to damage the dust seals.
9   Install the snap-rings. If difficulty is encountered in seating the snap-rings, strike the driveshaft yoke sharply with a hammer. This will spring the yoke ears slightly and allow the snap-rings to seat in the groove **(see illustration)**.
10   Install the grease fitting and fill the joint with grease. Be careful not to overfill the joint, as this could blow out the grease seals.
11   Install the driveshaft. Tighten the flange bolts to the specified torque.

Double cardan U-joint (upper joint on some front driveshafts)

12   Use the above procedure, but note that it will have to be repeated because the double cardan joint is made up of two single cardan joints. **Note:** *The double cardan joint must be replaced as an assembly - don't attempt to replace half of it, even if only one cross-and-roller assembly is worn.*

Double offset constant velocity joint (upper joint on some front driveshafts)

13   As of the time of writing, no information was available on this type of joint. Consult your local Jeep dealer service department.

---

14   Pinion oil seal - replacement

Refer to illustrations 14.6a and 14.6b
1   A pinion shaft oil seal failure results in the leakage of differential gear lubricant past the seal and onto the driveshaft yoke or

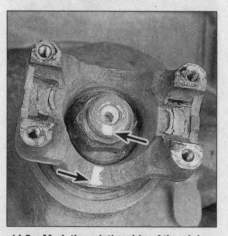

14.6a Mark the relationship of the pinion shaft to the pinion shaft yoke

flange. The seal is replaceable without removing or disassembling the differential.
2   Loosen the rear wheel lug nuts, raise the vehicle and place it on jackstands.
3   Remove the rear wheels and brake drums.
4   Disconnect the driveshaft from the pinion shaft yoke (see Section 12).

Rear differential

5   Using an inch-pound torque wrench, slowly turn the pinion shaft nut and measure the torque required to turn the pinion.
6   Mark the relationship of the pinion shaft to the pinion shaft yoke **(see illustration)**. Hold the pinion shaft yoke with a large pair of adjustable pliers, then remove the nut **(see illustration)**.
7   Remove the pinion shaft yoke from the shaft, using a puller if necessary.
8   After noting what the visible side of the oil seal looks like, carefully pry it out of the differential with a screwdriver or pry bar. Be careful not to damage the splines on the pinion shaft.

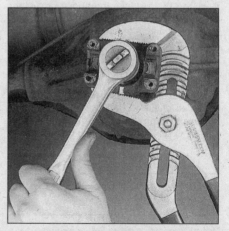

14.6b Hold the pinion shaft yoke stationary and loosen the nut

15.1 Before raising the vehicle, remove the cotter pin and nut lock, then loosen the axle hub nut

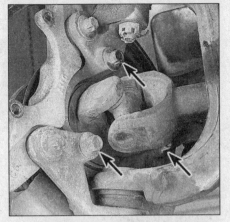

15.4a Remove the three bolts (arrows) (one is barely visible in this photo) that secure the hub assembly to the steering knuckle - a twelve-point socket will be necessary

15.4b Carefully tap the hub assembly out of the steering knuckle

9   Lubricate the new seal lip with multi-purpose grease or differential lubricant and carefully install it in position in the differential. Using a seal driver or a short section of pipe of the proper diameter and a hammer, carefully drive the seal into place.

10   Clean the sealing lip contact surface of the pinion shaft yoke. Apply a thin coat of multi-purpose grease to the seal contact surface and the shaft spines and, using a soft-faced hammer, tap the pinion shaft yoke onto the shaft, making sure the match-marks line up.

11   Install a new pinion shaft nut and, using the holder to hold the flange, tighten the nut just enough to eliminate all end-play in the pinion shaft.

12   Turn the pinion shaft yoke several times to seat the bearing.

13   Using a torque wrench, see how much torque is required to turn the pinion shaft. The desired preload is the previously recorded torque value plus five inch-pounds. If the preload is less than desired, tighten the nut in small increments until the desired preload is reached. **Caution:** *Don't exceed the recommended preload torque; if you do, the collapsible spacer on the pinion shaft will have to be replaced. Also, don't loosen the pinion nut in an attempt to reduce preload.* After the preload is properly adjusted, proceed to Step 17.

Front differential

14   Using the techniques described in Steps 6, 7 and 8, remove the pinion shaft nut and washer, pull the yoke from the shaft and remove the seal.

15   Following Steps 9 and 10, install the seal and pinion shaft yoke.

16   Lubricate the pinion shaft nut with multi-purpose grease. Install the washer and nut, then tighten the nut to the specified torque.

Front and rear differentials

17   Connect the driveshaft to the pinion shaft yoke (see Section 12).

18   Install the brake drums and wheels,

lower the vehicle to the ground and tighten the lug nuts to the specified torque (rear only).

19   Test drive the vehicle and check around the differential pinion shaft yoke for evidence of leakage.

---

15   Front axle hub and bearings - removal, service and installation

---

Refer to illustrations 15.1, 15.4a and 15.4b
Note: *This procedure applies to 4WD vehicles only. For the 2WD bearing servicing procedure, refer to Chapter 1.*

Removal

1   Pry the hubcap from the wheel, remove the cotter pin and nut lock then loosen the axle hub nut **(see illustration)**.

2   Loosen the front wheel lug nuts, raise the front of the vehicle and support it securely on jackstands. Remove the wheel.

3   Remove the disc brake caliper and disc (see Chapter 9).

4   Remove the axle hub nut. Unbolt the hub assembly from the steering knuckle and tap it out of the knuckle bore **(see illustrations)**. If the axle sticks in the hub splines, push it out of the hub with a puller.

Service

5   Due to the special tools and expertise required to separate and reassemble the hub and bearings, this job should be left to a dealer service department or repair shop.

Installation

6   Using sandpaper or emery cloth, clean the opening in the steering knuckle to remove any rust or dirt that may be present. Lubricate the axleshaft splines with wheel bearing grease. Smear the opening in the steering knuckle with wheel bearing grease and install the hub assembly, tightening the bolts to the specified torque.

7   Install the washer and hub nut, tightening the nut securely. Install the brake disc and caliper (see Chapter 9), mount the wheel and lower the vehicle. Tighten the lug nuts to the specified torque.

8   Tighten the axle hub nut to the specified torque and install the nut lock and a new cotter pin. Install the hubcap.

---

16   Front axleshafts - removal, overhaul and installation

Note: *This procedure applies to 4WD models only.*

Removal

Left or right outer axle

Refer to illustration 16.4

1   Following the procedure described in Section 15, remove the front axle hub and bearing assembly.

2   Remove the disc brake splash shield.

3   If the right side axle is being removed, follow the procedure described in Section 17 and remove the front axle shift motor.

4   Slide the axle straight out of the axle housing **(see illustration)**.

16.4 Once the hub and bearing assembly has been removed, the axleshaft can be pulled straight out of the housing

8

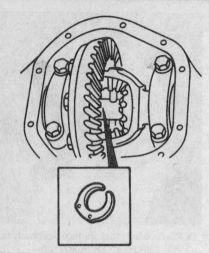

16.7 The right side intermediate axleshaft is held in the differential by a retaining clip

16.9a Exploded view of the front axleshaft U-joint assembly

Right side intermediate axle

Refer to illustration 16.7

5   Remove the right side outer axle (see above Steps).

6   Remove the differential cover (see the differential lubricant changing procedure in Chapter 1).

7   Remove the intermediate shaft retaining clip in the differential case **(see illustration)**.

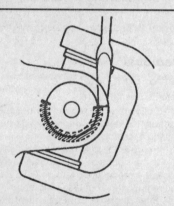

16.9b Push the snap-rings out of the groove in the U-joint bearings with a small screwdriver

8   Slide the intermediate shaft out of the axle housing.

U-joint overhaul

Single cardan design

Refer to illustrations 16.9a and 16.9b

9   Follow the U-joint replacement procedure in Section 13, but note that the snap-rings are inboard of the yoke ears and fit into grooves in the bearing caps **(see illustration)**. They are removed by driving them out with a screwdriver **(see illustration)**.

Double offset (constant velocity) design

Refer to illustrations 16.11a, 16.11b, 16.12a, 16.12b, 16.13, 16.14, 16.15a, 16.15b, 16.18 and 16.19

Note: *The constant velocity outer front axle joint is not repairable - if it becomes worn, noisy or breaks, the entire joint must be replaced. The joint can and should, however, be disassembled, cleaned and repacked with grease in the event of a boot failure.*

10   Cut both boot retaining clamps and discard them. Slide the boot back on the axleshaft.

11   Mount the axleshaft in a vise. The jaws of the vise must be lined with wood or rags to avoid damage to the axle. Using a hammer and a brass punch positioned on the inner race, knock the joint off the axleshaft **(see illustrations)**.

12   Remove the axleshaft from the vise. Place the joint assembly in the vise, with the stub axle pointing down. Again, wood or rags must line the vise jaws to prevent marring the

stub axle surface. Tap the inner race with the brass punch to angle it far enough to allow a ball bearing to be removed **(see illustration)**. Repeat this until all the balls are removed. If the balls are stuck, a screwdriver can be used to pry them from the cage **(see illustration)**.

13   With all of the balls removed, tilt the inner race and cage assembly 90-degrees, align the cage windows with the outer race (joint housing) lands and remove the assembly from the outer race **(see illustration)**.

14   Align the inner race lands with the cage windows and rotate the inner race out of the cage **(see illustration)**.

15   Wash all of the components in solvent and blow them dry with compressed air, if available. Inspect the cage and races for pitting, score marks, cracks and other signs of wear and damage **(see illustration)**. Shiny, polished spots are normal and don't affect CV joint operation **(see illustration)**.

16   Install the inner race in the cage by reversing the technique described in Step 14.

17   Install the inner race and cage assembly in the outer race by reversing the removal method used in Step 13. The small diameter of the cage must face out and the stopping groove in the inner race must face in.

18   Press the balls into the cage windows **(see illustration)**.

19   Pack the CV joint assembly with half of the lubricant supplied in the boot/joint kit through the inner splined hole. Force the grease into the bearing by inserting a wooden dowel through the splined hole and pushing it to the bottom of the joint **(see illustration)**. Repeat this procedure until the joint is thor-

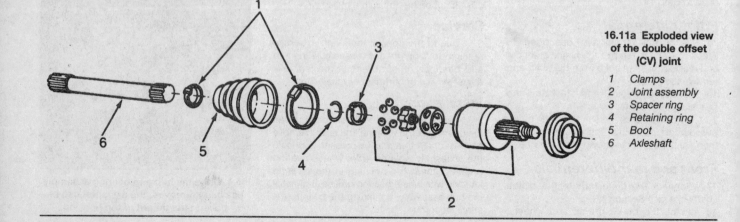

16.11a Exploded view of the double offset (CV) joint

1   Clamps
2   Joint assembly
3   Spacer ring
4   Retaining ring
5   Boot
6   Axleshaft

16.11b Dislodge the CV joint assembly with a brass punch and hammer (be careful not to let the joint fall!)

16.12a Tilt the inner race far enough to allow ball removal - a brass punch can be used if the inner race is difficult to move

16.12b If necessary, pry the ball bearings out with a screwdriver

16.13 Tilt the inner race and cage 90-degrees, then align the windows in the cage with the lands and rotate the inner race up and out of the outer race

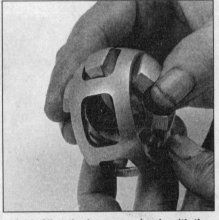

16.14 Align the inner race lands with the cage windows and rotate the inner race out of the cage

16.15a Check the inner race lands and grooves for pitting and score marks

oughly packed (the other half of the grease is to be placed in the boot).

20   Wrap the axleshaft splines with tape to avoid damaging the boot. Place the small boot clamp in the groove in the small end of the boot, then slide the boot onto the axleshaft. Remove the tape and smear the remainder of the grease into the boot.

21   Install the replacement retaining ring and spacer ring on the shaft.

22   Slide the CV joint onto the shaft until the inner race contacts the inner retaining ring.

23   Install the large boot clamp in its groove in the boot and slide the boot over the outer race (joint housing).

24   Tighten both boot clamps.

16.15b Check the cage for cracks, pitting and score marks (shiny spots are normal and don't affect operation)

16.18 Align the cage windows and the inner and outer race grooves, then tilt the cage and inner race to insert the balls

16.19 Apply grease through the splined hole, then insert a wooden dowel into the hole and push down - the dowel will force grease into the joint

8

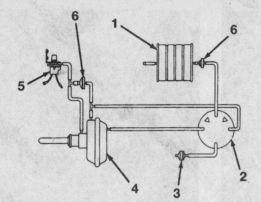

17.2a Command-Trac front axle shift motor vacuum diagram

1	Vacuum storage tank	5	Four-wheel drive indicator
2	Vacuum control switch		light and vacuum switch
3	Air vent filter	6	Vacuum check valves
4	Vacuum shift motor		

17.2b Selec-Trac front axle shift motor vacuum diagram

1	Mode selector (in	3	Vacuum shift motors
	console)	4	Vacuum check valves
2	Vacuum storage tank		

17.8 Unplug the vacuum harness connectors (1), then remove the four bolts that retain the shift motor housing (2)

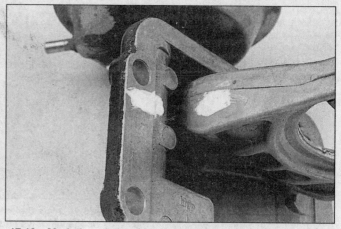

17.10a Mark the relationship of the shift fork to the housing so it will be reinstalled correctly

Installation

25   Installation is the reverse of the removal procedure. If the intermediate axle was removed, use a new gasket on the differential cover and fill the differential housing with the specified type of gear lubricant (see Chapter 1).

17   Front axle shift motor - check, removal and installation

Refer to illustrations 17.2a, 17.2b, 17.8, 17.10a and 17.10b

1   Raise the front of the vehicle and support it securely on jackstands.

Check

2   Disconnect the vacuum line from the shift motor **(see illustration)** and connect a hand-held vacuum pump to the front port of the motor.

3   Apply vacuum to the motor and rotate the right front wheel to ensure that the axle is fully disengaged. The shift motor should hold vacuum for at least 30 seconds. If it leaks,

replace it.

4   If the motor holds vacuum, connect the vacuum pump to the rear port on the shift motor. Plug the other ports and apply vacuum once again. It should hold vacuum for 30 seconds also. If not, replace the motor.

5   With vacuum still applied, remove the plug from the port where the vacuum line to the transfer case connects. Check for vacuum. If vacuum is not present, continue on to the next step.

6   Rotate the right front wheel to ensure that the axle has shifted completely. If it hasn't, remove the shift motor and check for freeness of the sliding shift collar.

7   Inspect the vacuum harness from the shift motor to the transfer case for kinks, cracks and other signs of damage. Check the vacuum harness connectors for a good, tight fit on the vacuum ports.

Removal

8   Unplug the vacuum harness from the shift motor **(see illustration)**.

9   Remove the shift motor housing-to-axle housing bolts and lift the shift motor housing, motor and fork from the axle housing.

Remove all traces of old gasket material.

10   Mark the relationship of the shift fork to the housing to return it to its original position upon reassembly **(see illustration)**. Rotate the shift motor and remove the shift fork and motor retaining snap-rings **(see illustration)**

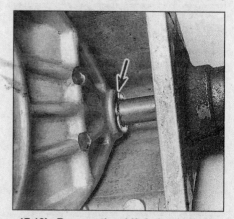

17.10b Remove the shift fork and motor retaining snap-rings (arrow is pointing to fork retaining snap-ring) and slide the motor out of the housing

19.2 Unscrew the four brake backing plate nuts using a socket on an extension passing through the hole in the axle flange

19.3 Pull the axle from the housing using a slide hammer and axle flange adapter

by pushing them down with two screwdrivers. Pull the motor out of the housing.

Installation

11   Remove the O-ring from the shift motor and install a new one if the same motor is to be reinstalled.

12   Slide the shift motor into the housing and install the shift fork, lining up the previously applied match-marks. Install the snaprings, making sure they are completely seated in their grooves.

13   Liberally coat the shift collar and axleshaft splines with wheel bearing grease. Engage the shift fork with the shift collar and set the assembly into position on the axle housing. Be sure to use a new gasket. Install the bolts and tighten them to the specified torque.

14   Connect the vacuum harness and check the front differential lubricant level (see Chapter 1).

18   Front axle assembly - removal and installation

Note: *This procedure is applicable to both 4WD and 2WD vehicles. If you are working on a 2WD vehicle, ignore references to the driveshaft, differential and shift motor.*

Removal

1   Loosen the front wheel lug nuts, raise the front of the vehicle and support it securely on jackstands positioned under the frame rails. Remove the front wheels.

2   Unbolt the front brake calipers and hang them out of the way with pieces of wire - don't let the calipers hang by the brake hose (see Chapter 9).

3   Remove the brake pads, anchor plates and brake discs (see Chapter 9).

4   Disconnect the vacuum harness from the front axle shift motor (see Section 17).

5   Mark the relationship of the front driveshaft to the front differential pinion shaft yoke, then disconnect the driveshaft from the yoke (see Section 12). Discard the U-joint straps.

6   Disconnect the stabilizer bar link from the bracket on the axle (see Chapter 10).

7   Disconnect the tie-rod and center link from the steering knuckle arms (see Chapter 10). Position them out of the way and hang them with pieces of wire from the underbody.

8   Unbolt the lower ends of the shock absorbers from the axle.

9   Remove the steering dampener (see Chapter 10).

10   Remove the ABS sensor (if equipped) (see Chapter 9 for general information).

11   Position a hydraulic jack under the differential. If two jacks are available, place one under the right side axle tube to balance the assembly (4WD models). On 2WD models, position the jack in the center of the axle.

12   Unbolt the upper and lower suspension arms from the axle (see Chapter 10).

13   Slowly lower the assembly to the ground.

Installation

14   Installation is the reverse of the removal procedure. When raising the axle into position, make sure the coil springs seat properly. Be sure to use new U-joint straps. For fasteners with specified torque values, be sure they are tightened to the specified torque.

19.5a Remove the inner axle seal from the housing using a slide hammer and internal puller jaw attachment

19   Rear axleshaft and bearing assembly - removal and installation

1989 and earlier models

Removal

Refer to illustrations 19.2, 19.3, 19.5a and 19.5b

1   Loosen the wheel lug nuts, raise the rear of the vehicle and support it securely on jackstands. Remove the wheel.

2   Remove the brake drum and brake backing plate retaining nuts **(see illustration)**.

3   Connect a slide hammer and adapter to the axle flange and pull the axle from the housing **(see illustration)**.

4   If the bearing must be replaced, take the assembly to a dealer service department or a repair shop to have the old bearing removed and a new one pressed on. If a new bearing is installed, be sure to also replace the outer axle seal (see next Step).

5   If the brake backing plate shows evidence of leaking differential lubricant, the outer axleshaft seal should be replaced. Remove the seal using the slide hammer with an internal puller jaw attachment **(see illus-**

8

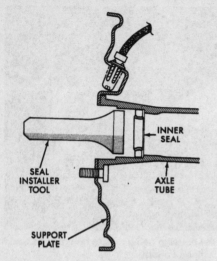

19.5b Lubricate the seal lip, then drive the seal into position with an installation tool or a piece of pipe with an outside diameter slightly smaller than the outside diameter of the seal

19.14 Push the axle flange in, then remove the C-lock from the inner end of the axleshaft

tration). Install the new seal using a seal driver or piece of pipe **(see illustration)**.

Installation

6   Wipe the bearing bore in the axle housing clean. Pack the bearing with wheel bearing grease and apply a thin coat of grease to the outer surface of the bearing.
7   Smear the lips of the inner seal with grease, then guide the axleshaft straight into the axle housing, being careful not to damage the seal.
8   Install the brake backing plate nuts, tightening them to the specified torque. Slide the brake drum over the axle flange.
9   Install the wheel and lug nuts, tightening the lug nuts securely. Lower the vehicle and tighten the lug nuts to the torque specified in Chapter 1.

1990 and later models
Removal
Refer to illustrations 19.12, 19.13 and 19.14
10   Loosen the wheel lug nuts, raise the

19.12 Remove the pinion shaft lock bolt

vehicle and support it securely on jackstands. Remove the wheel and brake drum (refer to Chapter 9).
11   Remove the cover from the differential carrier and allow the oil to drain into a container (see Chapter 1).
12   Remove the lock bolt from the differential pinion shaft **(see illustration)**.
13   Remove the pinion shaft **(see illustration)**. **Caution:** *Do not rotate the differential with the pinion shaft removed or the pinion gears will be dislodged.*
14   Push the outer (flanged) end of the axleshaft in and remove the C-lock from the inner end of the shaft **(see illustration)**. Withdraw the axleshaft.
15   Pry the oil seal out of the end of the axle housing with a seal removal tool or small prybar.
16   A bearing puller which grips the bearing from behind will be required for bearing removal. Attach a slide hammer to the puller and extract the bearing from the axle housing.

Installation

17   Clean out the bearing recess and drive in the new bearing with a bearing installer or a large socket positioned against the outer bearing race. Make sure the bearing is tapped in to the full depth of the recess and the numbers on the bearing are visible from the outer end of the axle housing.
18   Apply high-temperature grease to the oil seal recess and tap the new seal evenly into place with a hammer and seal installation tool, large socket or piece of pipe so the lips are facing in and the metal face is visible from the end of the axle housing. When correctly installed, the face of the oil seal should be flush with the end of the axle housing.
19   Lubricate the seal lip and the bore in the axle bearing with clean differential lubricant.
20   Install the axleshaft in the reverse of removal. Apply a non-hardening thread-locking compound to the threads of the lock bolt, then tighten the lock bolt to 14 ft-lbs. Tighten the wheel lug nuts to the torque listed in the Chapter 1 Specifications.
21   Clean off all traces of old gasket material from the differential cover and rear axle housing, then apply a bead of RTV sealant to the

19.13 Carefully remove the pinion shaft from the differential case (don't turn the differential case after the shaft has been removed, or the pinion gears may fall out)

cover. Install the cover and bolts, tightening the bolts in a criss-cross pattern to 35 ft-lbs.
22   Refill the axle with the correct quantity and grade of lubricant (see Chapter 1).
23   The remainder of installation is the reverse of removal.

20   Rear axle assembly - removal and installation

1   Loosen the rear wheel lug nuts, raise the vehicle and support it securely on jackstands placed underneath the frame. Remove the wheels.
2   Support the rear axle assembly with a floor jack placed underneath the differential.
3   Remove the shock absorber lower mounting nuts and compress the shocks to get them out of the way (see Chapter 10).
4   Disconnect the driveshaft from the differential pinion shaft yoke and hang the rear of the driveshaft from the underbody with a piece of wire (see Section 12).
5   Unbolt the stabilizer bar from the stabilizer bar link, if so equipped (see Chapter 10).
6   Disconnect the parking brake cables from the equalizer (see Chapter 9). Disconnect the height sensing proportioning valve control rod from the axle housing (Comanche models only).
7   Disconnect the flexible brake hose from the junction block on the rear axle housing. Plug the end of the hose or wrap a plastic bag tightly around it to prevent excessive fluid loss and contamination.
8   Remove the U-bolt nuts from the leaf spring tie plates (see Chapter 10).
9   Raise the rear axle assembly slightly, then unbolt the springs from the shackles (see Chapter 10) and lower the rear ends of the springs to the floor (only on models with the axle housing mounted above the spring).
10   Lower the jack and move the axle assembly out from under the vehicle.
11   Installation is the reverse of the removal procedure. Be sure to tighten the U-bolt nuts and the U-joint strap bolts to the specified torque.

Chapter 9   Brakes

Contents

Specifications

General

Brake fluid type	See Chapter 1

Disc brakes

Disc minimum thickness*	
1989 and earlier	0.815 in (20.7 mm)
1990	
4-wheel drive models	0.94 in (24.0 mm)
2-wheel drive models	0.866 in (22.0 mm)
1991 and later	
4-wheel drive models	0.89 in (22.7 mm)
2-wheel drive models	0.866 in (22.0 mm)
Disc runout (maximum)	0.004 in (0.10 mm)
Brake pad minimum thickness	See Chapter 1

Refer to marks cast in the disc (they supersede information printed here)

Drum brakes

Drum standard diameter	10.000 in (254 mm)
Drum maximum diameter*	10.060 in (255.5 mm)
Minimum brake lining thickness	See Chapter 1

Refer to marks cast in the drum (they supersede information printed here)

Torque specifications

	Ft-lbs (unless otherwise indicated)
Brake pedal pivot pin nut	
Manual transmission	20
Automatic transmission	26
Power brake booster pushrod nuts	
Inner nut	25
Outer nut	75 in-lbs
Power brake booster mounting nuts	30
Master cylinder mounting nuts	15
Caliper mounting pins	
1989 and earlier	30
1990 through 1993	
2WD	30
4WD	11
1994 and later	11
Caliper anchor plate bolts	77
Brake hose-to-caliper inlet fitting bolt	23
Wheel cylinder mounting bolts	90 in-lbs
Brake backing plate bolts	32

9

1　General information

Refer to illustration 1.2

The vehicles covered by this manual are equipped with hydraulically operated front and rear brake systems. The front brakes are disc type and the rear brakes are drum type. Both the front and rear brakes are self adjusting. The front disc brakes automatically compensate for pad wear, while the rear drum brakes incorporate an adjustment mechanism which is activated as the brakes are applied when the vehicle is driven in reverse.

Hydraulic system

The hydraulic system consists of two separate circuits **(see illustration)**. The master cylinder has separate reservoirs for the two circuits and in the event of a leak or failure in one hydraulic circuit, the other circuit will remain operative. A visual warning of circuit failure or air in the system is given by a warning light activated by displacement of the piston in the pressure differential switch portion of the combination valve from its normal "in balance" position.

Combination valve

A combination valve, located in the engine compartment below the master cylinder, consists of three sections providing the following functions. The metering section limits pressure to the front brakes until a predetermined front input pressure is reached and until the rear brakes are activated. There is no restriction at inlet pressures below 3 psi, allowing pressure equalization during non-braking periods. The proportioning section proportions outlet pressure to the rear brakes after a predetermined rear input pressure has been reached, preventing early rear wheel lock-up under heavy brake loads. The valve is also designed to assure full pressure to one brake system should the other system fail. The pressure differential warning switch incorporated into the combination valve is designed to continuously compare the front and rear brake pressure from the master cylinder and energize the dash warning light in the event of either a front or rear brake system failure. The design of the switch and valve are such that the switch will stay in the Warning position once a failure has occurred. The only way to turn the light off is to repair the cause of the failure and apply a brake pedal force of 450 psi.

Power brake booster (non-ABS models)

The power brake booster, utilizing engine manifold vacuum and atmospheric pressure to provide assistance to the hydraulically operated brakes, is mounted on the firewall in the engine compartment.

Parking brake

The parking brake operates the rear brakes only, through cable actuation. It's

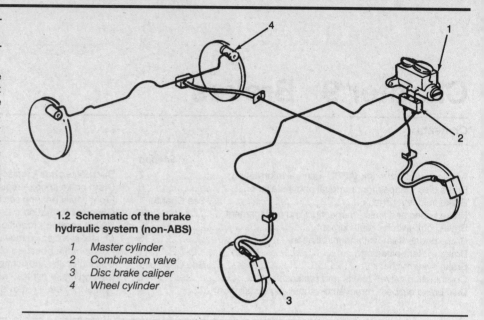

1.2 Schematic of the brake hydraulic system (non-ABS)

1　*Master cylinder*
2　*Combination valve*
3　*Disc brake caliper*
4　*Wheel cylinder*

activated by a lever mounted in the center console (Cherokee models) or a pedal mounted on the left side kick panel (Comanche models).

Service

After completing any operation involving disassembly of any part of the brake system, always test drive the vehicle to check for proper braking performance before resuming normal driving. When testing the brakes, perform the tests on a clean, dry flat surface. Conditions other than these can lead to inaccurate test results.

Test the brakes at various speeds with both light and heavy pedal pressure. The vehicle should stop evenly without pulling to one side or the other. Avoid locking the brakes because this slides the tires and diminishes braking efficiency and control of the vehicle.

Tires, vehicle load and front-end alignment are factors which also affect braking performance.

2　Anti-lock brake system (ABS) - general information

Refer to illustrations 2.2, 2.4 and 2.5

The anti-lock brake system was introduced in 1989 and is designed to maintain vehicle steerability, directional stability and optimum deceleration under severe braking conditions and on most road surfaces. It does so by monitoring the rotational speed of each wheel and controlling the brake line pressure to each wheel during braking. This prevents the wheel from locking-up and provides maximum vehicle controllability.

Components

Actuation assembly

The actuation assembly consists of the master cylinder/power booster unit, an elec-

tric booster pump and motor assembly (which contains an accumulator), a pressure modulator and an accumulator and pressure switch assembly **(see illustration)**.

a) *The electric pump provides hydraulic pressure to charge the accumulator, which supplies pressure to the braking system. The pump and accumulator are mounted to the firewall on the righthand side.*

b) *The pressure modulator mounts to the side of the master cylinder and power booster unit and modulates brake line pressure during ABS operation. The valve body contains three valves - one for each front wheel and one for the both of the rear wheels combined.*

Wheel sensors

These sensors are located at each wheel and generate small electrical pulsations when the toothed sensor rings are turning, sending a signal to the electronic controller, indicating wheel rotational speed.

The front wheel sensors **(see illustration)** are mounted to the steering knuckles in close relationship to the tone wheels, which are integral with the front axleshafts.

The rear wheel sensors bolt to the drum brake backing plates **(see illustration)**. The tone wheels are integral with the rear axleshafts.

Electronic controller

The electronic controller is mounted under the rear seat and is the "brain" for the ABS system. The function of the control module - consisting of microprocessors, a mercury switch and the related circuits needed for their operation - is to accept and process information received from the wheel speed sensors in conjunction with the signal from the internal mercury switch to control the hydraulic line pressure, avoiding wheel lock-up. The controller also constantly monitors the system, even under normal driving

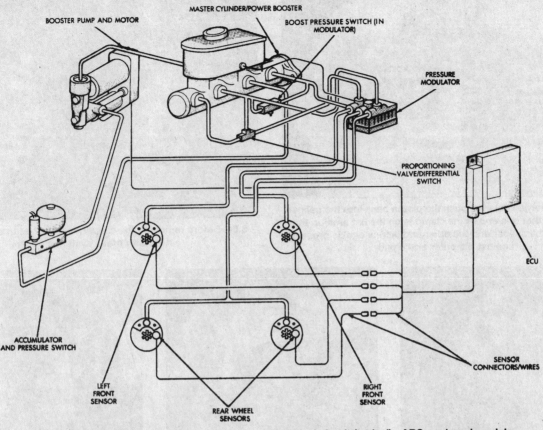

BOOSTER PUMP AND MOTOR

MASTER CYLINDER/POWER BOOSTER

BOOST PRESSURE SWITCH (IN MODULATOR)

PRESSURE MODULATOR

PROPORTIONING VALVE/DIFFERENTIAL SWITCH

ECU

SENSOR CONNECTORS/WIRES

ACCUMULATOR AND PRESSURE SWITCH

LEFT FRONT SENSOR

REAR WHEEL SENSORS

RIGHT FRONT SENSOR

2.2 Schematic of the brake hydraulic system and control circuit (typical) - ABS-equipped models

conditions, to find faults within the system.

If a problem develops within the system, a yellow or red light will glow on the dashboard. A diagnostic code will also be stored in the controller, which, when retrieved by a service technician, will indicate the problem area or component.

Diagnosis and repair

If a dashboard warning light comes on and stays on while the vehicle is in operation, the ABS system requires attention. Although a special electronic ABS diagnostic tester is necessary to properly diagnose the system, the home mechanic can perform a few preliminary checks before taking the vehicle to a dealer who is equipped with this tester.

a) *Check the brake fluid level in the reservoir.*

b) *Lift up the rear seat cushion and check that the controller electrical connector is securely connected.*

c) *Check the electrical connectors at the pump motor, accumulator pressure switch and pressure modulator.*

d) *Check the fuses.*

e) *Follow the wiring harness to each wheel and check that all connections are secure and that the wiring is not damaged.*

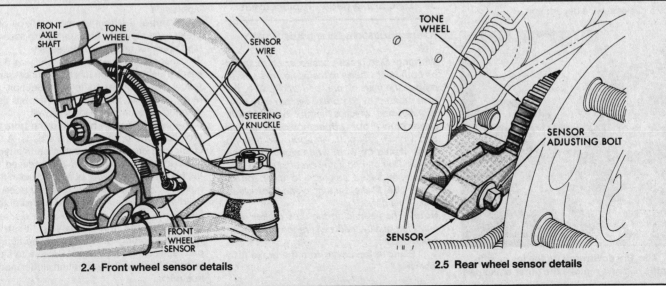

FRONT AXLE SHAFT

TONE WHEEL

SENSOR WIRE

STEERING KNUCKLE

FRONT WHEEL SENSOR

2.4 Front wheel sensor details

TONE WHEEL

SENSOR ADJUSTING BOLT

SENSOR

2.5 Rear wheel sensor details

9

3.5 Using a large C-clamp, push the piston back into the caliper bore - note that one end of the clamp is on the flat area on the backside of the caliper and the other end (screw end) is pressing against the outer brake pad

3.6a Before removing the caliper, wash off all traces of brake dust with brake system cleaner

3.6b Using an Allen wrench, unscrew the two caliper mounting pins

3.6c Swing the upper end of the caliper out of the anchor plate, then remove the caliper completely. Take this opportunity to check for fluid leakage around the caliper piston boot, which would indicate the need to overhaul the calipers (see Section 4)

3.6d Once the caliper is removed from the anchor plate, hang it from the coil spring with a piece of wire - DON'T let it hang by the brake hose!

If the above preliminary checks do not rectify the problem, the vehicle should be diagnosed by a dealer service department. Due to the rather complex nature of this system and the high operating pressures involved, all actual repair work must be done by a dealer service department.

3.6e Pry downward on the lower anti-rattle clip and remove the outer brake pad

3   Disc brake pads - replacement

Refer to illustrations 3.5 and 3.6a through 3.6k

Warning: Disc brake pads must be replaced on both front wheels at the same time - never replace the pads on only one wheel. Also, the dust created by the brake system may contain asbestos, which is harmful to your health. Never blow it out with compressed air and don't inhale any of it. An approved filtering mask should be worn when working on the brakes. Do not, under any circumstances, use petroleum based solvents to clean brake parts. Use brake system cleaner or clean brake fluid only!

Note: When servicing the disc brakes, use only high quality, nationally recognized name brand pads.

1   Remove the cover from the brake fluid reservoir.

2   Loosen the wheel lug nuts, raise the front of the vehicle and support it securely on jackstands.

3   Remove the front wheels. Work on one brake assembly at a time, using the assembled brake for reference if necessary.

4   Inspect the brake disc carefully as outlined in Section 5. If machining is necessary, follow the information in that Section to remove the disc, at which time the pads can be removed from the calipers as well.

5   Push the piston back into the bore to provide room for the new brake pads. A C-clamp can be used to accomplish this **(see illustration)**. As the piston is depressed to the bottom of the caliper bore, the fluid in the master cylinder will rise. Make sure it doesn't overflow. If necessary, siphon off some of the fluid.

6   Follow **the accompanying illustrations, beginning with 3.6a**, for the actual pad replacement procedure. Be sure to stay in order and read the caption under each illustration.

3.6f Pull the anti-rattle clips away from the pad with your index fingers and force the inner brake pads out with your thumbs

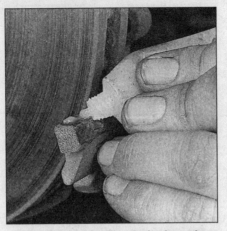

3.6g Before installing the brake pads, clean the sliding surfaces on the caliper and anchor plate and apply a thin coat of high-temperature grease to the anchor plate in the area shown

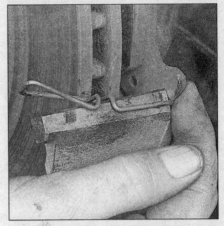

3.6h Position the anti-rattle clips on the anchor plate and install the inner brake pad

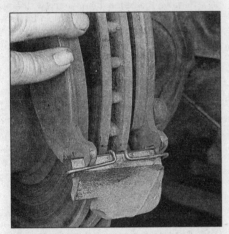

3.6i Place the lower end of the outer pad on the anchor plate and push it down against the anti-rattle clip, then lift up on the upper anti-rattle clip and swing the pad into position

3.6j Engage the notch in the lower end of the caliper with the anchor plate, then rotate it over the pads

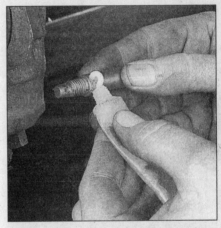

3.6k Lubricate the mounting pins with high-temperature grease, push them into the caliper and tighten them to the specified torque

7   When reinstalling the caliper, be sure to tighten the mounting pins to the specified torque. After the job has been completed, firmly depress the brake pedal a few times to bring the pads into contact with the disc.

8   Check for fluid leakage and make sure the brakes operate normally before driving in traffic.

---

4   Disc brake caliper - removal, overhaul and installation

Warning: *Dust created by the brake system may contain asbestos, which is harmful to your health. Never blow it out with compressed air and don't inhale any of it. An approved filtering mask should be worn when working on the brakes. Do not, under any circumstances, use petroleum-based solvents to clean brake parts. Use brake system cleaner or clean brake fluid only!*

Note: *If an overhaul is indicated (usually because of fluid leakage) explore all options before beginning the job. New and factory rebuilt calipers are available on an exchange basis, which makes this job quite easy. If it's decided to rebuild the calipers, make sure a rebuild kit is available before proceeding. Always rebuild the calipers in pairs - never rebuild just one of them.*

Removal

Refer to illustration 4.4

1   Remove the cover from the brake fluid reservoir, siphon off two thirds of the fluid into a container and discard it.

2   Loosen the wheel lug nuts, raise the front of the vehicle and support it securely on jackstands. Remove the front wheels.

3   Bottom the piston in the caliper bore **(see illustration 3.5)**.

4   **Note:** *Do not remove the brake hose from the caliper if you are only removing the caliper. Remove the brake hose inlet fitting bolt and detach the hose* **(see illustration)**. *Have a rag handy to catch spilled fluid and wrap a plastic bag tightly around the end of the hose to pre-*

4.4 Location of the brake hose inlet fitting bolt (arrow) - when reinstalling the bolt, be sure to use new sealing washers on each side of the fitting to prevent fluid leaks

9

4.7 With the caliper padded to catch the piston, use compressed air to force the piston out of its bore - make sure your hands or fingers are not between the piston and caliper frame!

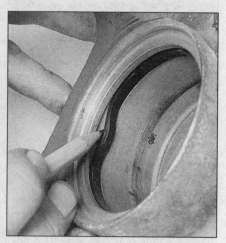

4.9 The piston seal should be removed with a plastic or wooden tool to avoid damage to the bore and seal groove. A pencil will do the job

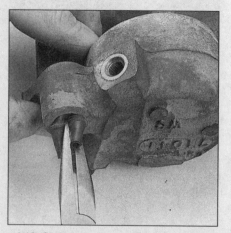

4.10 Grab the ends of the mounting pin bushings with needle-nose pliers and, using a twisting motion, push them through the caliper ears

4.8a Carefully pry the dust boot out of the caliper

vent fluid loss and contamination.

5   Unscrew the two caliper pins and detach the caliper from the vehicle (refer to Section 3 if necessary).

Overhaul

Refer to illustrations 4.7, 4.8a, 4.8b, 4.9, 4.10, 4.15 and 4.16

6   Clean the exterior of the caliper with brake cleaner or clean brake fluid. **Never use gasoline, kerosene or petroleum-based cleaning solvents. Place the caliper on a clean workbench.**

7   Position a wooden block or several shop rags in the caliper as a pad, then use compressed air to remove the piston from the caliper **(see illustration)**. Use only enough air pressure to ease the piston out of the bore. If the piston is blown out, even with the cushion in place, it may be damaged. **Warning:** *Never place your fingers in front of the piston in an attempt to catch or protect it when applying compressed air, as serious injury could occur.*

8   Carefully pry the dust boot out of the caliper bore **(see illustrations)**.

9   Using a wood or plastic tool, remove the

4.15 When installing the piston, make sure it doesn't become cocked in the caliper bore while pushing it down

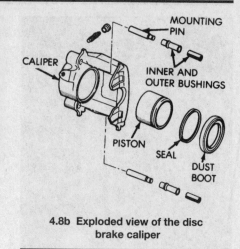

4.8b Exploded view of the disc brake caliper

piston seal from the groove in the caliper bore **(see illustration)**. Metal tools may cause bore damage.

10   Remove the caliper bleeder screw, then remove and discard the mounting pin bushings **(see illustration)**. Discard all rubber parts.

11   Clean the remaining parts with brake system cleaner or clean brake fluid, then blow them dry with compressed air.

12   Carefully examine the piston for nicks and burrs and loss of plating. If surface defects are present, the parts must be replaced.

13   Check the caliper bore in a similar way. Light polishing with crocus cloth is permissible to remove light corrosion and stains. Discard the caliper pins if they're corroded or damaged.

14   When assembling the caliper, lubricate the piston bore and seal with clean brake fluid. Position the seal in the caliper bore groove.

15   Lubricate the piston with clean brake fluid, install it squarely in the bore and apply pressure to bottom it in the caliper **(see illustration)**.

16   Stretch the dust boot over the groove in

4.16 If the correct seal driver tool isn't available, a drift punch can be used to tap around the circumference until the dust boot is seated

5.4a Use a dial indicator to check disc runout - if the reading exceeds the specified allowable runout limit, the disc will have to be machined or replaced

5.4b Using a swirling motion, remove the glaze from the disc surface with sandpaper or emery cloth

5.5a The minimum thickness is cast into the inside of the disc

5.5b Use a micrometer to measure disc thickness at several points

the piston, then carefully seat it in the caliper bore **(see illustration)**.

17 Install the bleeder screw.

18 Install new inner and outer mounting pin bushings.

Installation

19 Inspect the caliper pins for excessive corrosion, replacing them if necessary.

20 Clean the sliding surfaces of the caliper and the anchor plate **(see illustration 3.6g)**.

21 Install the caliper as described in **illustrations 3.6j and 3.6k**.

22 Install the brake hose and inlet fitting bolt, using new copper washers, then tighten the bolt to the specified torque.

23 If the line was disconnected, be sure to bleed the brakes (see Section 11).

24 Install the wheels and lower the vehicle.

25 After the job has been completed, firmly depress the brake pedal a few times to bring the pads into contact with the disc.

26 Check brake operation before driving the vehicle in traffic.

5 Brake disc - inspection, removal and installation

Inspection

Refer to illustrations 5.4a, 5.4b, 5.5a and 5.5b

1 Loosen the wheel lug nuts, raise the vehicle and support it securely on jackstands. Remove the wheel.

2 Remove the brake caliper as described in Section 3. It's not necessary to disconnect the brake hose. After removing the caliper mounting pins, suspend the caliper out of the way with a piece of wire as shown in illustrations 3.6d. Don't let the caliper hang by the hose and don't stretch or twist the hose. Reinstall two of the lug nuts to hold the disc in place (4WD models only).

3 Visually check the disc surface for score marks and other damage. Light scratches

and shallow grooves are normal after use and may not always be detrimental to brake operation, but deep score marks - over 0.015-inch (0.38 mm) - require disc removal and refinishing by an automotive machine shop. Be sure to check both sides of the disc. If pulsating has been noticed during application of the brakes, suspect disc runout. Be sure to check the wheel bearings to make sure they're properly adjusted.

4 To check disc runout, place a dial indicator at a point about 1/2-inch from the outer edge of the disc **(see illustration)**. Set the indicator to zero and turn the disc. The indicator reading should not exceed the specified allowable runout limit. If it does, the disc should be refinished by an automotive machine shop. **Note:** *Professionals recommend resurfacing of brake discs regardless of the dial indicator reading (to produce a smooth, flat surface that will eliminate brake pedal pulsations and other undesirable symptoms related to questionable discs). At the very least, if you elect not to have the discs resurfaced, deglaze them with sandpaper or emery cloth (use a swirling motion to ensure a nondirectional finish)* **(see illustration)**.

5 The disc must not be machined to a thickness less than the specified minimum refinish thickness. The minimum (or discard) thickness is cast into the inside of the disc **(see illustration)**. The disc thickness can be checked with a micrometer **(see illustration)**.

Removal

6 If the vehicle is a 4WD or 1993 and later 2WD, remove the two lug nuts that were installed to hold the disc in place, then slide the disc off the hub. If the vehicle is a 1992 and earlier 2WD, refer to Chapter 1, Front wheel bearing check, repack and adjustment for the hub/disc removal procedure.

Installation

7 On 2WD models, install the disc and hub assembly and adjust the wheel bearing (see Chapter 1). If the disc has been machined on a 2WD model, thoroughly wash

out the hub with solvent, then clean and repack the wheel bearing (see Chapter 1). On 4WD models simply place the disc over the wheel studs on the hub.

8 Install the caliper and brake pad assembly over the disc and position it on the steering knuckle (refer to Section 4 for the caliper installation procedure, if necessary). Tighten the caliper mounting pins to the specified torque.

9 Install the wheel, then lower the vehicle to the ground. Depress the brake pedal a few times to bring the brake pads into contact with the disc. Bleeding of the system will not be necessary unless the brake hose was disconnected from the caliper. Check the operation of the brakes carefully before placing the vehicle into normal service.

6 Drum brake shoes - replacement

Refer to illustrations 6.4a through 6.4x, 6.5, and 6.8

Warning: *Drum brake shoes must be replaced on both wheels at the same time - never replace the shoes on only one wheel. Also, the dust created by the brake system*

9

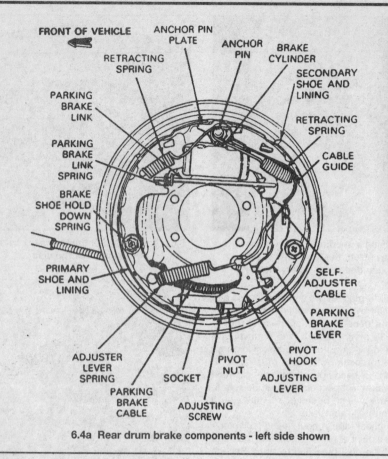

FRONT OF VEHICLE

ANCHOR PIN PLATE
ANCHOR PIN
BRAKE CYLINDER
RETRACTING SPRING
SECONDARY SHOE AND LINING
PARKING BRAKE LINK
RETRACTING SPRING
PARKING BRAKE LINK SPRING
CABLE GUIDE
BRAKE SHOE HOLD DOWN SPRING
PRIMARY SHOE AND LINING
SELF-ADJUSTER CABLE
PARKING BRAKE LEVER
ADJUSTER LEVER SPRING
PIVOT NUT
PIVOT HOOK
SOCKET
PARKING BRAKE CABLE
ADJUSTING SCREW
ADJUSTING LEVER

6.4a Rear drum brake components - left side shown

6.4b Before removing any internal drum brake components, wash them off with brake cleaner and allow them to dry - position a drain pan under the brake to catch the residue - DO NOT USE COMPRESSED AIR TO BLOW THE BRAKE DUST FROM THE PARTS!

brake shoes must be replaced at the same time, but to avoid mixing up parts, work on only one brake assembly at a time.

4   Follow the **accompanying illustrations (6.4a through 6.4x)** for the inspection and replacement of the brake shoes. Be sure to stay in order and read the caption under each illustration. **Note:** *If the brake drum cannot be easily pulled off the axle and shoe assembly, make sure that the parking brake is completely released, then apply some penetrating oil at the hub-to-drum joint. Allow the oil to soak in and try to pull the drum off. If the drum still cannot be pulled off, the brake shoes will have to be retracted. This is accomplished by first removing the plug from the backing plate. With the plug removed, pull the lever off the adjusting star wheel with one narrow screwdriver while turning the adjusting wheel with another narrow screwdriver, moving the shoes away from the drum* **(see illustration 6.7**, *if necessary). The drum should now come off.*

may contain asbestos, which is harmful to your health. Never blow it out with compressed air and don't inhale any of it. An approved filtering mask should be worn when working on the brakes. Do not, under any circumstances, use petroleum based solvents to clean brake parts. Use brake system cleaner or clean brake fluid only!
Caution: *Whenever the brake shoes are replaced, the retracting and hold-down springs should also be replaced. Due to the continuous heating/cooling cycle that the*

springs are subjected to, they lose their tension over a period of time and may allow the shoes to drag on the drum and wear at a much faster rat than normal. When replacing the rear brake shoes, use only high quality nationally recognized brand name parts.

1   Loosen the wheel lug nuts, raise the rear of the vehicle and support it securely on jackstands. Block the front wheels to keep the vehicle from rolling.
2   Release the parking brake.
3   Remove the wheel. **Note:** *All four rear*

6.4c Pull outward on the adjuster lever and turn the star wheel to retract the brake shoes

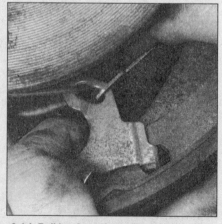

6.4d Pull back on the self-adjuster cable and push the adjusting lever toward the rear, unhooking it from the secondary brake shoe

6.4e Remove the primary and secondary shoe retracting springs - the spring removal tool shown here can be purchased at most auto parts stores and greatly simplifies this step

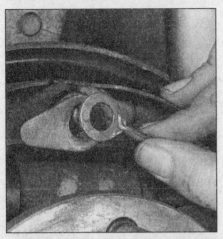

6.4f Remove the self-adjuster cable and anchor pin plate from the anchor pin

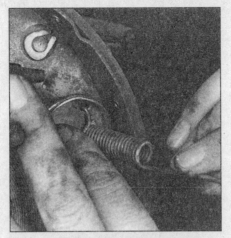

6.4g Remove the secondary shoe retracting spring and cable guide from the secondary shoe

6.4h Remove the primary shoe hold down spring and pin . . .

6.4i . . . then lift the primary shoe and adjusting screw from the backing plate

6.4j Remove the parking brake link

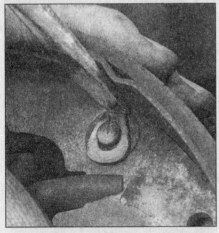

6.4k Remove the secondary shoe holddown spring and pin then lift the shoe from the backing plate - pry the parking brake lever retaining clip off of the pivot pin . . .

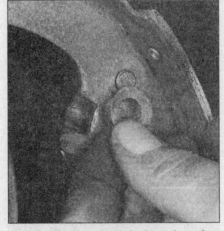

6.4l . . . then separate the lever from the secondary shoe - be careful not to lose the spring washer

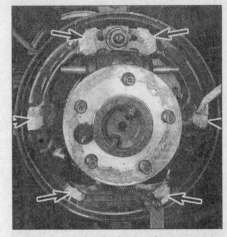

6.4m Lubricate the brake shoe contact areas (arrows) with high temperature grease

9

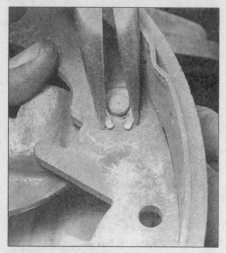

6.4n Attach the new shoe to the parking brake lever, install the spring washer and retaining clip on the pivot pin then crimp the clip closed with a pair of pliers

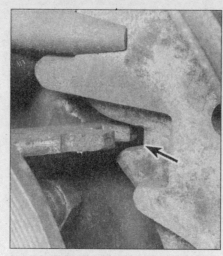

6.4o Install the secondary shoe and hold down spring to the backing plate, then position the end of the parking brake link into the notch (arrow)

6.4p Lubricate the adjusting screw with high temperature multi-purpose grease

6.4q Place the primary shoe against the backing plate, then install the holddown spring - make sure the parking brake strut and wheel cylinder pushrods engage in the brake shoe slots

6.4r Install the anchor pin plate . . .

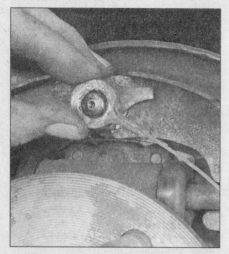

6.4s . . . and the self-adjuster cable

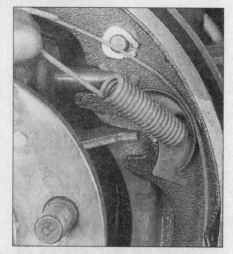

6.4t Hook the end of the secondary shoe retractor spring through the cable guide and into the hole in the shoe, then stretch the spring over the anchor pin

6.4u Install the primary shoe retractor spring - the tool shown here is available at most auto parts stores and makes this step much easier and safer

6.4v Hook the adjuster lever spring into the hole at the bottom of the primary shoe

6.4w Hook the adjuster lever spring and cable into the adjuster lever and pull the cable down and to the rear, inserting the hook on the lever into the hole in the secondary shoe

6.4x Wiggle the assembly to ensure the shoes are centered on the backing plate

6.5 The maximum allowable drum diameter is cast into the drum

6.7 Using a screwdriver or brake adjuster tool, turn the adjuster wheel in the direction shown until the shoes drag on the brake drum; then, insert a small screwdriver through the backing plate to move the adjuster lever away from the adjuster wheel and turn the wheel in the opposite direction until the drum turns freely

5   Before reinstalling the drum it should be checked for cracks, score marks, deep scratches and hard spots, which will appear as small discolored areas. If the hard spots cannot be removed with fine emery cloth or if any of the other conditions listed above exist, the drum must be taken to an automotive machine shop to have it turned. **Note:** *Professionals recommend resurfacing the drums whenever a brake job is done. Resurfacing will eliminate the possibility of out-of-round drums. If the drums are worn so much that they can't be resurfaced without exceeding the maximum allowable diameter (stamped into the drum)* **(see illustration)**, *then new ones will be required. At the very least, if you elect not to have the drums resurfaced, remove the glazing from the surface with medium-grit emery cloth using a swirling motion.*

6   Once the new shoes are in place, install the drums on the axle flanges. Remove the rubber plugs from the brake backing plates.

7   Insert a narrow screwdriver or brake adjusting tool through the adjustment hole and turn the star wheel until the brakes drag slightly as the drum is turned **(see illustration)**.

8   Turn the star wheel in the opposite direction until the drum turns freely. Keep the adjuster lever from contacting the star wheel or it won't turn. This can be done by pushing on it with a narrow screwdriver.

9   Repeat the adjustment on the opposite wheel and install the backing plate plugs.

10   Mount the wheel, install the lug nuts, then lower the vehicle. Tighten the lug nuts to the torque specified in Chapter 1.

11   Make a number of forward and reverse stops to allow the brakes to further adjust themselves.

12   Check brake operation before driving the vehicle in traffic.

7   Wheel cylinder - removal, overhaul and installation

Note: *If an overhaul is indicated (usually because of fluid leakage or sticky operation) explore all options before beginning the job. New wheel cylinders are available, which makes this job quite easy. If it's decided to rebuild the wheel cylinder, make sure that a rebuild kit is available before proceeding. Never overhaul only one wheel cylinder always rebuild both of them at the same time.*

Removal

Refer to illustration 7.4

1   Raise the rear of the vehicle and support it securely on jackstands. Block the front wheels to keep the vehicle from rolling.

2   Remove the brake shoe assembly (see Section 6).

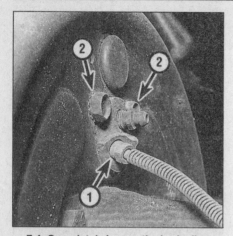

7.4 Completely loosen the brake line fitting (1), then remove the two wheel cylinder mounting bolts (2) (one is barely visible in this photo)

3   Remove all dirt and foreign material from around the wheel cylinder.
4   Completely loosen the brake line fitting **(see illustration)**. Don't pull the brake line away from the wheel cylinder.
5   Remove the wheel cylinder mounting bolts **(see illustration 7.4)**.
6   Detach the wheel cylinder from the brake backing plate and place it on a clean workbench. Immediately plug the brake line to prevent fluid loss and contamination. **Note:** *If the brake shoe linings are contaminated with brake fluid, install new brake shoes.*

Overhaul

Refer to illustration 7.7
7   Remove the bleeder screw, piston cups, pistons, boots and expander assembly from the wheel cylinder body **(see illustration)**.
8   Clean the wheel cylinder with brake fluid or brake system cleaner. **Warning:** *Do not, under any circumstances, use petroleum based solvents to clean brake parts!*
9   Use compressed air to remove excess fluid from the wheel cylinder and to blow out the passages.
10   Check the cylinder bore for corrosion and score marks. Crocus cloth can be used to remove light corrosion and stains, but the cylinder must be replaced with a new one if the defects cannot be removed easily, or if the bore is scored.
11   Lubricate the new cups with brake fluid.
12   Assemble the wheel cylinder components. Make sure the recessed sides of the cups face in.

Installation

13   Place the wheel cylinder in position and install the bolts.
14   Connect the brake line and tighten the fitting. Install the brake shoe assembly.
15   Bleed the brakes (see Section 11).
16   Check brake operation before driving the vehicle in traffic.

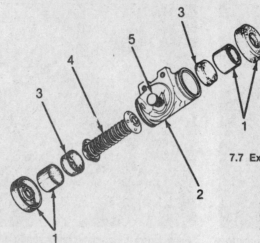

7.7 Exploded view of the wheel cylinder

1   *Piston and boot*
2   *Wheel cylinder body*
3   *Piston cups*
4   *Expander assembly*
5   *Bleeder screw*

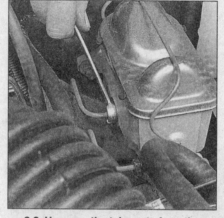

8.2 Unscrew the tube nuts from the master cylinder - use a flare-nut wrench (if one is available)

8   Master cylinder - removal, overhaul and installation

Note: *Before deciding to overhaul the master cylinder, investigate the availability and cost of a new or factory rebuilt unit and also the availability of a rebuild kit. This procedure does not apply to models equipped with ABS.*

Removal

Refer to illustration 8.2
1   Place rags under the brake line fittings and prepare caps or plastic bags to cover the ends of the lines once they are disconnected. **Caution:** *Brake fluid will damage paint. Cover all body parts and be careful not to spill fluid during this procedure.*
2   Loosen the tube nuts at the ends of the brake lines where they enter the master cylinder. To prevent rounding off the flats on these nuts, a flare-nut wrench, which wraps around the nut, should be used **(see illustration)**.
3   Pull the brake lines away from the master cylinder slightly and plug the ends to prevent contamination.
4   Remove the two master cylinder mounting nuts and remove the master cylinder from

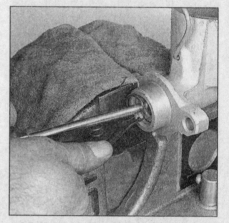

8.7 Use a Phillips head screwdriver to push the primary piston into the cylinder, then remove the snap-ring with a pair of snap-ring pliers

the vehicle.
5   Remove the reservoir cover, then discard any fluid remaining in the reservoir.

Overhaul

Refer to illustrations 8.7, 8.8, 8.9, and 8.13
6   Mount the master cylinder in a vise with the vise jaws clamping on the mounting flange.
7   Remove the primary piston snap-ring by

8.8 Remove the primary piston assembly from the cylinder

8.9 Eject the secondary piston assembly by applying compressed air to the secondary brake line port. While doing this, cover the opening in the cylinder with a piece of wood to serve as a shield - only apply enough air pressure to ease the piston out!

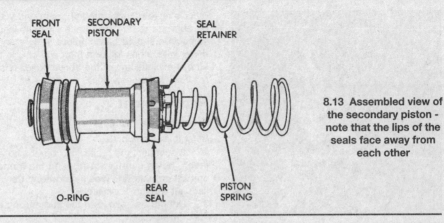

8.13 Assembled view of the secondary piston - note that the lips of the seals face away from each other

depressing the piston and extracting the ring with a pair of snap-ring pliers **(see illustration)**.

8   Remove the primary piston assembly from the cylinder bore **(see illustration)**. Discard it, since it can only be serviced as an assembly.

9   Reinstall the reservoir cover and remove the secondary piston assembly from the cylinder bore by applying compressed air to the secondary brake line port **(see illustration)**.

10   Inspect the cylinder bore for corrosion and damage. If any corrosion or damage is found, replace the master cylinder body with a new one, as abrasives cannot be used on the bore.

11   Remove the seals, O-ring, piston spring and seal retainer from the secondary piston. Discard them.

12   Clean the master cylinder body and secondary piston with clean brake fluid or brake cleaner.

13   Assemble the O-rings, seals, piston spring and seal retainer to the secondary piston **(see illustration)**. Make sure the seal lips face away from each other.

14   Lubricate the cylinder bore and primary and secondary piston assemblies with clean brake fluid. Install the secondary piston assembly into the cylinder.

15   Install the primary piston assembly in the cylinder bore, depress it and install the snap-ring.

16   Inspect the reservoir cover and diaphragm for cracks and deformation. Replace it if it's damaged in any way.

17   **Note:** *Whenever the master cylinder is removed, the complete hydraulic system must be bled. The time required to bleed the system can be reduced if the master cylinder is filled with fluid and bench bled (refer to Steps 18 through 22) before the master cylinder is installed on the vehicle.*

18   Insert threaded plugs of the correct size into the cylinder outlet holes and fill the reservoirs with brake fluid. The master cylinder should be supported in such a manner that brake fluid will not spill during the bench bleeding procedure.

19   Loosen one plug at a time, starting with the rear outlet port first, and push the piston assembly into the bore to force air from the master cylinder. To prevent air from being drawn back into the cylinder, the appropriate plug must be replaced before allowing the piston to return to its original position.

20   Stroke the piston three or four times for each outlet to ensure that all air has been expelled.

21   Since high pressure is not involved in the bench bleeding procedure, an alternative to the removal and replacement of the plugs with each stroke of the piston assembly is available. Before pushing in on the piston assembly, remove one of the plugs completely. Before releasing the piston, however, instead of replacing the plug, simply put your finger tightly over the hole to keep air from being drawn back into the master cylinder. Wait several seconds for the brake fluid to be drawn from the reservoir to the piston bore, then repeat the procedure. When you push down on the piston it will force your finger off the hole, allowing the air inside to be expelled. When only brake fluid is being ejected from the hole, replace the plug and go on to the other port.

22   Refill the master cylinder reservoir and install the diaphragm an cover assembly.

Installation

23   Carefully install the master cylinder by reversing the removal steps, then bleed the brakes (refer to Section 11).

9   Combination valve - check and replacement

Refer to illustration 9.1

Check

1   Disconnect the wire connector from the pressure differential switch **(see illustration)**. **Note:** *When unplugging the connector,*

squeeze the side lock releases, moving the inside tabs away from the switch, then pull up. Pliers may be used as an aid if necessary.

2   Using a jumper wire, connect the switch wire to a good ground, such as the engine block.

3   Turn the ignition key to the On position. The warning light in the instrument panel should light.

4   If the warning light does not light, either the bulb is burned out or the electrical circuit is defective. Replace the bulb (refer to Chapter 12) or repair the electrical circuit as necessary.

5   When the warning light functions correctly, turn the ignition switch off, disconnect the jumper wire and reconnect the wire to the switch terminal.

6   Make sure the master cylinder reservoirs are full, then attach a bleeder hose to one of the rear wheel bleeder valves and immerse the other end of the hose in a container partially filled with clean brake fluid.

7   Turn the ignition switch on.

8   Open the bleeder valve while a helper applies moderate pressure to the brake pedal. The brake warning light on the instrument panel should light.

9   Close the bleeder valve before the helper releases the brake pedal.

9.1 Location of the combination valve - master cylinder removed for clarity - the arrow shows the location of the wire connector

9

10.2 To remove the brake hose, unscrew the brake line from the hose fitting, using a flare-nut wrench; then unbolt the hose from the frame, using a Torx socket or screwdriver

10 Reapply the brake pedal with moderate to heavy pressure. The brake warning light should go out.
11 Attach the bleeder hose to one of the front brake bleeder valves and repeat Steps 8 through 10. The warning light should react in the same manner as in Steps 8 and 10.
12 Turn the ignition switch off.
13 If the warning light did not come on in Steps 8 and 11, but does light when a jumper is connected to ground, the warning light switch portion of the combination valve is defective and the combination valve must be replaced with a new one since the components of the combination valve are not individually serviceable.

Replacement

14 Place a container under the combination valve and protect all painted surfaces with newspapers or rags.
15 Disconnect the hydraulic lines at the combination valve, using a flare-nut wrench, if available. Plug the lines to prevent further loss of fluid and to protect the lines from contamination.
16 Disconnect the electrical connector from the pressure differential switch.
17 Remove the bolt holding the valve to the mounting bracket and remove the valve from the vehicle.
18 Installation is the reverse of the removal procedure.
19 Bleed the entire brake system.

10   Brake hoses and lines - inspection and replacement

Inspection

1 About every six months, with the vehicle raised and supported securely on jackstands, the rubber hoses which connect the steel brake lines with the front and rear brake assemblies should be inspected for cracks, chafing of the outer cover, leaks, blisters and other damage. These are important and vulnerable parts of the brake system and inspection should be complete. A light and mirror will be helpful for a thorough check. If a hose exhibits any of the above conditions, replace it with a new one.

Replacement

Front brake hose

Refer to illustration 10.2
2 Disconnect the brake line from the hose fitting, being careful not to bend the frame bracket or brake line **(see illustration)**. Use a flare nut wrench, if available.
3 Unbolt the hose from the frame using a Torx drive socket.
4 Remove the inlet fitting bolt from the brake caliper **(see illustration 4.4**, if necessary) and separate the hose from the caliper. Discard the sealing washers.
5 To install the hose, first attach it to the caliper, using new sealing washers on both sides of the fitting. Tighten the inlet fitting bolt to the specified torque.
6 Without twisting the hose, connect the brake line to the hose fitting, but don't tighten it yet.
7 Install the bolt retaining the hose to the frame, tightening it securely.
8 Tighten the brake line fitting securely.
9 When the brake hose installation is complete, there should be no kinks in the hose. Make sure the hose doesn't contact any part of the suspension. Check this by turning the wheels to the extreme left and right positions. If the hose makes contact, remove it and correct the installation as necessary. Bleed the system (Section 11).

Rear brake hose

10 Using a back-up wrench, disconnect the hose at the frame bracket, being careful not to bend the bracket or steel lines.
11 Remove the U-clip with a pair of pliers and separate the female fitting from the bracket.
12 Disconnect the two hydraulic lines at the junction block, then unbolt and remove the hose.
13 Bolt the junction block to the axle housing and connect the lines, tightening them securely. Without twisting the hose, install the female end of the hose in the frame bracket.
14 Install the U-clip retaining the female end to the bracket.
15 Using a back-up wrench, attach the steel line fittings to the female fittings. Again, be careful not to bend the bracket or steel line.
16 Make sure the hose installation did not loosen the frame bracket. Tighten the bracket if necessary.
17 Fill the master cylinder reservoir and bleed the system (refer to Section 11).

Metal brake lines

18 When replacing brake lines be sure to use the correct parts. Don't use copper tubing for any brake system components. Purchase steel brake lines from a dealer or auto parts store.
19 Prefabricated brake line, with the tube ends already flared and fittings installed, is available at auto parts stores and dealers. These lines are also bent to the proper shapes.
20 When installing the new line, make sure it's securely supported in the brackets and has plenty of clearance between moving or hot components.
21 After installation, check the master cylinder fluid level and add fluid as necessary. Bleed the brake system as outlined in the next Section and test the brakes carefully before driving the vehicle in traffic.

11   Brake system bleeding

Refer to illustrations 11.8 and 11.15
Warning: *Wear eye protection when bleeding the brake system. If the fluid comes in contact with your eyes, immediately rinse them with water and seek medical attention.*
Note: *Bleeding the hydraulic system is necessary to remove any air that manages to find its way into the system when it's been opened during removal and installation of a hose, line, caliper or master cylinder.*

Vehicles with conventional (non-ABS) brakes

1 It will probably be necessary to bleed the system at all four brakes if air has entered the system due to low fluid level, or if the brake lines have been disconnected at the master cylinder.
2 If a brake line was disconnected only at a wheel, then only that caliper or wheel cylinder need be bled.
3 If a brake line is disconnected at a fitting located between the master cylinder and any of the brakes, that part of the system served by the disconnected line must be bled.
4 Remove any residual vacuum from the brake power booster by applying the brake several times with the engine off.
5 Remove the master cylinder reservoir cover and fill the reservoir with brake fluid. Reinstall the cover. **Note:** *Check the fluid level often during the bleeding operation and add fluid as necessary to prevent the fluid level from falling low enough to allow air bubbles into the master cylinder.*
6 Have an assistant on hand, as well as a supply of new brake fluid, a clear container partially filled with clean brake fluid, a length of 3/16-inch plastic, rubber or vinyl hose to fit over the bleeder valve and a wrench to open and close the bleeder valve.
7 Beginning at the right rear wheel, loosen the bleeder valve slightly, then tighten it to a point where it is snug but can still be loosened quickly and easily.
8 Place one end of the hose over the bleeder valve and submerge the other end in brake fluid in the container **(see illustration)**.
9 Have the assistant pump the brakes slowly a few times to get pressure in the sys-

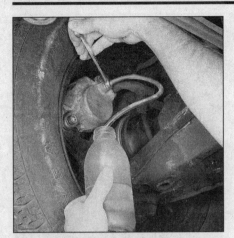

11.8 When bleeding the brakes, a hose is connected to the bleeder valve at the caliper or wheel cylinder and then submerged in brake fluid. Air will be seen as bubbles in the tube and container. All air must be expelled before moving to the next wheel.

tem, then hold the pedal firmly depressed.

10   While the pedal is held depressed, open the bleeder valve just enough to allow a flow of fluid to leave the valve. Watch for air bubbles to exit the submerged end of the tube. When the fluid flow slows after a couple of seconds, close the valve and have your assistant release the pedal.

11   Repeat Steps 9 and 10 until no more air is seen leaving the tube, then tighten the bleeder valve and proceed to the left rear wheel, the right front wheel and the left front wheel, in that order, and perform the same procedure. Be sure to check the fluid in the master cylinder reservoir frequently.

12   Never use old brake fluid. It contains moisture which will deteriorate the brake system components.

13   Refill the master cylinder with fluid at the end of the operation.

14   Check the operation of the brakes. The pedal should feel solid when depressed, with no sponginess. If necessary, repeat the entire process. **Warning:** *Do not operate the vehicle if you are in doubt about the effectiveness of the brake system.*

Vehicles with ABS brakes

15   If equipped with ABS brakes, begin by bleeding the system at the master cylinder/modulator **(see illustration)**. This is done by loosening and tightening the brake line fittings, one at a time, while an assistant operates the brake pedal. The ignition must be on.

16   The remaining bleeding operations are the same as for conventional brakes. However, note the following:

a) *It is imperative that no dirt enter the system with the reservoir cap off.*

b) *Insure that the fluid in the master cylinder does not drop too low during bleeding, since this will lead to severe pump damage.*

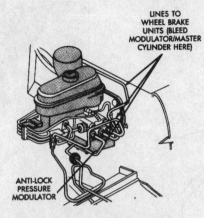

11.15 Master cylinder/modulator bleeding points

12   Power brake booster - check, removal and installation

Refer to illustrations 12.2, 12.3 and 12.9

Operating check

1   Depress the brake pedal several times with the engine off and make sure that there is no change in the pedal reserve distance.

2   Depress the pedal and start the engine. If the pedal goes down slightly, operation is normal **(see illustration)**.

Air tightness check

3   Start the engine and turn it off after one or two minutes. Depress the brake pedal several times slowly. If the pedal goes down farther the first time but gradually rises after the second or third depression, the booster is air tight **(see illustration)**.

4   Depress the brake pedal while the engine is running, then stop the engine with the pedal depressed. If there is no change in the pedal reserve travel after holding the pedal for 30 seconds, the booster is air tight.

Removal

5   Power brake booster units should not

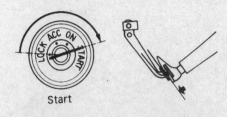

12.2 Push down on the brake pedal, then start the engine - the brake pedal should go down slightly, indicating normal booster operation

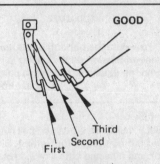

12.3 With the engine turned off the pedal should gradually rise with each pump if the booster is functioning properly

be disassembled. They require special tools not normally found in most service stations or shops. They are fairly complex and because of their critical relationship to brake performance it is best to replace a defective booster unit with a new or rebuilt one.

6   To remove the booster, first unbolt the brake master cylinder from the booster and carefully pull it forward.

7   Disconnect the hose leading from the engine to the booster. Be careful not to damage the hose when removing it from the booster fitting.

8   Locate the pushrod connecting the booster to the brake pedal. This is accessible from the interior in front of the driver's seat.

9   Disconnect the brake light switch wire

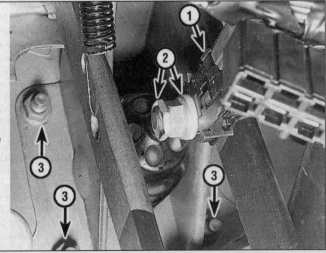

12.9 Disconnect the wire connector from the brake light switch (1), then remove the two booster pushrod nuts (2) and slide out the pushrod bolt - the booster is fastened to the firewall by four nuts; three of which are visible in this photo (3)

9

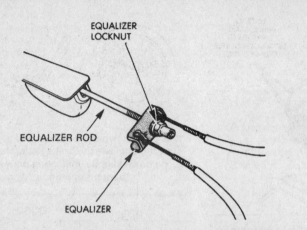

13.4 To adjust the parking brake, turn the equalizer locknut until the desired tension on the cables is reached; it may be necessary to hold the equalizer rod with a pair of locking pliers to prevent it from turning

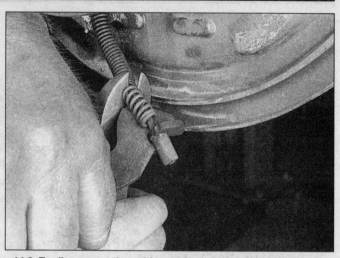

14.3 To disconnect the cable end from the parking brake lever, pull back on the return spring and maneuver the cable out of the slot in the lever

connector, remove the pushrod bolt nuts and pull out the bolt **(see illustration)**. Discard the bolt and nuts and obtain new ones of the same part number for installation.

10   Remove the four nuts holding the brake booster to the firewall. You may need a light to see these, as they are up under the dash area **(see illustration 12.9)**.

11   Slide the booster straight out from the firewall until the studs clear the holes and pull the booster, brackets and gaskets from the engine compartment area.

Installation

Warning: *When installing the booster pushrod to brake pedal bolt, insert it from the right side on manual transmission models and from the left side on automatic transmission models. In either case, the bolt must first pass through the brake pedal, THEN the pushrod* **(see illustration 15.3 if necessary)**.

12   Installation procedures are basically the reverse of those for removal. Tighten the pushrod nuts, booster mounting nuts and the master cylinder mounting nuts to the specified torque.

13 Parking brake - adjustment

Refer to illustration 13.4

1   The adjustment of the parking brake, often overlooked or put off by many motorists, is actually a fairly critical adjustment. If the parking brake cables are too slack, the brake won't hold the vehicle on an incline - if they're too tight, the brakes may drag, causing them to wear prematurely. Another detrimental side effect of a tightly adjusted parking brake cable is the restriction of the automatic adjuster assembly on the rear drum brakes, which will not allow them to function properly.

2   The first step in adjusting slack parking

14.4 Compress the retainer tangs (arrows) to free the cable and housing from the backing plate

brake cables is to ensure the correct adjustment of the rear drum brakes. This can be accomplished by making a series of forward and reverse stops (approximately 10 of them), which will bring the brake shoes into proper relationship with the brake drums.

3   Fully apply and release the parking brake four or five times, then set the parking brake lever or pedal to the fifth notch.

4   Raise the rear of the vehicle and support it securely on jackstands. While holding the equalizer rod with a pair of locking pliers, turn the equalizer locknut **(see illustration)** until the cables are fairly taut.

5   Release the parking brake and apply it, making sure it travels five to seven clicks. If it travels too far, tighten the equalizer locknut a little more. If the travel is less than five clicks, the locknut will have to be loosened.

6   After the parking brake has been properly adjusted, place the handle or pedal in the released position and rotate the rear wheels, making sure the brakes don't drag.

7   Lower the vehicle and test the operation of the parking brake on an incline.

14 Parking brake cables - replacement

Refer to illustrations 14.3, 14.4 and 14.5

1   Release the parking brake. Loosen the rear wheel lug nuts, raise the rear of the vehicle and support it securely on jackstands.

2   Remove the rear wheel and brake drum. Loosen the equalizer nut fully **(see illustration 13.4)**.

3   Following the procedure in Section 6, remove the brake shoes, then disconnect the parking brake cable end from the parking brake lever **(see illustration)**.

4   Compress the cable housing retainer tangs at the brake backing plate **(see illustration)** and push the cable and housing through the backing plate.

5   Remove the spring and clip from the cable housing **(see illustration)** and pry the housing out of the frame bracket, then disconnect the forward end of the cable from the slot in the equalizer **(see illustration 13.4)**.

6   Installation is the reverse of the removal

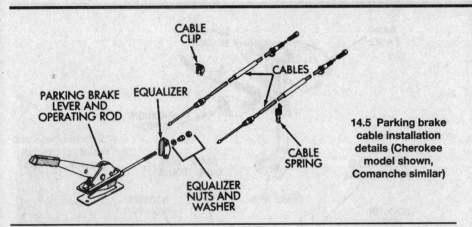

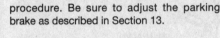

14.5 Parking brake cable installation details (Cherokee model shown, Comanche similar)

and skills required to diagnose and service the height sensing proportioning valve, it is not recommended that the home mechanic attempt the procedures. If servicing the system becomes necessary, take the vehicle to a dealer service department or other qualified repair shop.

Warning: *Any suspension modifications that alter the distance between the rear axle and the frame (such as the installation of air shocks, heavy duty springs or lift kits) will furnish the height sensing proportioning valve with a false reading. This could lead to inadequate braking characteristics, possibly resulting in an accident.*

procedure. Be sure to adjust the parking brake as described in Section 13.

15   Brake pedal - removal and installation

Refer to illustration 15.3

Removal

1   Disconnect the cable from the negative battery terminal.
2   Disconnect the power brake booster pushrod from the brake pedal (see Section 12).
3   Remove the nut from the right end of the pivot pin and slide the pin to the left just far enough to allow removal of the brake pedal **(see illustration)**.
4   Check the bushings in the top of the pedal for excessive wear, replacing them if necessary.

Installation

5   Apply a light coat of multi-purpose grease to the sleeve and bushings and push them into the top of the pedal.
6   Position the pedal in the bracket and slide the pivot pin to the right, making sure it passes through the pedal bushings. Install the nut and tighten it to the specified torque.
7   Connect the booster pushrod to the brake pedal, inserting a new bolt from the right side on models with a manual transmission and from the left side on automatic transmission models **(see illustration 15.3)**. Install new nuts and tighten them to the specified torques.
8   Connect the negative battery cable and test the brakes for proper operation.

16   Height sensing proportioning valve - general information

Note: *Only the Comanche models are equipped with this valve.*

The height sensing proportioning valve regulates hydraulic pressure to the rear brakes in accordance with the amount of weight present in the bed of the truck. When

the bed is empty, pressure to the rear brakes will be decreased to avoid locking up the wheels. When the bed contains a load, the valve senses the lower ride height and allows more hydraulic pressure to the rear brakes.

Due to the special tools, test equipment

17   Brake light switch - replacement

Refer to illustration 17.2
1   Disconnect the cable from the negative terminal of the battery.
2   From under the dash, unplug the brake

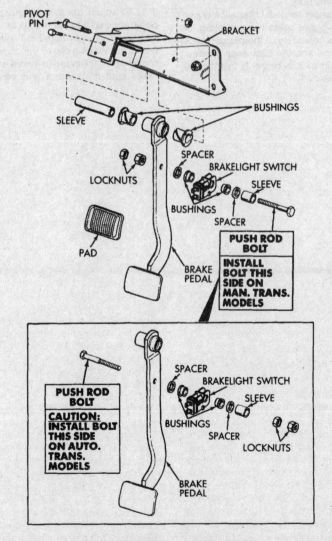

15.3 Brake pedal mounting details

light switch electrical connector and remove the pushrod bolt **(see accompanying illustration and illustration 15.3)**. Remove the brake light switch, noting the positions of the bushings, spacers and sleeves.

3   Lubricate the switch bushings, spacers and sleeves with a light coat of multi-purpose grease and position the switch against the brake pedal. Install the pushrod bolt from the correct side **(see illustration 15.3)** and install the nuts, tightening them to the specified torque.

4   Attach the electrical connector and reconnect the battery. The switch is not adjustable.

5   On pre-1991 models, the switch is not adjustable. On 1991 and later models, it is adjustable.

6   To check the switch adjustment on 1991 and later models, move the brake pedal forward by hand and watch the switch plunger. It should be fully extended at the point at which pedal freeplay is taken up and brake application begins. A clearance of about 1/8-inch should exist between the plunger and the pedal at this point.

7   If the plunger-to-pedal clearance is correct and the brake lights are operating, no adjustment is necessary. If the plunger doesn't fully extend and the clearance between the pedal and switch barrel is insufficient, adjust the switch.

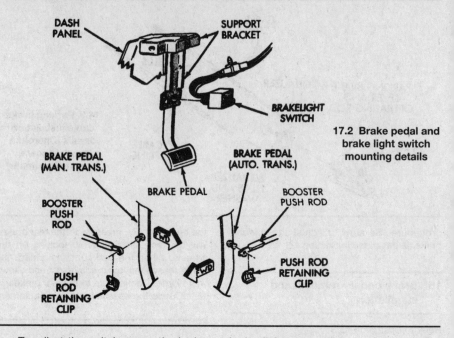

17.2 Brake pedal and brake light switch mounting details

8   To adjust the switch, grasp the brake pedal and pull it to the rear as far as possible. The switch plunger barrel will "ratchet" to the rear in its retaining clip to the correct position. Measure the plunger-to-pedal clearance to make sure it's correct and verify that the brake lights are still operating correctly. **Warning:** *Make SURE the brake pedal returns to its fully released position after adjustment. The switch can interfere with full pedal return if it's too far forward, resulting in dragging brakes caused by partial brake application.*

Chapter 10
Suspension and steering systems

Contents

Specifications

Torque Specifications
Ft-lbs (unless otherwise indicated)

Front Suspension
Track bar ballstud nut	62
Track bar-to-axle bracket nut	74
Front axle suspension arm bolts and nuts	
Upper arm	
Axle end	55
Frame end	66
Lower arm (both ends)	133
Steering knuckle-to-axle housing balljoint stud nuts	75
Brake caliper anchor plate bolts	77

Rear suspension
Leaf spring	
Front and rear mounting bolts	109
Tie plate U-bolt nuts	
Cherokee	52
Comanche	82

Steering
Airbag module	
Retaining nuts (1995 and 1996 models)	100 in-lbs
Retaining screws (1997 and later models)	90 in-lbs
Steering wheel-to-steering shaft hub nut	25
Intermediate shaft pinch bolts	33
Steering gear mounting bolts	65
Pitman arm nut	185
Center link-to-Pitman arm nut	35
Center link-to-steering knuckle nut	35
Tie-rod-to-center link nut	35
Tie-rod-to-steering knuckle nut	35
Steering damper-to-center link nut	35
Steering damper-to-axle bracket bolt	55
Wheel lug nuts	See Chapter 1

10

1.1 Front suspension and steering components

1	Stabilizer bar	6	Shock absorber (barely visible in this	9	Steering knuckle
2	Upper suspension arms		photo)	10	Lower suspension arms
3	Center link	7	Coil spring	11	Front axle housing
4	Track bar	8	Tie rod	12	Steering damper
5	Pitman arm				

1   General information

Refer to illustrations 1.1 and 1.2

All of the vehicles covered by this manual utilize a solid front axle, suspended by two coil springs. Four control arms allow the axle to move vertically, and lateral movement is prevented by a track bar. A telescopic dual-action shock absorber is mounted on each side. The steering knuckles pivot on balljoints and a stabilizer bar controls body roll **(see illustration)**.

The rear axle on all models is suspended by two semi-elliptic leaf springs and two dual-action telescopic shock absorbers **(see illustration)**. A stabilizer bar is installed on most models. As an option, some Comanche models are equipped with an automatic load leveling system.

Steering is either manual or power assisted. A recirculating ball type steering gearbox transmits the turning force through the steering linkage to the steering knuckle

arms. A steering damper is mounted between the frame and the center link to reduce unwanted bump steer. An intermediate shaft connects the steering gear to the steering column. The steering column is designed to collapse in the event of an accident.

Frequently, when working on the suspension or steering system components, you may come across fasteners which seem impossible to loosen. These fasteners on the underside of the vehicle are continually subjected to water, road grime, mud, etc., and can become rusted or *"frozen,"* making them extremely difficult to remove. In order to unscrew these stubborn fasteners without damaging them (or other components), be sure to use lots of penetrating oil and allow it to soak in for a while. Using a wire brush to clean exposed threads will also ease removal of the nut or bolt and prevent damage to the threads. Sometimes a sharp blow with a hammer and punch is effective in breaking the bond between a nut and bolt threads, but care must be taken to prevent the punch

from slipping off the fastener and ruining the threads. Heating the stuck fastener and surrounding area with a torch sometimes helps too, but isn't recommended because of the obvious dangers associated with fire. Long breaker bars and extension, or "cheater," pipes will increase leverage, but never use an extension pipe on a ratchet - the ratcheting mechanism could be damaged. Sometimes, turning the nut or bolt in the tightening (clockwise) direction first will help to break it loose. Fasteners that require drastic measures to unscrew should always be replaced with new ones.

Since most of the procedures that are dealt with in this chapter involve jacking up the vehicle and working underneath it, a good pair of jackstands will be needed. A hydraulic floor jack is the preferred type of jack to lift the vehicle, and it can also be used to support certain components during various operation. **Warning:** *Never, under any circumstances, rely on a jack to support the vehicle while working on it. Whenever any of the sus-*

1.2 Rear suspension components

1	Stabilizer bar	3	Shock absorber	5	Leaf spring shackle
2	U-bolt	4	Leaf spring	6	Rear axle housing

pension or steering fasteners are loosened or removed they must be inspected and, if necessary, be replaced with new ones of the same part number or of original equipment quality and design. Torque specifications must be followed for proper reassembly and component retention. Never attempt to heat or straighten any suspension or steering components. Instead, replace any bent or damaged part with a new one.

2   Front stabilizer bar and bushings - removal and installation

Refer to illustrations 2.2 and 2.3

Removal

1   Apply the parking brake. Raise the front of the vehicle and support it securely on jackstands.

2   Remove the stabilizer bar-to-link nuts, noting how the spacers, washers and bushings are positioned **(see illustration)**. If it is necessary to remove the links, simply unbolt them from the axle brackets.

3   Remove the stabilizer bar bracket bolts

and detach the bar from the vehicle **(see illustration)**.

4   Pull the brackets off the stabilizer bar and inspect the bushings for cracks, hardness and other signs of deterioration. If the bushings are damaged, replace them.

Installation

5   Position the stabilizer bar bushings on the bar.

6   Push the brackets over the bushings and raise the bar up to the frame. Install the bracket bolts but don't tighten them completely at this time.

7   Install the stabilizer bar-to-link nuts, washers, spacers and rubber bushings and tighten the nuts securely.

8   Tighten the bracket bolts.

3   Front shock absorber - removal and installation

Refer to illustrations 3.2 and 3.3

Removal

1   Loosen the wheel lug nuts, raise the

2.2 Note the positions of the bushings and washers before removing the stabilizer bar-to-link nuts - if the link turns while loosening the nut, hold it with a pair of locking pliers

vehicle and support it securely on jackstands. Apply the parking brake. Remove the wheel.

2   Remove the upper shock absorber stem

10

2.3 The stabilizer bar is attached to the frame with two brackets like this - remove the bolts (arrows) to detach the bar from the frame - the rubber bushings should be replaced if they are hard, cracked or otherwise deformed

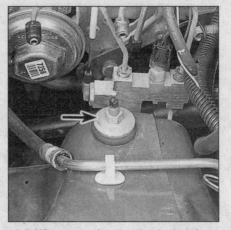

3.2 When removing the shock absorber stem nut (arrow), it may be necessary to hold the stem with an open end wrench or locking pliers to prevent it from turning - the brake master cylinder has been removed for clarity

3.3 The lower end of the shock absorber is connected to the front axle housing by two bolts and nuts

nut **(see illustration)**. Use an open end wrench to keep the stem from turning. If the nut won't loosen because of rust, squirt some penetrating oil on the stem threads and allow it to soak in for awhile. It may be necessary to keep the stem from turning with a pair of locking pliers, since the flats provided for a wrench are quite small.

3　Remove the two lower shock mount nuts and bolts **(see illustration)** and pull the shock absorber out from the wheel well. Remove the washers and the rubber grommets from the top of the shock absorber.

Installation

4　Extend the new shock absorber as far as possible. Position a new washer and rubber grommet on the stem and guide the shock up into the upper mount.

5　Install the upper rubber grommet and washer and wiggle the set back-and-forth to ensure the grommets are centered in the mount. Tighten the stem nut securely.

6　Install the lower mounting bolts and nuts and tighten them securely.

4　Track bar - removal and installation

Refer to illustration 4.2

Warning: *Whenever any of the suspension or steering fasteners are loosened or removed they must be inspected and, if necessary, replaced with new ones of the same part number or of original equipment quality and design. Torque specifications must be followed for proper reassembly and component retention.*

1　Raise the front of the vehicle and support it securely on jackstands.

2　Remove the cotter pin and retaining nut from the ballstud at the frame rail bracket **(see illustration)**. Discard the cotter pin.

3　Remove the bolt and nut from the axle bracket end of the bar and remove the bar from the vehicle.

4　Installation is the reverse of the removal procedure, but be sure to tighten the fasteners to the specified torque.

5　Front axle suspension arms - removal and installation

Refer to illustration 5.2

Warning: *Whenever any of the suspension or steering fasteners are loosened or removed they must be inspected and, if necessary, replaced with new ones of the same part number or of original equipment quality and design. Torque specifications must be followed for proper reassembly and component retention.*

Note: *Remove and install only one suspension arm at a time to avoid the possibility of the axle housing shifting out of position, which would make reassembly much more difficult. If it is absolutely necessary to remove more than one at a time, support the axle with a floor jack.*

1　Raise the front of the vehicle and support it securely on jackstands.

2　Remove the nut and bolt securing the suspension arm to the axle housing **(see illustration)**.

3　Remove the arm-to-frame bracket bolt and remove the arm from the vehicle.

4　Check the arm for distortion and cracks,

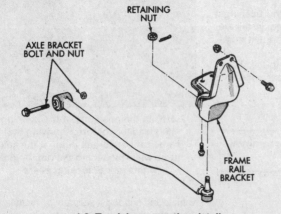

4.2 Track bar mounting details

5.2 Remove the nut (arrow), then slide out the bolt that secures the suspension arm to the front axle housing

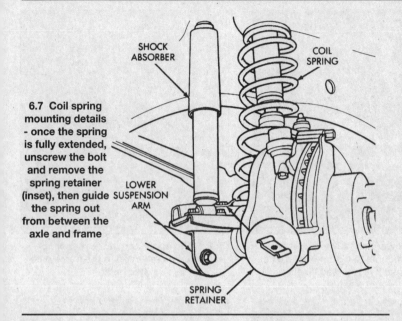

6.7 Coil spring mounting details - once the spring is fully extended, unscrew the bolt and remove the spring retainer (inset), then guide the spring out from between the axle and frame

7.5 Remove the cotter pins and loosen the castellated nuts on the upper (A) and lower (not visible in this photo) balljoint stud nuts - if a new steering knuckle is being installed, first remove the anchor plate bolts (B), then remove the anchor plate

replacing it if either of these conditions are noted.

5   Inspect the bushing in the axle housing (upper arm only) for cracking, hardness and general deterioration. If it is in need of replacement, reinstall the suspension arm and take the vehicle to a dealer service department or other qualified shop to have it replaced, as special tools are required to perform the job successfully.

6   Installation is the reverse of the removal procedure, but be sure to tighten the fasteners to the specified torque.

6   Front coil spring(s) - removal and installation

Refer to illustration 6.7

Removal

1   Loosen the front wheel lug nuts, raise the front of the vehicle and support it securely on jackstands. Remove the wheels.

2   Referring to Chapter 8, mark and disconnect the front driveshaft from the front differential pinion shaft yoke. Hang the driveshaft out of the way with a piece of wire.

3   Support the axle assembly with a floor jack (under the differential), or preferably two jacks (one at each end of the axle), to provide better balance. Unbolt the lower suspension arms from the axle (see Section 5).

4   Unbolt the stabilizer bar links and the shock absorbers at the front axle housing (see Sections 2 and 3).

5   Disconnect the track bar at the axle bracket (see Section 4).

6   Separate the steering center link from the Pitman arm (see Section 15).

7   Slowly lower the axle assembly until the coil springs are fully extended. If you are using only one jack, have an assistant support the right side of the axle as it's lowered.

Remove the spring retainer **(see illustration)** and remove the spring from the vehicle.

8   Check the spring for deep nicks and corrosion, which will cause premature failure of the spring. Replace the spring if these or any other questionable conditions are evident.

Installation

9   Position the coil spring on the axle housing, place the spring retainer over the bottom coil of the spring and tighten the spring retainer bolt securely.

10   Raise the axle up into position and connect the lower suspension arms to the axle housing (see Section 5). Tighten the fasteners to the specified torque.

11   Connect the center link to the Pitman arm (see Section 15).

12   Connect the track bar to its bracket on the axle (see Section 4).

13   Connect the stabilizer bar links and the shock absorbers to the front axle housing (see Sections 2 and 3).

14   Connect the front driveshaft to the differential pinion shaft yoke (see Chapter 8).

15   Install the wheels and lug nuts and lower the vehicle. Tighten the lug nuts to the torque specified in Chapter 1.

7   Steering knuckle - removal and installation

Refer to illustrations 7.5, 7.6 and 7.9

Warning: *Whenever any of the suspension or steering fasteners are loosened or removed they must be inspected and, if necessary, replaced with new ones of the same part number or of original equipment quality and design. Torque specifications must be followed for proper reassembly and component retention.*

Note: *This procedure applies to 4WD and*

2WD models. If you are working on a 2WD model, simply ignore references to the components particular to 4WD vehicles.

Removal

1   Loosen the wheel lug nuts, raise the vehicle and support it securely on jackstands. Remove the wheel.

2   Disconnect the tie-rod from the steering knuckle (see Section 16).

3   Remove the disc brake caliper and disc (see Chapter 9).

4   Remove the front axle hub and bearing assembly (4WD only) and remove the axleshaft (see Chapter 8).

5   Remove the cotter pins and loosen the castellated nuts on the balljoint studs **(see illustration)**.

6   Using a large hammer, tap the steering knuckle at the top to separate it from the balljoint studs **(see illustration)**.

7.6 Strike the steering knuckle at the top to separate it from the balljoint studs - note that the castellated nuts are loosened, but not removed - this will prevent the knuckle from falling when it suddenly breaks loose

10

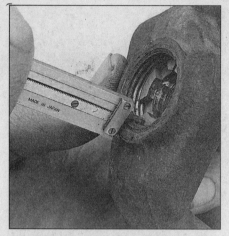

7.9 The split ring seat must be screwed in to a depth of 0.206 in (5.23 mm), measured from the spindle boss to the flats of the seat; this can be measured with a depth micrometer or a vernier caliper, as shown here

7   Carefully check the steering knuckle for cracks, especially around the steering arm and spindle mounting area. Check for elongated balljoint stud holes. Replace the steering knuckle if any of these conditions are found.

Installation

8   If the steering knuckle is being replaced, unscrew the split ring seat from the lower mount and transfer it to the new knuckle. Also remove the disc anchor plate from the old knuckle **(see illustration 7.5)**.

9   Check the depth of the split ring seat in the knuckle using a depth micrometer or vernier caliper **(see illustration)**. Screw the split ring seat in or out to attain the proper depth of 0.206 in (5.23 mm).

10   Position the steering knuckle on the axle housing, inserting the balljoint studs into the holes in the knuckle. Install the nuts and tighten them to the specified torque. Remember to use new cotter pins.

11   Install the axleshaft (see Chapter 8).

12   Install the front axle hub and bearing assembly (see Chapter 8).

13   Install the disc anchor plate (if removed previously) and tighten the bolts to the specified torque. Then install the brake disc and caliper (see Chapter 9).

14   Connect the tie-rod to the steering knuckle (see Section 16).

15   Install the wheel and lug nuts. Lower the vehicle and tighten the lug nuts to the torque specified in Chapter 1.

8   Balljoints - replacement

The balljoints are a press fit in the front axle housing, which necessitates the use of a special tool to remove and install them. Since the tool is not normally available to the home mechanic, it is recommended that the vehicle

9.3 The lower end of the rear shock absorber mounts to a stud on the axle housing and is retained by a nut and washer

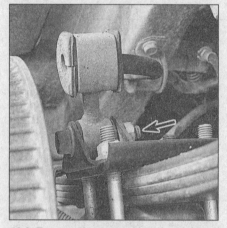

10.2 Remove the nut and bolt (arrow) that retain the stabilizer bar link to the spring tie plate

be taken to a dealer service department or other qualified repair shop to have the balljoints replaced.

9   Rear shock absorber - removal and installation

Refer to illustrations 9.3 and 9.4

Warning: *Whenever any of the suspension or steering fasteners are loosened or removed they must be inspected and, if necessary, replaced with new ones of the same part number or of original equipment quality and design.*

1   Raise the rear of the vehicle and support it securely on jackstands.

2   Support the rear axle assembly with a floor jack placed under the differential. Raise the jack just enough to take the spring pressure off the shock absorbers (the shock absorbers limit downward travel of the suspension).

3   Remove the shock absorber lower retaining nut and washer **(see illustration)**.

9.4 The upper end of the rear shock absorber is mounted to the frame with two bolts

10.3 The stabilizer bar is connected to the frame rails with two brackets like this

4   Remove the two upper mounting bolts **(see illustration)** and slide the shock off the lower mounting stud.

5   Installation is the reverse of the removal procedure.

10   Rear stabilizer bar - removal and installation

Refer to illustrations 10.2 and 10.3

Warning: *Whenever any of the suspension or steering fasteners are loosened or removed they must be inspected and, if necessary, replaced with new ones of the same part number or of original equipment quality and design.*

Removal

1   Raise the rear of the vehicle and support it securely on jackstands.

2   Unbolt the stabilizer bar link at the spring tie plate **(see Illustration)**.

3   Unbolt the stabilizer bar brackets from the frame rails **(see illustration)** and remove the bar from the vehicle.

11.5 The rear axle must be supported before removing the U-bolt nuts (arrows); otherwise the axle will fall

11.6 Unscrew the nut (arrow), then remove the rear spring-to-shackle bolt

11.7 Remove the spring front eye-to-frame bracket bolt (arrow)

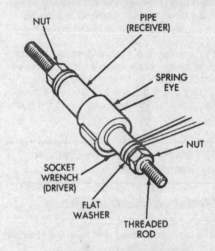

12.8 A threaded rod and two appropriately-sized sockets can be used to remove and install the spring eye bushings

4   Pull the brackets and bushings off the bar and inspect them for wear, cracking and general deterioration, replacing them if necessary.

Installation

5   Push the bushings and brackets onto the stabilizer bar. Position the bar against the frame and install the bracket bolts, tightening them securely.
6   Connect the links to the spring tie plates and tighten them securely. Lower the vehicle,

11   Rear leaf spring - removal and installation

Refer to illustrations 11.5, 11.6 and 11.7
Warning: *Whenever any of the suspension or steering fasteners are loosened or removed they must be inspected and, if necessary, replaced with new ones of the same part number or of original equipment quality and design. Torque specifications must be followed for proper reassembly and component retention.*

Removal

1   Loosen the rear wheel lug nuts, raise the rear of the vehicle and support it securely on jackstands. Remove the wheel.
2   Support the rear axle assembly with a floor jack positioned underneath the differential. Raise the axle just enough to take the spring pressure off of the shock absorbers.
3   Disconnect the shock absorber from the axle bracket (see Section 9).
4   Unbolt the stabilizer bar links from the spring tie plate (see Section 10).
5   Support the axle, then unscrew the U-bolt nuts **(see illustration)**. Remove the spring tie plate.
6   Remove the rear spring-to-shackle bolt **(see illustration)**.
7   Remove the spring front eye-to-frame bracket bolt **(see illustration)** and remove the spring from the vehicle.

Installation

8   Installation is the reverse of the removal procedure. Be sure to tighten the spring mounting bolts and the spring tie plate U-bolt nuts to the specified torque. **Note:** *The vehicle must be standing at normal ride height before tightening the front and rear mounting bolts.*

12   Rear leaf spring bushing - check and replacement

Refer to illustration 12.8

Check

1   All models are equipped with rubber bushings which are pressed into the spring eyes. The bushings should be inspected for cracks, damage and looseness indicating excessive wear. To check for wear, jack up the frame until the weight is removed from the spring bushing. Pry the spring eye up-and-down to check for movement. If there is considerable movement, the bushing is worn and should be replaced.

Replacement

2   Remove the spring (see Section 11).
3   The bushings are of two different sizes and tools can be fabricated from threaded rod for pressing them out. For small diameter bushings, cut an eight inch length of 3/8-inch diameter threaded rod and, for the large diameter bushings, cut an eleven-inch length of 1/2-inch diameter threaded rod.
4   Insert the threaded rod through the bushing.
5   Place a socket over one end of the rod with the open end toward the bushing to serve as a driver. The socket must be large enough to bear against the bushing outer sleeve and small enough to pass through the spring eye.
6   Install a flat washer and hex nut on the rod behind the socket.
7   On the opposite end of the threaded

rod, install a piece of pipe to serve as a receiver. The inside diameter of the pipe must be large enough to accommodate the bushing while still seating against the spring eye surface. It must also be long enough to accept the entire bushing.
8   Secure the pipe section on the rod with a flat washer and nut **(see illustration)**. The washer must be large enough to properly support the pipe.
9   Tighten the nuts finger tight to align the components. The socket must be positioned in the spring eye and aligned with the bushing and the pipe must butt against the eye surface so the bushing can pass through it.
10   Press the bushing out of the spring eye by tightening the nut at the socket end of the rod.
11   Remove the bushing and tool from the spring eye.
12   Install the new bushing on the threaded rod and assemble and align the tools as previously described.
13   Line up the bushing with the spring eye and press the new bushing into position.

10

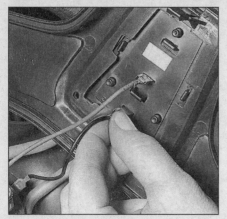

13.2 After removing the horn pad, disconnect the horn switch wires; the horn pad is retained to the steering wheel by three screws, accessible from the dashboard side of the wheel

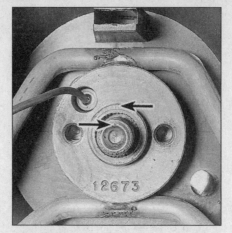

13.3 Before removing the steering wheel, check to see if any relationship marks exist (arrows) - if not, use a sharp scribe or white paint to make your own marks

13.4 Remove the wheel from the shaft with a puller - DO NOT HAMMER ON THE SHAFT!

14   Loosen the nuts and check to make sure the bushing is centered in the spring eye with the ends of the bushing flush with or slightly below the sides of the eye. If necessary, reinstall the tools and adjust the bushing position.

13   Steering wheel - removal and installation

Models without an airbag

Refer to illustrations 13.2, 13.3 and 13.4

1   Disconnect the cable from the negative terminal of the battery.

2   Detach the horn pad from the steering wheel and disconnect the wires to the horn switch **(see illustration)**.

3   Remove the steering wheel retaining nut, then mark the relationship of the steering shaft to the hub (if marks don't already exist or don't line up) to simplify installation and ensure steering wheel alignment **(see illustration)**.

4   Use a puller to detach the steering wheel from the shaft **(see illustration)**. Don't hammer on the shaft to dislodge the steering wheel.

5   To install the wheel, align the mark on the steering wheel hub with the mark on the shaft and slip the wheel onto the shaft. Install the nut and tighten it to the specified torque.

6   Connect the horn wire and install the horn pad.

7   Connect the negative battery cable.

Models with an airbag

Refer to illustrations 13.10 and 13.11

8   If you're working on a 1995 or 1996 model, roll down the driver's side window.

9   Disconnect the cable from the negative terminal of the battery. Wait at least two minutes before proceeding (this will disable the airbag system on 1997 and later models; proceed to the next Step for the airbag disarming procedure on 1995 and 1996 models).

10   If you're working on a 1995 or 1996

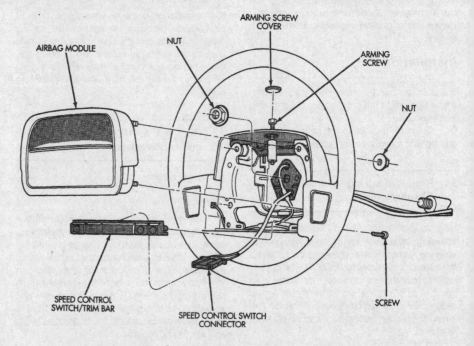

13.10 Airbag module and steering wheel details - 1995 and 1996 models

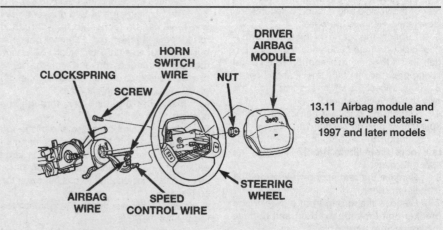

13.11 Airbag module and steering wheel details - 1997 and later models

14.2 Using white paint, mark the relationship of the upper and lower universal joints to the steering shaft and steering gear input shaft, respectively; then remove the upper pinch bolt (arrow) . . .

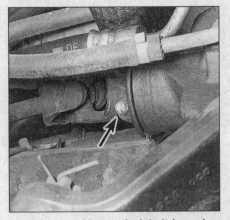

14.3 . . . and lower pinch bolt (arrow)

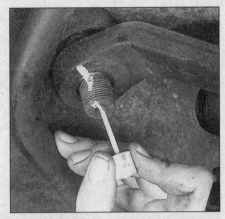

15.4 Before removing the Pitman arm, mark its relationship to the steering gear shaft

model, pry off the small cover from the top of the steering wheel hub. Get out of the vehicle, reach into the vehicle and, using an 8 mm socket, unscrew the arming screw until it stops (it should protrude approximately one inch from the surface of the steering wheel trim) **(see illustration)**. The airbag is now disarmed.

11   Remove the airbag module fasteners from the backside of the steering wheel. Lift the airbag module away from the wheel and, on 1997 and later models, unplug the electrical connectors from the module **(see the accompanying illustration and illustration 13.10)**. Set the module aside in an isolated location. **Warning:** *Carry the module with the trim side facing away from your body, and set it down with the trim side facing up.*

12   Perform Steps 3 and 4 of this Section.

13   To install the wheel, refer to Step 5. **Warning:** *If the clockspring for the airbag system was inadvertently turned, refer to Chapter 12, Section 20, for the centering procedure. Be sure to carefully thread the wiring harness for the cruise control, airbag (1997 and later models) and horn through the proper openings in the steering wheel.*

14   Position the airbag module over the steering wheel, plug in the electrical connectors and install the module. Tighten the module fasteners to the torque listed in this Chapter's Specifications.

15   Connect the negative battery cable.

14   Intermediate shaft - removal and installation

Refer to illustrations 14.2 and 14.3

Warning: *On 1997 and later models, make sure the steering shaft is not turned while the intermediate shaft is removed or you could damage the airbag system. To prevent the shaft from turning, position the wheels straight ahead, thread the seat belt through the steering wheel and fasten it into its latch.*

If the steering wheel is inadvertently turned, remove the steering wheel and center the clockspring (see Chapter 12).

1   Turn the front wheels to the straight ahead position.

2   Using white paint, mark the relationship of the upper universal joint to the steering shaft and the lower universal joint to the steering gear input shaft **(see illustration)**.

3   Remove the upper and lower universal joint pinch bolts **(see illustration)**. Some designs require the steering gear to be loosened (bolts removed) and repositioned to allow shaft removal.

4   Pry the intermediate shaft out of the steering shaft universal joint with a large screwdriver, then pull the shaft from the steering gearbox.

5   Installation is the reverse of the removal procedure. Be sure to align the marks and tighten the pinch bolts to the specified torque.

15   Steering gear - removal and installation

Refer to illustrations 15.4, 15.5 and 15.6

Warning: *On 1997 and later models, make sure the steering shaft is not turned while the steering gear is removed or you could dam-*

age the airbag system. To prevent the shaft from turning, position the wheels straight ahead, thread the seat belt through the steering wheel and fasten it into its latch. If the steering wheel is inadvertently turned, remove the steering wheel and center the clockspring (see Chapter 12).

Removal

1   Raise the front of the vehicle and support it securely on jackstands. Apply the parking brake.

2   Place a drain pan under the steering gear (power steering only). Remove the hoses/lines and cap the ends to prevent excessive fluid loss and contamination. If available, use a flare-nut wrench to remove the hoses/lines.

3   Mark the relationship of the intermediate shaft lower universal joint to the steering gear input shaft. Remove the intermediate shaft lower pinch bolt **(see illustration 14.3)**.

4   Remove the Pitman arm nut and washer. Mark the relationship of the Pitman arm to the shaft so it can be installed in the same position **(see illustration)**.

5   Separate the Pitman arm from the shaft with a two-jaw puller **(see illustration)**.

6   Support the steering gear and remove

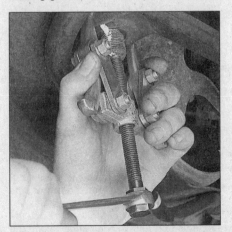

15.5 Use a two-jaw puller to separate the Pitman arm from the steering gear shaft

15.6 The steering gear is mounted to the frame rail with three bolts (arrows)

10

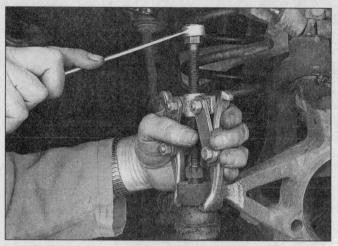

16.8 Use a two-jaw puller to detach the tie-rod end from the steering knuckle - notice the nut has been loosened but not removed; this will prevent the components from separating violently

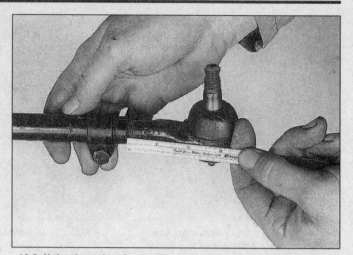

16.9 If the tie-rod ends must be replaced, measure the distance from the end of the tie rod tube to the center of the ballstud so the new tie-rod end can be set to this dimension

the mounting bolts **(see illustration)**. Lower the unit, separate the intermediate shaft from the steering gear input shaft and remove the steering gear from the vehicle.

Installation

7   Raise the steering gear into position and connect the intermediate shaft, aligning the marks.

8   Install the mounting bolts and washers and tighten them to the specified torque.

9   Slide the Pitman arm onto the shaft. Make sure the marks are aligned. Install the washer and nut and tighten the nut to the specified torque.

10   Install the intermediate shaft lower pinch bolt and tighten it to the specified torque.

11   Connect the power steering hoses/lines to the steering gear and fill the power steering pump reservoir with the recommended fluid (see Chapter 1).

12   Lower the vehicle and bleed the steering system as discussed in Section 18.

---

16   Steering linkage - inspection, removal and installation

Warning: *Whenever any of the suspension or steering fasteners are loosened or removed they must be inspected and, if necessary, replaced with new ones of the same part number or of original equipment quality and design. Torque specifications must be followed for proper reassembly and component retention. Never attempt to heat, straighten or weld any suspension or steering component. Instead, replace any bent or damaged part with a new one.*
Caution: *DO NOT use a "pickle fork" type balljoint separator - it may damage the balljoint seals.*

Inspection

1   The steering linkage connects the steer-

ing gear to the front wheels and keeps the wheels in proper relation to each other **(see illustration 1.1)**. The linkage consists of the Pitman arm, fastened to the steering gear shaft, which moves the center link back and forth. The back and-forth motion of the center link is transmitted to the right steering knuckle. A tie-rod assembly connects the center link to the left knuckle. The tie-rod is made up of a tube, clamps and two tie-rod ends. A steering damper, connected between the center link and the frame reduces shimmy and unwanted forces to the steering gear.

2   Set the wheels in the straight ahead position and lock the steering wheel.

3   Raise one side of the vehicle until the tire is approximately 1-inch off the ground.

4   Mount a dial indicator with the needle resting on the outside edge of the wheel. Grasp the front and rear of the tire and using light pressure, wiggle the wheel back-and-forth and note the dial indicator reading. The gauge reading should be less than 0.108-inch. If the play in the steering system is more than specified, inspect each steering linkage pivot point and ball stud for looseness and replace parts if necessary.

5   Raise the vehicle and support it on jackstands. Check for torn ball stud boots, frozen joints and bent or damaged linkage components.

Removal and installation

Refer to illustrations 16.8 and 16.9

Tie-rod

6   Loosen the wheel lug nuts, raise the vehicle and support it securely on jackstands. Apply the parking brake. Remove the wheel.

7   Remove the cotter pins and loosen, but do not remove, the castellated nuts from the ball studs on each end of the tie-rod.

8   Using a two-jaw puller, separate the tie-rod ends from the steering knuckle and the center link **(see illustration)**. Remove the castellated nuts and pull the tie-rod ends

from the knuckle and center link.

9   If the tie-rod ends must be replaced, measure the distance from the end of the tie-rod tube to the center of the ball stud and record it **(see illustration)**. Loosen the adjuster tube clamp and unscrew the tie-rod end.

10   Lubricate the threaded portion of the tie-rod end with chassis grease. Screw the new tie-rod end into the adjuster tube and adjust the distance from the tube to the ball stud to the previously measured dimension. The number of threads showing on the inner and outer tie-rod ends should be equal within three threads. Don't tighten the clamp yet.

11   To install the tie-rod, insert the tie-rod end ball studs into the steering knuckle and center link until they're seated. Install the nuts and tighten them to the specified torque. If a ball stud spins when attempting to tighten the nut, force it into the tapered hole with a large pair of pliers.

12   Install new cotter pins. If necessary, tighten the nut slightly to align a slot in the nut with the hole in the ball stud.

13   Tighten the clamp nuts. The bolts should be nearly horizontal.

14   Install the wheel and lug nuts, lower the vehicle and tighten the lug nuts to the torque specified in Chapter 1. Drive the vehicle to an alignment shop to have the front end alignment checked and, if necessary, adjusted.

Center link

15   Raise the front of the vehicle and support it securely on jackstands. Apply the parking brake.

16   Loosen, but do not remove, the nuts securing the center link ball studs to the tie-rod and steering knuckle. Separate the joints with a two-jaw puller, then remove the nuts **(see illustration 16.8)**.

17   Separate the center link from the Pitman arm using the same technique. If the tie-rod end on the left end of the center link is in need of replacement, follow Steps 9 and 10

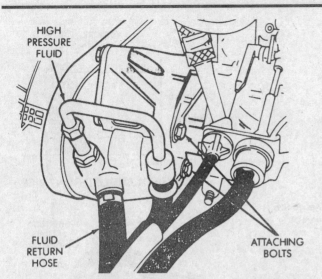

17.2 Power steering pump line and hose connections and rear attaching bolts

17.3 Power steering pump front attaching bolts

of this Section.
18   Installation is the reverse of the removal procedure. If the ball studs spin when attempting to tighten the nuts, force them into the tapered holes with a large pair of pliers. Be sure to tighten all of the nuts to the specified torque.

Pitman arm

19   Refer to Section 15 of this Chapter for the Pitman arm removal procedure.

Steering damper

20   Raise the front of the vehicle and support it securely on jackstands.
21   Separate the steering damper from the center link using the technique described in Step 16.
22   Unbolt the damper from the frame and remove it from the vehicle.
23   Installation is the reverse of the removal procedure.

17 Power steering pump - removal and installation

Refer to illustration 17.2, 17.3, 17.4 and 17.5
Warning: *Whenever any of the suspension or steering fasteners are loosened or removed they must be inspected and, if necessary, replaced with new ones of the same part number or of original equipment quality and design.*

1   Loosen the pump drivebelt and slip the belt over the pulley (see Chapter 1).
2   Using a suction gun, suck out as much power steering fluid from the reservoir as possible. Position a drain pan under the pump and disconnect the high pressure line and fluid return hose **(see illustration)**. It may be necessary to remove the air cleaner housing for access to the return hose (if so, see Chapter 4). Cap the ends of the lines to prevent excessive fluid leakage and the entry of

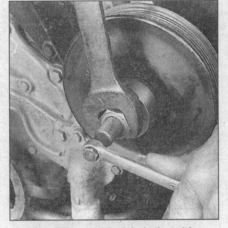

17.4 This special tool, designed for removing power steering pump pulleys, is available at most auto parts stores

contaminants.
3   Remove the front attaching bolts **(see illustration)** and the rear attaching bolts **(see illustration 17.2)** and nuts, then lift the pump from the engine.
4   If it is necessary to remove the pulley from the pump, first measure how far the pump shaft protrudes from the face of the pulley hub. Remove the pulley from the shaft with a special power steering pump pulley removal tool **(see illustration)**. This tool can be purchased at most auto parts stores.
5   A special pulley installation tool is available for pressing the pulley back onto the pump shaft, but an alternate tool can be fabricated from a long bolt, nut, washer and a socket of the same diameter as the pulley hub **(see illustration)**. Push the pulley onto the shaft until the shaft protrudes from the hub the previously recorded amount.
6   Installation of the power steering pump is the reverse of the removal procedure. Be sure to bleed the power steering system following the procedure in Section 18.

17.5 A long bolt with the same thread pitch as the internal threads of the power steering pump shaft, a nut, washer and a socket that is the same diameter as the pulley hub can be used to install the pulley on the shaft

18 Power steering system - bleeding

1   Following any operation in which the power steering fluid lines have been disconnected, the power steering system must be bled to remove all air and obtain proper steering performance.
2   With the front wheels in the straight ahead position, check the power steering fluid level and, if low, add fluid until it reaches the Cold (C) mark on the dipstick.
3   Start the engine and allow it to run at fast idle. Recheck the fluid level and add more if necessary to reach the Cold (C) mark on the dipstick.
4   Bleed the system by turning the wheels from side-to-side, without hitting the stops. This will work the air out of the system. Keep the reservoir full of fluid as this is done.
5   When the air is worked out of the sys-

10

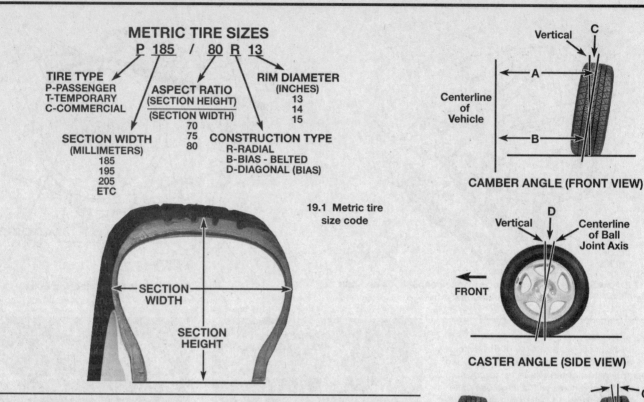

METRIC TIRE SIZES

P 185 / 80 R 13

TIRE TYPE
P-PASSENGER
T-TEMPORARY
C-COMMERCIAL

ASPECT RATIO
(SECTION HEIGHT)
———————————
(SECTION WIDTH)
70
75
80

RIM DIAMETER
(INCHES)
13
14
15

SECTION WIDTH
(MILLIMETERS)
185
195
205
ETC

CONSTRUCTION TYPE
R-RADIAL
B-BIAS - BELTED
D-DIAGONAL (BIAS)

SECTION WIDTH

SECTION HEIGHT

19.1 Metric tire size code

CAMBER ANGLE (FRONT VIEW)

Vertical
C
A
Centerline of Vehicle
B

CASTER ANGLE (SIDE VIEW)

Vertical   D   Centerline of Ball Joint Axis

FRONT

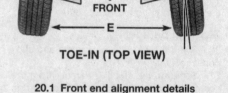

TOE-IN (TOP VIEW)

G
F
FRONT
E

20.1 Front end alignment details

A minus B = C (degrees camber)
D = Degrees caster
E minus F = toe-in (measured in inches)
G - toe-in (expressed in degrees)

tem, return the wheels to the straight ahead position and leave the vehicle running for several more minutes before shutting it off.

6   Road test the vehicle to be sure the steering system is functioning normally and noise free.

7   Recheck the fluid level to be sure it is up to the Hot (H) mark on the dipstick while the engine is at normal operating temperature. Add fluid if necessary (see Chapter 1).

19   Wheels and tires - general information

Refer to illustration 19.1

All vehicles covered by this manual are equipped with metric-sized fiberglass or steel belted radial tires **(see illustration)**. Use of other size or type of tires may affect the ride and handling of the vehicle. Don't mix different types of tires, such as radials and bias belted, on the same vehicle as handling may be seriously affected. It's recommended that tires be replaced in pairs on the same axle, but if only one tire is being replaced, be sure it's the same size, structure and tread design as the other.

Because tire pressure has a substantial effect on handling and wear, the pressure on all tires should be checked at least once a month or before any extended trips (see Chapter 1).

Wheels must be replaced if they are bent, dented, leak air, have elongated bolt holes, are heavily rusted, out of vertical symmetry or if the lug nuts won't stay tight. Wheel repairs that use welding or peening are not recommended.

Tire and wheel balance is important to the overall handling, braking and performance of the vehicle. Unbalanced wheels can adversely affect handling and ride characteristics as well as tire life. Whenever a tire is installed on a wheel, the tire and wheel should be balanced by a shop with the proper equipment.

20   Front end alignment - general information

Refer to illustration 20.1

A front end alignment refers to the adjustments made to the front wheels so they are in proper angular relationship to the suspension and the ground. Front wheels that are out of proper alignment not only affect steering control, but also increase tire wear. The only front end adjustments possible on these vehicles are caster and toe-in **(see illustration)**.

Getting the proper front wheel alignment is a very exacting process, one in which complicated and expensive machines are necessary to perform the job properly. Because of this, you should have a technician with the proper equipment perform these tasks. We will, however, use this space to give you a basic idea of what is involved with front end alignment so you can better understand the process and deal intelligently with the shop that does the work.

Toe-in is the turning in of the front wheels. The purpose of a toe specification is to ensure parallel rolling of the front wheels. In a vehicle with zero toe-in, the distance between the front edges of the wheels will be

the same as the distance between the rear edges of the wheels. The actual amount of toe-in is normally only a fraction of an inch. Toe-in adjustment is controlled by the tie-rod end position on the tie-rod. Incorrect toe-in will cause the tires to wear improperly by making them scrub against the road surface.

Caster is the tilting of the top of the front steering axis from the vertical. A tilt toward the rear is positive caster and a tilt toward the front is negative caster. This angle is adjusted by adding or subtracting shims at the rear of the lower suspension arms.

Camber (the tilting of the front wheels from vertical when viewed from the front of the vehicle) is factory present at 0-degrees and cannot be adjusted. If the camber angle isn't correct, the components causing the problem must be replaced. NEVER ATTEMPT TO ADJUST THE CAMBER ANGLE BY HEATING OR BENDING THE AXLE OR ANY OTHER SUSPENSION COMPONENT!

Chapter 11   Body

Contents

Specifications

Torque specifications

	Ft-lb
Door	
Glass stud nuts	4
Hinge bolts	26
Hood-to-hinge bolts	23
Liftgate striker screws	22
Front seat retaining nuts	18
Seatback cushion pivot bolts	33
Rear seat-to-bottom cushion hinge striker bolt	33

1   General information

The vehicles covered in this manual have a separate frame and body.

As with other parts of the vehicle, proper maintenance of body components plays an important part in preserving the vehicle's market value. It's far less costly to handle small problems before they grow into larger ones. Information in this Chapter will tell you all you need to know to keep seals sealing, body panels aligned and general appearance up to par.

Major body components which are particularly vulnerable in accidents are removable. These include the hood, fenders, grille and doors. It's often cheaper and less time consuming to replace an entire panel than it is to attempt a restoration of the old one. However, this must be decided on a case-by-case basis.

2   Body - maintenance

1   The condition of your vehicle's body is very important, because the resale value depends a great deal on it. It's much more difficult to repair a neglected or damaged body than it is to repair mechanical components. The hidden areas of the body, such as the wheel wells, the frame and the engine compartment, are equally important, although they don't require as frequent attention as the rest of the body.
2   Once a year, or every 12,000 miles, it's a good idea to have the underside of the body steam cleaned. All traces of dirt and oil will be removed and the area can then be inspected carefully for rust, damaged brake lines, frayed electrical wires, damaged cables and other problems. The front suspension components should be greased after completion of this job.
3   At the same time, clean the engine and

the engine compartment with a steam cleaner or water soluble degreaser.
4   The wheel wells should be given close attention, since undercoating can peel away and stones and dirt thrown up by the tires can cause the paint to chip and flake, allowing rust to set in. If rust is found, clean down to the bare metal and apply an anti-rust paint.
5   The body should be washed about once a week. Wet the vehicle thoroughly to soften the dirt, then wash it down with a soft sponge and plenty of clean soapy water. If the surplus dirt is not washed off very carefully, it can wear down the paint.
6   Spots of tar or asphalt thrown up from the road should be removed with a cloth soaked in solvent.
7   Once every six months, wax the body and chrome trim. If a chrome cleaner is used to remove rust from any of the vehicle's plated parts, remember that the cleaner also removes part of the chrome, so use it sparingly.

11

These photos illustrate a method of repairing simple dents. They are intended to supplement *Body repair - minor damage* in this Chapter and should not be used as the sole instructions for body repair on these vehicles.

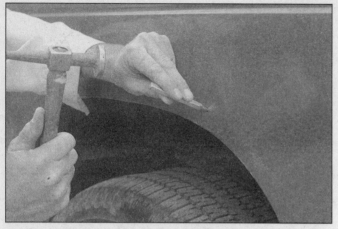

1   If you can't access the backside of the body panel to hammer out the dent, pull it out with a slide-hammer-type dent puller. In the deepest portion of the dent or along the crease line, drill or punch hole(s) at least one inch apart . . .

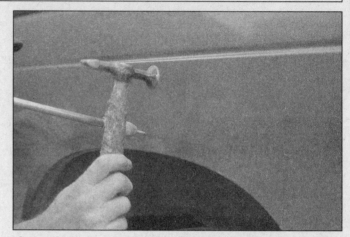

2   . . . then screw the slide-hammer into the hole and operate it. Tap with a hammer near the edge of the dent to help 'pop' the metal back to its original shape. When you're finished, the dent area should be close to its original contour and about 1/8-inch below the surface of the surrounding metal

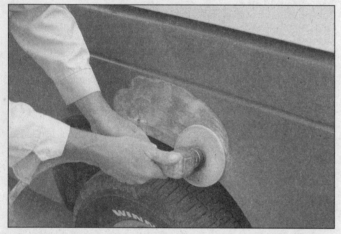

3   Using coarse-grit sandpaper, remove the paint down to the bare metal. Hand sanding works fine, but the disc sander shown here makes the job faster. Use finer (about 320-grit) sandpaper to feather-edge the paint at least one inch around the dent area

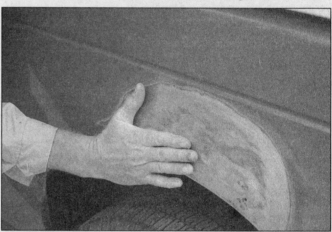

4   When the paint is removed, touch will probably be more helpful than sight for telling if the metal is straight. Hammer down the high spots or raise the low spots as necessary. Clean the repair area with wax/silicone remover

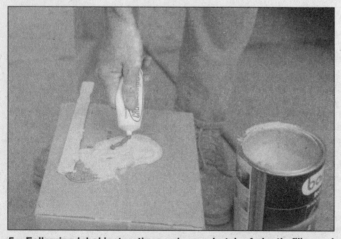

5   Following label instructions, mix up a batch of plastic filler and hardener. The ratio of filler to hardener is critical, and, if you mix it incorrectly, it will either not cure properly or cure too quickly (you won't have time to file and sand it into shape)

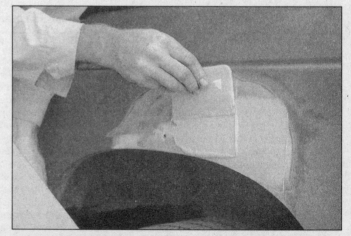

6   Working quickly so the filler doesn't harden, use a plastic applicator to press the body filler firmly into the metal, assuring it bonds completely. Work the filler until it matches the original contour and is slightly above the surrounding metal

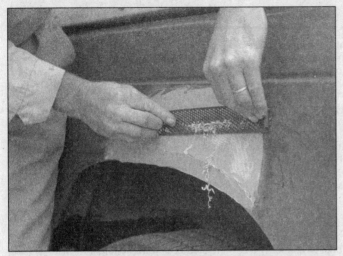

7   Let the filler harden until you can just dent it with your fingernail. Use a body file or Surform tool (shown here) to rough-shape the filler

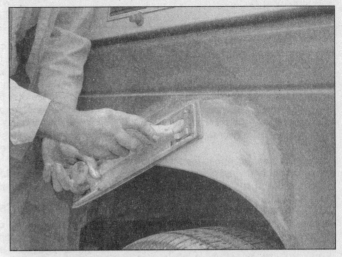

8   Use coarse-grit sandpaper and a sanding board or block to work the filler down until it's smooth and even. Work down to finer grits of sandpaper - always using a board or block - ending up with 360 or 400 grit

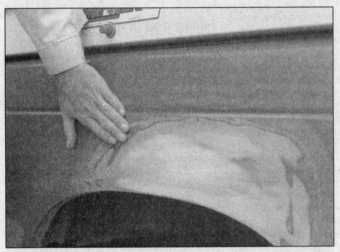

9   You shouldn't be able to feel any ridge at the transition from the filler to the bare metal or from the bare metal to the old paint. As soon as the repair is flat and uniform, remove the dust and mask off the adjacent panels or trim pieces

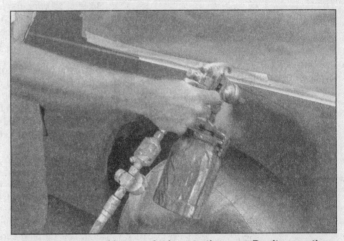

10   Apply several layers of primer to the area. Don't spray the primer on too heavy, so it sags or runs, and make sure each coat is dry before you spray on the next one. A professional-type spray gun is being used here, but aerosol spray primer is available inexpensively from auto parts stores

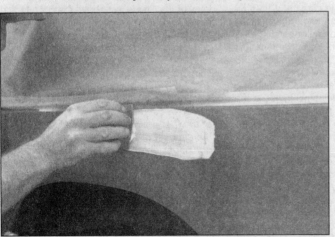

11   The primer will help reveal imperfections or scratches. Fill these with glazing compound. Follow the label instructions and sand it with 360 or 400-grit sandpaper until it's smooth. Repeat the glazing, sanding and respraying until the primer reveals a perfectly smooth surface

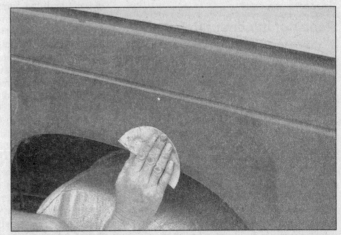

12   Finish sand the primer with very fine sandpaper (400 or 600-grit) to remove the primer overspray. Clean the area with water and allow it to dry. Use a tack rag to remove any dust, then apply the finish coat. Don't attempt to rub out or wax the repair area until the paint has dried completely (at least two weeks)

3   Vinyl trim - maintenance

Don't clean vinyl trim with detergents, caustic soap or petroleum based cleaners. Plain soap and water works just fine, with a soft brush to clean dirt that may be ingrained. Wash the vinyl as frequently as the rest of the vehicle.

After cleaning, application of a high quality rubber and vinyl protectant will help prevent oxidation and cracks. The protectant can also be applied to weatherstripping, vacuum lines and rubber hoses, which often fail as a result of chemical degradation, and to the tires.

4   Upholstery and carpets - maintenance

1   Every three months remove the carpets or mats and clean the interior of the vehicle (more frequently if necessary). Vacuum the upholstery and carpets to remove loose dirt and dust.

2   Leather upholstery requires special care. Stains should be removed with warm water and a very mild soap solution. Use a clean, damp cloth to remove the soap, then wipe again with a dry cloth. Never use alcohol, gasoline, nail polish remover or thinner to clean leather upholstery.

3   After cleaning, regularly treat leather upholstery with a leather wax. Never use car wax on leather upholstery.

4   In areas where the interior of the vehicle is subject to bright sunlight, cover leather seats with a sheet if the vehicle is to be left out for any length of time.

5   Body repair - minor damage

See photo sequence

Repair of minor scratches

1   If the scratch is superficial and does not penetrate to the metal of the body, repair is very simple. Lightly rub the scratched area with a fine rubbing compound to remove loose paint and built up wax. Rinse the area with clean water.

2   Apply touch-up paint to the scratch, using a small brush. Continue to apply thin layers of paint until the surface of the paint in the scratch is level with the surrounding paint. Allow the new paint at least two weeks to harden, then blend it into the surrounding paint by rubbing with a very fine rubbing compound. Finally, apply a coat of wax to the scratch area.

3   If the scratch has penetrated the paint and exposed the metal of the body, causing the metal to rust, a different repair technique is required. Remove all loose rust from the bottom of the scratch with a pocket knife, then apply rust inhibiting paint to prevent the formation of rust in the future. Using a rubber

or nylon applicator, coat the scratched area with glaze-type filler. If required, the filler can be mixed with thinner to provide a very thin paste, which is ideal for filling narrow scratches. Before the glaze filler in the scratch hardens, wrap a piece of smooth cotton cloth around the tip of a finger. Dip the cloth in thinner and then quickly wipe it along the surface of the scratch. This will ensure that the surface of the filler is slightly hollow. The scratch can now be painted over as described earlier in this section.

Repair of dents

4   When repairing dents, the first job is to pull the dent out until the affected area is as close as possible to its original shape. There is no point in trying to restore the original shape completely as the metal in the damaged area will have stretched on impact and cannot be restored to its original contours. It is better to bring the level of the dent up to a point which is about 1/8-inch below the level of the surrounding metal. In cases where the dent is very shallow, it is not worth trying to pull it out at all.

5   If the back side of the dent is accessible, it can be hammered out gently from behind using a soft-face hammer. While doing this, hold a block of wood firmly against the opposite side of the metal to absorb the hammer blows and prevent the metal from being stretched.

6   If the dent is in a section of the body which has double layers, or some other factor makes it inaccessible from behind, a different technique is required. Drill several small holes through the metal inside the damaged area, particularly in the deeper sections. Screw long, self tapping screws into the holes just enough for them to get a good grip in the metal. Now the dent can be pulled out by pulling on the protruding heads of the screws with locking pliers.

7   The next stage of repair is the removal of paint from the damaged area and from an inch or so of the surrounding metal. This is easily done with a wire brush or sanding disk in a drill motor, although it can be done just as effectively by hand with sandpaper. To complete the preparation for filling, score the surface of the bare metal with a screwdriver or the tang of a file or drill small holes in the affected area. This will provide a good grip for the filler material. To complete the repair, see the Section on filling and painting.

Repair of rust holes or gashes

8   Remove all paint from the affected area and from an inch or so of the surrounding metal using a sanding disk or wire brush mounted in a drill motor. If these are not available, a few sheets of sandpaper will do the job just as effectively.

9   With the paint removed, you will be able to determine the severity of the corrosion and decide whether to replace the whole panel, if possible, or repair the affected area. New body panels are not as expensive as most people think and it is often quicker to install a

new panel than to repair large areas of rust.

10   Remove all trim pieces from the affected area except those which will act as a guide to the original shape of the damaged body, such as headlight shells, etc. Using metal snips or a hacksaw blade, remove all loose metal and any other metal that is badly affected by rust. Hammer the edges of the hole inward to create a slight depression for the filler material.

11   Wire brush the affected area to remove the powdery rust from the surface of the metal. If the back of the rusted area is accessible, treat it with rust inhibiting paint.

12   Before filling is done, block the hole in some way. This can be done with sheet metal riveted or screwed into place, or by stuffing the hole with wire mesh.

13   Once the hole is blocked off, the affected area can be filled and painted. See the following subsection on filling and painting.

Filling and painting

14   Many types of body fillers are available, but generally speaking, body repair kits which contain filler paste and a tube of resin hardener are best for this type of repair work. A wide, flexible plastic or nylon applicator will be necessary for imparting a smooth and contoured finish to the surface of the filler material. Mix up a small amount of filler on a clean piece of wood or cardboard (use the hardener sparingly). Follow the manufacturer's instructions on the package, otherwise the filler will set incorrectly.

15   Using the applicator, apply the filler paste to the prepared area. Draw the applicator across the surface of the filler to achieve the desired contour and to level the filler surface. As soon as a contour that approximates the original one is achieved, stop working the paste. If you continue, the paste will begin to stick to the applicator. Continue to add thin layers of paste at 20-minute intervals until the level of the filler is just above the surrounding metal.

16   Once the filler has hardened, the excess can be removed with a body file. From then on, progressively finer grades of sandpaper should be used, starting with a 180-grit paper and finishing with a 600-grit wet or-dry paper. Always wrap the sandpaper around a flat rubber or wooden block, otherwise the surface of the filler will not be completely flat. During the sanding of the filler surface, the wet-or-dry paper should be periodically rinsed in water. This will ensure that a very smooth finish is produced in the final stage.

17   At this point, the repair area should be surrounded by a ring of bare metal, which in turn should be encircled by the finely feathered edge of good paint. Rinse the repair area with clean water until all of the dust produced by the sanding operation is gone.

18   Spray the entire area with a light coat of primer. This will reveal any imperfections in the surface of the filler. Repair the imperfections with fresh filler paste or glaze filler and once more smooth the surface with sandpa-

per. Repeat this spray-and-repair procedure until you are satisfied that the surface of the filler and the feathered edge of the paint are perfect. Rinse the area with clean water and allow it to dry completely.

19 The repair area is now ready for painting. Spray painting must be carried out in a warm, dry, windless and dust free atmosphere. These conditions can be created if you have access to a large indoor work area, but if you are forced to work in the open, you will have to pick the day very carefully. If you are working indoors, dousing the floor in the work area with water will help settle the dust which would otherwise be in the air. If the repair area is confined to one body panel, mask off the surrounding panels. This will help minimize the effects of a slight mismatch in paint color. Trim pieces such as chrome strips, door handles, etc., will also need to be masked off or removed. Use masking tape and several thicknesses of newspaper for the masking operations.

20 Before spraying, shake the paint can thoroughly, then spray a test area until the spray painting technique is mastered. Cover the repair area with a thick coat of primer. The thickness should be built up using several thin layers of primer rather than one thick one. Using 600-grit wet-or-dry sandpaper, rub down the surface of the primer until it is very smooth. While doing this, the work area should be thoroughly rinsed with water and the wet-or-dry sandpaper periodically rinsed as well. Allow the primer to dry before spraying additional coats.

21 Spray on the top coat, again building up the thickness by using several thin layers of paint. Begin spraying in the center of the repair area and then, using a circular motion, work out until the whole repair area and about two inches of the surrounding original paint is covered. Remove all masking material 10 to 15 minutes after spraying on the final coat of paint. Allow the new paint at least two weeks to harden, then use a very fine rubbing compound to blend the edges of the new paint into the existing paint. Finally, apply a coat of wax.

6 Body repair - major damage

1 Major damage must be repaired by an auto body shop specifically equipped to perform unibody repairs. These shops have the specialized equipment required to do the job properly.

2 If the damage is extensive, the body must be checked for proper alignment or the vehicle's handling characteristics may be adversely affected and other components may wear at an accelerated rate.

3 Due to the fact that all of the major body components (hood, fenders, etc.) are separate and replaceable units, any seriously damaged components should be replaced rather than repaired. Sometimes the components can be found in a wrecking yard that specializes in used vehicle components,

9.1 The grille is held in place by eight screws

often at considerable savings over the cost of new parts.

7 Hinges and locks - maintenance

Once every 3000 miles, or every three months, the hinges and latch assemblies on the doors, hood and liftgate or tailgate should be given a few drops of light oil or lock lubricant. The door latch strikers should also be lubricated with a thin coat of grease to reduce wear and ensure free movement. Lubricate the door and liftgate locks with spray-on graphite lubricant.

8 Fixed glass - replacement

Replacement of the windshield and fixed glass requires the use of special fast-setting adhesive/caulk materials and some specialized tools and techniques. These operations should be left to a dealer service department or a shop specializing in glass work.

9 Radiator grille - removal and installation

Refer to illustration 9.1

1 Remove the phillips head mounting screws and detach the grille from the vehicle **(see illustration)**.

2 Installation is the reverse of removal.

10 Hood - removal, installation and adjustment

Refer to illustrations 10.2, 10.4 and 10.12

Note: *The hood is heavy and somewhat awkward to remove and install - at least two people should perform this procedure.*

Removal and installation

1 Use blankets or pads to cover the cowl area of the body and the fenders. This will protect the body and paint as the hood is

10.2 Use white paint or a marker to make alignment marks around the hood bolt heads - note the padding used to protect the body in case the hood swings back

10.4 Be sure to remove and keep track of the adjusting shims before lifting the hood off - it's a good idea to mark them so you know which side to return them to on installation

lifted off.

2 Scribe or paint alignment marks around the bolt heads to insure proper alignment during installation **(see illustration)**.

3 Disconnect any cables or wire harnesses which will interfere with removal.

4 Have an assistant support the weight of the hood. Remove the hinge-to-hood bolts and adjusting shims **(see illustration)**.

11

10.12 The hood bumpers, located at the front of the hood, can be screwed up or down to adjust the closed position of the hood

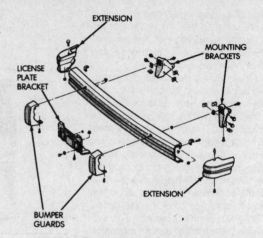

11.3a Typical front bumper details

5   Lift off the hood.
6   Installation is the reverse of removal. Be sure to reinstall the adjusting shims and tighten the hinge-to-hood bolts to the specified torque.

Adjustment

7   Fore-and-aft and side-to-side adjustment of the hood is done by moving the hood in relation to the hinge plate after loosening the bolts.
8   Scribe or paint around the hood bolt heads so you can judge the amount of movement **(see illustration 10.2)**.
9   Loosen the bolts and move the hood into correct alignment. Move it only a little at a time. Tighten the hood bolts and carefully lower the hood to check the alignment.
10   If necessary after installation, the entire hood latch assembly can be adjusted up-and-down as well as from side-to-side on the radiator support so the hood closes securely and the front of the hood is flush with the fenders. To do this, scribe a line around the hood latch mounting bolts to provide a reference point. Then loosen the bolts and reposition the latch assembly as necessary. Following adjustment, retighten the mounting bolts.
11   The rear of the hood can also be adjusted up-and-down so it is flush with the fenders. To do this, loosen the hood-to-hinge bolt and add or subtract adjusting shims, which are available at your dealer.
12   Finally, adjust the hood bumpers so the hood, when closed, is flush with the fenders **(see illustration)**.
13   The hood latch assembly, as well as the hinges, should be periodically lubricated with white lithium-base grease to prevent sticking and wear.

11   Bumpers - removal and installation

Refer to illustrations 11.3a, 11.3b and 11.3c

1   Disconnect any wiring or other components that would interfere with bumper removal.

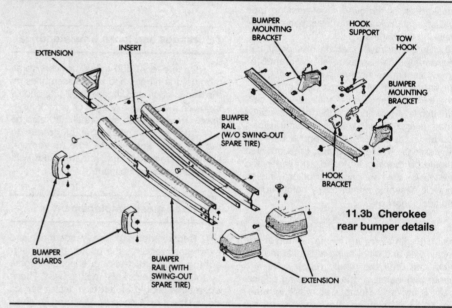

11.3b Cherokee rear bumper details

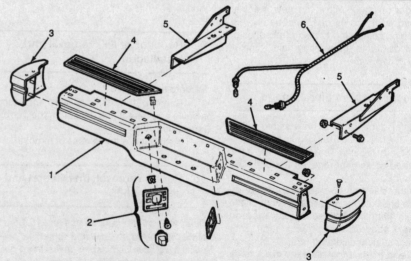

11.3c Typical Comanche rear bumper details

1 Bumper rail	4 Anti-skid plates
2 License plate light bulb housings	5 Bumper mounting brackets
3 Bumper extensions	6 License plate lights and wires

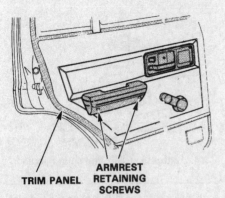

TRIM PANEL

ARMREST
RETAINING
SCREWS

**12.2a Remove the two armrest
retaining screws . . .**

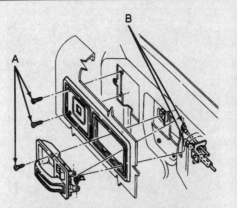

**12.2b . . . remove the three door
handle/bezel screws (A), detach the rod
links (B), then remove the door handle**

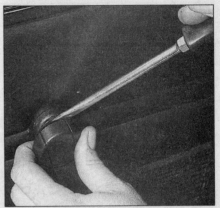

12.3 Pry off the crank handle

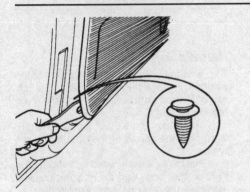

**12.4 A flat, forked tool such as this one
makes door panel removal much easier
because the long ribbed plastic retainers
are difficult to pry out**

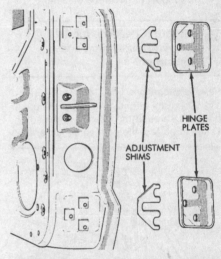

13.1 Remove the door check pin (arrow)

HINGE
PLATES

ADJUSTMENT
SHIMS

**13.6 Forward-and-backward adjustments
are made by adding or subtracting
adjustment shims between the door and
hinge plates**

2   Support the bumper with a jack or jack-stand. Alternatively, have an assistant support the bumper as the bolts are removed.
3   Remove the bolts that secure the bumper mounting brackets to the frame, then detach the bumper **(see illustrations)**.
4   Installation is the reverse of removal.
5   Tighten the retaining bolts securely.
6   Install any components that were removed.

12 Door trim panel - removal and installation

Refer to illustrations 12.2a, 12.2b, 12.3 and 12.4

1   Disconnect the negative cable from the battery.
2   Remove the armrest and door handle assemblies **(see illustrations)**.
3   On manual window regulator equipped models, remove the crank handle **(see illustration)**. On power regulator models, pry out the control switch assembly and unplug it.
4   Insert a flat, forked tool or a putty knife between the trim panel and the door and disengage the ribbed plastic retainers **(see illustration)**. Work around the outer edge until the

panel is free.
5   Once all of the retainers are disengaged, detach the trim panel, unplug any wire harness connectors and remove the trim panel from the vehicle.
6   For access to the inner door, carefully peel back the plastic watershield.
7   Prior to installation of the door panel, be sure to reinstall any retainers in the panel which may have come out during the removal procedure and remain in the door itself.
8   Plug in the wire harness connectors and place the panel in position in the door. Press the door panel into place until the clips are seated and install the handle and armrest assemblies. Install the manual regulator window crank or power window switch assembly.

13 Door - removal, installation and adjustment

Refer to illustrations 13.1 and 13.6

1   Remove the door trim panel (see Section 12). Disconnect any wire harness connectors and push them through the door opening so they won't interfere with door removal. Remove the door check pin **(see illustration)**.

2   Place a jack or jackstand under the door or have an assistant on hand to support it when the hinge bolts are removed. **Note:** *If a jack or jackstand is used, place a rag between it and the door to protect the door's painted surfaces.*
3   Scribe around the door hinges.
4   Remove the hinge-to-door bolts (a #30 Torx head tool is required) and carefully lift off the door.
5   Installation is the reverse of removal. Tighten the hinge-to-door bolts to the specified torque.
6   Following installation of the door, check the alignment and adjust it if necessary as follows:

a) *Forward-and-backward adjustments are made by adding or removing as necessary* **(see illustrations)**.
b) *The door lock striker can also be adjusted both up-and-down and sideways to provide positive engagement with the lock mechanism. This is done by loosening and moving the striker as necessary.*

11

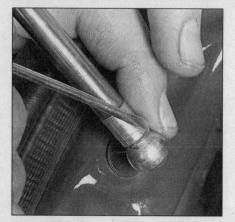

14.4 Detach the wire retainer with a small screwdriver and pry the end of the support strut off the ballstud

14.5 A #30 Torx head tool will be required to remove the hinge bolts

14.8 Liftgate striker and shim details

14 Liftgate (Cherokee models) - removal, installation and adjustment

Refer to illustrations 14.4, 14.5 and 14.8

1   Open the liftgate and cover the upper body area around the opening with pads or cloths to protect the painted surfaces when the liftgate is removed.

2   Detach the left side cargo area trim panel and unplug the electrical connector for the liftgate wiring harness.

3   Paint or scribe around the hinge flanges so the liftgate can be easily adjusted when installed.

4   While an assistant supports the liftgate, detach the support struts **(see illustration)**.

5   Remove the hinge bolts (a Torx head tool is necessary) and detach the liftgate from the vehicle **(see illustration)**.

6   Installation is the reverse of removal. Tighten the hinge bolts securely.

7   After installation, close the liftgate and make sure it's in proper alignment with the surrounding body panels. Adjustments are made by adding or removing shims (available at your dealer) between the hinge and liftgate or roof panel.

8   The engagement of the liftgate can be adjusted by removing the screws and the striker and adding or removing shims underneath to raise or lower it **(see illustration)**. If you add or remove shims, tighten the striker screws to the specified torque.

15 Tailgate (Comanche models) - removal and installation

Refer to illustrations 15.1a and 15.1b

Removal

1   Lower the tailgate and pull the centers of the supports up **(see illustration)**. Push

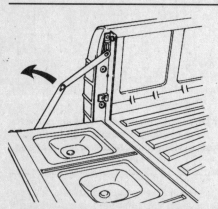

15.1a Pull the center of each tailgate support up . . .

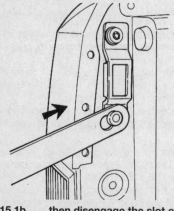

15.1b . . . then disengage the slot on the end of the support from the retaining dowel

each support forward to disengage the slot on the end of the support from the retaining dowel **(see illustration)**.

2   Pull the right side of the tailgate rearward to disengage the right hinge, then pull the tailgate to the right to disengage the left hinge.

Installation

3   With the tailgate in the open (horizontal) position, engage the female half of the left hinge (located on the tailgate) with the male half (on the cargo box).

4   Force the right side of the tailgate forward until the halves of the right hinge are engaged.

5   Pull the supports upward and engage the slots on their ends with the retaining dowels, then force the supports down to seat them.

16 Dashboard panels - removal and installation

Refer to illustrations 16.2, 16.4, 16.7 and 16.10

Warning: *Some models covered by this manual are equipped with Supplemental Restraint Systems, more commonly known as airbags. Always disable the airbag system before working in the vicinity of any airbag system components, to avoid the possibility of accidental deployment of the airbag(s), which could cause personal injury (see Chapter 12).*

1   Disconnect the negative cable at the battery.

1984 through 1996 models

Lower instrument panel

2   Remove the retaining screws, lower the

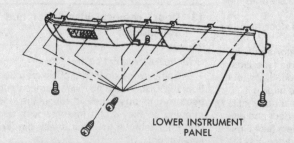

16.2 Remove the retaining screws from the lower instrument panel (1984 through 1996 models)

LOWER INSTRUMENT PANEL

panel, unplug any electrical connectors and remove the panel **(see illustration)**.

3   Installation is the reverse of removal.

Instrument cluster bezel

4   Remove the attaching screws and unsnap the bezel at the snap fit locations in the dash **(see illustration)**.

5   Remove the bezel.

6   Installation is the reverse of removal.

Radio and heater control panel

7   Remove the attaching screws and pull the panel assembly out **(see illustration)**.

8   Unplug the electrical connectors and remove the assembly.

9   Installation is the reverse of removal.

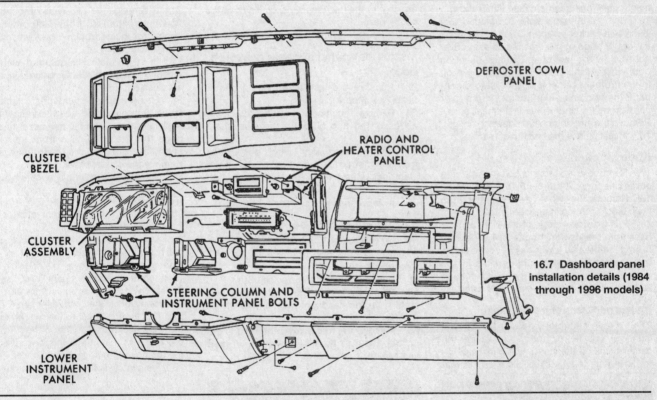

16.4 The instrument cluster bezel is held in place by four screws and is a snap fit (1984 through 1996 models)

16.7 Dashboard panel installation details (1984 through 1996 models)

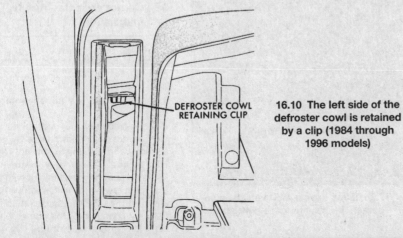

16.10 The left side of the defroster cowl is retained by a clip (1984 through 1996 models)

Defroster cowl panel

10   Remove the screws, detach the retaining clip on the driver's side and lift the cowl panel out **(see illustration)**.

11   Installation is the reverse of removal.

1997 and later models

Steering column opening cover

Refer to illustration 16.14

12   Disconnect the cable from the negative battery terminal.

13   If the vehicle is equipped with a tilt steering column, raise the column to its highest position.

14   Remove the three screws that attach the lower edge of the cover to the lower instru-

11

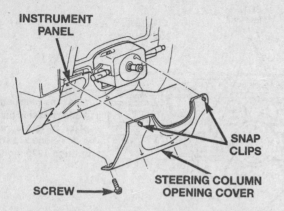

16.14 Steering column opening cover mounting details (1997 and later models)

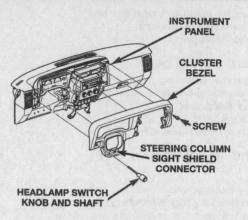

16.30 Instrument cluster bezel mounting details (1997 and later models)

ment panel reinforcement **(see illustration)**.

15   Using a suitable wide, flat-bladed tool (such as a gasket scraper), carefully pry loose the upper edge of the steering column cover right below the instrument cluster bezel on both sides of the steering column. Pry it off the instrument panel just enough to disengage the two snap-clip retainers from their corresponding slots in the instrument panel. Remove the steering column opening cover.

16   Installation is the reverse of removal.

Knee blocker

17   Remove the steering column opening cover (see Steps 12 through 15).

18   Remove the two screws that attach the knee blocker to the instrument panel.

19   Pull the upper edge of the knee blocker back from the instrument panel far enough to disengage the two lower mounting tabs from their corresponding mounting slots in the instrument panel. Remove the knee blocker.

20   Installation is the reverse of removal.

Instrument panel center bezel

21   Using a suitable wide, flat-bladed tool (such as a gasket scraper), carefully pry the center bezel off the instrument panel enough to disengage the six snap-clip retainers that

attach it to the instrument panel. Remove the center bezel.

22   Installation is the reverse of removal.

Instrument panel accessory switch bezel

23   Remove the instrument panel center bezel (see Step 21).

24   Remove the three screws that attach the accessory switch bezel to the instrument panel.

25   Pull the accessory switch bezel back from the instrument panel and unplug the electrical connectors. Remove the switch bezel.

26   Installation is the reverse of removal.

Instrument cluster bezel

Refer to illustration 16.30

27   Remove the knee blocker (see Steps 17 through 19).

28   Remove the center bezel (see Step 21).

29   Remove the headlight switch knob and shaft from the headlight switch (see Chapter 12).

30   Remove the four screws that attach the cluster bezel to the instrument panel **(see illustration)**.

31   Disengage the two ends of the steering column sight shield from each other at the connector located below the lower steering column shroud.

32   If the vehicle is equipped with a tilt steering column, place the column in its lowest position.

33   Using a suitable wide, flat-bladed tool (such as a gasket scraper), carefully pry all the way around the cluster bezel to disengage the five snap-clips from their corresponding slots in the instrument panel. Remove the cluster bezel.

34   Installation is the reverse of removal.

Instrument panel top cover

35   Remove the instrument cluster bezel (see Steps 27 through 33).

36   Using a suitable wide, flat-bladed tool (such as a gasket scraper), carefully pry up the rear edge of the top cover (the edge closest to you) from the instrument panel and disengage the seven snap-clip retainers from their corresponding slots in the instrument panel.

37   Pull the top cover to the rear (away from the windshield) to disengage the four snap-clip retainers that attach the forward edge of the top cover. Remove the top cover.

38   Installation is the reverse of removal.

---

17   Center console - removal and installation

1984 through 1996 models

Refer to illustrations 17.3a, 17.3b, 17.4a and 17.4b

1   Disconnect the negative cable at the battery.

2   On vehicles with automatic transmission, remove the shift handle by grasping it securely and pulling straight up. On vehicles with manual transmission, remove the shift knob by loosening the locknut with a wrench, then unscrewing the shift knob by hand.

3   Remove the transmission shift boot or plate by prying up at the corner with a small screwdriver **(see illustration)**. On 4WD vehi-

17.3a Pry up on the corner of the transmission shift boot to detach it (manual transmission shown, automatic transmission similar)

17.3b If you have a 4WD vehicle, also detach the transfer case shift plate

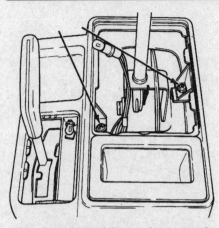

17.4a Locations of the two front console retaining screws (arrows)

cles, also remove the transfer case shift plate **(see illustration)**.
4   Remove the retaining screws or bolts **(see illustrations)**.
5   Detach the console and lift it from the vehicle.
6   Installation is the reverse of removal.

1997 and later models

7   On models with an automatic transmission, remove the shift lever handle by pulling it straight up. Then, carefully pry off the gear position indicator bezel with a suitably wide, flat-bladed tool.
8   On models with a manual transmission, loosen the locknut and unscrew the knob from the shift lever. Then, pull off the shift lever dust boot.
9   Open the console lid and remove the screws that attach the console to the floor-pan and mounting bracket.
10   Lift up the console and unplug all electrical connectors. Remove the console.
11   Installation is the reverse of removal.

18   Overhead console - removal and installation

Refer to illustrations 18.3 and 18.5
1   Disconnect the negative cable at the battery.

1984 through 1996 models

2   Remove the mounting screw ahead of the compass unit.
3   Using your hands, flex the console unit outward while pressing upward to disengage the unit from the rear mounting bracket **(see illustration)**.
4   Carefully slide the console toward the rear until the console is free from the front mounting bracket.
5   Press up on the rear of the console, then slide the console forward while holding it away from the headliner **(see illustration)**. Move the console forward until it is free from the rear mounting bracket.

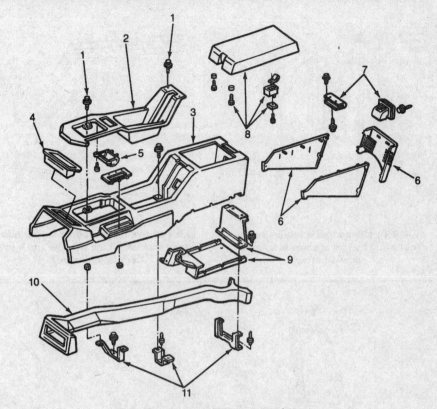

17.4b Center console component layout

1	*Center panel retaining screws*	*7*	*Ashtray and bracket*
2	*Center panel*	*8*	*Compartment lid*
3	*Console*	*9*	*Bottom panels*
4	*Storage pocket*	*10*	*Heater air ducts*
5	*Storage pocket*	*11*	*Brackets*
6	*Wing and rear panels*		

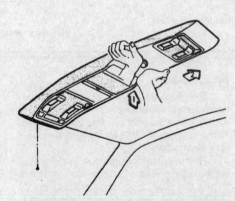

18.3 Flex the console unit outward (arrow 1), then press upward to disengage the unit from the rear mounting bracket

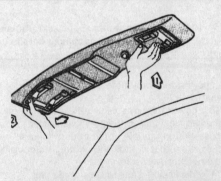

18.5 Press up on the rear of the console (arrow, 1), slide the console forward (arrow), hold it away from the headliner (arrow 2) and move the console forward until it is free from the rear mounting bracket

6   Carefully lower the console and disconnect the electrical connectors from the compass and remote keyless entry modules.
7   Installation is the reverse of removal. Flex the console housing outward near the remote keyless entry module until the console snaps onto the rear mounting bracket.

1997 and later models

8   Remove the two screws **(see illustration)** that attach the forward end of the overhead console to the headliner.
9   Using your fingertips, carefully pull the sides of the console housing out, near the rear, then slide the console forward to disen-

11

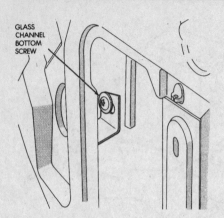

19.4 A Torx head tool is required when removing the glass channel bottom screw (arrow)

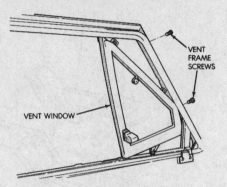

19.5 Remove the front door vent window frame screws, tilt the window and frame back and remove it

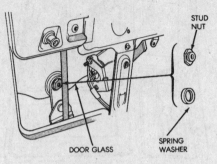

19.6 Front door glass attachment details

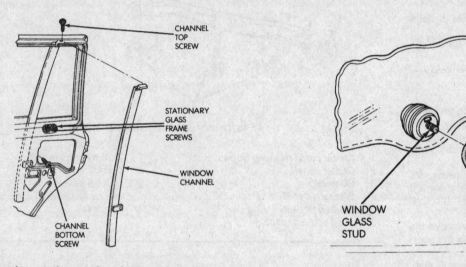

19.8 Rear door stationary glass frame and window channel attachment details

19.9 Rear door glass attachment details

gage the rear mounting tab from the headliner.

10   Lower the console slightly and unplug the two electrical connectors (one is near the push-button module at the front of the console, the other is near the center). Remove the overhead console.

11   Installation is the reverse of removal.

19   Door window glass - removal and installation

Refer to illustrations 19.4, 19.5, 19.6, 19.8 and 19.9

1   Remove the door trim panel and watershield (see Section 12).

2   Lower the window glass.

3   Remove the door window hardware, weather seal and moldings as necessary for access.

Front door

4   Remove the glass channel bottom screw **(see illustration)**.

5   Remove the vent window frame screws, tilt the window and frame back and remove it **(see illustration)**.

6   Remove the door glass attaching stud nut and spring washer, tilt the glass to detach it from the stud, then remove the glass by pulling it up and out of the door **(see illustration)**.

7   Installation is the reverse of removal. Tighten the vent window frame screws and stud nut securely.

Rear door

8   Remove the window channel screws and stationary glass frame screws, tilt the channel and frame forward and remove it from the door **(see illustration)**.

9   Remove the glass stud nut and wave washer **(see illustration)**.

10   Remove the window glass by tilting it to detach the stud, then sliding the glass up and out of the door.

11   Installation is the reverse of removal. Tighten the stationary glass frame screws and stud nut securely.

20   Door latch, lock cylinder and handle - removal and installation

Refer to illustrations 20.3, 20.4, 20.6 and 20.8

1   Remove the door trim panel and watershield (see Section 12).

2   Remove the door window glass (see Section 19).

Door latch

3   Remove the access plug (if equipped) in the end of the door **(see illustration)**.

4   Remove the three door latch retaining screws from the end of the door, disconnect

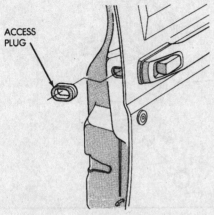

ACCESS
PLUG

20.3 On some models it is necessary to pry out the plug in the end of the door for access to the door handle

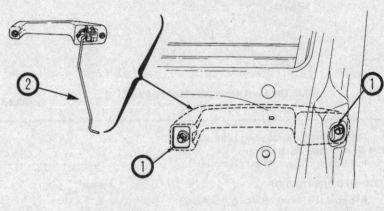

20.4 Door latch and lock details

1 *Latch retaining screws*
2 *Latch assembly*
3 *Power lock solenoid*
4 *Lock cylinder and clip*

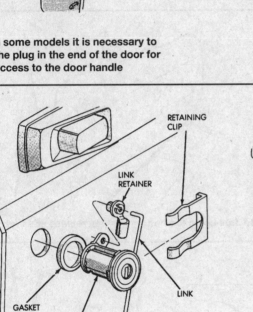

RETAINING
CLIP

LINK
RETAINER

LINK

GASKET

LOCK
CYLINDER

20.6 Lock cylinder installation details

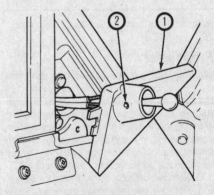

20.8 Remove the retaining nuts (1), pull the handle out, then detach the control rod (2)

the links, then remove the latch and (if equipped) solenoid **(see illustration)**. On power door locks, it will be necessary to drill out the rivets retaining the solenoid to the door.

5   Installation is the reverse of removal. Tighten the door latch retaining screws securely.

Lock cylinder

6   Disconnect the link from the lock cylinder **(see illustration)**.

7   Use pliers to slide the retaining clip off and remove the lock cylinder from the door. Installation is the reverse of removal. After installation, check the latch link freeplay by pressing the door handle release button in. Check that the latch link starts to move after the button moves about 1/64-inch. If necessary, loosen the adjusting screw on the latch, adjust the freeplay and then retighten it.

Handle

8   Remove the retaining nuts, pull the handle out, detach the control rod and remove the handle from the door **(see illustration)**.

9   Installation is the reverse of removal. Tighten the retaining nuts securely.

21   Outside mirror - removal and installation

Refer to illustration 21.2, 21.3 and 21.5

Standard mirror

1   Remove the door trim panel (see Section 12).

2   Remove the retaining screw, pull out the trim cover, loosen the adjustment knob set screw and pull off the trim cover **(see illustration)**.

21.2 Remove the retaining screw, pull out the trim cover (1), loosen the adjustment knob set screw (2) and detach the panel from the door

11

21.3 Three bolts (arrows) retain the mirror to the door

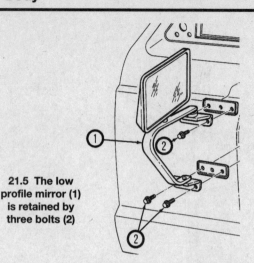

21.5 The low profile mirror (1) is retained by three bolts (2)

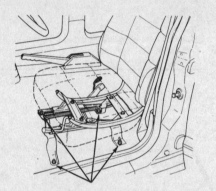

22.2 The front seats are retained by four nuts (arrows)

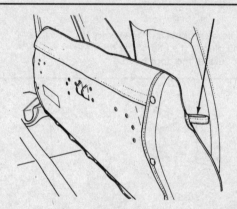

22.4a Pull up on the bottom cushion release strap (arrow) . . .

3   Remove the three retaining bolts and lift off the mirror **(see illustration)**.
4   Installation is the reverse of removal.

Low profile mirror

5   Remove the three retaining bolts and lower the mirror assembly from the vehicle **(see illustration)**.
6   Installation is the reverse of removal.

22   Seats - removal and installation

Refer to illustrations 22.2, 22.4a, 22.4b and 22.7

Front seat

Warning: *Some models covered by this manual are equipped with Supplemental Restraint Systems, more commonly known as airbags. Always disable the airbag system before working in the vicinity of any airbag system components, to avoid the possibility of accidental deployment of the airbag(s), which could cause personal injury (see Chapter 12).*
1   Remove the seat frame trim panels.
2   Remove the retaining nuts, unplug any electrical connectors and lift the seats from the vehicle **(see illustration)**.
3   Installation is the reverse of removal. Tighten the retaining nuts to the specified torque.

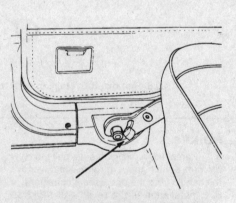

22.4b . . . tilt the cushion forward, then unlock the frame release lever (arrow)

Rear seat

Bottom cushion

4   Pull up on the bottom cushion release strap, tilt the cushion forward, then unlock the frame release lever and lift the assembly out **(see illustrations)**.
5   To install the cushion, place the cushion assembly in position, insert the left pivot shaft into the receptacle, snap the release lever in place and swing the seat down.

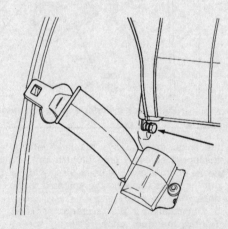

22.7 The seatback cushion pivot bolts (arrow) are located at the base of the cushion

Seatback cushion

6   Remove the bottom cushion (see above), remove the seatbelts from their elastic straps, then release the seatback lock.
7   Remove the pivot bolts and washers, move the cushion forward, then lift it from the vehicle **(see illustration)**.
8   To install the cushion, set it in place and install the pivot bolts and washers. Tighten

the bolts to the specified torque. Lock the seatback in position, insert the seatbelts through the elastic straps and install the bottom cushion.

23 Instrument panel - removal and installation

Refer to illustration 23.4

Warning: *Some models covered by this manual are equipped with Supplemental Restraint Systems, more commonly known as airbags. Always disable the airbag system before working in the vicinity of any airbag system components, to avoid the possibility of accidental deployment of the airbag(s), which could cause personal injury (see Section 20).*

1   Disconnect the negative cable at the battery.

1984 through 1996 models

2   Remove all the dashboard panels (see Chapter 11).

3   Remove the parking brake release handle (if equipped) and the lower heater/air conditioning duct, which is located below the steering column.

4   Loosen the mounting bolts, then rotate the instrument panel back so you can disconnect the tubes, wires, cables and harnesses **(see illustration).**

5   Make sure all instrument panel components are disconnected, then lift the instrument panel from the vehicle.

6   Installation is the reverse of removal. Make sure none of the wires or hoses get pinched between components when the instrument panel is rotated back into position.

1997 and later models

Refer to illustrations 23.9, 23.17 and 23.23

Warning: *Make sure that the airbag module is disarmed before servicing the instrument panel (see Chapter 12).*

7   Place the wheels in the straight-ahead position.

8   Disconnect the cable from the negative battery terminal. Wait at least two minutes before proceeding.

9   Remove the retaining screws from each sill trim plate **(see illustration)** and remove the left and right sill trim plates. Remove the fuse panel access door from the right cowl side trim panel. Remove the nut behind the access door. Remove the cowl side trim panel retaining screws and remove the left and right cowl side trim panels.

10 Remove the airbag module from the steering wheel (see Section 20 in Chapter 12).

11 Remove the steering wheel (see Chapter 10).

12 Remove the knee blocker cover, the knee blocker, the instrument panel center bezel and the instrument panel top cover (see Section 16).

13 Remove the center console (see Section 17).

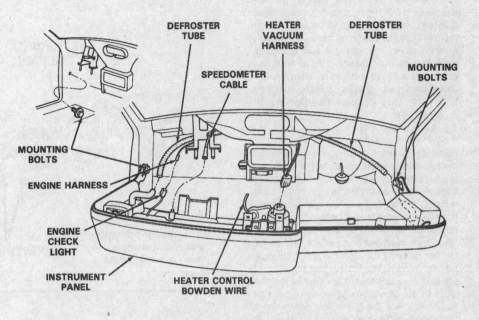

23.4 Loosen the mounting bolts, rotate the instrument panel back, then disconnect the tubes, wires, cables and harnesses

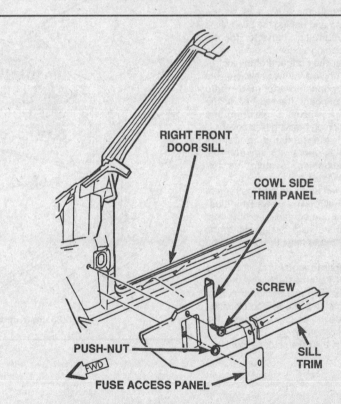

23.9 Sill trim plate and cowl side trim panel mounting details

14 Peel back the carpet from the front of the transmission tunnel. Remove the two upper bracket nuts that attach the instrument panel center support bracket to the studs on the instrument panel center support and remove the two lower nuts that attach the support bracket to the studs on the transmission tunnel. Remove the center support bracket.

15 Turn the ignition key to the On position and remove the ignition key lock cylinder (see Section 9 in Chapter 12).

16 Working below the steering column, remove the two screws from the lower steering column cover and remove the lower cover.

11

17   Remove the steering coupler bolt and the steering column mounting nuts **(see illustration)**. Lower the steering column from the mounting studs. Remove the upper steering column cover. Detach the wiring harness from the steering column. If the vehicle is equipped with an automatic transmission, disconnect the brake/transmission shift interlock cable (see Chapter 7B). Remove the steering column.

18   Locate the electrical connectors near the left cowl side inner panel. Remove the retaining screws from both connectors, detach the connectors from the body and unplug them. Locate the instrument panel-to-floor electrical connector on top of the transmission tunnel, remove the retaining screw from the connector and unplug the big connector and the two smaller ones.

19   Open the glove box. Locate the two rubber bumpers on the upper edge of the glove box opening in the instrument panel. Slide the bumpers downward and out of the their slots in the glove box opening upper reinforcement. Roll the glove box downward so that the stops molded into the glove box bin pass through the bumper slots in the upper reinforcement.

20   Reach through the inner side of the glove box opening and unplug the two halves of the vacuum harness connector for the heater/air conditioning system.

21   Reach under the right end of the instrument panel and unplug the two halves of the coaxial cable connector for the radio antenna. Also disengage the retainer on the radio half of the coaxial cable from the heater/air conditioning housing kick cover.

22   Disconnect the temperature control cable and blend-air door crank arm from the heater/air conditioning system control assembly (see Chapter 3).

23   Loosen the left and right side instrument panel cowl side roll down screws **(see illustration)**. Remove the screws and nuts that attach the top of the instrument panel to the top of the dash panel near the base of the windshield.

24   Using a helper to assist you, lift the top of the instrument panel assembly off the two dash panel studs, then pull the lower part of the instrument panel to the rear to clear the cowl side roll down screws. Remove the instrument panel assembly.

25   Installation is the reverse of removal.

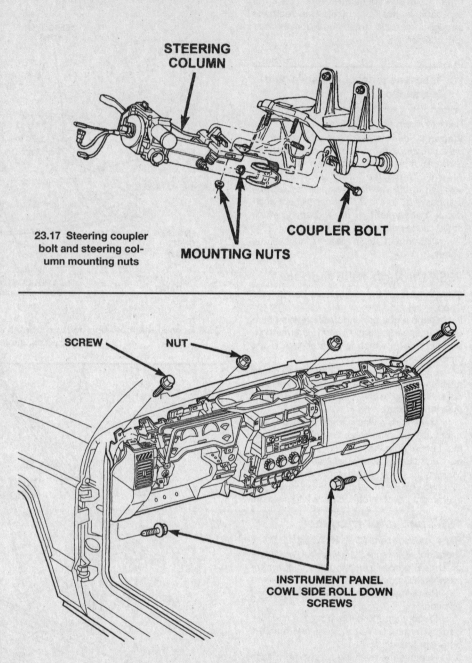

23.17 Steering coupler bolt and steering column mounting nuts

23.23 Instrument panel cowl fastener locations

Chapter 12
Chassis electrical system

Contents

Specifications

Light bulb application

	Number
Under hood light	
1984 through 1992	90
1993 on	105
Front park/turn signal light	2057NA
Headlight	
Single	
Up to 1992	H6052
1993 on	H6054
Dual	
High beam	H4701
Low beam	H4703
Fog light	H3
License plate light	
Cherokee	
With swing-out spare tire	168
Without swing-out spare tire	67
Comanche	
With step bumper	194
Without step bumper	67
Stop/tail light	2057
High-mounted brake light	922
Rear turn signal light	1156
Side marker light	97
Backup light	1156

1 General information

The electrical system is a 12-volt, negative ground type. Power for the lights and all electrical accessories is supplied by a lead/acid-type battery which is charged by the alternator.

This Chapter covers repair and service procedures for the various electrical components not associated with the engine. Information on the battery, alternator, distributor and starter motor can be found in Chapter 5.

It should be noted that when portions of the electrical system are serviced, the negative battery cable should be disconnected from the battery to prevent electrical shorts and/or fires.

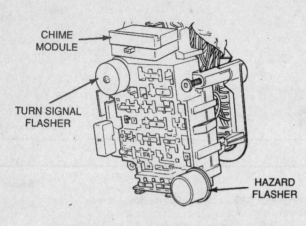

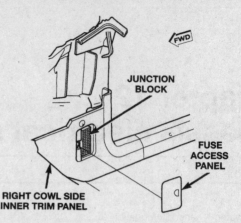

3.1a The fuse block is located under the left side of the instrument panel (1984 through 1996 models)

3.1b On 1997 and later models, the fuse panel is located behind a small access panel in the right cowl trim panel (this fuse panel is sometimes referred to as a "junction block"

2   Electrical troubleshooting - general information

A typical electrical circuit consists of an electrical component, any switches, relays, motors, fuses, fusible links or circuit breakers related to that component and the wiring and connectors that link the component to both the battery and the chassis. To help you pinpoint an electrical circuit problem, wiring diagrams are included at the end of this book.

Before tackling any troublesome electrical circuit, first study the appropriate wiring diagrams to get a complete understanding of what makes up that individual circuit. Trouble spots, for instance, can often be narrowed down by noting if other components related to the circuit are operating properly. If several components or circuits fail at one time, chances are the problem is in a fuse or ground connection, because several circuits are often routed through the same fuse and ground connections.

Electrical problems usually stem from simple causes, such as loose or corroded connections, a blown fuse, a melted fusible link or a bad relay. Visually inspect the condition of all fuses, wires and connections in a problem circuit before troubleshooting it.

If testing instruments are going to be utilized, use the diagrams to plan ahead of time where you will make the necessary connections in order to accurately pinpoint the trouble spot.

The basic tools needed for electrical troubleshooting include a circuit tester or voltmeter (a 12-volt bulb with a set of test leads can also be used), a continuity tester, which includes a bulb, battery and set of test leads, and a jumper wire, preferably with a circuit breaker incorporated, which can be used to bypass electrical components. Before attempting to locate a problem with test instruments, use the wiring diagram(s) to decide where to make the connections.

Voltage checks

Voltage checks should be performed if a circuit is not functioning properly. Connect one lead of a circuit tester to either the negative battery terminal or a known good ground. Connect the other lead to a connector in the circuit being tested, preferably nearest to the battery or fuse. If the bulb of the tester lights, voltage is present, which means that the part of the circuit between the connector and the battery is problem free. Continue checking the rest of the circuit in the same fashion. When you reach a point at which no voltage is present, the problem lies between that point and the last test point with voltage. Most of the time the problem can be traced to a loose connection. **Note:** *Keep in mind that some circuits receive voltage only when the ignition key is in the Accessory or Run position.*

Finding a short

One method of finding shorts in a circuit is to remove the fuse and connect a test light or voltmeter in its place to the fuse terminals. There should be no voltage present in the circuit. Move the wiring harness from side-to-side while watching the test light. If the bulb goes on, there is a short to ground somewhere in that area, probably where the insulation has rubbed through. The same test can be performed on each component in the circuit, even a switch.

Ground check

Perform a ground test to check whether a component is properly grounded. Disconnect the battery and connect one lead of a self-powered test light, known as a continuity tester, to a known good ground. Connect the other lead to the wire or ground connection being tested. If the bulb goes on, the ground is good. If the bulb does not go on, the ground is not good.

Continuity check

A continuity check is done to determine if there are any breaks in a circuit - if it is passing electricity properly. With the circuit off (no power in the circuit), a self-powered continuity tester can be used to check the circuit. Connect the test leads to both ends of the circuit (or to the "power" end and a good ground), and if the test light comes on the circuit is passing current properly. If the light doesn't come on, there is a break somewhere in the circuit. The same procedure can be used to test a switch, by connecting the continuity tester to the switch terminals. With the switch turned On, the test light should come on.

Finding an open circuit

When diagnosing for possible open circuits, it is often difficult to locate them by sight because oxidation or terminal misalignment are hidden by the connectors. Merely wiggling a connector on a sensor or in the wiring harness may correct the open circuit condition. Remember this when an open circuit is indicated when troubleshooting a circuit. Intermittent problems may also be caused by oxidized or loose connections.

Electrical troubleshooting is simple if you keep in mind that all electrical circuits are basically electricity running from the battery, through the wires, switches, relays, fuses and fusible links to each electrical component (light bulb, motor, etc.) and to ground, from which it is passed back to the battery. Any electrical problem is an interruption in the flow of electricity to and from the battery.

3   Fuses - general information

Refer to illustrations 3.1a, 3.1b and 3.3

The electrical circuits of the vehicle are protected by a combination of fuses, circuit

3.3 When a fuse blows, the element between the terminals melts - the fuse on the left is blown, the fuse on the right is good

Bad   Good

breakers and fusible links. The fuse block is located under the instrument panel on the left side of the dashboard **(see illustration)**.

Each of the fuses is designed to protect a specific circuit, and the various circuits are identified on the fuse panel itself.

Miniaturized fuses are employed in the fuse block. These compact fuses, with blade terminal design, allow fingertip removal and replacement. If an electrical component fails, always check the fuse first. A blown fuse is easily identified through the clear plastic body. Visually inspect the element for evidence of damage **(see illustration)**. If a continuity check is called for, the blade terminal tips are exposed in the fuse body.

Be sure to replace blown fuses with the correct type. Fuses of different ratings are physically interchangeable, but only fuses of the proper rating should be used. Replacing a fuse with one of a higher or lower value than specified is not recommended. Each electrical circuit needs a specific amount of protection. The amperage value of each fuse is molded into the fuse body.

If the replacement fuse immediately fails, don't replace it again until the cause of the problem is isolated and corrected. In most cases, the cause will be a short circuit in the wiring caused by a broken or deteriorated wire.

4   Fusible links - general information

Some circuits are protected by fusible links. The links are used in circuits which are not ordinarily fused, such as the ignition circuit.

Although the fusible links appear to be a heavier gauge than the wire they are protecting, the appearance is due to the thick insulation. All fusible links are four wire gauges smaller than the wire they are designed to protect.

Fusible links cannot be repaired, but a new link of the same size wire can be put in its place. The procedure is as follows:

a) *Disconnect the negative cable from the battery.*
b) *Disconnect the fusible link from the wiring harness.*
c) *Cut the damaged fusible link out of the wiring just behind the connector.*
d) *Strip the insulation back approximately 1/2-inch.*
e) *Position the connector on the new fusible link and crimp it into place.*
f) *Use rosin core solder at each end of the new link to obtain a good solder joint.*
g) *Use plenty of electrical tape around the*

soldered joint. No wires should be exposed.
h) *Connect the battery ground cable. Test the circuit for proper operation.*

5   Circuit breakers - general information

Circuit breakers protect components such as power windows, power door locks and headlights. Some circuit breakers are located in the fuse box.

On some models the circuit breaker resets itself automatically, so an electrical overload in a circuit breaker protected system will cause the circuit to fail momentarily, then come back on. If the circuit does not come back on, check it immediately. Once the condition is corrected, the circuit breaker will resume its normal function. Some circuit breakers must be reset manually.

6   Relays - general information

Refer to illustration 6.2

Several electrical accessories in the vehicle use relays to transmit the electrical signal to the component. If the relay is defective, that component will not operate properly.

The various relays are grouped together in several locations **(see illustration)**.

If a faulty relay is suspected, it can be removed and tested by a dealer service department or a repair shop. Defective relays must be replaced as a unit.

7   Turn signal and hazard flashers - check and replacement

1984 through 1995 models
Turn signal flasher

1   The turn signal flasher, a small canister-shaped unit located in the fuse block **(see**

6.2a Some of the relays (arrows) are located in the engine compartment adjacent to the battery, under a cover

LEFT FENDER

BATTERY   POWER DISTRIBUTION CENTER   80ba775b

6.2b On 1997 and later models, relays and fuses for the engine compartment are housed in the power distribution center, which is located behind the battery

12

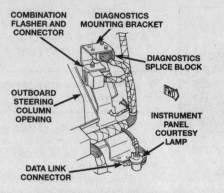

7.19 Combination flasher unit installation details (1997 and later models)

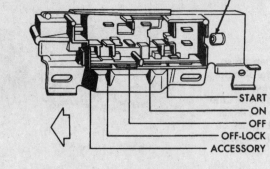

8.8a Standard steering column ignition switch details

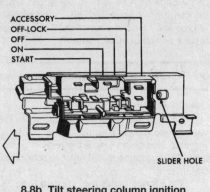

8.8b Tilt steering column ignition switch details

illustration 3.1), flashes the turn signals.

2 When the flasher unit is functioning properly, an audible click can be heard during its operation. If the turn signals fail on one side or the other and the flasher unit does not make its characteristic clicking sound, a faulty turn signal bulb is indicated.

3 If both turn signals fail to blink, the problem may be due to a blown fuse, a faulty flasher unit, a broken switch or a loose or open connection. If a quick check of the fuse box indicates that the turn signal fuse has blown, check the wiring for a short before installing a new fuse.

4 To replace the flasher, simply pull it out of the fuse block.

5 Make sure that the replacement unit is identical to the original. Compare the old one to the new one before installing it.

6 Installation is the reverse of removal.

Hazard flasher

7 The hazard flasher, a small canister-shaped unit located in the fuse block, flashes all four turn signals simultaneously when activated.

8 The hazard flasher is checked in a fashion similar to the turn signal flasher (see Steps 2 and 3).

9 To replace the hazard flasher, pull it from the back of fuse block.

10 Make sure the replacement unit is identical to the one it replaces. Compare the old

one to the new one before installing it.

11 Installation is the reverse of removal.

1996 and later models

12 On these models, a combination flasher serves as the turn signal flasher and as the hazard flasher.

13 Disconnect the cable from the negative battery cable.

1996 models

15 Remove the lower instrument panel (see Section 16 in Chapter 11).

16 Locate the relay center, on the lower instrument panel reinforcement, under the instrument panel, to the right of the steering column. Unplug the (blue) combination flasher from the relay center.

17 Installation is the reverse of removal.

1997 and later models

Refer to illustration 7.19

18 Remove the knee blocker from the instrument panel (see Section 16 in Chapter 11).

19 Reach through the left side of the steering column opening and disengage the retainer for the combination flasher electrical connector from the instrument panel diagnostics mounting bracket **(see illustration)**. Use a flashlight if necessary.

20 Pull out the combination flasher slightly, then unplug the flasher from the electrical connector. Remove the combination flasher.

21 Installation is the reverse of removal.

8 Ignition switch (1984 through 1994 models) - removal and installation

Warning: *Some models covered by this manual are equipped with Supplemental Restraint Systems, more commonly known as airbags. Always disable the airbag system before working in the vicinity of any airbag system components, to avoid the possibility of accidental deployment of the airbag(s), which could cause personal injury (see Section 20).*
Refer to illustrations 8.8a and 8.8b

1 Disconnect the negative cable at the battery.

Removal

2 The ignition switch is located on the lower end of the steering column and is connected by an actuating rod to the key lock assembly.

3 Remove the lower dashboard trim panel (see Chapter 11).

4 Set the key lock in the Off-Lock position.

5 Remove the two screws retaining the switch to the steering column and disconnect the actuating rod.

6 Unplug the black harness electrical connector from the white connector and remove the switch from the steering column.

Installation

7 Place the switch in position.

8 Move the slider on the switch to the Accessory position **(see illustrations)**. On standard columns, the Accessory position is to the extreme left. On tilt columns, the Accessory position is on the extreme right.

9 Connect the actuating rod to the slider hole and install the switch with the screws finger tight.

10 Hold the key lock in the Accessory position, push the switch toward the bottom of the column slightly to remove any slack in the rod and tighten the retaining screws securely.

11 Plug in the connector and install the dashboard trim panel.

9 Ignition switch and key lock cylinder (1995 and later models) - removal and installation

1995 and 1996 models

Ignition switch

Refer to illustration 9.6

1 Disconnect the cable from the negative battery terminal and wait at least two minutes before proceeding.

2 If the vehicle has a tilt column, unscrew the tilt lever (turn it counterclockwise) and remove it.

3 Remove the upper and lower steering column shroud retaining screws and remove the upper and lower shrouds.

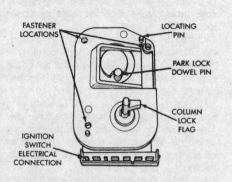

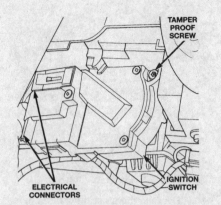

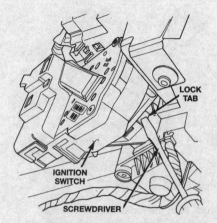

9.6 The park lock dowel pin and column lock flag must also be correctly indexed, as shown, before installing the switch

9.23 Remove the ignition switch retaining screw (the tamper-proof Torx type) (1997 and later models)

9.24 Using a small screwdriver, depress the locking tab and remove the ignition switch from the steering column (1997 and later models)

4   Remove the ignition switch mounting screws. (These screws are tamper-proof Torx type screws, so you'll need the correct Torx bit.)

5   Carefully detach the switch from the steering column, unlock the locking tabs on the two electrical connectors and unplug the connectors. Remove the switch.

6   Install the ignition switch as follows:

a) *Before attaching the ignition switch to a tilt steering column, make sure that the shift lever is in the PARK position. The park lock dowel pin and column lock flag must also be correctly indexed* **(see illustration)** *before installing the switch.*

b) *Put the ignition switch in the LOCK position (the switch is in the LOCK position when the column lock flag is parallel to the ignition switch terminals* **(see illustration 9.6)**.

c) *Position the ignition switch par lock dowel pin so that it will engage the steering column park lock slider linkage.*

d) *Place the ignition switch against the lock housing opening on the steering column. Make sure that the ignition switch park lock dowel pin enters the slot in the park lock slider linkage in the steering column.*

7   Installation is otherwise the reverse of removal.

Key lock cylinder

8   Remove the ignition switch/key lock cylinder assembly (see Steps 1 through 5).

9   Insert the key in the ignition switch and turn the key to the LOCK position.

10   Using the correct size Torx bit, remove the key lock cylinder retaining screw and bracket.

11   Rotate the key clockwise to the OFF position. The key lock cylinder will unseat itself, i.e. it will come out about 1/8-inch. DO NOT TRY TO REMOVE THE KEY CYLINDER YET, or you will damage the ignition switch

halo light ring.

12   With the key cylinder in this unseated position, rotate the key counterclockwise to the LOCK position and remove the key.

13   Remove the key lock cylinder.

14   Installation is the reverse of removal.

1997 and later models
Key lock cylinder

15   Disconnect the negative battery cable and wait at least two minutes before proceeding.

16   If the vehicle is equipped with an automatic transmission, put the shift lever in the PARK position.

17   Turn the key to the ON position.

18   Insert a small screwdriver or pin punch through the access hole in the lower steering column cover and depress the release tang on the bottom of the key lock cylinder. While pushing up on the release tang, pull out the key lock cylinder.

19   Installation is the reverse of removal. Be sure to push in the key lock cylinder until it snaps into place.

Ignition switch

Refer to illustrations 9.23, 9.24 and 9.25

20   Remove the ignition key lock cylinder (see Section 18).

21   Remove the lower steering column cover screws and remove the cover.

22   Unplug the two electrical connectors from the ignition switch.

23   Remove the ignition switch retaining screw **(see illustration)**. (The screw is a tamper-proof Torx type.)

24   Using a small screwdriver, depress the locking tab **(see illustration)** and remove the ignition switch from the steering column.

25   Installation is the reverse of removal. Before installing the switch, rotate the slot in the switch to the ON position **(see illustration)**.

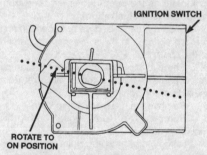

9.25 Before installing the switch, be sure to rotate the slot in the switch to the ON position (1997 and later models)

10   Turn signal switch - removal and installation

Warning: *Some models covered by this manual are equipped with Supplemental Restraint Systems, more commonly known as airbags. Always disable the airbag system before working in the vicinity of any airbag system components, to avoid the possibility of accidental deployment of the airbag(s), which could cause personal injury (see Section 20).*

1984 through 1995 models

Refer to illustrations 10.4 and 10.9

1   Disconnect the negative battery cable and remove the steering wheel as detailed in Chapter 10.

2   Remove the lower trim cover located at the base of the dashboard (see Chapter 11).

3   At the end of the steering column, some models have a plastic cover plate which should be pried out of the column using a screwdriver in the slots provided.

4   The lock plate will now have to be removed from the steering column. This is

12

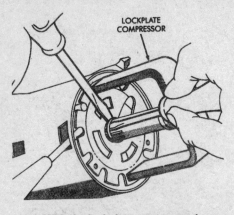

10.4 With the lockplate compressed, pry out the snap-ring with a screwdriver

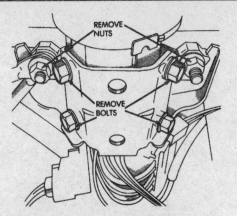

10.9 Remove the steering column nuts and bolts so the column can be lowered

held in place with a snap-ring which fits into a groove in the steering shaft. The lock plate must be depressed to relieve pressure on the snap-ring. A special U-shaped tool which fits on the shaft should be used to depress the lock plate as the snap-ring is removed from its groove **(see illustration)**.

5   Slide the cancelling cam, upper bearing preload spring and thrust washer off the end of the shaft.

6   Remove the turn signal lever attaching screw and withdraw the turn signal lever from the side of the column.

7   Remove the hazard warning knob.

8   Remove the turn signal assembly mounting screws.

9   Remove the nuts and bolts retaining the steering column to the bracket **(see illustration)**. Loosen the column brace mounting nut at the driver's side kick panel. This will allow the column to drop, providing clearance for switch and wiring removal.

10   Pull the switch wiring connector out of the bracket on the steering column jacket. Tape the connector terminals to prevent damage. Feed the wiring connector up through the column support bracket and pull the switch, wiring harness and connectors out the top of the steering column.

11   Installation is a reversal of removal; however, make sure the wiring harness is in the protector as it is pulled into position. Before installing the thrust washer, upper bearing preload spring and cancelling cam, make sure the switch is in the neutral position and the warning knob is pulled out. Always use a new snap-ring on the shaft for the lock plate.

1996 and later models

12   On these models, the headlight dimmer, turn signal and hazard flasher and windshield wiper/washer switches are integrated into one switch assembly on the steering column. This switch is known as a multi-function switch.

13   Disconnect the cable from the negative battery terminal and then wait at least two minutes before proceeding.

1996 models

14   Remove the tilt steering column lever, if applicable.

15   Remove the lower instrument panel/knee blocker (see Section 23 in Chapter 11).

16   Remove the upper and lower steering column shroud retaining screws and remove

the shrouds. (Note that there are two upper and two lower shrouds; you can't remove one of the upper shrouds until the steering column upper bracket mounting nuts have been loosened.)

17   Loosen - but do not remove - the steering column upper bracket mounting nuts.

18   Move the other upper shroud out of the way to gain access to the multi-function switch.

19   Remove the tamper-proof Torx retaining screws from the multi-function switch, carefully detach the switch from the steering column and loosen the electrical connector retaining screw (the screw stays in the connector).

20   Unplug the electrical connector from the multi-function switch and remove the switch.

21   Installation is the reverse of removal.

1997 and later models

Refer to illustrations 10.27a and 10.27b

22   Remove the knee blocker (see Section 23 in Chapter 11).

23   Place the tilt column, if equipped, in the fully raised position.

24   Insert a small screwdriver or pin punch through the access hole in the lower steering column cover and push in the ignition key lock cylinder retaining tumbler (see Section 9). While depressing the retaining tumbler, pull the ignition lock cylinder and key out of the ignition lock housing.

25   Remove the three screws that attach the lower steering column cover to the upper cover.

26   On vehicles with a non-tilt steering column, loosen the two nuts that secure the column upper mounting bracket to the dash panel steering column bracket studs. Lower the column far enough to remove the upper steering column cover.

27   Remove the three screws (two on the top and one on the side of the switch water shield) that attach the switch water shield and bracket to the steering column **(see illustrations)**.

28   Carefully pull the lower mounting tab on

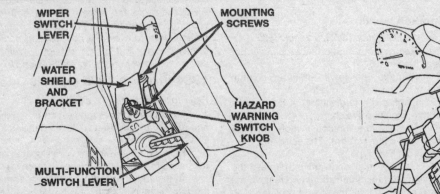

10.27a To detach the switch watershield and bracket from the steering column, remove the two screws (arrows) on the top of the switch shield . . .

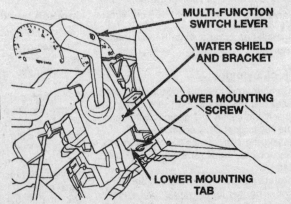

10.27b . . . and the screw (arrow) on the bottom of the bracket (1997 and later models)

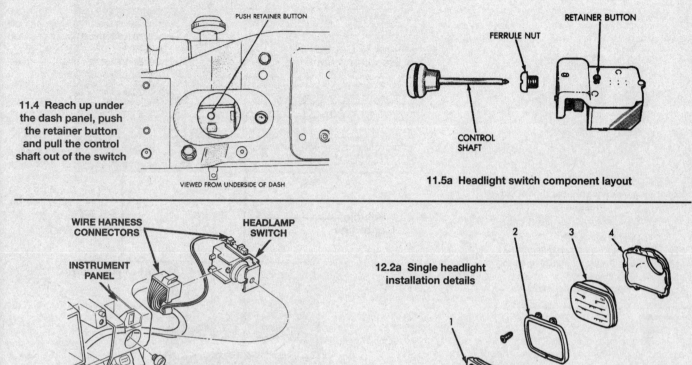

11.4 Reach up under the dash panel, push the retainer button and pull the control shaft out of the switch

VIEWED FROM UNDERSIDE OF DASH

PUSH RETAINER BUTTON

RETAINER BUTTON

FERRULE NUT

CONTROL SHAFT

11.5a Headlight switch component layout

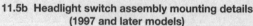

WIRE HARNESS CONNECTORS

HEADLAMP SWITCH

INSTRUMENT PANEL

NUT

11.5b Headlight switch assembly mounting details (1997 and later models)

12.2a Single headlight installation details

1   Headlight bezel
2   Headlight retainer
3   Headlight
4   Headlight bucket

the switch water shield bracket away from the steering column far enough to clear the screw boss below the multi-function switch lever.

29   Remove the water shield and bracket and the multi-function switch from the steering column and unplug the electrical connectors. (On vehicles equipped with a tilt steering column, lifting up on the tilt release lever will provide extra clearance for switch removal.)

30   Carefully remove the water shield by pulling it off over the hazard warning switch knob and the multi-function switch lever. This left half of the multi-function switch includes the headlight dimmer switch and the hazard warning flasher and turn signal switch.

31   To remove the windshield wiper and washer switch (the right half of the multi-function switch), carefully pull the windshield wiper and washer switch up and away from the right side of the steering column and unplug the electrical connector from the switch.

32   Installation is the reverse of removal.

11 Headlight switch - removal and installation

Refer to illustrations 11.4, 11.5a and 11.5b

1   Disconnect the negative cable at the battery.

2   On 1997 and later models, remove the knee blocker (see Section 16 in Chapter 11).

3   Pull the headlight switch out to the On position.

4   Reach up under the dash panel, push the retainer button, then pull the control shaft out of the switch **(see illustration)**.

5   Unplug the electrical connector, unscrew the ferrule nut and remove the switch from the instrument panel **(see illustrations)**.

6   To install the switch, plug in the electrical connector, place the switch in position and install the ferrule nut. Insert the shaft into the switch until it is seated.

12 Headlights - removal and installation

Refer to illustrations 12.2a and 12.2b

1   Disconnect the negative cable from the battery.

2   Remove the retaining screws and detach the headlight bezel **(see illustrations)**.

3   Remove the headlight retainer screws, taking care not to disturb the adjustment screws.

4   Remove the retainer and pull the headlight out enough to allow the connector to be unplugged.

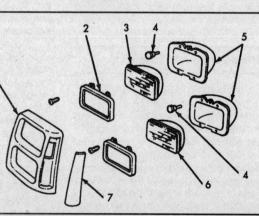

12.2b Dual headlight installation details

1   Headlight bezel
2   Headlight retainer
3   Headlight (low beam)
4   Headlight adjusters
5   Headlight buckets
6   Headlight (high beam)
7   Side marker light

12

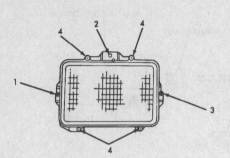

13.1a Single headlight adjustment screw details

1   *Left-and-right adjusting screw (right headlight)*
2   *Up-and-down adjusting screw*
3   *Left-and-right adjusting screw (left headlight)*
4   *Headlight retainer screws*

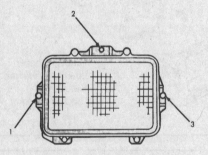

13.1b Dual headlight adjustment screw details

1   *Left-and-right adjusting screw (right headlight)*
2   *Up-and-down adjusting screw*
3   *Left-and-right adjusting screw (left headlight)*

5   Remove the headlight.
6   To install the headlight, plug the connector in, place the headlight in position and install the retainer and screws. Tighten the screws securely.
7   Place the headlight bezel in position and install the retaining screws.

13   Headlights - adjustment

Refer to illustrations 13.1a, 13.1b and 13.2
Note: *The headlights must be aimed correctly. If adjusted incorrectly they could blind the driver of an oncoming vehicle and cause a serious accident or seriously reduce your ability to see the road. The headlights should be checked for proper aim every 12 months and any time a new headlight is installed or front end body work is performed. It should be emphasized that the following procedure*

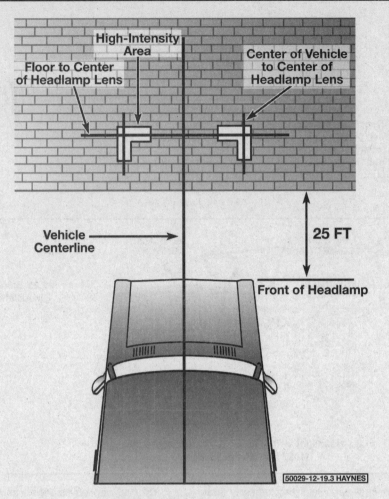

13.2 Headlight aiming details

is only an interim step which will provide temporary adjustment until the headlights can be adjusted by a properly equipped shop.
1   Headlights have two spring loaded adjusting screws, one on the top controlling up-and-down movement and one on the side controlling left-and-right movement **(see illustrations)**.
2   There are several methods of adjusting the headlights. The simplest method requires a blank wall 25 feet in front of the vehicle and a level floor **(see illustration)**.
3   Position masking tape vertically on the wall in reference to the vehicle centerline and the centerlines of both headlights.
4   Position a horizontal tape line in reference to the centerline of all the headlights. **Note:** *It may be easier to position the tape on the wall with the vehicle parked only a few inches away.*
5   Adjustment should be made with the vehicle sitting level, the gas tank half-full and no unusually heavy load in the vehicle.
6   Starting with the low beam adjustment, position the high intensity zone so it is two inches below the horizontal line and two inches to the right of the headlight vertical line. Adjustment is made by turning the top

adjusting screw clockwise to raise the beam and counterclockwise to lower the beam. The adjusting screw on the side should be used in the same manner to move the beam left or right.
7   With the high beams on, the high intensity zone should be vertically centered with the exact center just below the horizontal line. **Note:** *It may not be possible to position the headlight aim exactly for both high and low beams. If a compromise must be made, keep in mind that the low beams are the most used and have the greatest effect on driver safety.*
8   Have the headlights adjusted by a dealer service department or service station at the earliest opportunity.

14   Bulb replacement

Refer to illustrations 14.1a, 14.1b, 14.1c, 14.1d, 14.1e, 14.1f, 14.1g, 14.1h and 14.3
1   The lenses of many lights are held in place by screws, which makes it a simple procedure to gain access to the bulbs **(see illustrations)**.
2   On some lights the lenses are held in

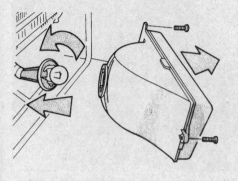

14.1a Front turn signal/parking light bulb replacement details

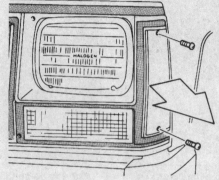

14.1b Side marker light housing removal details

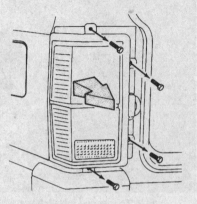

14.1c Cherokee tail light housing removal details

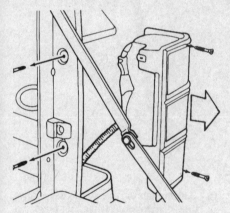

14.1d Comanche tail light housing removal details

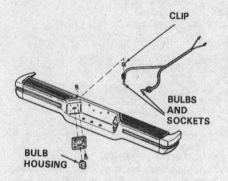

14.1e Comanche license plate bulb replacement details

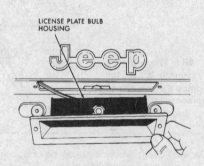

14.1f Cherokee license plate bulb replacement details (without swing-out spare tire)

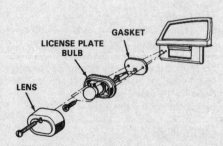

14.1g Cherokee license plate bulb replacement details (with swing-out spare tire)

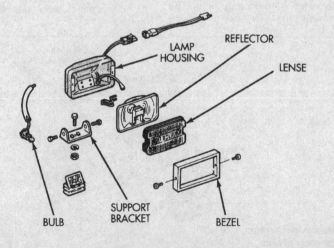

14.1h Fog lamp installation details

14.3 To remove a bulb housing, rotate the housing counterclockwise about 60-degrees and pull it out

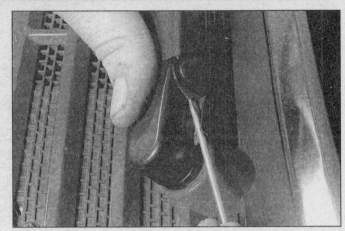

15.2 Pry up on the locking lever with a small screwdriver, then pull the wiper arm off the pivot

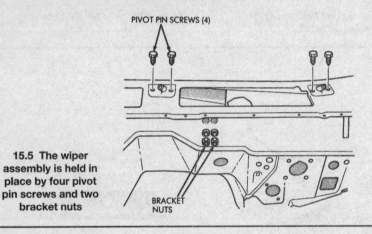

PIVOT PIN SCREWS (4)

BRACKET NUTS

15.5 The wiper assembly is held in place by four pivot pin screws and two bracket nuts

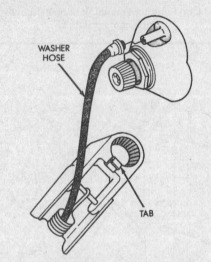

WASHER HOSE

TAB

15.7 Lift the tab to release the wiper arm, then disconnect the washer hose

place by clips. The lenses can be removed either by unsnapping them or by using a small screwdriver to pry them off.

3   Some bulbs, such as the tail light bulbs, are located in bulb housings. To remove a bulb housing, rotate the housing counter-clockwise about 60-degrees and pull it out **(see illustration)**.

4   Several types of bulbs are used. Some are removed by pushing in and turning them counterclockwise. Others can simply be unclipped from the terminals or pulled

straight out of the socket.

5   To gain access to the instrument cluster lights, the instrument cluster will have to be removed first (see Section 16).

15 Wiper motors - removal and installation

Refer to illustrations 15.2, 15.5, 15.7 and 15.8

1   Disconnect the negative cable at the battery.

Windshield wiper

2   Use as small screwdriver to pry the locking levers up, then pull the wiper arms off the pivots while still pulling the locking tab up **(see illustration)**.

3   Remove the cowl trim cover retaining screws, then detach the cover for access to the motor.

4   Disconnect the washer hoses.

5   Remove the bracket nuts and the pivot pin screws **(see illustration)**. Unplug the electrical connector and remove the assembly from the vehicle. Be careful not to damage the rubber waterproof boot shroud when handling the motor assembly.

6   Installation is the reverse of removal.

Rear wiper

7   Remove the wiper arm and disconnect the washer hose **(see illustration)**.

8   Remove the pivot pin retaining nut and washers. Note the order in which the washers are arranged for ease of installation **(see illustration)**.

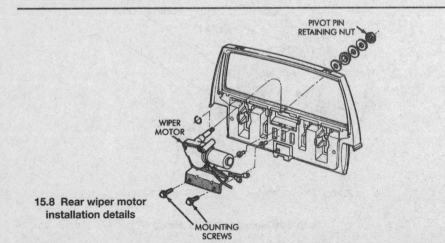

PIVOT PIN RETAINING NUT

WIPER MOTOR

MOUNTING SCREWS

15.8 Rear wiper motor installation details

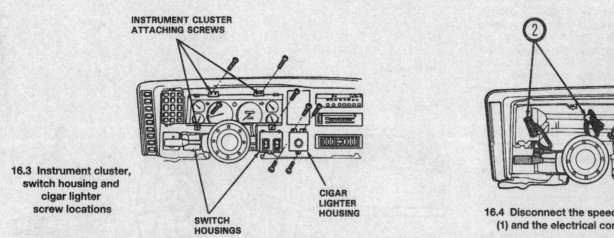

16.3 Instrument cluster, switch housing and cigar lighter screw locations

INSTRUMENT CLUSTER ATTACHING SCREWS

SWITCH HOUSINGS

CIGAR LIGHTER HOUSING

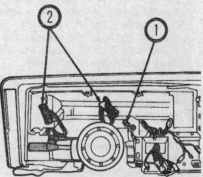

16.4 Disconnect the speedometer cable (1) and the electrical connectors (2)

9   Open the liftgate and remove the trim panel.
10   Unplug the electrical connector, remove the two mounting screws and lower the motor from the liftgate (see illustration).
11   Installation is the reverse of removal.

16   Instrument cluster - removal and installation

Refer to illustration 16.3 and 16.4
Warning: *Some models covered by this manual are equipped with Supplemental Restraint Systems, more commonly known as airbags. Always disable the airbag system before working in the vicinity of any airbag system components, to avoid the possibility of accidental deployment of the airbag(s), which could cause personal injury (see Section 20).*
1   Disconnect the negative cable at the battery.

1984 through 1996 models

2   Remove the instrument cluster bezel (see Section 23 in Chapter 11).
3   Remove the screws attaching the instrument cluster, cigar lighter and (if equipped) switch housings (see illustration).
4   Pull the cluster out and disconnect the speedometer cable and the electrical connectors (see illustration). On some later models it may be necessary to remove the nut under the vehicle on the left side upper suspension arm and disconnect the speedometer cable bracket to provide sufficient slack so the cluster can be pulled out. You can pull the speedometer cable out of the instrument cluster after you press the spring clip on the instrument cluster and squeeze the clip attached to the cable (if equipped).
5   Detach the cluster from the instrument panel.
6   Installation is the reverse of removal.

1997 and later models

Refer to illustration 16.8
7   Remove the cluster bezel (see Section

23 in Chapter 11).
8   Remove the four cluster retaining screws (see illustration).
9   Pull the cluster straight back and disengage it from the two "self-docking" (fixed) electrical connectors.
10   Installation is the reverse of removal.

17   Cruise control system - description and check

The cruise control system maintains vehicle speed with a vacuum actuated servo motor located in the engine compartment, which is connected to the throttle linkage by a cable. The system consists of the servo motor, clutch switch, brake switch, control switches, a relay and associated vacuum hoses.

Diagnosis can usually be limited to simple checks of the wiring and vacuum connections for minor faults which can be easily repaired. These include:

a) *Inspect the cruise control actuating switches for broken wires and loose connections.*
b) *Check the cruise control fuse.*

c) *The cruise control system is operated by vacuum so it's critical that all vacuum switches, hoses and connections are secure. Check the hoses in the engine compartment for tight connections, cracks and obvious vacuum leaks.*

18   Power door lock system - description and check

The power door lock system operates the door lock actuators mounted in each door. The system consists of the switches, actuators and associated wiring. Diagnosis can usually be limited to simple checks of the wiring connections and actuators for minor faults which can be easily repaired. These include:

a) *Check the system fuse and/or circuit breaker.*
b) *Check the switch wires for damage and loose connections. Check the switches for continuity.*
c) *Remove the door panel(s) and check the actuator wiring connections to see if they're loose or damaged. Inspect the actuator rods (if equipped) to make sure*

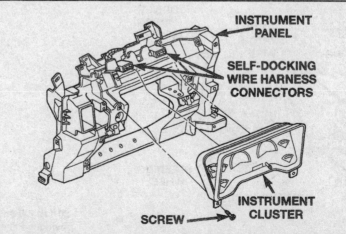

INSTRUMENT PANEL

SELF-DOCKING WIRE HARNESS CONNECTORS

INSTRUMENT CLUSTER

SCREW

16.8 Instrument cluster assembly installation details (1997 and later models)

12

they aren't bent or damaged. Inspect the actuator wiring for damaged or loose connections. The actuator can be checked by applying battery power momentarily. A clicking sound indicates the solenoid is operating properly.

19 Power window system - description and check

The power window system operates the electric motors mounted in the doors which lower and raise the windows. The system consists of the control switches, the motors (regulators), glass mechanisms and associated wiring.

Diagnosis can usually be limited to simple checks of the wiring connections and motors for minor faults which can be easily repaired. These include:

a) *Inspect the power window actuating switches for broken wires and loose connections.*

b) *Check the power window fuse/and or circuit breaker.*

c) *Remove the door panel(s) and check the power window motor wires to see if they're loose or damaged. Inspect the glass mechanisms for damage which could cause binding.*

20 Airbag - general information

1 1995 and later models are equipped with a Supplemental Restraint System (SRS), more commonly known as an airbag. This system is designed to protect the driver and, on 1997 and later models, the front seat passenger, from serious injury in the event of a head-on or frontal collision. It consists of an airbag module in the center of the steering wheel and, on models so equipped, the right side of the

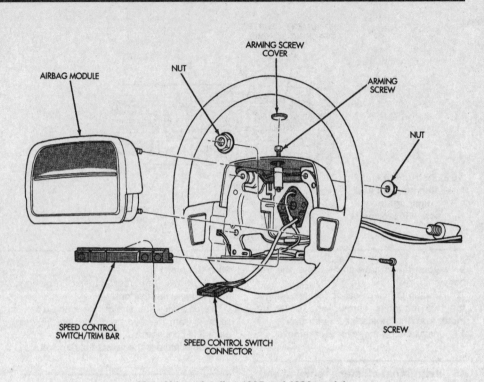

20.4 Airbag details - 1995 and 1996 models

instrument panel. 1995 and 1996 models have a self-contained system, in which the impact sensor is housed within the airbag module. The system on 1997 and later models utilizes a control module mounted under the driver's seat (which contains the impact sensor), a "clockspring" under the steering wheel which can transmit current to the driver's side airbag module regardless of steering wheel position, and a passenger's side airbag module mounted above the glove box.

2 Whenever working in the vicinity of any airbag system components it is important to disable the system to prevent accidental deployment of the airbag(s), which could cause personal injury.

1995 and 1996 models

Refer to illustration 20.4

Disarming the system

3 Roll down the driver's side window.

4 Pry off the small cover from the top of the steering wheel hub. Get out of the vehicle, reach into the vehicle and, using an 8 mm socket, unscrew the arming screw until it

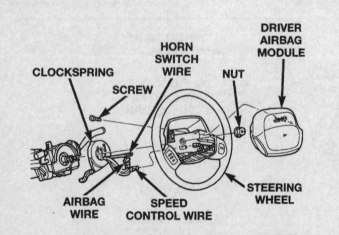

20.11a Airbag details - 1997 and later models

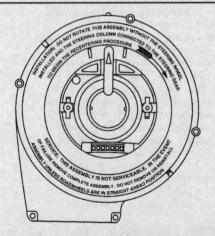

20.11b When properly centered, the airbag clockspring will be positioned like this, with the arrow on the rotor pointing straight up, the flats in the center of the rotor will be horizontal, and the wiring harness will be positioned at the bottom

stops (it should protrude approximately one inch from the surface of the steering wheel trim) **(see illustration)**. The airbag is now disarmed.

Arming the system

5   Reach through the driver's window and turn the arming screw until it is seated (tighten it to approximately 15 in-lbs).
6   Install the trim cover on the steering wheel hub.

1997 and later models

Refer to illustrations 20.11a and 20.11b

Disarming the system

7   Disconnect the cable from the negative terminal of the battery, then wait at least two minutes before proceeding with any repairs (the system has a back-up capacitor that must fully discharge before it can be considered disabled). Make sure the battery cable cannot accidentally come into contact with the battery terminal.

Arming the system

8   To arm the system, simply reconnect the cable to the negative terminal of the battery.
9   Turn the ignition key to the On position and observe the airbag light on the dash - it should glow for a few seconds, then go out. If it stays on, have the system diagnosed by a dealer service department or other qualified repair facility.

Centering the clockspring

10   If the airbag clockspring becomes uncentered (during steering wheel or steering gear removal, for example), it must be centered to prevent it from breaking.
11   Turn the clockspring rotor clockwise until it stops (don't apply too much force) **(see illustrations)**.
12   Turn the clockspring rotor counterclock-

wise 2-1/2 turns; the flats on the rotor should be horizontal and the arrow on the rotor should be pointing straight up. Also, the wiring harness should be positioned at the bottom; if it isn't, turn the clockspring an additional 1/2 turn.
13   Install the steering wheel following the procedure described in Chapter 10.

21   Wiring diagrams - general information

Since it isn't possible to include all wiring diagrams for every year covered by

this manual, the following diagrams are those that are typical and most commonly needed.

Prior to troubleshooting any circuits, check the fuse and circuit breakers (if equipped) to make sure they're in good condition. Make sure the battery is properly charged and check the cable connections (Chapter 1).

When checking a circuit, make sure that all connectors are clean, with no broken or loose terminals. When unplugging a connector, do not pull on the wires. Pull only on the connector housings themselves.

Refer to the accompanying chart for the wire color codes applicable to your vehicle.

WIRE COLOR CODE CHART					
COLOR CODE	COLOR	STANDARD TRACER COLOR	COLOR CODE	COLOR	STANDARD TRACER CODE
BK	BLACK	WT	PK	PINK	BK OR WH
BR	BROWN	WT	RD	RED	WT
DB	DARK BLUE	WT	TN	TAN	WT
DG	DARK GREEN	WT	VT	VIOLET	WT
GY	GRAY	BK	WT	WHITE	BK
LB	LIGHT BLUE	BK	YL	YELLOW	BK
LG	LIGHT GREEN	BK	*	WITH TRACER	
OR	ORANGE	BK			

21.4 This chart will help you identify wire colors from the wiring diagrams that follow - the term tracer is used to designate a different-colored stripe on a wire

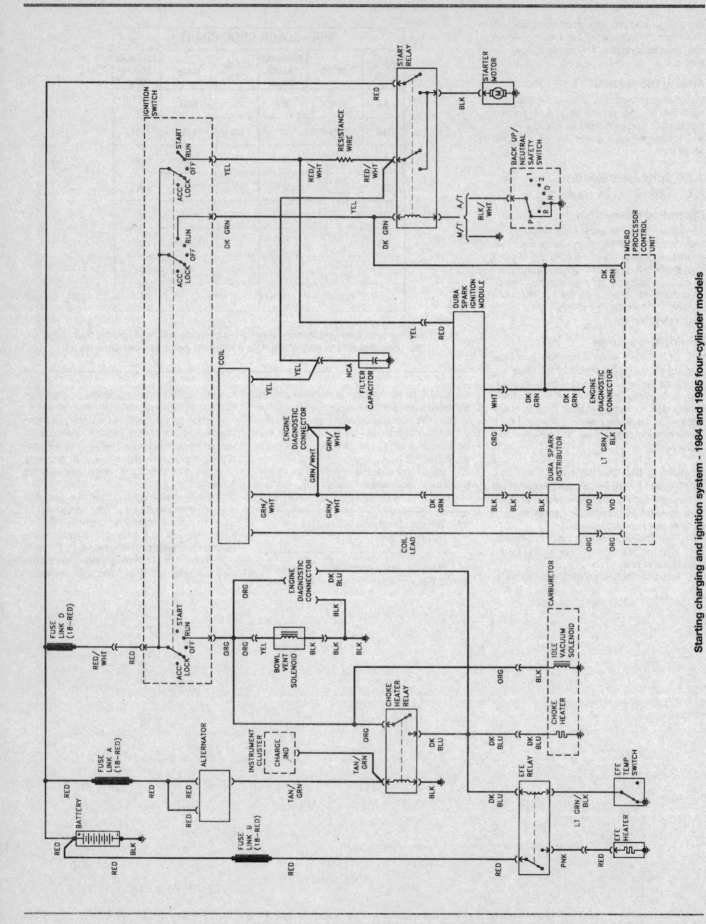

Starting charging and ignition system - 1984 and 1985 four-cylinder models

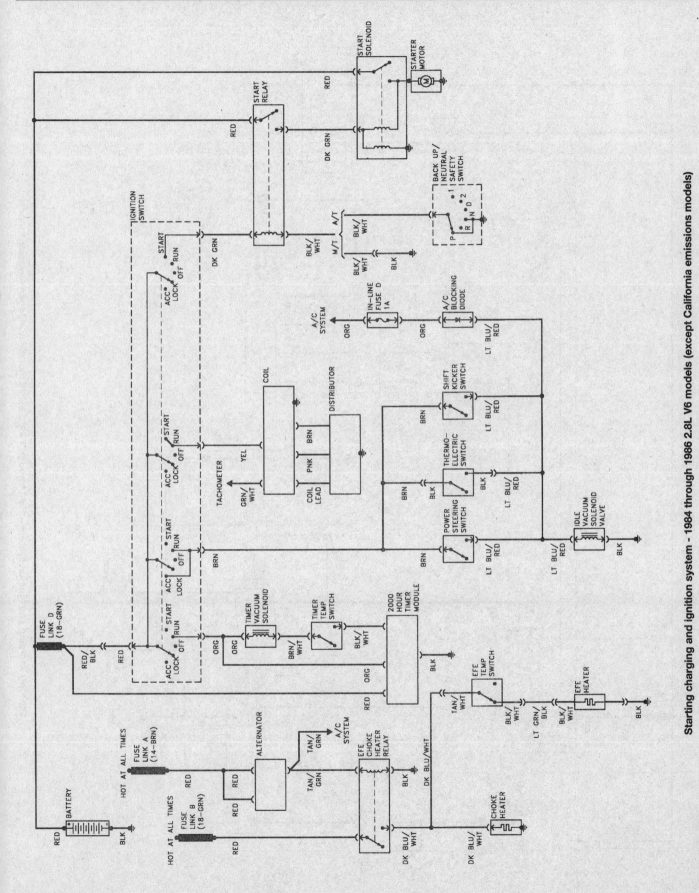

Starting charging and ignition system - 1984 through 1986 2.8L V6 models (except California emissions models)

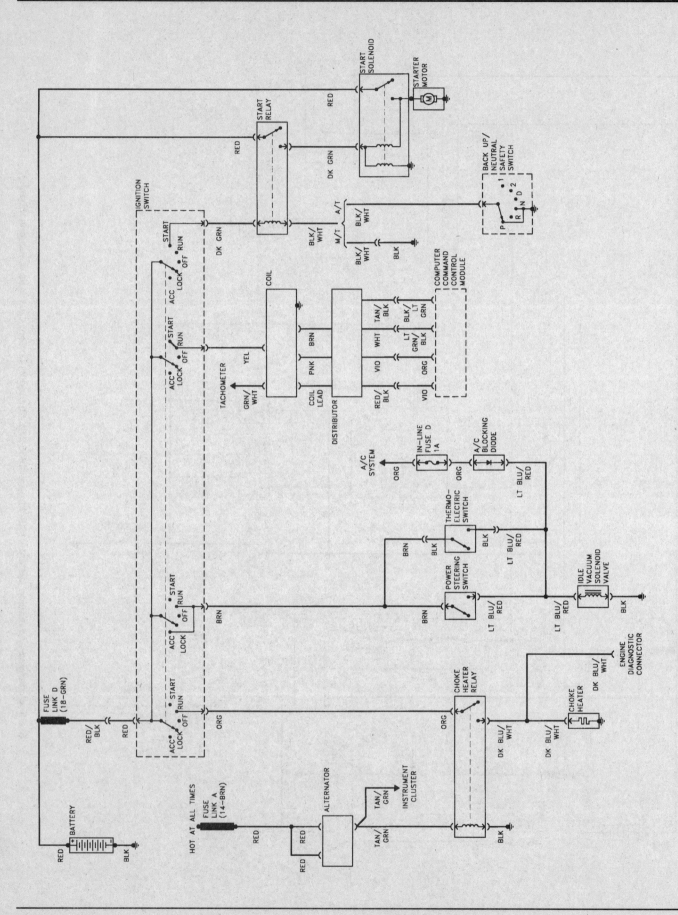

Starting charging and ignition system – 1984 through 1986 2.8L V6 models (California emissions models)

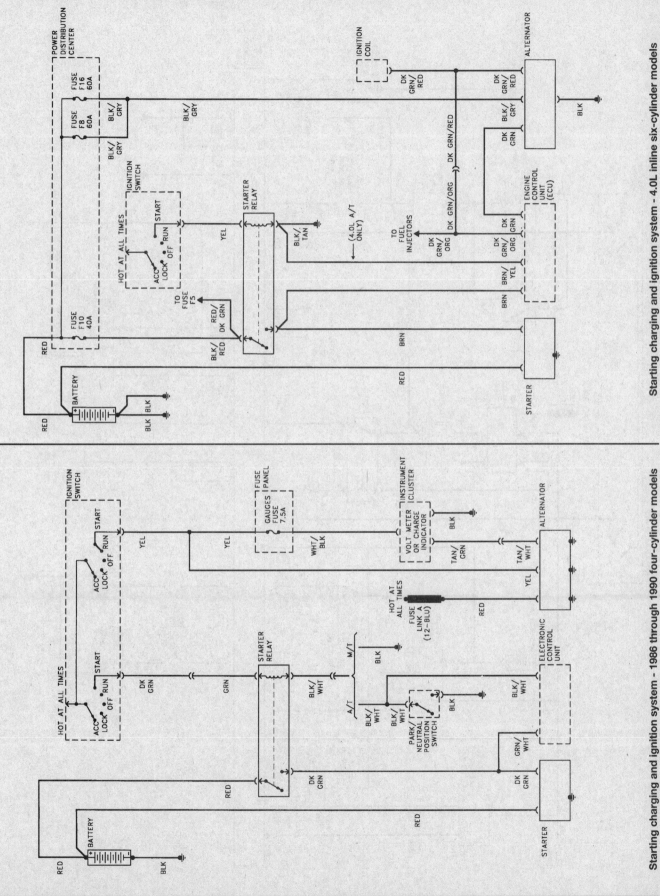

Starting charging and ignition system - 4.0L inline six-cylinder models

Starting charging and ignition system - 1986 through 1990 four-cylinder models

12

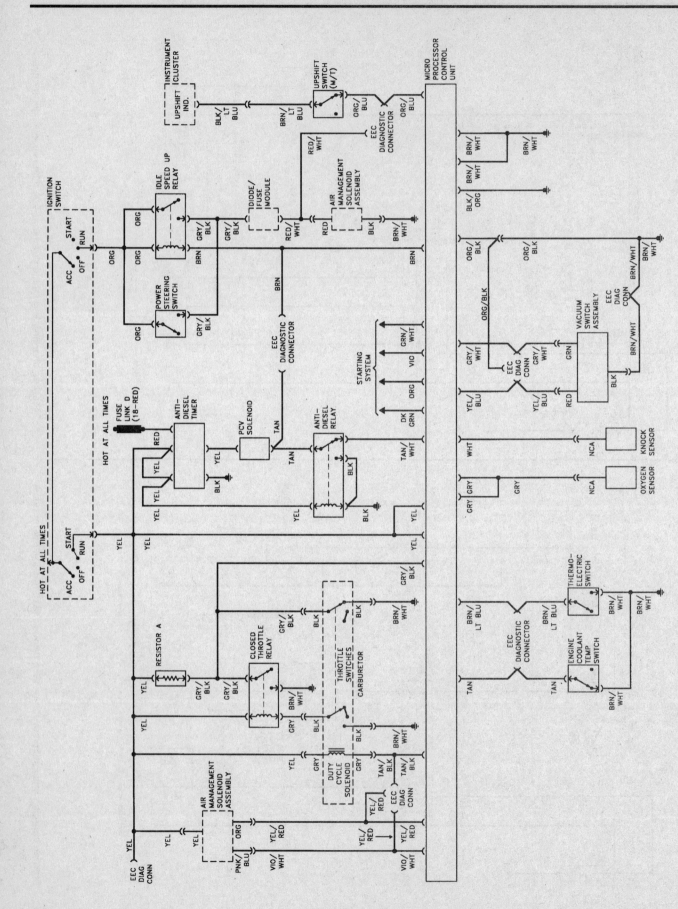

Engine control system - 1984 and 1985 four-cylinder models

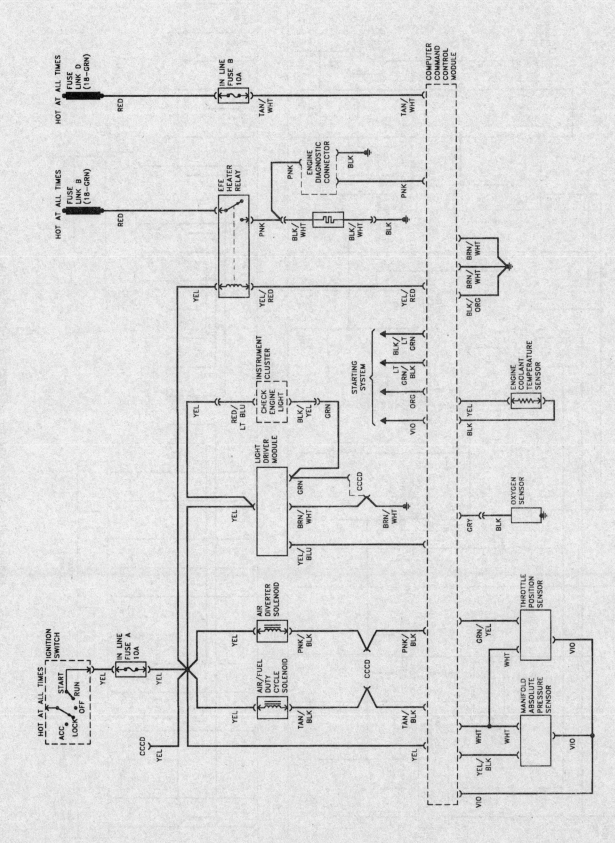

Engine control system - 1984 through 1986 2.8L V6 models

12

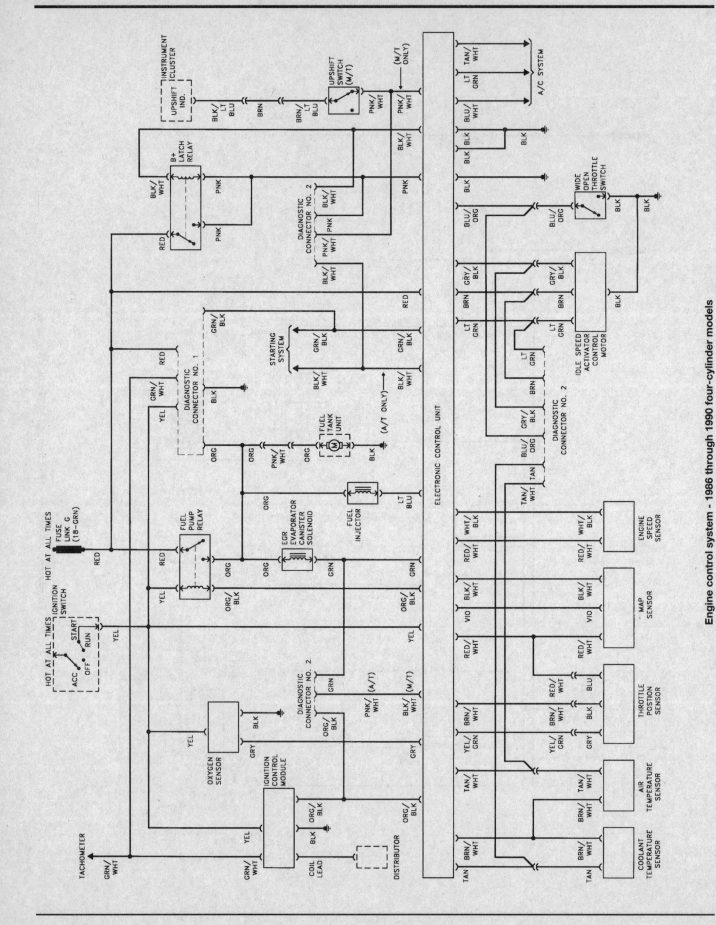

Engine control system - 1986 through 1990 four-cylinder models

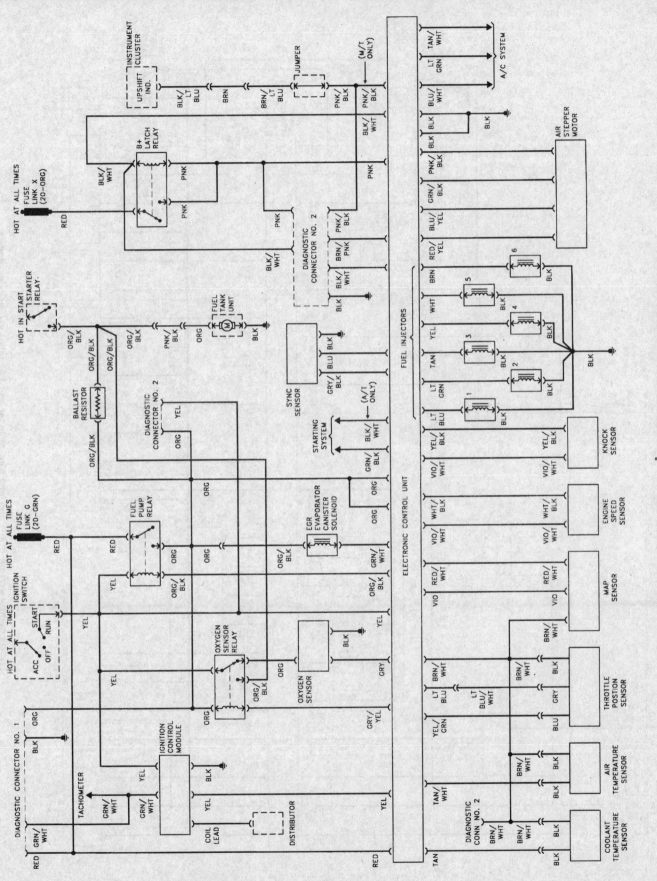

Engine control system - 1987 through 1990 4.0L inline six-cylinder models

12

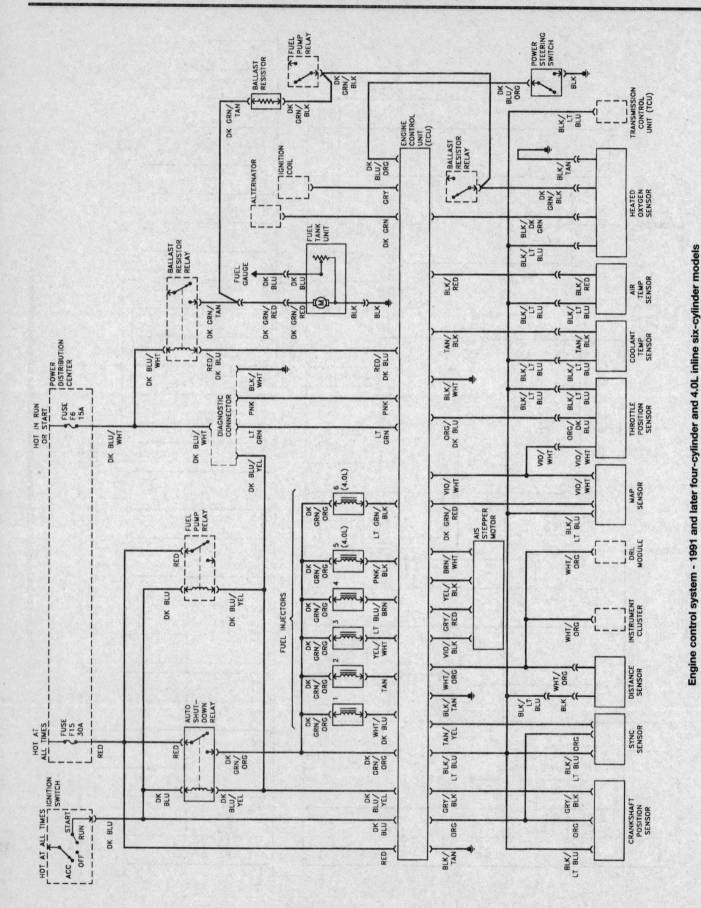

Engine control system - 1991 and later four-cylinder and 4.0L inline six-cylinder models

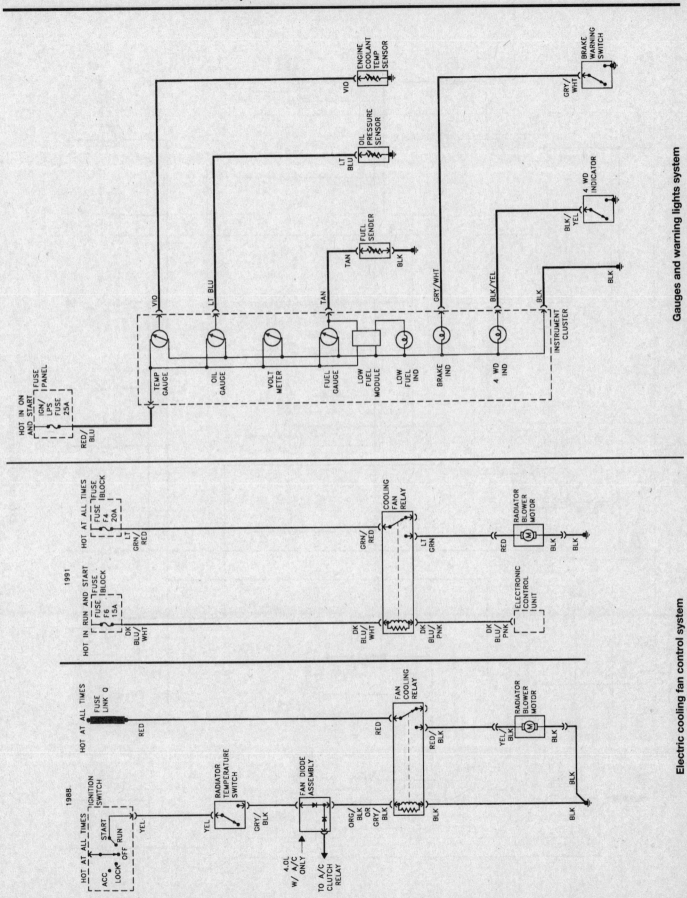

Gauges and warning lights system

Electric cooling fan control system

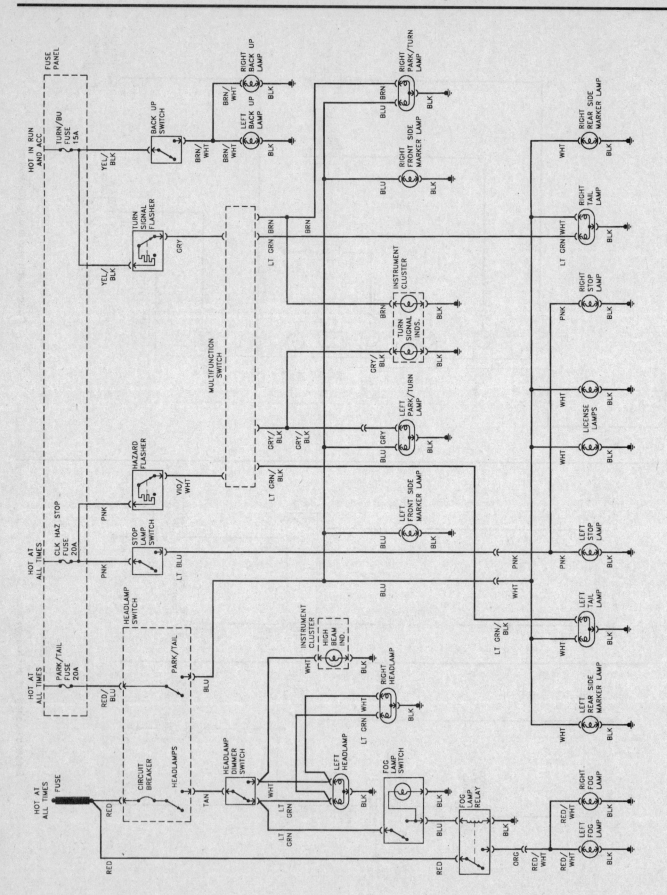

Exterior lighting system

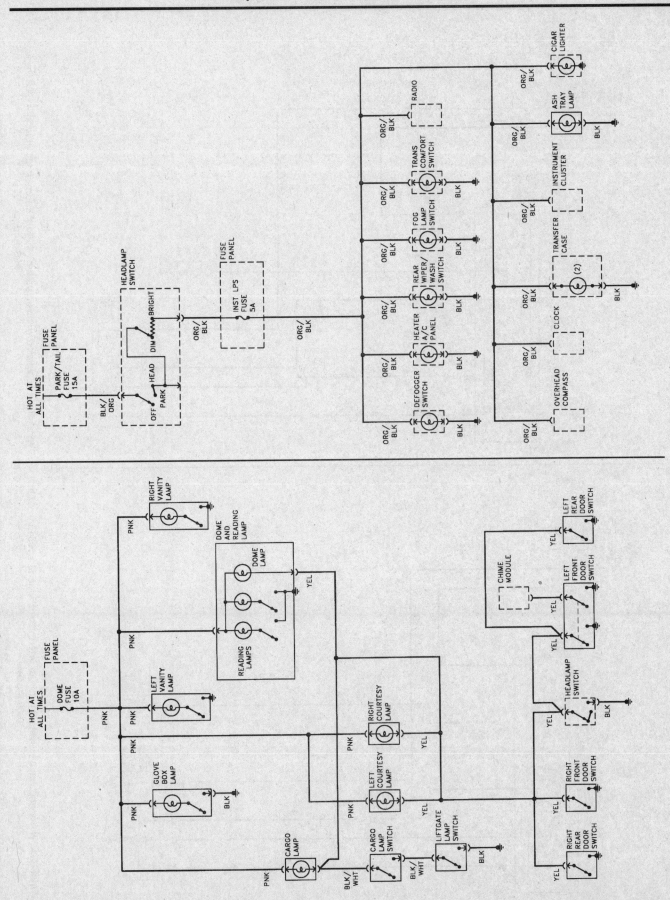

Interior lighting system

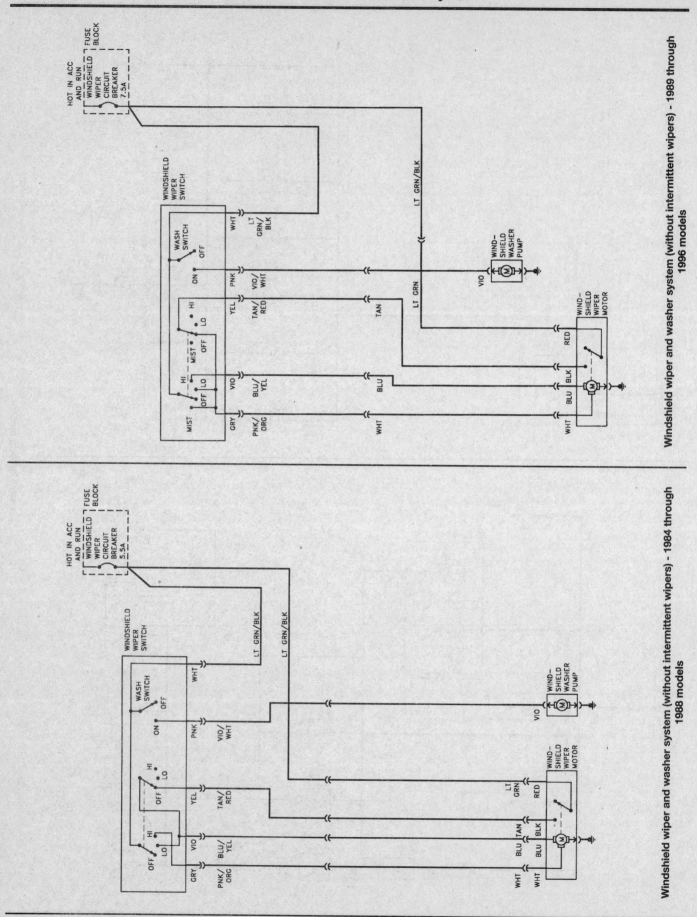

Windshield wiper and washer system (without intermittent wipers) - 1989 through 1996 models

Windshield wiper and washer system (without intermittent wipers) - 1984 through 1988 models

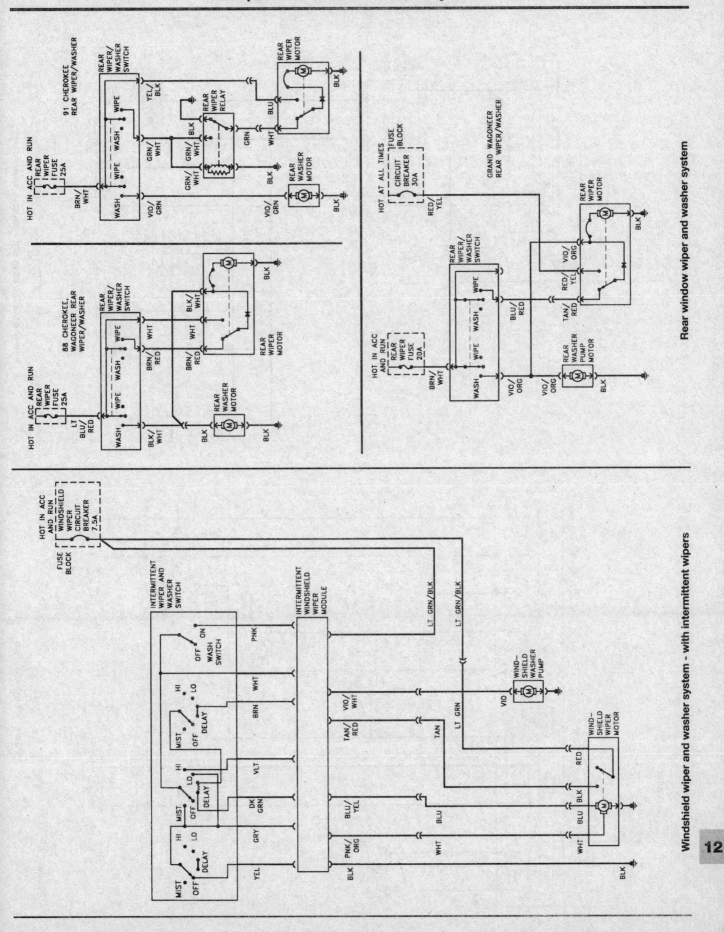

Rear window wiper and washer system

Windshield wiper and washer system - with intermittent wipers

12

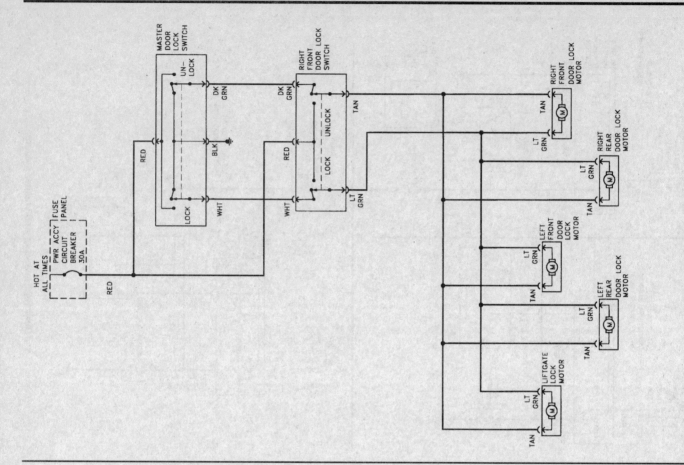

Power door lock system - without keyless entry

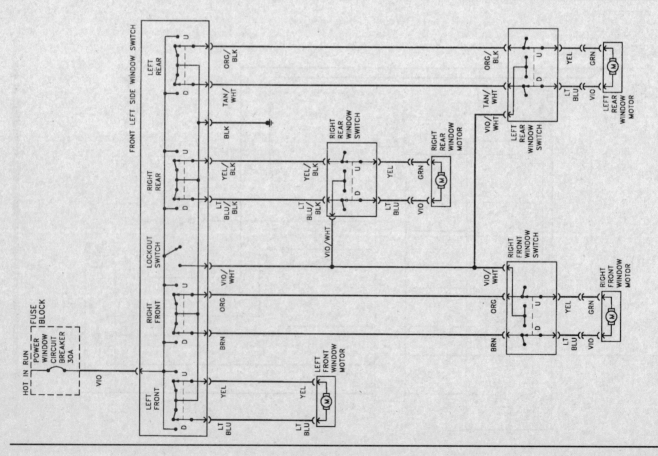

Power window system

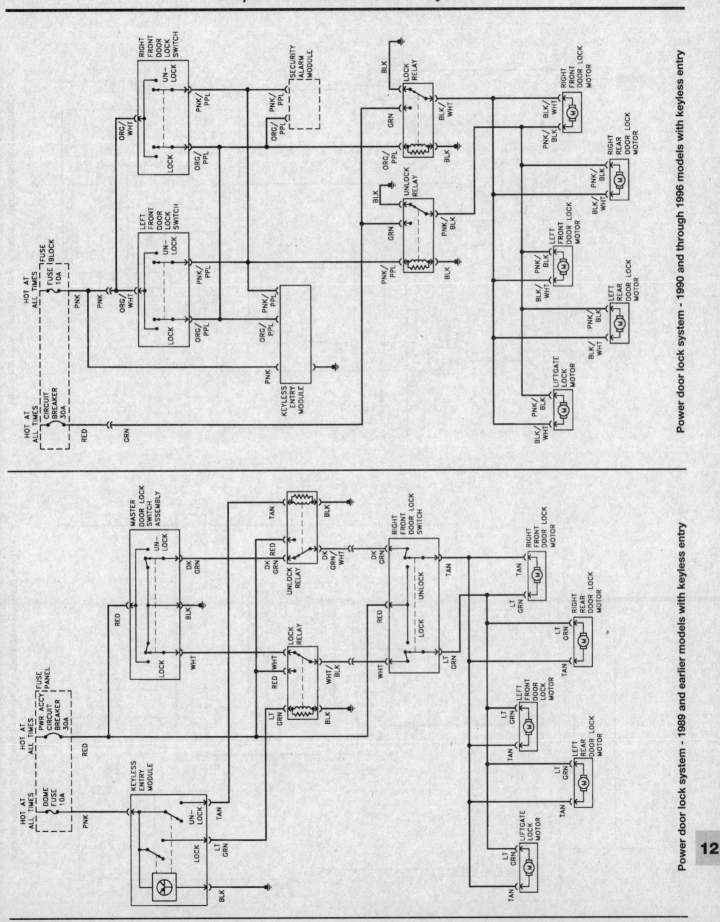

Power door lock system - 1990 and through 1996 models with keyless entry

Power door lock system - 1989 and earlier models with keyless entry

12

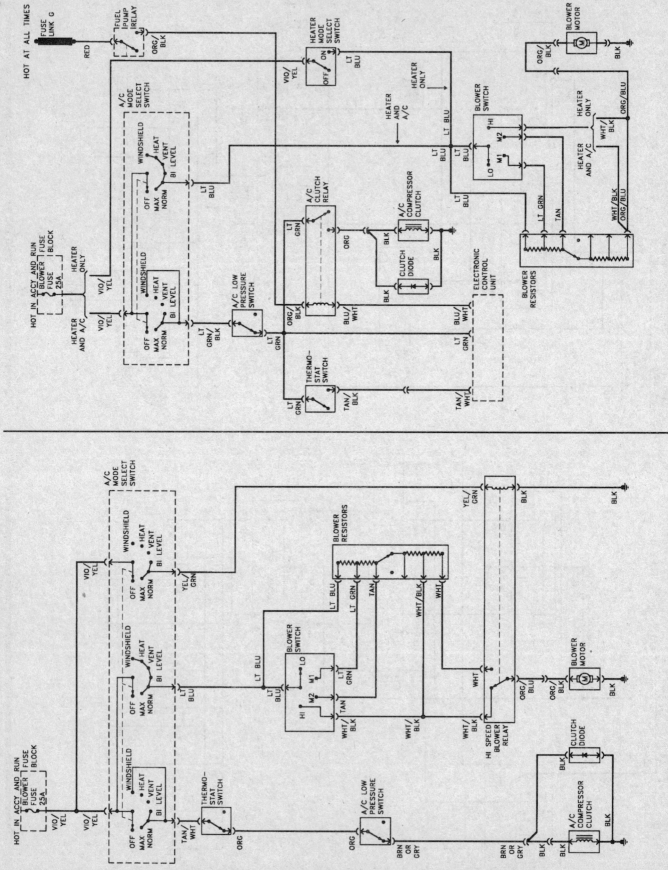

Heater and air conditioning system - 1987 through 1990 models

Heater and air conditioning system - 1984 through 1986 models

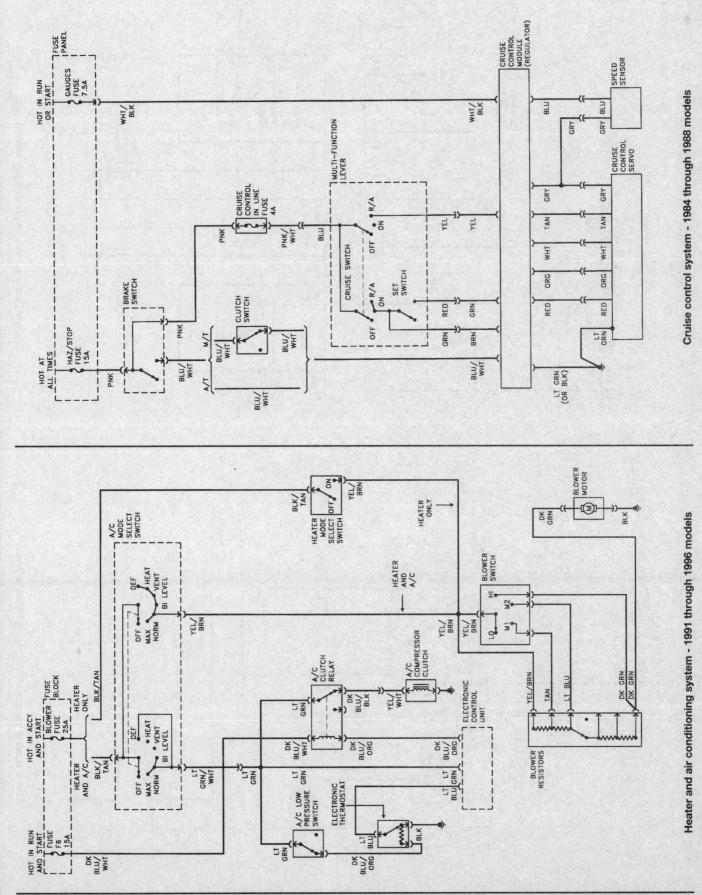

Cruise control system - 1984 through 1988 models

Heater and air conditioning system - 1991 through 1996 models

12

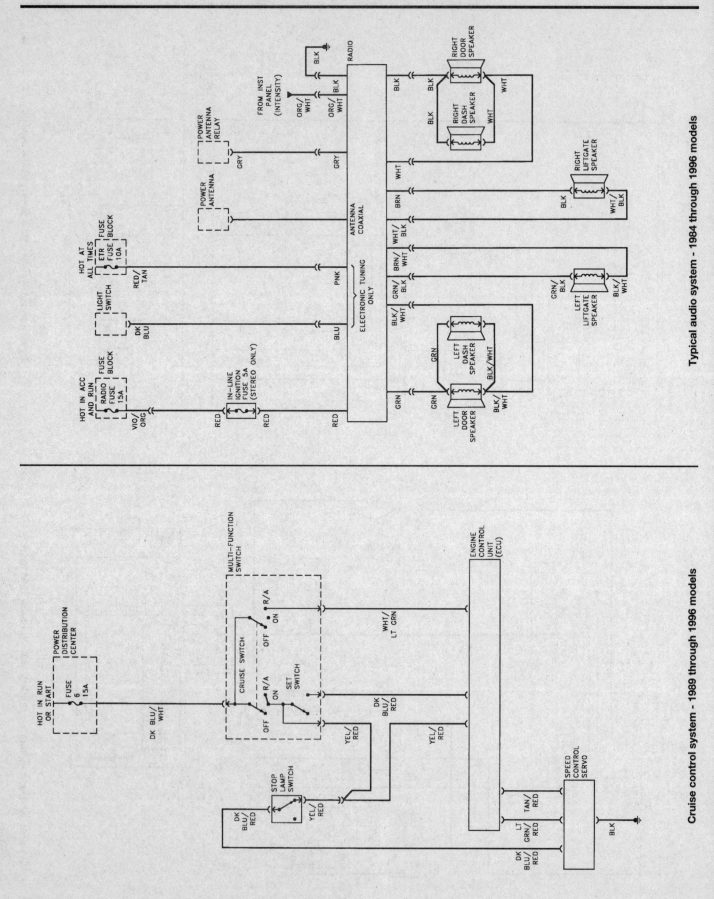

Typical audio system - 1984 through 1996 models

Cruise control system - 1989 through 1996 models

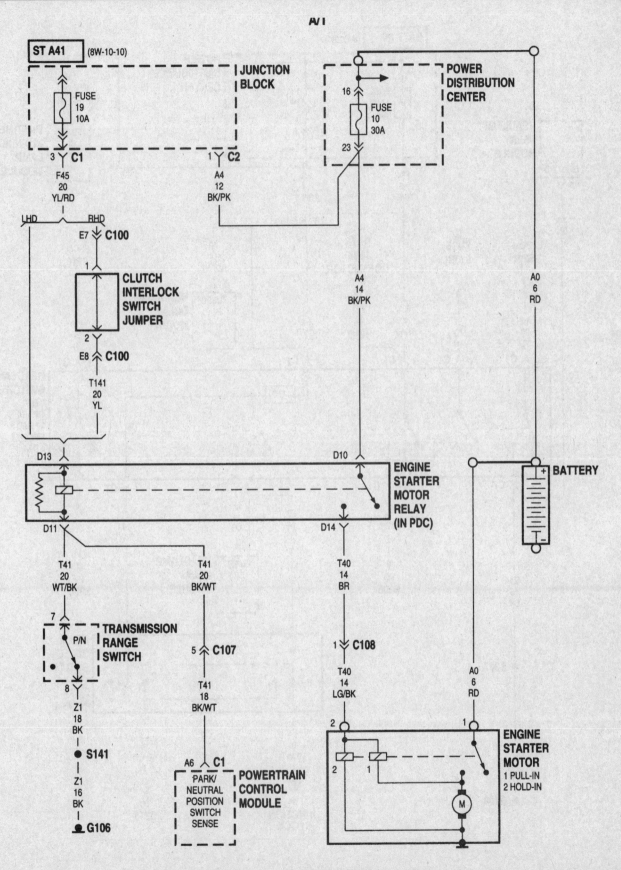

Typical starting system circuit (1997 and later models)

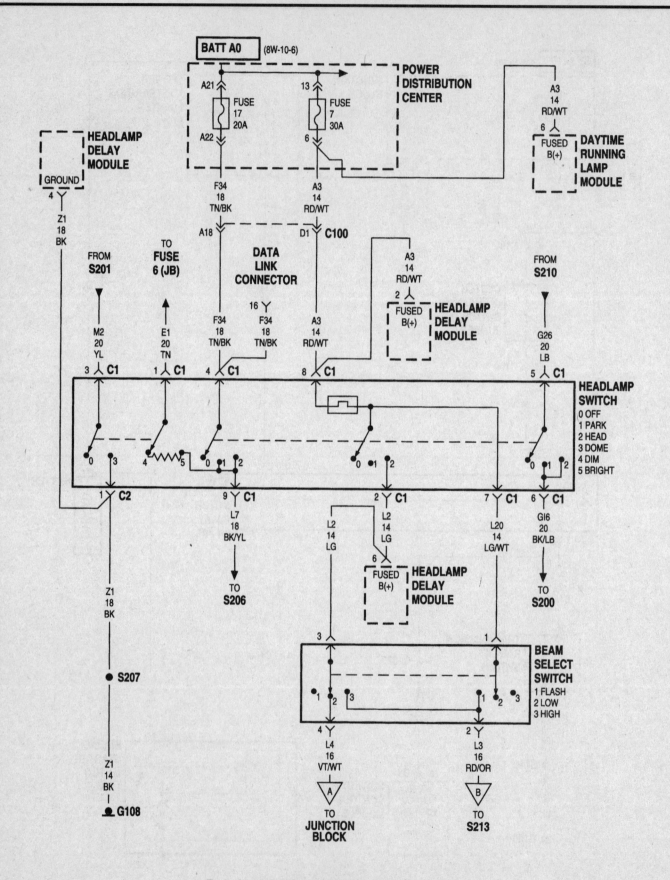

Typical front lighting circuit (1997 and later models)

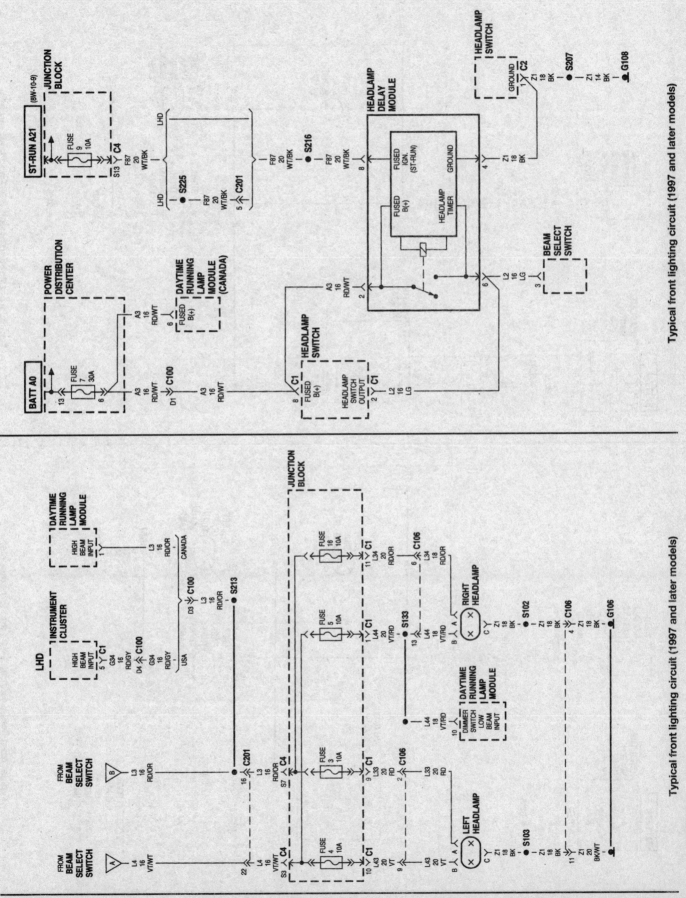

Typical front lighting circuit (1997 and later models)

Typical front lighting circuit (1997 and later models)

12

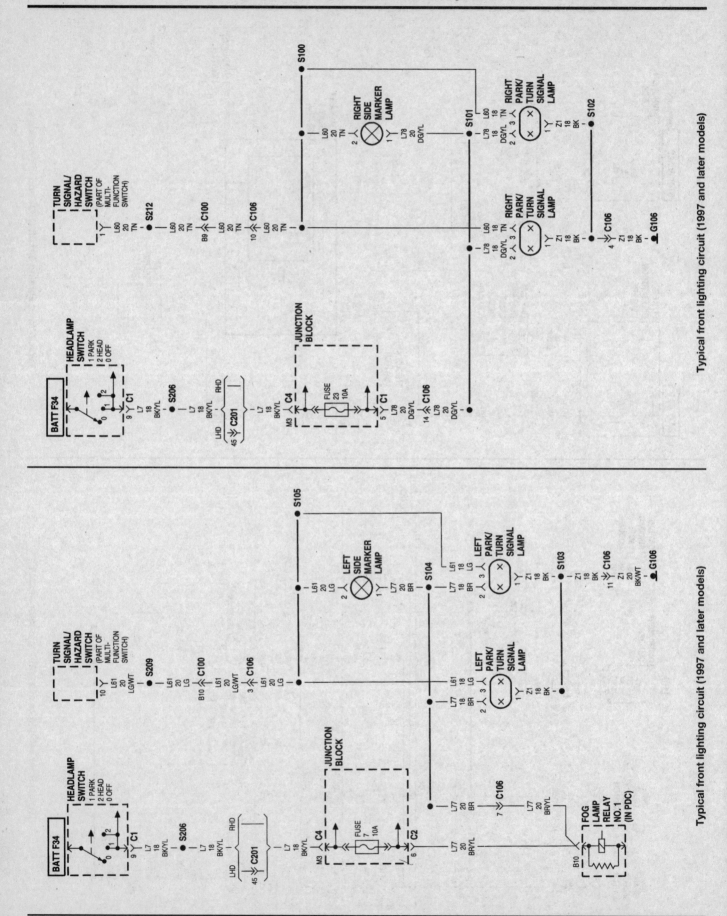

Typical front lighting circuit (1997 and later models)

Typical front lighting circuit (1997 and later models)

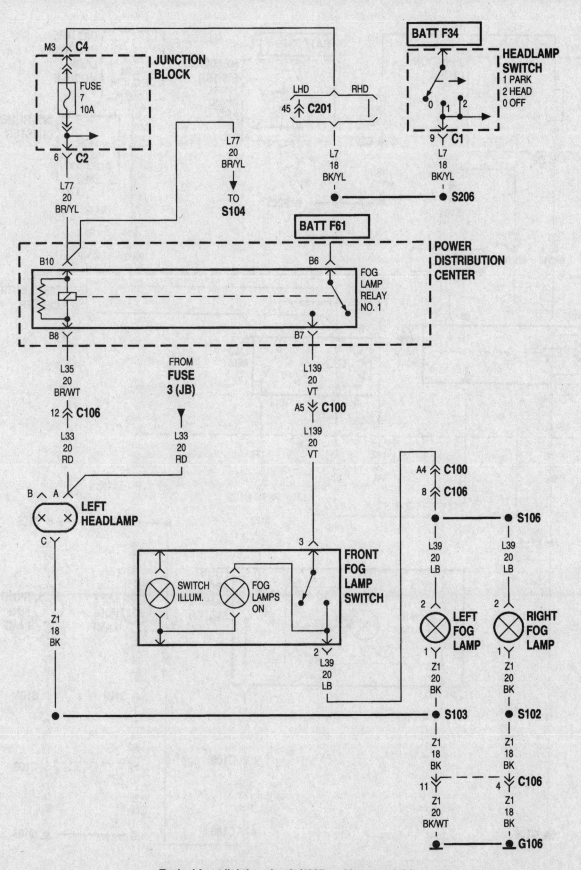

Typical front lighting circuit (1997 and later models)

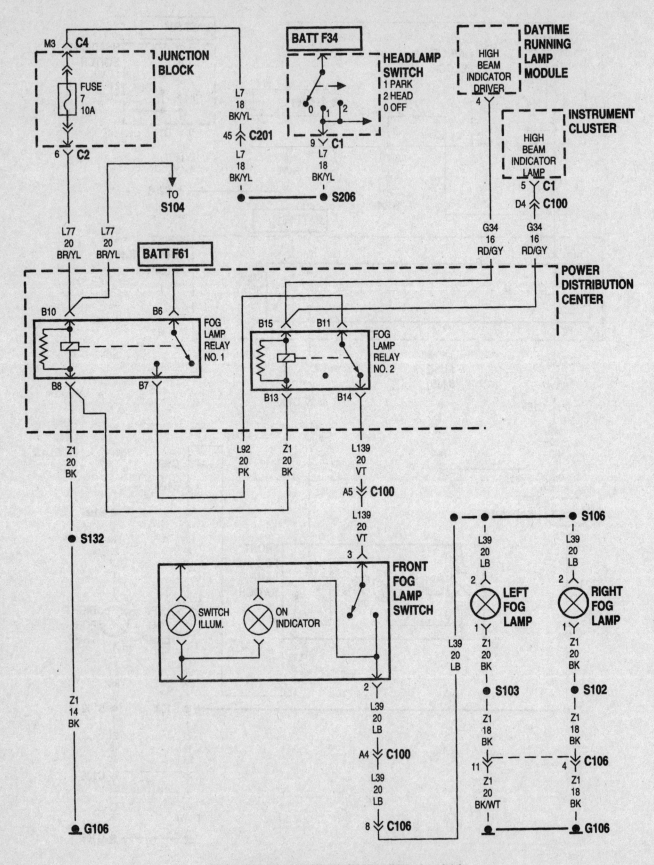

Typical front lighting circuit (1997 and later models)

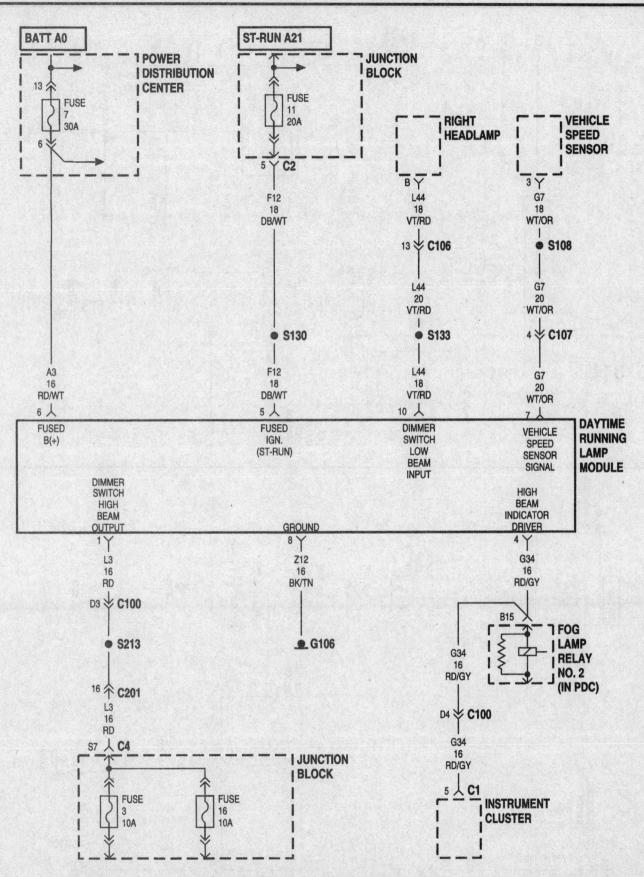

Typical front lighting circuit (1997 and later models)

12

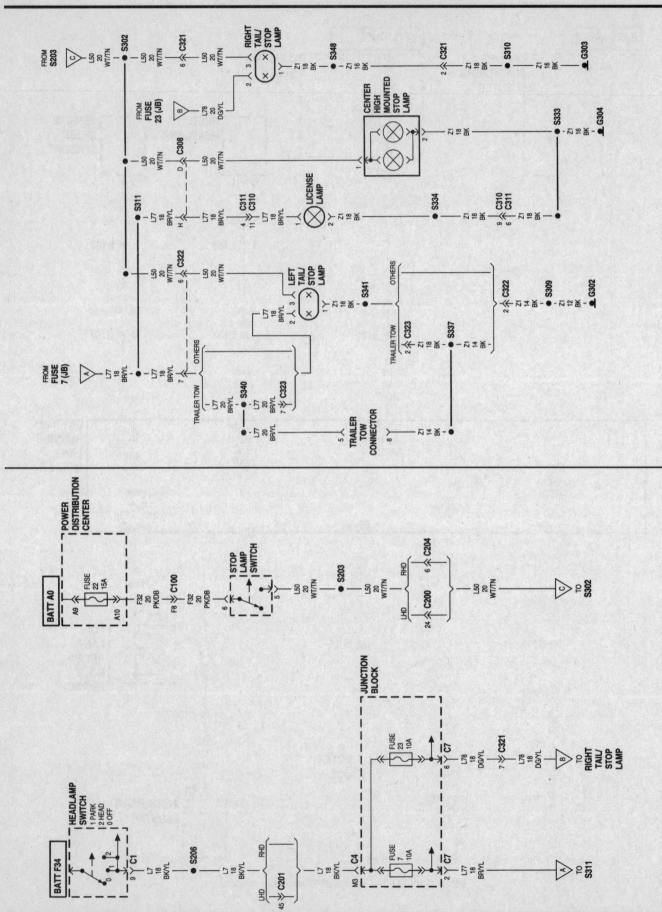

Typical rear lighting circuit (1997 and later models)

Typical rear lighting circuit (1997 and later models)

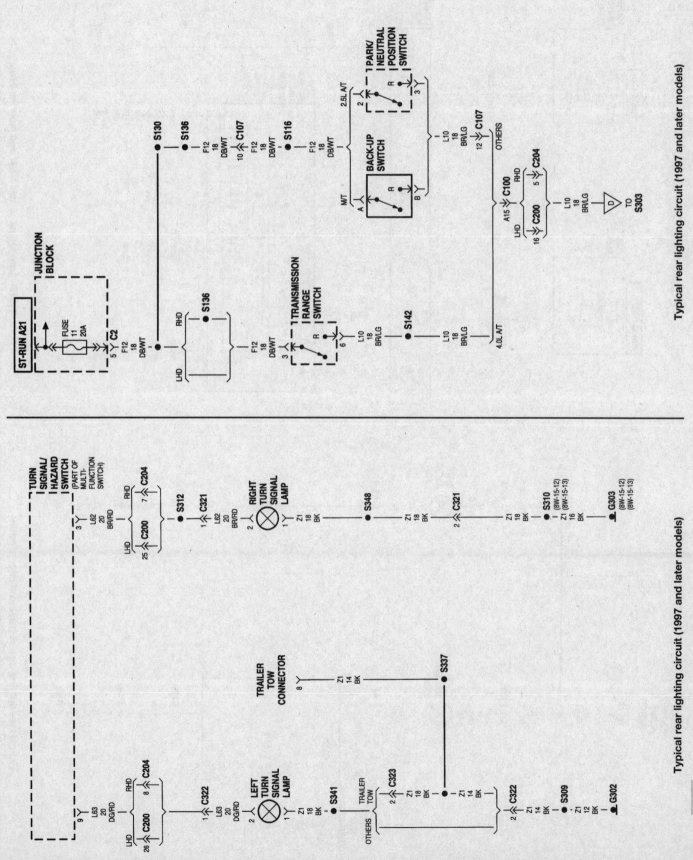

Typical rear lighting circuit (1997 and later models)

Typical rear lighting circuit (1997 and later models)

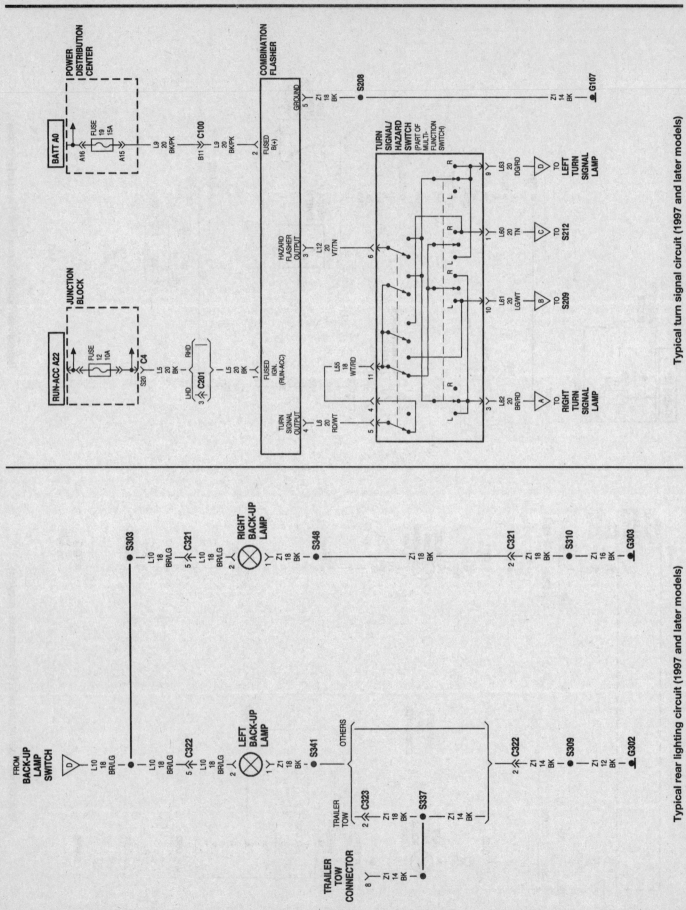

Typical turn signal circuit (1997 and later models)

Typical rear lighting circuit (1997 and later models)

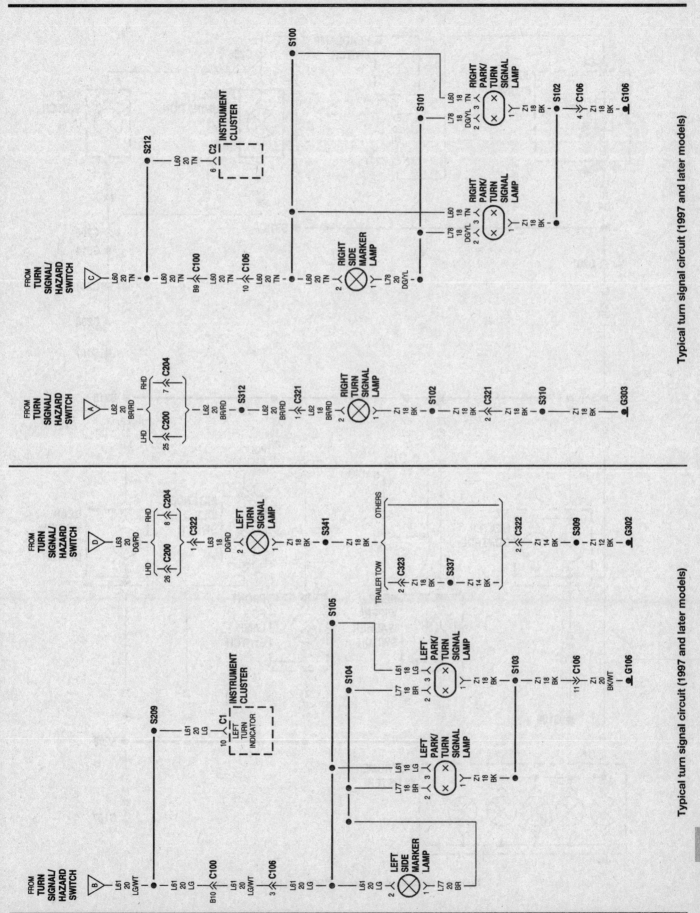

Typical turn signal circuit (1997 and later models)

Typical turn signal circuit (1997 and later models)

12

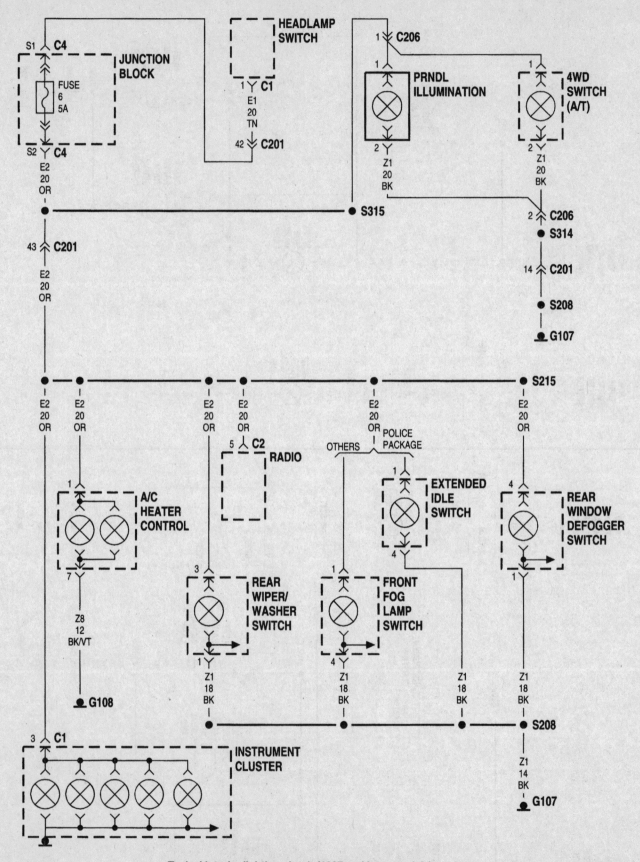

Typical interior lighting circuit (1997 and later models)

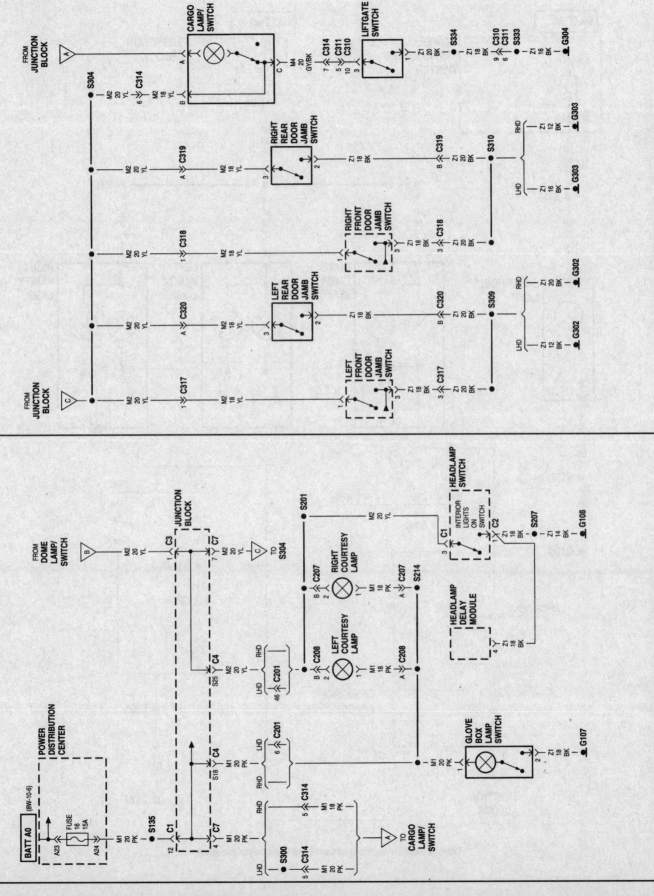

Typical interior lighting circuit (1997 and later models)

Typical interior lighting circuit (1997 and later models)

12

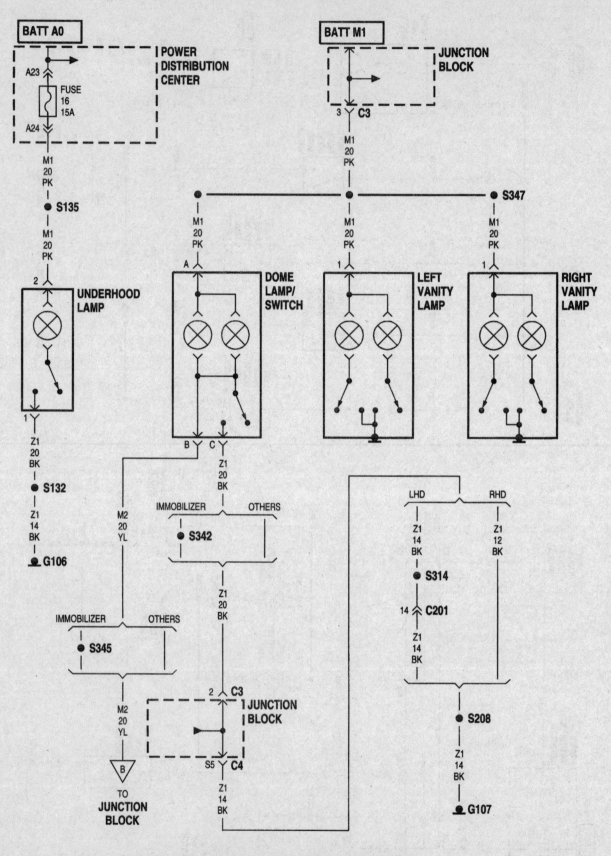

Typical interior lighting circuit (1997 and later models)

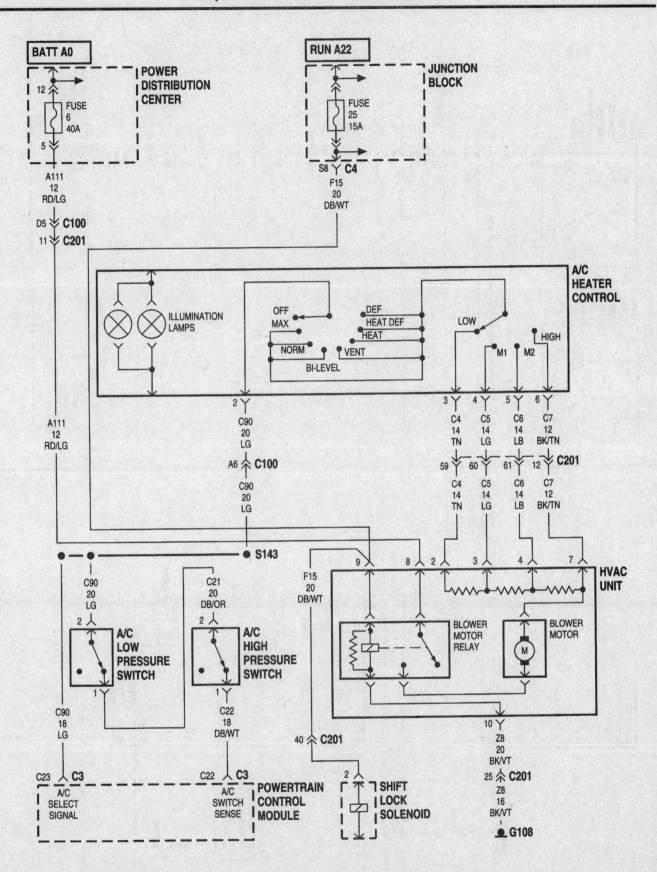

Typical air conditioning and heater circuit (1997 and later models)

12

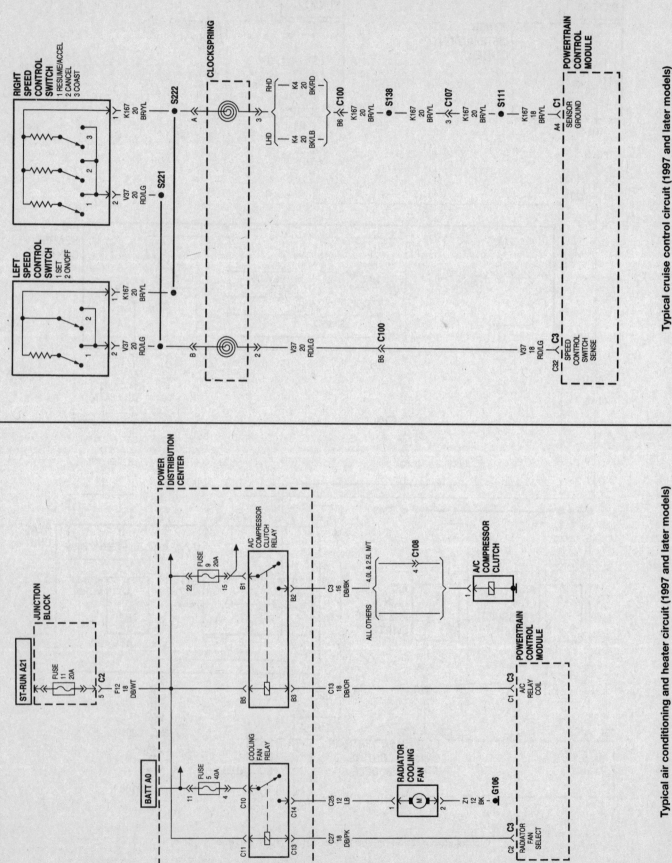

Typical cruise control circuit (1997 and later models)

Typical air conditioning and heater circuit (1997 and later models)

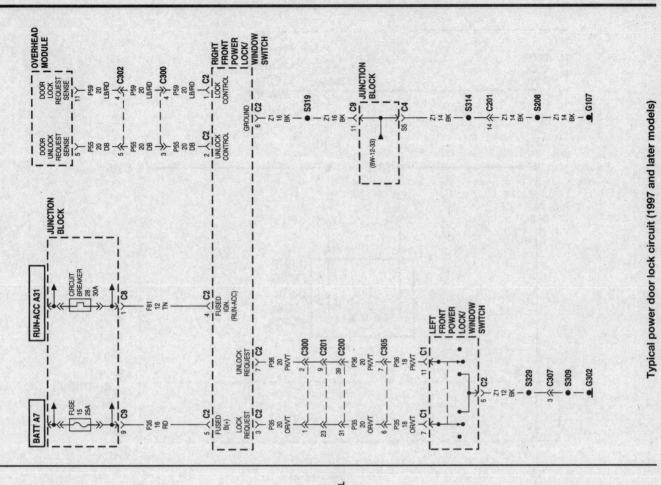

Typical power door lock circuit (1997 and later models)

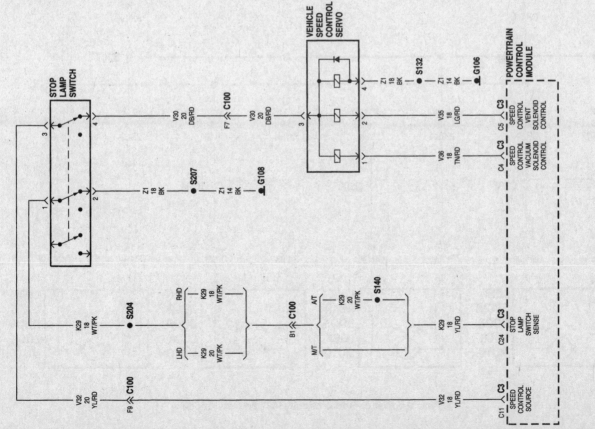

Typical cruise control circuit (1997 and later models)

12

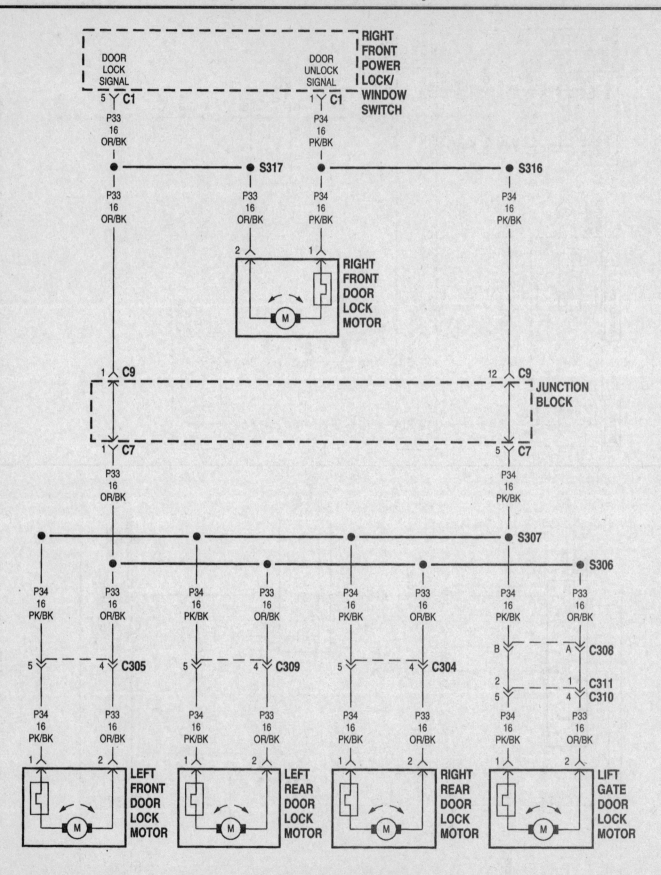

Typical power door lock circuit (1997 and later models)

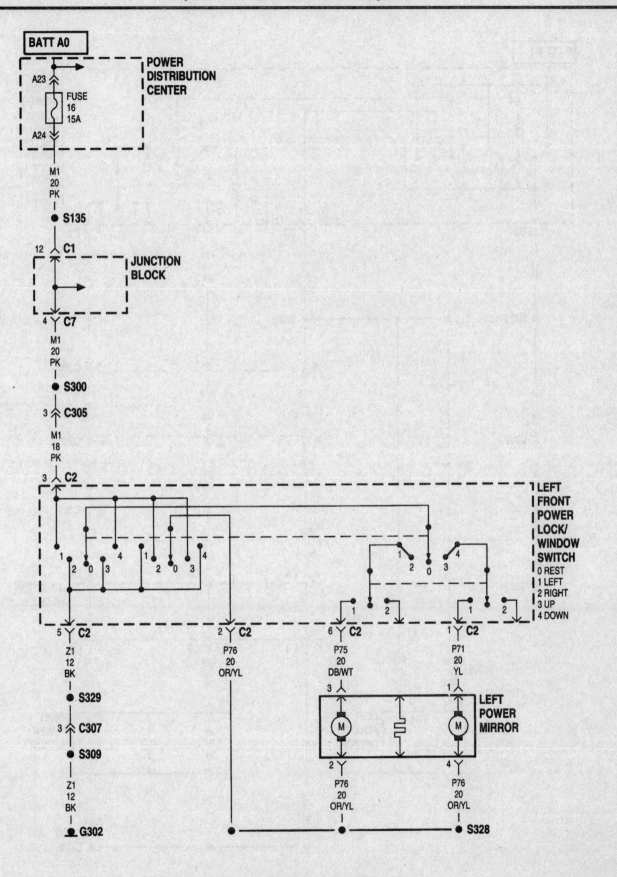

Typical power mirror circuit (1997 and later models)

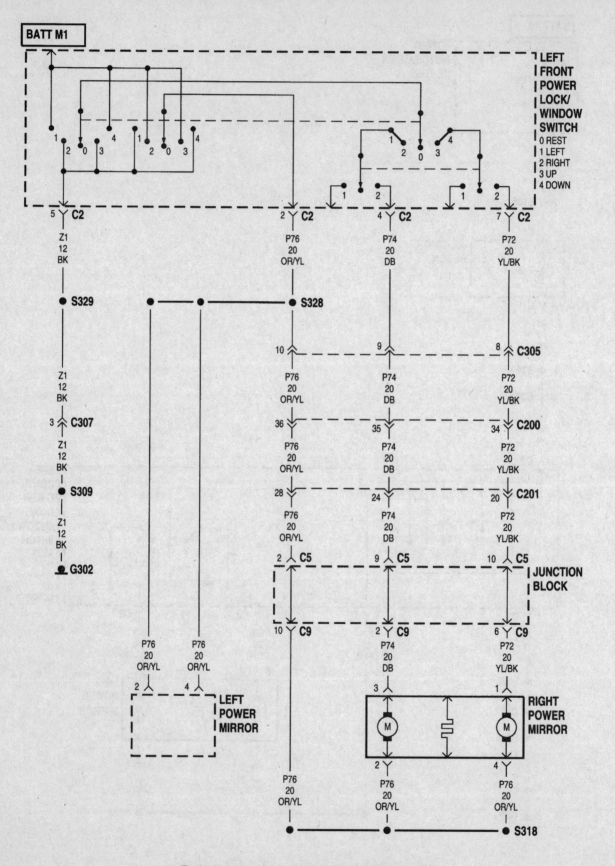

Typical power mirror circuit (1997 and later models)

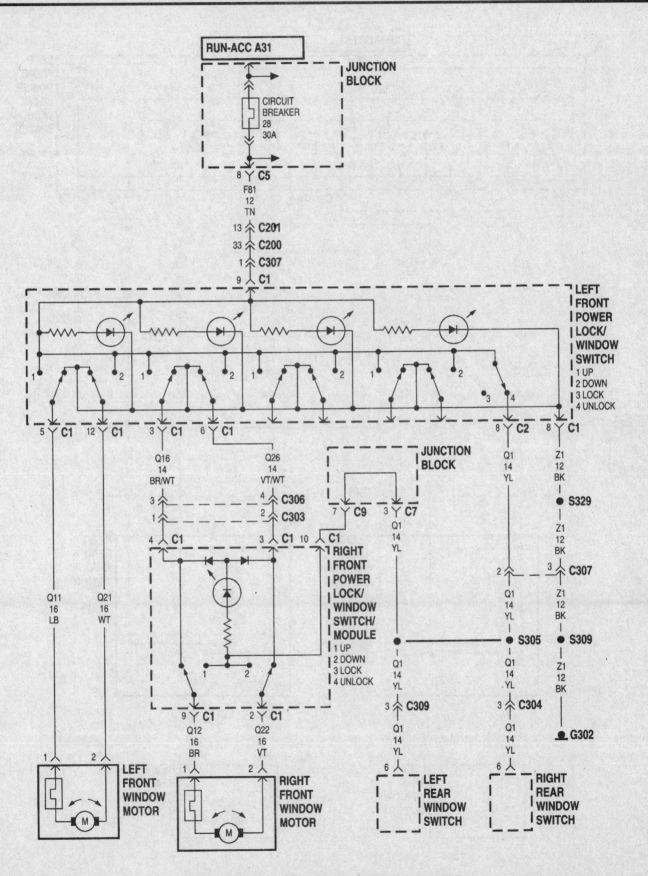

Typical power window circuit (1997 and later models)

12

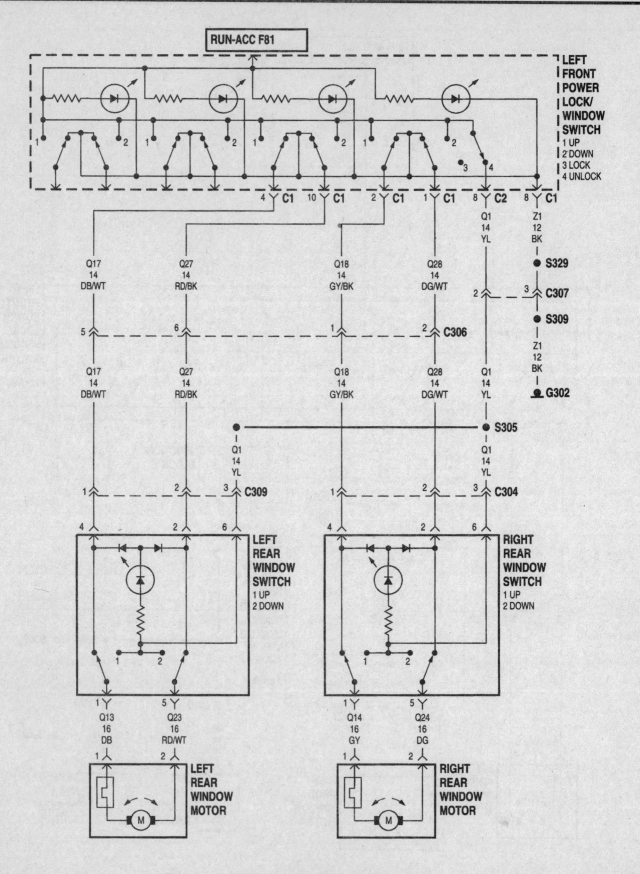

Typical power window circuit (1997 and later models)

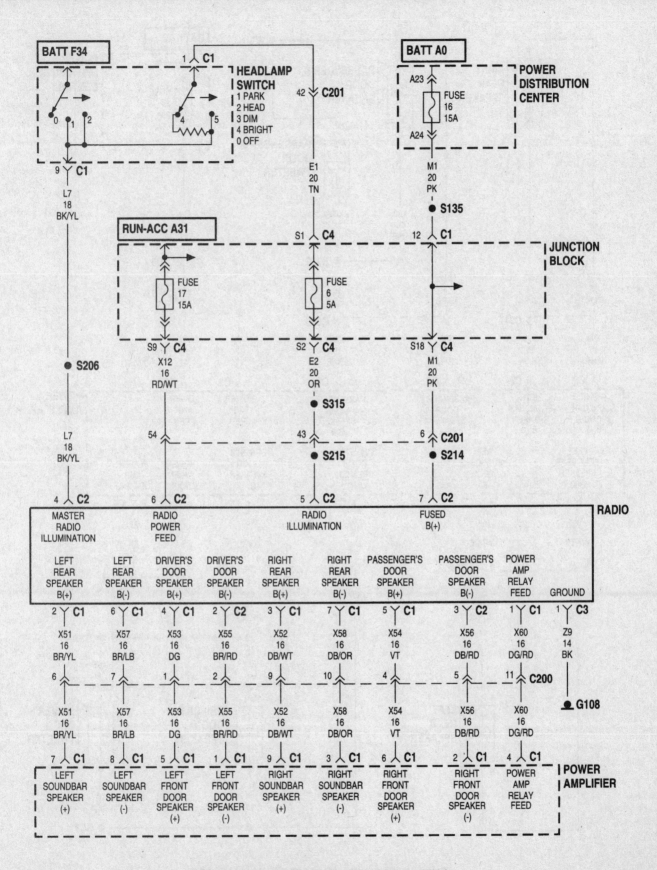

Typical audio system circuit (1997 and later models)

12

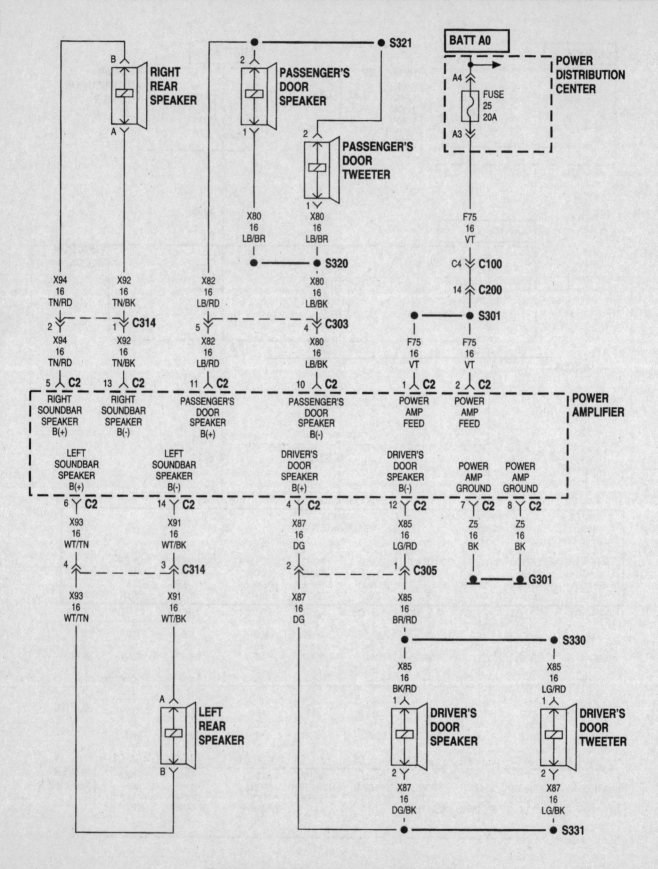

Typical audio system circuit (1997 and later models)

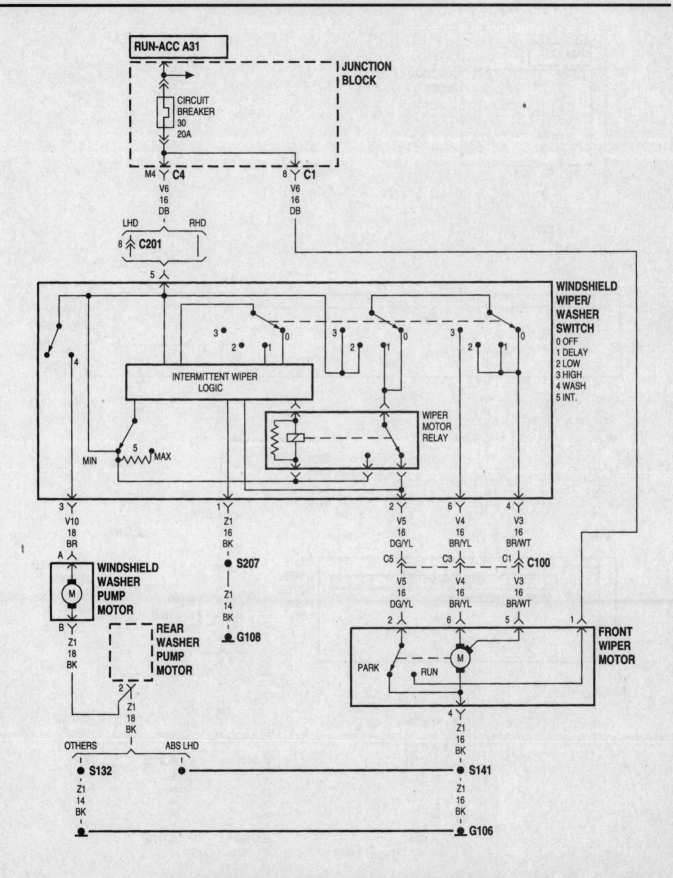

Typical windshield wiper and washer circuit (1997 and later models)

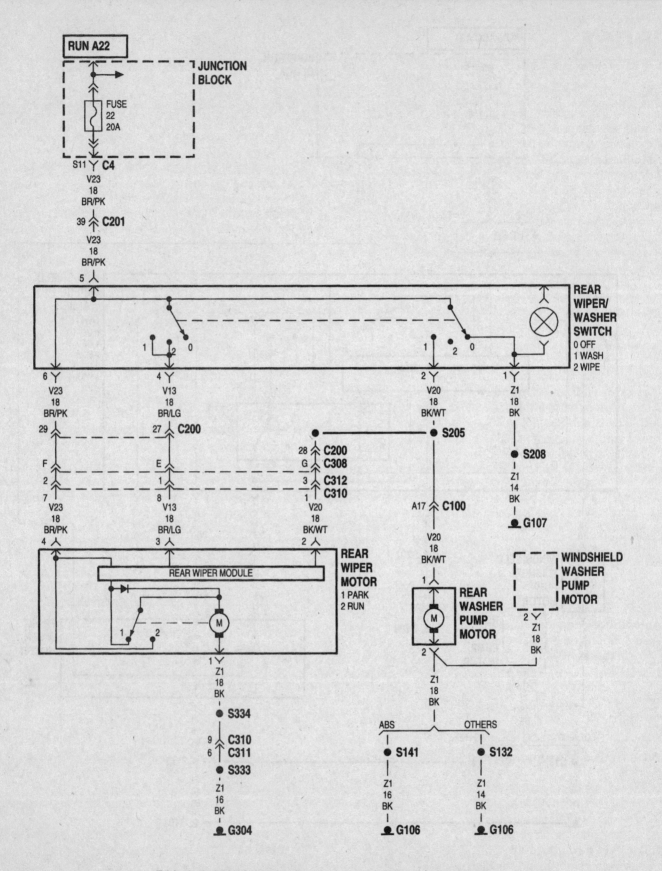

Typical windshield wiper and washer circuit (1997 and later models)

Index

Haynes Automotive Manuals

ACURA
12020	**Integra** '86 thru '89 & **Legend** '86 thru '90
12021	**Integra** '90 thru '93 & **Legend** '91 thru '95

AMC
	Jeep CJ - see JEEP (50020)
14020	**Mid-size models** '70 thru '83
14025	**(Renault) Alliance & Encore** '83 thru '87

AUDI
15020	**4000** all models '80 thru '87
15025	**5000** all models '77 thru '83
15026	**5000** all models '84 thru '88

AUSTIN-HEALEY
	Sprite - see MG Midget (66015)

BMW
*18020	**3/5 Series** not including diesel or all-wheel drive models '82 thru '92
18021	**3-Series** incl. Z3 models '92 thru '98
18025	**320i** all 4 cyl models '75 thru '83
18050	**1500 thru 2002** except Turbo '59 thru '77

BUICK
*19010	**Buick Century** '97 thru '02
	Century (front-wheel drive) - see GM (38005)
*19020	**Buick, Oldsmobile & Pontiac Full-size (Front-wheel drive)** '85 thru '02 **Buick** Electra, LeSabre and Park Avenue; **Oldsmobile** Delta 88 Royale, Ninety Eight and Regency; **Pontiac** Bonneville
19025	**Buick Oldsmobile & Pontiac Full-size (Rear wheel drive)** **Buick** Estate '70 thru '90, Electra '70 thru '84, LeSabre '70 thru '85, Limited '74 thru '79 **Oldsmobile** Custom Cruiser '70 thru '90, Delta 88 '70 thru '85, Ninety-eight '70 thru '84 **Pontiac** Bonneville '70 thru '81, Catalina '70 thru '81, Grandville '70 thru '75, Parisienne '83 thru '86
19030	**Mid-size Regal & Century** all rear-drive models with V6, V8 and Turbo '74 thru '87 **Regal** - see GENERAL MOTORS (38010) **Riviera** - see GENERAL MOTORS (38030) **Roadmaster** - see CHEVROLET (24046) **Skyhawk** - see GENERAL MOTORS (38015) **Skylark** - see GM (38020, 38025) **Somerset** - see GENERAL MOTORS (38025)

CADILLAC
21030	**Cadillac Rear Wheel Drive** all gasoline models '70 thru '93 **Cimarron** - see GENERAL MOTORS (38015) **DeVille** - see GM (38031 & 38032) **Eldorado** - see GM (38030 & 38031) **Fleetwood** - see GM (38031) **Seville** - see GM (38030, 38031 & 38032)

CHEVROLET
*24010	**Astro & GMC Safari Mini-vans** '85 thru '02
24015	**Camaro V8** all models '70 thru '81
24016	**Camaro** all models '82 thru '92
24017	**Camaro & Firebird** '93 thru '00 **Cavalier** - see GENERAL MOTORS (38016) **Celebrity** - see GENERAL MOTORS (38005)
24020	**Chevelle, Malibu & El Camino** '69 thru '87
24024	**Chevette & Pontiac T1000** '76 thru '87 **Citation** - see GENERAL MOTORS (38020)
24032	**Corsica/Beretta** all models '87 thru '96
24040	**Corvette** all V8 models '68 thru '82
24041	**Corvette** all models '84 thru '96
10305	**Chevrolet Engine Overhaul Manual**
24045	**Full-size Sedans** Caprice, Impala, Biscayne, Bel Air & Wagons '69 thru '90
24046	**Impala SS & Caprice and Buick Roadmaster** '91 thru '96 **Impala** - see LUMINA (24048) **Lumina** '90 thru '94 - see GM (38010)
*24048	**Lumina & Monte Carlo** '95 thru '01 **Lumina APV** - see GM (38035)
24050	**Luv Pick-up** all 2WD & 4WD '72 thru '82 **Malibu** '97 thru '00 - see GM (38026)

24055	**Monte Carlo** all models '70 thru '88 **Monte Carlo** '95 thru '01 - see LUMINA (24048)
24059	**Nova** all V8 models '69 thru '79
24060	**Nova and Geo Prizm** '85 thru '92
24064	**Pick-ups '67 thru '87** - Chevrolet & GMC, all V8 & in-line 6 cyl, 2WD & 4WD '67 thru '87; Suburbans, Blazers & Jimmys '67 thru '91
24065	**Pick-ups '88 thru '98** - Chevrolet & GMC, full-size pick-ups '88 thru '98, C/K Classic '99 & '00, Blazer & Jimmy '92 thru '94; Suburban '92 thru '99; Tahoe & Yukon '95 thru '99
*24066	**Pick-ups '99 thru '01** - Chevrolet Silverado & GMC Sierra full-size pick-ups '99 thru '01, Suburban/Tahoe/Yukon/Yukon XL '00 thru '01
24070	**S-10 & S-15 Pick-ups** '82 thru '93, Blazer & Jimmy '83 thru '94,
*24071	**S-10 & S-15 Pick-ups** '94 thru '01, Blazer & Jimmy '95 thru '01, Hombre '96 thru '01
24075	**Sprint** '85 thru '88 & Geo Metro '89 thru '01
24080	**Vans - Chevrolet & GMC** '68 thru '96

CHRYSLER
25015	**Chrysler Cirrus, Dodge Stratus, Plymouth Breeze** '95 thru '00
10310	**Chrysler Engine Overhaul Manual**
25020	**Full-size Front-Wheel Drive** '88 thru '93 **K-Cars** - see DODGE Aries (30008) **Laser** - see DODGE Daytona (30030)
25025	**Chrysler LHS, Concorde, New Yorker, Dodge Intrepid, Eagle Vision,** '93 thru '97
*25026	**Chrysler LHS, Concorde, 300M, Dodge Intrepid,** '98 thru '03
25030	**Chrysler & Plymouth Mid-size** front wheel drive '82 thru '95 **Rear-wheel Drive** - see Dodge (30050)
*25035	**PT Cruiser** all models '01 thru '03
*25040	**Chrysler Sebring, Dodge Avenger** '95 thru '02

DATSUN
28005	**200SX** all models '80 thru '83
28007	**B-210** all models '73 thru '78
28009	**210** all models '79 thru '82
28012	**240Z, 260Z & 280Z** Coupe '70 thru '78
28014	**280ZX** Coupe & 2+2 '79 thru '83 **300ZX** - see NISSAN (72010)
28016	**310** all models '78 thru '82
28018	**510 & PL521 Pick-up** '68 thru '73
28020	**510** all models '78 thru '81
28022	**620 Series Pick-up** all models '73 thru '79 **720 Series Pick-up** - see NISSAN (72030)
28025	**810/Maxima** all gasoline models, '77 thru '84

DODGE
	400 & 600 - see CHRYSLER (25030)
30008	**Aries & Plymouth Reliant** '81 thru '89
30010	**Caravan & Plymouth Voyager** '84 thru '95
*30011	**Caravan & Plymouth Voyager** '96 thru '02
30012	**Challenger/Plymouth Saporro** '78 thru '83
30016	**Colt & Plymouth Champ** '78 thru '87
30020	**Dakota Pick-ups** all models '87 thru '96
*30021	**Durango** '98 & '99, Dakota '97 thru '99
30025	**Dart, Demon, Plymouth Barracuda, Duster & Valiant** 6 cyl models '67 thru '76
30030	**Daytona & Chrysler Laser** '84 thru '89 **Intrepid** - see CHRYSLER (25025, 25026)
*30034	**Neon** all models '95 thru '99
30035	**Omni & Plymouth Horizon** '78 thru '90
30040	**Pick-ups** all full-size models '74 thru '93
*30041	**Pick-ups** all full-size models '94 thru '01
30045	**Ram 50/D50 Pick-ups & Raider and Plymouth Arrow Pick-ups** '79 thru '93
30050	**Dodge/Plymouth/Chrysler RWD** '71 thru '89
30055	**Shadow & Plymouth Sundance** '87 thru '94
30060	**Spirit & Plymouth Acclaim** '89 thru '95
*30065	**Vans - Dodge & Plymouth** '71 thru '03

EAGLE
	Talon - see MITSUBISHI (68030, 68031)
	Vision - see CHRYSLER (25025)

FIAT
34010	**124 Sport Coupe & Spider** '68 thru '78
34025	**X1/9** all models '74 thru '80

FORD
10355	**Ford Automatic Transmission Overhaul**
36004	**Aerostar Mini-vans** all models '86 thru '97
36006	**Contour & Mercury Mystique** '95 thru '00
36008	**Courier Pick-up** all models '72 thru '82
*36012	**Crown Victoria & Mercury Grand Marquis** '88 thru '00
10320	**Ford Engine Overhaul Manual**
36016	**Escort/Mercury Lynx** all models '81 thru '90
36020	**Escort/Mercury Tracer** '91 thru '00
36024	**Explorer & Mazda Navajo** '91 thru '01
36028	**Fairmont & Mercury Zephyr** '78 thru '83
36030	**Festiva & Aspire** '88 thru '97
36032	**Fiesta** all models '77 thru '80
*36034	**Focus** all models '00 and '01
36036	**Ford & Mercury Full-size** '75 thru '87
36040	**Granada & Mercury Monarch** '75 thru '80
36044	**Ford & Mercury Mid-size** '75 thru '86
36048	**Mustang V8** all models '64-1/2 thru '73
36049	**Mustang II** 4 cyl, V6 & V8 models '74 thru '78
36050	**Mustang & Mercury Capri** all models Mustang, '79 thru '93; Capri, '79 thru '86
*36051	**Mustang** all models '94 thru '03
36054	**Pick-ups & Bronco** '73 thru '79
36058	**Pick-ups & Bronco** '80 thru '96
*36059	**F-150 & Expedition** '97 thru '02, F-250 '97 thru '99 & Lincoln Navigator '98 thru '02
*36060	**Super Duty Pick-ups, Excursion** '97 thru '02
36062	**Pinto & Mercury Bobcat** '75 thru '80
36066	**Probe** all models '89 thru '92
36070	**Ranger/Bronco II** gasoline models '83 thru '92
*36071	**Ranger** '93 thru '00 & Mazda Pick-ups '94 thru '00
36074	**Taurus & Mercury Sable** '86 thru '95
*36075	**Taurus & Mercury Sable** '96 thru '01
36078	**Tempo & Mercury Topaz** '84 thru '94
36082	**Thunderbird/Mercury Cougar** '83 thru '88
36086	**Thunderbird/Mercury Cougar** '89 and '97
36090	**Vans** all V8 Econoline models '69 thru '91
*36094	**Vans** full size '92 thru '01
*36097	**Windstar Mini-van** '95 thru '03

GENERAL MOTORS
10360	**GM Automatic Transmission Overhaul**
38005	**Buick Century, Chevrolet Celebrity, Oldsmobile Cutlass Ciera & Pontiac 6000** all models '82 thru '96
*38010	**Buick Regal, Chevrolet Lumina, Oldsmobile Cutlass Supreme & Pontiac Grand Prix (FWD)** '88 thru '02
38015	**Buick Skyhawk, Cadillac Cimarron, Chevrolet Cavalier, Oldsmobile Firenza & Pontiac J-2000 & Sunbird** '82 thru '94
*38016	**Chevrolet Cavalier & Pontiac Sunfire** '95 thru '01
38020	**Buick Skylark, Chevrolet Citation, Olds Omega, Pontiac Phoenix** '80 thru '85
38025	**Buick Skylark & Somerset, Oldsmobile Achieva & Calais and Pontiac Grand Am** all models '85 thru '98
*38026	**Chevrolet Malibu, Olds Alero & Cutlass, Pontiac Grand Am** '97 thru '00
38030	**Cadillac Eldorado** '71 thru '85, **Seville** '80 thru '85, **Oldsmobile Toronado** '71 thru '85, Buick Riviera '79 thru '85
*38031	**Cadillac Eldorado & Seville** '86 thru '91, **DeVille** '86 thru '93, Fleetwood & Olds Toronado '86 thru '92, Buick Riviera '86 thru '93
38032	**Cadillac DeVille** '94 thru '02 & **Seville** - '92 thru '02
38035	**Chevrolet Lumina APV, Olds Silhouette & Pontiac Trans Sport** all models '90 thru '96
*38036	**Chevrolet Venture, Olds Silhouette, Pontiac Trans Sport & Montana** '97 thru '01 **General Motors Full-size Rear-wheel Drive** - see BUICK (19025)

GEO
	Metro - see CHEVROLET Sprint (24075)
	Prizm - '85 thru '92 see CHEVY (24060), '93 thru '02 see TOYOTA Corolla (92036)

(Continued on other side)

Haynes Automotive Manuals (continued)

NOTE: New manuals are added to this list on a periodic basis. If you do not see a listing for your vehicle, consult your local Haynes dealer for the latest product information.

40030 Storm all models '90 thru '93
Tracker - *see SUZUKI Samurai (90010)*

GMC
Vans & Pick-ups - *see CHEVROLET*

HONDA
42010 Accord CVCC all models '76 thru '83
42011 Accord all models '84 thru '89
42012 Accord all models '90 thru '93
42013 Accord all models '94 thru '97
*42014 **Accord** all models '98 and '99
42020 Civic 1200 all models '73 thru '79
42021 Civic 1300 & 1500 CVCC '80 thru '83
42022 Civic 1500 CVCC all models '75 thru '79
42023 Civic all models '84 thru '91
42024 Civic & del Sol '92 thru '95
*42025 **Civic** '96 thru '00, **CR-V** '97 thru '00, **Acura Integra** '94 thru '00
42040 Prelude CVCC all models '79 thru '89

HYUNDAI
*43010 **Elantra** all models '96 thru '01
43015 Excel & Accent all models '86 thru '98

ISUZU
Hombre - *see CHEVROLET S-10 (24071)*
*47017 **Rodeo** '91 thru '02; **Amigo** '89 thru '94 and '98 thru '02; **Honda Passport** '95 thru '02
47020 Trooper & Pick-up '81 thru '93

JAGUAR
49010 XJ6 all 6 cyl models '68 thru '86
49011 XJ6 all models '88 thru '94
49015 XJ12 & XJS all 12 cyl models '72 thru '85

JEEP
50010 Cherokee, Comanche & Wagoneer Limited all models '84 thru '00
50020 CJ all models '49 thru '86
*50025 **Grand Cherokee** all models '93 thru '00
50029 Grand Wagoneer & Pick-up '72 thru '91 Grand Wagoneer '84 thru '91, Cherokee & Wagoneer '72 thru '83, Pick-up '72 thru '88
*50030 **Wrangler** all models '87 thru '00

LEXUS
ES 300 - *see TOYOTA Camry (92007)*

LINCOLN
Navigator - *see FORD Pick-up (36059)*
59010 Rear-Wheel Drive all models '70 thru '01

MAZDA
61010 GLC Hatchback (rear-wheel drive) '77 thru '83
61011 GLC (front-wheel drive) '81 thru '85
61015 323 & Protegé '90 thru '00
*61016 **MX-5 Miata** '90 thru '97
61020 MPV all models '89 thru '94
Navajo - *see Ford Explorer (36024)*
61030 Pick-ups '72 thru '93
Pick-ups '94 thru '00 - *see Ford Ranger (36071)*
61035 RX-7 all models '79 thru '85
61036 RX-7 all models '86 thru '91
61040 626 (rear-wheel drive) all models '79 thru '82
61041 626/MX-6 (front-wheel drive) '83 thru '91
61042 626 '93 thru '01, **MX-6/Ford Probe** '93 thru '97

MERCEDES-BENZ
63012 123 Series Diesel '76 thru '85
63015 190 Series four-cyl gas models, '84 thru '88
63020 230/250/280 6 cyl sohc models '68 thru '72
63025 280 123 Series gasoline models '77 thru '81
63030 350 & 450 all models '71 thru '80

MERCURY
64200 Villager & Nissan Quest '93 thru '01
All other titles, see FORD Listing.

MG
66010 MGB Roadster & GT Coupe '62 thru '80
66015 MG Midget, Austin Healey Sprite '58 thru '80

MITSUBISHI
68020 Cordia, Tredia, Galant, Precis & Mirage '83 thru '93
68030 Eclipse, Eagle Talon & Ply. Laser '90 thru '94
*68031 **Eclipse** '95 thru '01, **Eagle Talon** '95 thru '98
68040 Pick-up '83 thru '96 & **Montero** '83 thru '93

NISSAN
72010 300ZX all models including Turbo '84 thru '89
72015 Altima all models '93 thru '01
72020 Maxima all models '85 thru '92
*72021 **Maxima** all models '93 thru '01
72030 Pick-ups '80 thru '97 **Pathfinder** '87 thru '95
*72031 **Frontier Pick-up** '98 thru '01, **Xterra** '00 & '01, **Pathfinder** '96 thru '01
72040 Pulsar all models '83 thru '86
Quest - *see MERCURY Villager (64200)*
72050 Sentra all models '82 thru '94
72051 Sentra & 200SX all models '95 thru '99
72060 Stanza all models '82 thru '90

OLDSMOBILE
73015 Cutlass V6 & V8 gas models '74 thru '88
For other OLDSMOBILE titles, see BUICK, CHEVROLET or GENERAL MOTORS listing.

PLYMOUTH
For PLYMOUTH titles, see DODGE listing.

PONTIAC
79008 Fiero all models '84 thru '88
79018 Firebird V8 models except Turbo '70 thru '81
79019 Firebird all models '82 thru '92
79040 Mid-size Rear-wheel Drive '70 thru '87
For other PONTIAC titles, see BUICK, CHEVROLET or GENERAL MOTORS listing.

PORSCHE
80020 911 except Turbo & Carrera 4 '65 thru '89
80025 914 all 4 cyl models '69 thru '76
80030 924 all models including Turbo '76 thru '82
80035 944 all models including Turbo '83 thru '89

RENAULT
Alliance & Encore - *see AMC (14020)*

SAAB
*84010 **900** all models including Turbo '79 thru '88

SATURN
*87010 **Saturn** all models '91 thru '02

SUBARU
89002 1100, 1300, 1400 & 1600 '71 thru '79
89003 1600 & 1800 2WD & 4WD '80 thru '94

SUZUKI
90010 Samurai/Sidekick & Geo Tracker '86 thru '01

TOYOTA
92005 Camry all models '83 thru '91
92006 Camry all models '92 thru '96
*92007 **Camry, Avalon, Solara, Lexus ES 300** '97 thru '01
92015 Celica Rear Wheel Drive '71 thru '85
92020 Celica Front Wheel Drive '86 thru '99
92025 Celica Supra all models '79 thru '92
92030 Corolla all models '75 thru '79
92032 Corolla all rear wheel drive models '80 thru '87
92035 Corolla all front wheel drive models '84 thru '92
92036 Corolla & Geo Prizm '93 thru '02
92040 Corolla Tercel all models '80 thru '82
92045 Corona all models '74 thru '82
92050 Cressida all models '78 thru '82
92055 Land Cruiser FJ40, 43, 45, 55 '68 thru '82
92056 Land Cruiser FJ60, 62, 80, FZJ80 '80 thru '96
92065 MR2 all models '85 thru '87
92070 Pick-up all models '69 thru '78
92075 Pick-up all models '79 thru '95
*92076 **Tacoma** '95 thru '00, **4Runner** '96 thru '00, & **T100** '93 thru '98
*92078 **Tundra** '00 thru '02 & **Sequoia** '01 thru '02

92080 Previa all models '91 thru '95
*92082 **RAV4** all models '96 thru '02
92085 Tercel all models '87 thru '94

TRIUMPH
94007 Spitfire all models '62 thru '81
94010 TR7 all models '75 thru '81

VW
96008 Beetle & Karmann Ghia '54 thru '79
*96009 **New Beetle** '98 thru '00
96016 Rabbit, Jetta, Scirocco & Pick-up gas models '74 thru '91 & Convertible '80 thru '92
96017 Golf, GTI & Jetta '93 thru '98 & **Cabrio** '95 thru '98
*96018 **Golf, GTI, Jetta & Cabrio** '99 thru '02
96020 Rabbit, Jetta & Pick-up diesel '77 thru '84
96023 Passat '98 thru '01, **Audi A4** '96 thru '01
96030 Transporter 1600 all models '68 thru '79
96035 Transporter 1700, 1800 & 2000 '72 thru '79
96040 Type 3 1500 & 1600 all models '63 thru '73
96045 Vanagon all air-cooled models '80 thru '83

VOLVO
97010 120, 130 Series & 1800 Sports '61 thru '73
97015 140 Series all models '66 thru '74
97020 240 Series all models '76 thru '93
97040 740 & 760 Series all models '82 thru '88
97050 850 Series all models '93 thru '97

TECHBOOK MANUALS
10205 Automotive Computer Codes
10210 Automotive Emissions Control Manual
10215 Fuel Injection Manual, 1978 thru 1985
10220 Fuel Injection Manual, 1986 thru 1999
10225 Holley Carburetor Manual
10230 Rochester Carburetor Manual
10240 Weber/Zenith/Stromberg/SU Carburetors
10305 Chevrolet Engine Overhaul Manual
10310 Chrysler Engine Overhaul Manual
10320 Ford Engine Overhaul Manual
10330 GM and Ford Diesel Engine Repair Manual
10340 Small Engine Repair Manual, 5 HP & Less
10341 Small Engine Repair Manual, 5.5 - 20 HP
10345 Suspension, Steering & Driveline Manual
10355 Ford Automatic Transmission Overhaul
10360 GM Automatic Transmission Overhaul
10405 Automotive Body Repair & Painting
10410 Automotive Brake Manual
10411 Automotive Anti-lock Brake (ABS) Systems
10415 Automotive Detaiing Manual
10420 Automotive Eelectrical Manual
10425 Automotive Heating & Air Conditioning
10430 Automotive Reference Manual & Dictionary
10435 Automotive Tools Manual
10440 Used Car Buying Guide
10445 Welding Manual
10450 ATV Basics

SPANISH MANUALS
98903 Reparación de Carrocería & Pintura
98905 Códigos Automotrices de la Computadora
98910 Frenos Automotriz
98915 Inyección de Combustible 1986 al 1999
99040 Chevrolet & GMC Camionetas '67 al '87 Incluye Suburban, Blazer & Jimmy '67 al '91
99041 Chevrolet & GMC Camionetas '88 al '98 Incluye Suburban '92 al '98, Blazer & Jimmy '92 al '94, Tahoe y Yukon '95 al '98
99042 Chevrolet & GMC Camionetas Cerradas '68 al '95
99055 Dodge Caravan & Plymouth Voyager '84 al '95
99075 Ford Camionetas y Bronco '80 al '94
99077 Ford Camionetas Cerradas '69 al '91
99083 Ford Modelos de Tamaño Grande '75 al '87
99088 Ford Modelos de Tamaño Mediano '75 al '86
99091 Ford Taurus & Mercury Sable '86 al '95
99095 GM Modelos de Tamaño Grande '70 al '90
99100 GM Modelos de Tamaño Mediano '70 al '88
99110 Nissan Camioneta '80 al '96, **Pathfinder** '87 al '95
99118 Nissan Sentra '82 al '94
99125 Toyota Camionetas y 4Runner '79 al '95

Over 100 Haynes motorcycle manuals also available

8-03

* *Listings shown with an asterisk (*) indicate model coverage as of this printing. These titles will be periodically updated to include later model years - consult your Haynes dealer for more information.*

Haynes North America, Inc., 861 Lawrence Drive, Newbury Park, CA 91320-1514 • (805) 498-6703